T0310075

BIOPROCESSING PIPING AND EQUIPMENT DESIGN

Wiley-ASME Press Series List

Introduction to Dynamics and Control in Mechanical Engineering Systems	To	March 2016
Fundamentals of Mechanical Vibrations	Cai	May 2016
Nonlinear Regression Modeling for Engineering Applications	Rhinehart	September 2016
Modeling, Model Validation, and Enabling Design of Experiments Stress in ASME Pressure Vessels	Jawad	November 2016
Combined Cooling, Heating, and Power Systems	Shi	January 2017

BIOPROCESSING PIPING AND EQUIPMENT DESIGN

A COMPANION GUIDE FOR THE ASME BPE STANDARD

William M. (Bill) Huitt

This Work is a co-publication between ASME Press and John Wiley & Sons, Inc.

WILEY

Published by John Wiley & Sons, Inc., Hoboken, New Jersey

Published simultaneously in Canada

Library of Congress Cataloging-in-Publication Data

Names: Huitt, William M., 1943– author.
Title: Bioprocessing piping and equipment design : a companion guide for the ASME BPE standard / William M. (Bill) Huitt.
Description: Hoboken, New Jersey : John Wiley & Sons, Inc., [2017] | Includes bibliographical references and index.
Identifiers: LCCN 2016024930| ISBN 9781119284239 (cloth) | ISBN 9781119284253 (ePub) | ISBN 9781119284246 (Adobe PDF)
Subjects: LCSH: Biochemical engineering–Equipment and supplies–Standards–Handbooks, manuals, etc. | Chemical plants–Piping–Standards–Handbooks, manuals, etc.
Classification: LCC TP157 .H\87 2017 | DDC 660.6/3–dc23
LC record available at https://lccn.loc.gov/2016024930

Printed in the United States of America

Set in 10/12pt Times by SPi Global, Pondicherry, India

10 9 8 7 6 5 4 3 2 1

ASME BPE 2014

Its Organization and Roster of Members

Organization

The ASME Bioprocessing Equipment (BPE) Standards Committee membership in 2014 was made up, in whole, of 195 members holding membership to anywhere from one to five committee/subcommittee memberships. The ASME BPE Standards Committee is, as self-described, considered a "committee," referring to itself as the ASME BPE Standards Committee, or simply Standards Committee. As indicated in Figure 1, organizational chart, the ASME BPE Standards Committee reports to the ASME Board on Pressure Technology Codes and Standards (BPTCS). Aside from the BPE Standards Committee, reporting also to the BPTCS are the Boiler and Pressure Vessel Code (BPVC) Committees, the B16 and B31 Committees, and other committees related to pressure containing subject matter.

The ASME BPE Committee is divided into a set of subtier groups of interest referred to as subcommittees. In other Standards Committees these subtier groups are referred to as "subgroups," not so with the BPE Standards Committee. Among this group of subcommittees there is no hierarchy. They are simply divided by and focused on the various subject matter interests of the BPE Standard and report directly to the BPE Standards Committee. These subject matter interests are referred to as Parts with the following identifiers as referenced in Table 1.

Referring to Figure 2, it is apparent that each of the subcommittee groups reports to the BPE Standards Committee. The work these subcommittees do, whether it's maintaining an existing part in the standard, respective of the subcommittee's part title, or in developing a new part for the standard, there is an ongoing liaison effort that takes place between all of the subcommittees. This helps in diverting conflicts among the various subcommittees and in improving content of the standard as a whole.

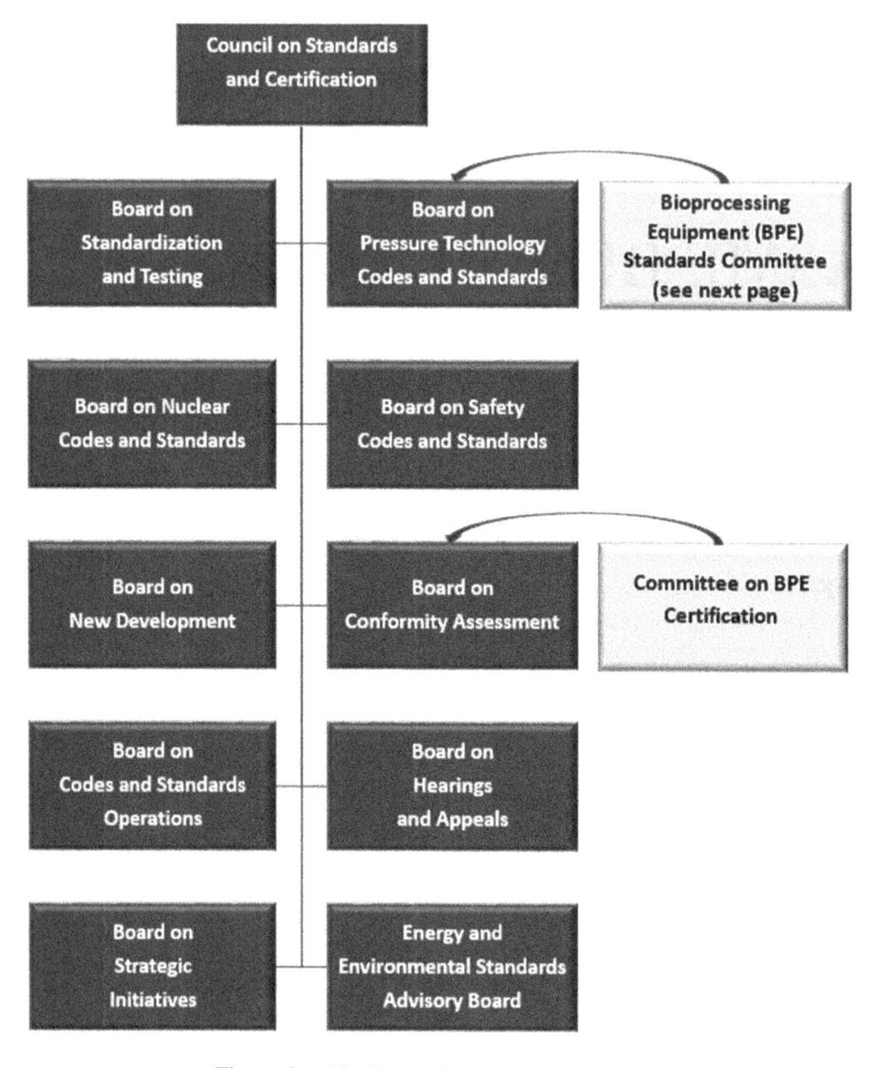

Figure 1 ASME boards and governing groups

Table 1 Subcommittee subject matter part identifiers

Part	Title	Part	Title
GR	General requirements	SG	Sealing components
SD	Systems design	PM	Polymeric materials
DT	Dimensions and tolerances	CR	Certification
MJ	Materials joining	MM	Metallic materials
SF	Surface finishes	PI	Process instrumentation

Each of the subcommittees is made up of a balanced membership wherein each member is assigned an interest category as follows:

- Designer/constructor (AC)—An organization performing design or design-related services, fabrication or erection, or both
- General interest (AF)—Consultants, educators, research and development organization personnel, and public interest persons
- Manufacturer (AK)—An organization producing components or assemblies
- Material manufacturer (AM)—An organization producing materials or ancillary material-related accessories or component parts
- User (AW)—An organization utilizing processes and/or facilities covered by the applicable standards

No one classification shall be represented by more than one-third of the subcommittee membership. By maintaining this balanced membership, no single interest group, whether it be a manufacturer or end user, or any other group, can monopolize the decisions made and the topics discussed in the subcommittee meetings.

Heading up the committees and subcommittees are elected officers of those groups. Each committee, subcommittee, and task group will have a chair and vice-chair. And depending on the size and complexity of any subcommittee, they may also have multiple vice-chairs and a secretary. The secretary for the BPE Standards Committee is an ASME staff secretary that not only provides a direct in-house link to ASME but also helps the entire membership maneuver through the procedural maze now and then when such procedural questions arise.

A subtier of groups under that of the subcommittees are the task groups. These are ad hoc groups that are assembled for a specific task and report to a particular subcommittee. These groups are where the majority of work gets done in standards development and maintenance. Some projects these groups are tasked to do are relatively small. But rather than take up time trying to resolve an issue during a subcommittee meeting, the issue or task will be assigned a temporary number, and volunteers are asked to work on resolving such issues offline or outside the confines of the subcommittee meeting.

Other task group issues are much more complex and involved. These tasks may take years to resolve and prepare for the balloting process. The balloting process itself is rigorous in that a proposal has to obtain consensus approval at multiple stages of the balloting process. That is, a proposal is balloted at the subcommittee level, then at the standards committee level, then to the board level, and finally to ANSI for procedural approval.

At each step of the process, a consensus has to be reached and each negative response has to be responded to with an attempt made to resolve all objections. But a consensus does not require unanimous approval. It does require approval by a simple majority of all of those voting. And to document all of this, ASME uses a system titled C&S Connect, the C&S standing for Codes and Standards. The basis for these procedures is consistent with the principles established for the World Trade Organization's Technical Barriers to Trade Committee.

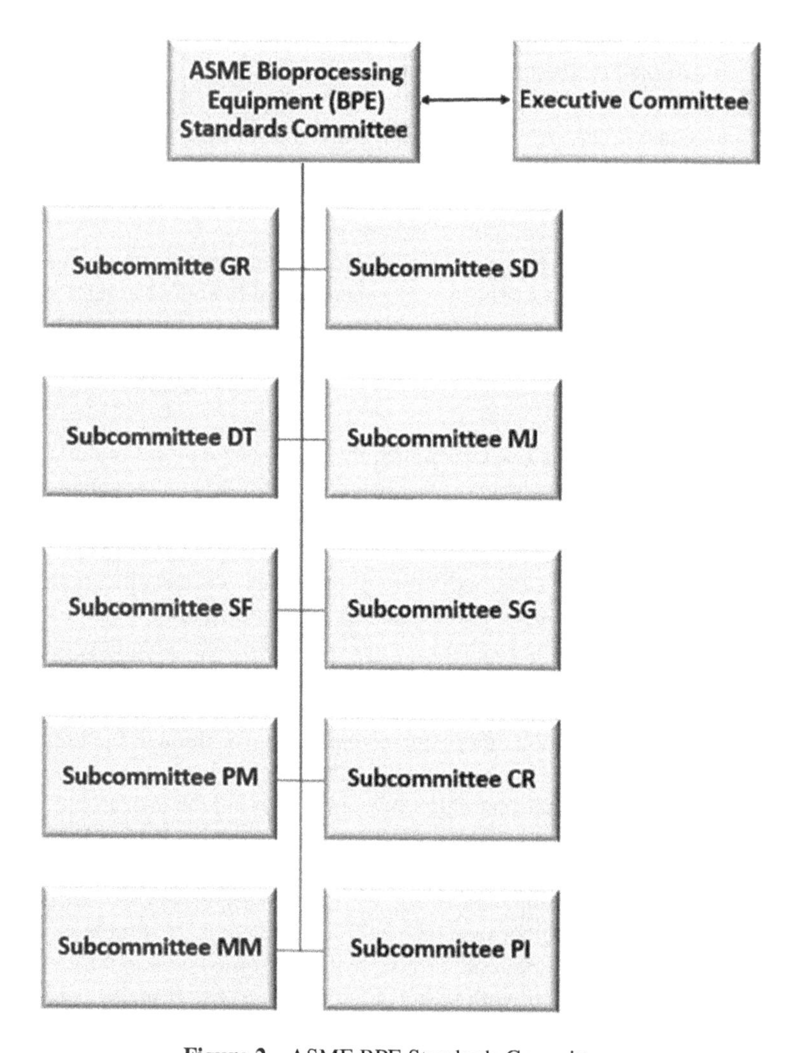

Figure 2 ASME BPE Standards Committee

The BPE Executive Committee, as seen in Figure 2, is a direct subset of the Standards Committee. This group is responsible for recommending approval or discharge of personnel and the governing of administrative items or actions as they relate to ASME policy and procedures. The vice-chair of the Standards Committee automatically serves as chair of the Executive Committee, and the chair of the Standards Committee automatically serves as vice-chair of the Executive Committee. Subcommittee chairs automatically hold membership to the Executive Committee, but membership on the Executive Committee beyond that does not require being a member of the Standards Committee.

In referring to Figure 1, there is an adjunct committee that is closely related to the Part CR subcommittee with the title of "Committee on BPE Certification" (CBPEC). This is a committee on its own and reports, as shown, to the Board on Conformity

Assessment (BCA). I will refer you to Section 1.2 of this book for a very brief synopsis of the scope of Part CR and all of the other subcommittees. But to clarify here how these two groups, Part CR and the CBPEC, work together is relatively simple.

The Part CR subcommittee is the group that developed and maintains Part CR in the standard, which intrinsically defines what BPE Certification is and how it interacts with the standard. It defines what the requirements are for BPE Certification and provides guidance on how to become a BPE Certificate Holder.

The CBPEC is the assessment and enforcement arm of the certification process. This is the group that, in working with the BCA, performs audits of those applying for BPE Certification; they review the subsequent auditor's assessment report and then make a determination, based on the auditor's report and deliberation, whether or not to recommend approval of the applicant. The final decision on that point is made by the BCA.

BPE Standards Committee Meetings

All committee and subcommittee meetings of the BPE Standard are open to the public. The only meetings not open to the public are those meetings in which discussions and decisions regarding personnel are held. The CBPEC meetings are closed, but these are conformity assessment meetings that typically follow the meetings of the CR subcommittee and are not BPE meetings.

The BPE Standards meetings follow an evolved schedule that runs for four days, Monday through Thursday. The Monday meetings typically include subcommittees CR and GR and task group meetings for any active task groups that need to discuss and finalize any outstanding issue relating to a task group's work.

Most task group activity takes place between the three committee meetings each year via conference calls and e-mail communication. Depending on the complexity and scope of a task group, some discussions and resolution need to take place in a face-to-face setting. These are the meetings that are scheduled for the Monday task group meetings. Tuesday and Wednesday are the two days in which the balance of subcommittees will meet. Thursday is the Standards Committee meeting at which the Standards Committee reports and all subcommittees report on what transpired at each of their meetings during the week.

As mentioned, all meetings are generally open to the public. New attendees should know that they are free to visit any meeting at any time except as explained previously. Only members are permitted to vote on subcommittee business. But any visitor is free to voice an opinion or make a point during a tabled discussion. It should also be known that visitors to these meetings are eligible to participate in task group work. If a visitor is considering membership, their work on task groups elevates their possibility of being approved.

Up until 2016 the BPE Standards Committee met three times each year. These meetings were held each year in January, May/June, and September/October. As a trial run it was voted on and planned that the committee hold only two meetings in 2016, one meeting in January and a second meeting in September. This was to test the waters to see if the committee could maintain the same level of efficiency and production of work on the standard with only two meetings per year.

A decision as to whether or not to remain with a two meeting per year format would not be decided upon until possibly the January 2017 meeting. That decision, I suspect, will be based largely on what is accomplished during 2016 and to what extent, good or bad, did the missing third meeting play a part.

Roster of Members

The following is a listing of all members of the ASME BPE Standards Committee and members of subcommittees reporting to the Standards Committee. The names are in alphabetical order and indicate if that person is a chair, vice-chair, or secretary of a committee or subcommittee and which subcommittees they are members of:

BPE STANDARDS COMMITTEE AND SUBCOMMITTEE MEMBERS

Abbreviations and Terminology used below:

EC = Executive Committee
DT = Dimensions and Tolerances
GR = General Requirements
MJ = Materials Joining
MM = Metallic Materials
PI = Process Instrumentation
PM = Polymeric Materials
SC = Standards Committee
SD = Systems Design
SF = Surface Finishes
SG = Sealing Components

Notes:

1. Contributing Member: an individual non-voting member whose contribution to a committee is through reviews and comments on proposals. Contributing members shall possess the technical qualifications necessary for individual voting members.
2. This individual is considered a "Delegate," which implies that they are an individual selected by the Standards committee to represent a group of experts outside of the U.S. and Canada. Each group represented has provided a clearly defined interest in participating on BPE subcommittees.

NAME	AFFILIATION	CHAIR	V. CHAIR	SECRETARY	SUBCOMMITTEES				
Allard, Michael	NewAge Industries				PM				
Anant, Janmeet	EMD Millipore				PM				
Anderson, Paul	Northland Stainless Inc.				MM				
Andrews, Jacob	Zenpure Americas, Inc.				PM				
Andrews, Todd	Colder Products Company				PM				
Ankers, Jay	Ocean Alloys LLC	SC	EC		SD	PI[1]			
Anton, George	Qualtech Inc.				PI				
Avery, Richard E.	Nickel Institute				MM	SF			
Balmer, Melissa L.	Sanofi Pasteur		SD		SC				
Banes, Patrick H.	Astro Pak Corp.		SF						
Baram, David	Clifton Enterprises				SC[1]	SG[1]			
Benway, Ernest A.	Ironwood Specialist Inc.				SC[1]	GR[1]	MJ[1]		
Bhaila, Kadeem	ITT Engineered Valves, LLC		MJ						
Bickel, Neill	Genentech				MJ	SF			
Billmyer, Bryan A.	Central States Industrial Equip.				SC	CR	DT	SD	
Blumenthal, Joel	Perceptual Focus LLC				PI	SG			
Bond, Richard	Anderson Instrument Co.				PI				
Bradley, Jeffrey L.	Eli Lilly and Co.				SD[1]	GR[1]	MJ[1]		
Bragg. Chuck J.	Burns Engineering, Inc.				PI				
Brockmann, Dan	Alfa Laval Inc.				CR	DT	SF		
Burg, William P.	DECCO Inc.			MJ	GR	MJ			
Cagne, William H.	JSG, LLC				SC	EC	GR		
Campbell, Dr. Richard D., PE	Bechtel	MJ			SC	EC	GR	CR	MM
Canty, Thomas	J M Canty Inc.		PI		SD[1]				
Carl A. Johnson	Genentech Inc.				SD				
Chapman, Chuck	Gemu Valves				DT	SD			
Chih-Feng, Kuo	King Lai International				SF				
Cirillo, Anthony P.	Cirillo Consulting Services LLC				SC[1]	EC[1]	GR[1]		
Cohen, Donald K.	Michigan Metrology, LLC				SF[1]				
Conley, Indumathi	DPS Engineering				SD				
Conn, Carlyle C.	Top Line Process Equipment Co.				SF				
Cook, Todd J.	T & C Stainless, Inc.				MJ	SF			
Cooper, Mark	United Stainless				SF				
Cosentino, Rodolfo	Giltec Ltda				DT	PI			
Cotter, Randolph A.	Cotter Brothers Corporation				SC	MJ	SD		
Crawley, Jere	Jacobs Engineering Group, Inc.				SD				
Daly, James	BSI Engineering				SD				
Daniels, James R., PE	ITT Engineered Valves, LLC				SG	SF			
Davis, Kenneth R.	Nordson Medical				DT	PM			
Defeo, John W.	Hoffer Flow Controls Inc.				PI				
Defusco, Sean J.	Integra Companies Inc.				PM	SG			
Dubiel, Robert J.	Parker Hannifin				SG				
Dunbar, Peter M.	VNE Corporation				CR	DT			
Dvorscek, James	Abbott Laboratories	MJ			SC	EC	CR	SD[1]	
Dyrness, Albert D., PE	Advent Engineering Inc.		SD		SD				
Elbich, Robert	Exigo Manufacturing				CR	DT			

(*Continued*)

Elkins, Curtis W.	Central States Industrial Equip.				MJ	SF			
Embury, Mark	ASEPCO	GR			EC	EC	SD		
Esbensen, Preben	Alfa Laval Kolding A/S				SG				
Evans, Greg	Ace Sanitary				PM				
Featherston, Jan-Marc	Weed Instrument Co.				PI				
Feldman, Jason	Yula Corporation				SD				
Fisher, E. Burrell	Fisher Engineering				SC	SD			
Fitts, Robert B.	Spraying Systems Co				DT	GR			
Foley, Gerard P.	PBM, Inc.				SG	SD			
Foley, Raymond F.	DPS Engineering				DT	SD			
Fortin, Jonathan	Lonza Group				SD				
Franks, John W.	Electrol Specialties Company				MM	SD			
Fridman Tamara	Vanasyl LLC.		PM	GR					
Fritz, James	TMR Stainless				MJ	MM			
Gallagher, Eoghan	Alkermes Pharma Ireland Ltd				ES				
Galvin, Paul G.	GF Piping Systems LLC			PM	PM				
Gayer, Ms. Evelyn L.	Holloway America				CR	MJ	SF		
George, Daryl	Hallam ICS				SD				
Gerra, Ronn	Shire Pharmaceuticals				SD				
Giffen, Jay	PBM Inc.				SG	SF			
Gillespie, David A.	BMW Constructors				CR	MJ	MM		
Gleeson, John	Hamilton Company				PI				
Gonzalez, Michelle M., PE	Engineering Consultant				SC[1]	CR[1]	SF[1]		
Gorbis, Vladimir	Genentech / Roche		CPI						
Govaert, Roger	Mettler Toledo Process Anal.				PI				
Gregg, Bradley D.	Top Line Process Equipment				SG				
Gu, Mr. Zhenghui[2]	Shanghai Morimatsu Pharma				SC	SD			
Guttzeit, Maik	GEA Lyophil GmbH				SD				
Haman, Scott	Fristam Pumps				SG				
Hamilton, Jody	RathGibson		SF						
Hanselka, Reinhard, PhD, PE	CRB Engineers				SC	MJ	SD		
Harper, Larry	Wika Instruments, Ltd				GR	SG			
Harrison, S. Tom	Harrison Electropolishing, LP				MM	SF			
Hartner, Scott M., PE	Baxalta US, Inc.				SD				
Harvey, Tom	Gemu Valves, Inc.				SG				
Helmke, Dennis R.	Flow Products LLC				CR	SG			
Henon, Dr. Barbara K.	Magnatech LLC				SC[1]	MJ[1]	SF[1]		
Hobick, Troy L.	Holland Applied Technologies		CR		SD				
Hogenson Dr. David	Amgen				SD				
Hohmann, Michael A.	Quality Coalescence				SC	CR	GR	MJ	
Huitt William M.	W. M. Huitt Company				GR	CR	MJ	MM	
Hutton, L. T.	Arkema Inc.		PM		SC	CR	MJ		
Inoue, Mikio	Fujikin Inc.				SG	SD			
Irish, Declan	Carten-Fujikin				SG				
Jain, Mukesh K.	W.L. Gore & Associates				PM				
Janousek, John	Abbott				SD				
Jensen, Bo B. B.	Alfa Laval		SD		SD				
Johnson, Carl A.	Genentech Inc.				SC	SG			
Johnson, Michael W.	Entegris				PM				
Juritsch, Daniel	Zeta Biopharma Gmbh				MJ				
Kelleher, Ciaran	Janssen Supply Chain				SG	SD			
Kettermann, Carl	RathGibson	CR			SC	EC	MJ	MM	
Kimbrel, Kenneth D.	Ultraclean Electropolish Inc.	SF			SC	EC	CR		
Klees, Daniel T.	Magnetrol International, Inc.	PI			SC	EC			
Klitgaard, Lars Beck	NNE Pharmaplan				SD				
Knox, Marianne	W. L. Gore & Associates				PM				
Kollar, Csilla	Dow Corning Corporation				PM				
Kranzpiller, Johann	GEA Tuchenhagen GmbH				ES				
Kresge, Ms. Denise	CRB Consulting Engineers				PI				
Kroehnert, Gerhard	Neumo				DT	DT	SF		
Kubera, Paul M.	ABEC, Inc.				SG	SD			
Kwilosz, David	Elanco Global Engineering			PI	GR	PI			
Lamore, Andrew	Burkert Fluid Control Systems				PI				
Larkin, Thomas Jr.	Amgen				PM				
Larson, John D.	DCI, Inc.				SD				
Lisboa, Ivan[2]	RathGibson				SC	DT			
Mahar Jeffrey T.	3M Purification				SC	SD	PM		
Manfredi, Marcello	ZDL Componentes De Proc.				DT				
Manning, Frank	VNE Corp		DT		SC	SF[1]			
Manser, Rolf	DCI, Inc.				SD				
Marks, David M., PE	DME	SD			SC	EC			
Marshall, Jeff	Perrigo-Inc.				SG				
Matheis, Kenneth J. Sr.	Complete Automation Inc.				CR	MJ	MM		

(Continued)

Mathien, Daniel J.	Behringer Corp.	DT			SC	EC			
McCauley Nicholas S.	A&B Process Systems				MJ				
McClune, Paul L., Jr.	ITT Pure-Flo				DT				
McCune, Daniel P.	Allegheny Bradford Corp				MM	SD			
McFeeters, Milena	Steridose	SG			SC	PM			
McGonigle, Robert	Active Chemical Corp.				SF[1]				
Michalak, Ryan A.	Eli Lilly and Co.			SD	S	SG	SD		
Minor, John W., PE	Paul Mueller Co				GR	SD			
Mogul, Rehan	Crane Flow Technologies Ltd				PM	SG			
Monachello, John F.	SP INDUSTRIES				SD				
Mondello, Matthew	MECO				SF				
Montgomery. Gabe	Tank Components Industries				DT				
Mortensen, Michael	NNE Pharmaplan A/S				SD				
Muller, Scott R.	GE Healthcare Bio-Sciences				SD				
Murakami, Sei[2]	Hitachi, Ltd.				S				
Nerstad, Joseph Richard	SOR, Inc.				PI				
Norton, Vickie L.	T&C Stainless				GR				
Obertanec, Andrew R.	Clark-Reliance Corporation				CR	SG	SD		
O'Connor, Tom	Central States Industrial Equip				MJ	MM			
Ortiz, William	Eli Lilly and Co.				GR[1]	SD[1]			
Pacheco, Christopher N., PE	Amgen				SC	SG	SD		
Page, George W. Jr.	Page Solutions				SG	SD			
Parker, Alton K. Jr	W.L. Gore and Associates Inc.				SG				
Pelletier, Marc PHD	CRB Engineers	EC	SC		GR	SD			
Peterman, Lloyd J.	United Industries Inc.				SC	DT	SF		
Petrillo, Peter A.	Millennium Facilities Resources				PI	SF			
Pierre, Philippe R.	Pierre Guerin SAS				ES				
Pitchford, Ernie	Parker Hannifin Corp.				PM				
Pitolaj, Steve	Garlock Sealing Technologies				SG				
Placide, Gilbert	Crosspoint Engineering				PI				
Pouliot, Jeffrey	Amgen				SG				
Powell, Alan L.	Merck & Co, Retired				SG	SD			
Priebe, Paul	Sartorius Stedim Biotech				PM				
Raney, Robert K.	Ultraclean Electropolish Inc.				SF				
Rau, Dr. Jan	Dockweiler AG		MM		SF				
Reinhold, Herman	AM Technical Solutions				MJ				
Rieger, Robert	John Crane Inc.				SG				
Roll, Daryl L., PE	Astro Pak Corporation				MM				
Roth, William L., PE	Procter & Gamble Company		MJ		SC	CR	MM		
Sams, William R.	Richards Industries				SG				
Schmidt, Neil A., PE	Boccard Life Sciences			MM	MM				
Schnell, Russell W.	DuPont Company				PM	SG			
Schroder, Richard	Newman Gasket				PM	SG			
Sedivy, Paul D.	RathGibson				SF				
Seibert, Kathy	Abec Inc.								
Seiler, David A.	Arkema Inc.				PM				
Shankar, Ravi	Endress + Hauser USA				PI				
Sharon, Steven	Genentech, Inc.				PI	SD			
Sisto, David P.	Purity Systems Inc.				MJ				
Smith, Robert A.	Flowserve Corp.				SG				
Snow, Robert A.	Sanofi Global				SC	PM	SD		
Solamon, Michael S.	Feldmeier Equipment Inc.				MJ	SF			
Stumpf, Paul D.	ASME			SC					
Sturgill, Paul	Sturgill Welding and Code Cnsltg.	MM			S	EC	GR	MJ	
Tabor, Glyn	Eli Lilly & Co				MJ				
Tamara Fridman	Vanasyl LLC.	PM			SC				
Tanner, Scott	Garlock Sealing Technologies				SG				
Tischler, Gregory	VEGA Americas				PI				
Trumbull, Christopher A.	Paul Mueller Company				S	MJ	SF		
Van Der Lans, Albert	Janssen Biologics BV				ES				
Villela, Fernando Garcia	Stockval Tecno Comercial Ltda				DT				
Vitti, John	CraneChemPharma Flow Solution				SG				
Vogel, James D.	The BioProcess Institute		SG		PM				
Wagner, Paul	Anderson Instrument				PI				
Warn, Robert A.	Commissioning Agents Inc.				SD				
Watson-Davies, Stuart J.	PBM Inc.				ES				
Weeks, Cullen	CRB Builders, LLC				MJ				
Westin, Karl-johan	Roplan Sales Inc.			SG	SD				
Wilson, Thomas G.	Consultant				DT[1]				
Winter, Thomas	Winter Technologies		GR		DT				
Wise, Daniel	Genentech, Inc.				SG				
Woods, Gary	Cross Point Engineering Grp.				PI				
Wu, Nanping	Fristam Pumps				SG				
Zinkowski, Richard J.	RJZ Alliances, LLC				S	EC	SG	SD	
Zuehlke, Dr. Simon	Endress & Hauser GmbH Co. KG				PI				
Zumbrum, Michael A.	Maztech, Inc.	PM			SC	EC	SG		

Table 1 – Subcommittee Subject Matter Part Identifiers

Part	Title	Part	Title
GR	General Requirements	SG	Sealing Components
SD	Systems Design	PM	Polymeric Materials
DT	Dimensions and Tolerances	CR	Certification
MJ	Materials Joining	MM	Metallic Materials
SF	Surface Finishes	PI	Process Instrumentation

To my wife
Doris
My children and their spouses
Monique and Michael
Robert and Daryl
And my grandchildren
Connor, Shayfer, and Willamina
I thank each and every one of you. Having your faith and trust inspires me to do more.

Contents

Appendices

List of Figures

List of Tables

List of Forms

Series Preface

The *Wiley-ASME Press Series in Mechanical Engineering* brings together two established leaders in mechanical engineering publishing to deliver high-quality, peer-reviewed books covering topics of current interest to engineers and researchers worldwide. The series publishes across the breadth of mechanical engineering, comprising research, design and development, and manufacturing. It includes monographs, references, and course texts. Prospective topics include emerging and advanced technologies in engineering design, computer-aided design, energy conversion and resources, heat transfer, manufacturing and processing, systems and devices, renewable energy, robotics, and biotechnology.

Preface

Scope and Intent of this Book with Early BPE History

Scope and Intent of this Book

This book is not meant to replace or act as a substitute for the American Society of Mechanical Engineers (ASME) Bioprocessing Equipment (BPE) Standard. It is instead a companion guide to the standard in providing clarification and to give basis and background for much of what is covered in the BPE Standard. And, in so doing, it has to be made clear that the dialogue and inferences made in this book are those of the author and not those of ASME. What is contained in this book are the results of decades of experience and insights from firsthand involvement in the field of industrial piping design, engineering, construction, and management, which includes the bioprocessing industry.

It is intended that this book both explain and go beyond the content of the ASME BPE Standard in helping to clarify much of its subject matter. Industry codes and standards are written in a manner that goes to the heart of a requirement or guideline without embellishment. They do not explain the reason why some statements in the standard are requirements while others are simply suggestions or recommendations. Neither does a code nor standard describe how something should be done. The reader is left with the requirement, but not the means to achieve it. This book is meant to close that gap of ambiguity to a large degree and make clear not only the standard itself but also its intent.

As various topics are discussed, you, the reader, will learn the reasons why certain things are done in a particular manner, such as electropolishing or orbital welding, and what those terms actually mean. Why are some materials passivated and others not and what does the term passivation really mean? Why mechanically polish tubing and why should piping be sloped? How much slope is sufficient and what is hold-up volume? These questions and more will be discussed and their answers made clear as we move through this book.

In an effort to make this book work in a somewhat logical manner, a manner that coincides more with the way a facility would be designed and constructed rather than the way the standard itself is structured, this book will flow in the following manner: (i) It will first of all provide information on the history of the BPE Standard immediately following this introduction to the book; (ii) following that it will describe the BPE Standard and then discuss codes and standards in general; (iii) the design and engineering aspect of the book will begin with materials of construction, both metallic and nonmetallic; and (iv) it will then touch on components. (v) After components it will get into fabrication, assembly, and installation of piping systems. (vi) It will then roll into examination, inspection, and testing; (vii) next it will discuss the ASME BPE Certification process. (viii) And finally it will bring it all together by discussing system design.

Much of the safety aspects of the BPE Standard are relegated to the ASME B31.3 Process Piping Code through references. This relates to such topics as leak testing, weld criteria, inspection, examination, and so on. Where such topics are touched on, B31.3 will be brought into the discussion. B31.3 will also be referenced in conjunction with Chapter X, the B31.3 high-purity safety arm of BPE.

Information contained in this book is based on content found in the 2014 edition of the ASME BPE Standard. It will also reveal some of the relevant supplemental data not included in the standard. Such information is residual to the large amount of accumulated data obtained during research, testing, and development as part of the process in qualifying content that ultimately finds its way into the standard through a long and arduous process, a process you can learn about in this chapter under Section C—Creating and Maintaining an American National Standard (ANS).

A small amount of essential and useable information is distilled from the macro-data that is accumulated throughout the ongoing process of improving the standard and in keeping current with ever-evolving technology. The compiled macro-data resulting from such research, testing, and development is distilled down to its elemental properties. That essential data and information destined to be included in the standard is then formed into a proposal and submitted to consensus committees in seeking approval to be added to the standard. The data and information that does not find its way into the standard is considered supplemental or residual to that found in the standard, and while it is good, viable data and information, it is simply not suitable or practical as content in an industry standard or for an industry code.

Also woven into the pages of this book are lessons learned by over five decades of work in the piping design and engineering field by the author. These are lessons not found in codes or standards, but are instead methods and procedures developed by the author to enhance the execution of a project.

ASME codes are typically not written to serve as design guides and are stated as such in their introduction. The BPE Standard though is very different in this respect as design of high-purity systems is at the very heart of the BPE Standard. This too will be made clear as you make your way through this book.

Early History and Development of the ASME BPE Standard

The foreword of ASME codes and standards contains a brief, key-point history for each code or standard. The history included in the foreword of these codes and standards contains only the essential elements of that document's creation and development with no narrative beyond that. The telling of the history of the BPE in this book will contain the names of the individuals responsible for its creation and development and will include some of the original documentation and communication of those early days. That documentation, as referenced in the following as Ref. 01, Ref. 03, and so on, can be found in Appendix K.

> *To recognize that the greatest error is not to have tried and failed, but that in trying, we did not give it our best effort*
>
> Gene Kranz—NASA Flight Director during the Gemini & Apollo programs from 1969 to 1974

Paul Kantner of Jefferson Airplane/Starship fame famously quipped that "If you can remember the 60's then you weren't really there." Well, you can take it from me that I was "really" there and I do remember those days. A huge part of the 1960s, that is, if you weren't spending most of your time stoned at some antiwar rally, was the space program. And one of the rock stars, yes, right up there with Jim Morrison and Joe Cocker, was a guy by the name of Gene Kranz. While the press and most of the public were fixated on the astronauts and the spectacle of the Apollo launches, many of us were transfixed on Gene Kranz. As NASA Flight Director (1969–1974) he was calling the shots and running the show on the ground from the Mission Operations Control Center in Houston, TX.

Gene was an unassuming technical rock star who came to full notoriety during the Apollo 13 crisis. And I still, to this day, am in awe of their very publically documented meetings and their live camera feeds during that nail-biting space emergency that placed their discussions and decision-making process, throughout that crisis, under the public microscope—talk about pressure. That was an era of exciting things, monumental engineering feats in which massive rockets were blasted into space and men actually walked on the moon. This was the time of Apollo 13 in which engineers and technicians, guided by the indefatigable Gene Kranz, created magic and saved lives with chewing gum and bailing wire changing certain defeat into a lifesaving victory.

In my mind Mr. Kranz has since become an analogy. One against which other people and other actions are measured. To say that there are no engineers, technicians, or managers among us that do not measure up to Gene Kranz today and in years past would be a mistake. There *are* men and women involved at the many levels and facets of our little niche in the world—bioprocessing—that do perform such work, but go, for the most part, unnoticed, doing work that perhaps may not rise to the grand scale of building and launching manned Apollo rockets, but nonetheless do take on daunting tasks that others choose to ignore, feel inadequate for the job, or simply shy away from such perceived challenges because the task seems too overwhelming.

But people analogous to Gene Kranz do seem to appear from nowhere and seemingly at the right moment in time. They are the same individuals who first founded ASME in 1880, who initiated discussion on the creation of the Boiler and Pressure Vessel Code (BPVC) first published in 1914/1915, and who have taken on tasks throughout the years to create the B31 Piping Codes, the B16 Standards, and many other meaningful documents used throughout the world. Typically these are unassuming but internally dynamic individuals who understand the critical need for something and have the intellect, work ethic, and conviction to step forward and be the driving force in creating that something from nothing.

Contained in the foreword of documents such as the B31 Piping Codes and the BPE Standard, there can be found a history of that particular document in timeline form. Meaning, it provides a very clinical interpretation of the history of the document in nothing more than an almost bulleted format. There are no names mentioned and there are certainly no real specifics, and justifiably so. The foreword in a technical standard or code is not the proper place to expound on a detailed history of how that document was conceived and developed. But that does not preclude the fact that the names and details of how such work was accomplished should be captured in some narrative written perhaps under separate cover. This is just such a narrative.

With the full support of their sponsoring company, many of the individuals who have spent countless hours of their own time and incurred the cost to spearhead the creation and development of the codes and standards we are all so familiar with will, most likely, never be known beyond their immediate associates. In that same context it must also be said that ASME and other such organizations do go to great lengths to recognize as many individuals as practical who have dedicated their time and effort in their ongoing work on codes and standards in a multitude of capacities.

For those that might otherwise go unrecognized along with the story behind the work they have done, a story to most likely be filed away in the back of a dusty file cabinet and shoved into the back room of history, their narrative, at least in this case, is herein documented before being lost through attrition. This narrative describes the inception and development of the BPE Standard with information taken from interviews with some of the key forward-thinking individuals involved in the difficult task of turning a need into an idea, into a plan, into a committee, and finally into the BPE Standard.

This narrative will take you through the very moments that the need was recognized, the idea came to light, a plan was formed, a committee was assembled, and the BPE Main Committee approved. In addition to those key moments in the history of the BPE Standard, we will also read about anecdotal moments that show both the resilience of its members and the development of friendships, friendships, in many cases, made for a lifetime, a quite unexpected bonus forged of hard work, long hours, and dedication to a singular objective.

Origins

In the mid-1980s Genentech had invested in a start-up by the name of Verax Corporation. Verax at the time was developing a design for a continuous fluidized-bed bioreactor. Bill Cagney, working for Genentech during that period, was assigned to support Verax in converting their bench research prototype into a pilot plant skid design suitable for scale-up GMP manufacturing.

Genentech had recently completed the design and construction of a large-scale cell culture manufacturing facility for recombinant human tissue plasminogen activator (rhtPA) that Bill had been involved in and had, like most of us at that time, gone through significant struggles to obtain the high-purity equipment and achieve the type of high-purity piping system design needed for a system of this type to function properly.

These were issues that anyone designing, engineering, or constructing high-purity systems had to deal with at the time. In order to obtain acceptable, high-quality welds on a consistent and acceptable basis, orbital welding, which was just coming into its own in those years, was the preferred method of welding. However, in order to achieve such consistent and acceptable welds on a repeatable basis between two welded components, the sulfur content of 316L stainless steel, the preferred material of construction then and now, had to be within a very narrow range, a range requirement that was nonexistent at the time. In the years prior to 1995, the maximum limit of 0.030% sulfur was indicated in the American Society for Testing and Materials (ASTM) material specification without option.

In order to dance around this issue, it was necessary to specify that all welded components had to be purchased from the same heat of material, meaning that suppliers would have to corral all weld fittings for a specific project with documentation verifying that they were manufactured from the same material heat. If this was not achievable, it was acceptable to alternately use weld fittings from heats of material in which the content of their sulfur was all reasonably close in comparison. "Reasonably close" was at the owner's direction; there were no industry recommended, much less standard values for such a range. Some companies also took the approach of developing techniques and procedures using mixed gases to overcome these same welding issues.

Aside from the welding issues, there were also issues with hygienic clamps not fitting properly over the ferrules. Poorly fitting clamps would leave installers with a false sense of having created a leak-tight seal. In many cases a clamp could be tightened against its own metal (metal to metal) and still spin loosely over the ferrules leaving less than a desirable compression load at the ferrule and gasket interface. In order to increase the odds of a clamp fitting properly over mated ferrules and gasket, the clamps were specified to come from the same manufacturer as that of the ferrules. This created logistical issues when having to purchase replacement parts in later years by having to return to the same manufacturer time and again, essentially eliminating competitive pricing and no options when delivery was an issue.

In addition there were no standard guidelines for high-purity pressure vessel design, no standards on fitting dimensions and tolerances, and no standard guidelines or requirements for slope, drainability, dead legs, and surface finishes, and the list goes on. The 3A Food and Dairy Standards somewhat filled the gap, but fell way short of what was really needed for the bioprocessing industry and pharmaceutical industry in general.

As Mr. Cagney worked with Verax, he conveyed his concern to them over these same issues. Those concerns reached the ear of Robert C. (Bob) Dean, Jr., founder of Verax. Bob had been very involved with ASME over the years and was actually an ASME Fellow, becoming a recipient of the ASME Gold Medal in 1996. But in 1988 he confided in Mr. Cagney that he was surprised to discover that there were no industry standards to guide the requirements for the design of equipment for pharmaceutical use.

Rather than keeping the issue internal to his company by choosing to simply develop their own set of specifications in order to rectify and move through the problems posed by Bill for that particular project and move on, like most of us did back then, Bob went the extra distance. He decided to take action by asking Bill if he would be interested in running an industry forum at the upcoming ASME 1988 Winter Annual Meeting in Chicago to test the waters and perhaps solicit industry support in developing a standard for the emerging bioprocessing industry. This singular action that Bob took in approaching Bill with this idea was the genesis of the BPE Standard. If you are looking for the spark of life for the BPE, this was the moment.

Bill Cagney did indeed accept the opportunity to chair the panel discussion at the 1988 ASME Winter Annual Meeting. In preparation for that meeting in Chicago, Bob had worked with the ASME meeting facilitators in getting the panel forum onto the WAM meeting agenda and invited the following individuals to participate in the discussion panel. They were:

William H. (Bill) Cagney (Panel Chair)—Genentech
Tony Wolk—US FDA
Ivy Logsdon—Eli Lilly
Dave Delucia—Verax
Dave Cowfer—Southwest Research Institute
Rick Zinkowski—ITT Grinnell
Cas Perkowski—Stearns Catalytic
Bob Greene—Fluor Daniel
Mel Green—ASME

Dave Delucia was one of the main facilitators at this meeting and helped drive the meeting. In addition, two individuals involved with the work being done at Verax were present as audience participants. They were:

Frank J. (Chip) Manning—TEK Supply, Inc.
Randy Cotter—Cotter Bros.

The 1988 Winter Annual Meeting was one in which the panel sat on an elevated dais and the audience were assembled in a theater-type seating arrangement. After much active discussion and debate from both the audience and the panel, the unanimous consensus, from all participants, was that a standard was most certainly needed and it was agreed that the process of developing a BPE Standard should move forward to the next step in its development.

At an interim meeting it was decided that an Ad Hoc BPE Committee would be formed. Dave Smith, who had recently guided a new ASME Standard for Reinforced Thermoset Plastic Corrosion-Resistant Equipment, ASME RTP-1, through development and approval, was tapped to serve as chair of this new Ad Hoc BPE Committee, and he accepted. On December 6, 1989, a letter notifying Dave (Ref. 01) that he had been appointed chair and a letter notifying Bill Cagney that he had been appointed vice-chair of the Ad Hoc BPE Committee had been sent out.

Leading up to the 1989 Winter Annual Meeting, various companies heavily invested in the pharmaceutical industry at the time had been contacted inquiring as to their assistance in recommending and supporting the nomination of employee delegates to participate as members in the new Ad Hoc BPE Committee.

Lloyd Peterman recalls that:

> *...someone, or some group, I don't recall which, solicited resume's of people who might be qualified for the first ad hoc group. I knew that several people within Tri-Clover, as well as several other companies, had applied. I believe I was selected because I started working with Upjohn and Eli Lilly in 1972 to replace their glass lines with stainless.*

At the 1989 Winter Annual Meeting in San Francisco, CA, resumés were received in response to the solicitation and were reviewed. From that review a list of potential members was developed. To that small but talented group of designees, acceptance letters then went out (Ref. 02). These were like-minded individuals who could bring experience and know-how to the committee in helping to create something from nothing. Many of them answered the call and agreed to participate in the development of the BPE Standard. On February 23, 1990, letters went out (Ref. 03) to those respondents, which included the following:

Hans Koning-Bastiaan—Genentech, Inc.
Frank J. (Chip) Manning—TEK Supply, Inc.
Richard E. Markovitz—Trinity Industries, Inc.
Theodore Mehalko—Kinetic Systems, Inc.
Joseph Van Houten—Merck & Company, Inc.
Ivy Logsdon—Eli Lilly and Company
Lloyd Peterman—Tri-Clover, Inc.
Frederick D. Zikas—Parker Hannifin Corp.

The letters notified them that they had been appointed as members to the Ad Hoc BPE Committee.

On March 1, 1990, the Ad Hoc BPE Committee held its first meeting in the ASME offices in New York. The minutes of that meeting (Ref. 04) reflected attendance of the following members:

David Smith (Chair)—United Engineers and Constructors
William H. (Bill) Cagney (V. Chair)—Biogen, Inc.

Hans Koning-Bastiaan—Genentech, Inc.
Frank J. (Chip) Manning—TEK Supply, Inc.
Richard E. Markovitz—Trinity Industries, Inc.
Theodore Mehalko—Kinetic Systems, Inc.
Joseph Van Houten—Merck & Company, Inc.
Ivy Logsdon—Eli Lilly and Company
Lloyd Peterman—Tri-Clover, Inc.
Mark E. Sheehan—ASME Secretary
Frederick D. Zikas—Parker Hannifin Corp.

Visitors attending that first meeting, as reflected in the minutes, included:

Pat Banes—Oakley Services, Inc.
David Baram—Vanasyl Valves, Inc.
Nigel Brooks—Fluor Daniel
Anthony (Tony) Cirillo—Jacobs Engineers
Randy Cotter—Cotter Corp.
Robert Daggett—Allegheny Bradford Corp.
William M. Dodson—Precision Stainless
Randolph Greasham—Merck & Co.
Barbara Henon—Arc Machines, Inc. (AMI)
Peter Leavesley—Membrex, Inc.
Julie Lee—ASME BPEP Manager
Tom Ransohoff—Dorr-Oliver
Arlene Spadafino—ASME Codes and Standards
Rick Zinkowski—ITT Engineered Valves

The purpose of that first meeting, as stated in the meeting minutes, was to:

> *...determine the need for a standard and to initiate preparation of a presentation to the Board on PTCS describing said need. Since these meetings are to be conducted under ASME policies, all meetings are open to the public and guests are invited to participate in discussions, but not in voted actions.*

The minutes went on to state, under Para. 90-3-5, that:

> *This Ad Hoc Committee has been charged with the development of a recommendation to the Board on PTCS with respect to the formation of a Main Committee to develop and administer Standards for bioprocessing equipment. The scope of such standards was stated by the Council as follows:*
>
> *This standard is intended for design, materials, construction, inspection, and testing of vessels, piping, and related accessories: such as pumps, valves, and fittings for use in the biopharmaceutical industry. The rules provide for adoption of other ASME and related national standards and when so referenced become a part of the standard.*

The minutes go on to state:

Assuming that this committee determines that there is a need for these standards and that the Board on PTCS approves the formation of a new Main Committee, the next task for this group would be to develop a proposed charter for the committee and subsequently develop preliminary operating procedures for the Main Committee. The development of the charter may involve some fine tuning of the scope of the standards to be developed.

The March 1990 issue of the Bioprocess Engineering Program (BPEP) Newsletter (Ref. 04-13) had captured the highlights of the pre-Ad Hoc BPE Committee meeting during the 1989 ASME Winter Annual Meeting at which the group discussed a path forward. The article describes the meeting as follows:

The annual session at the ASME Winter Annual Meeting has become a popular venue for informal public discussions between all interested parties involved with bioprocess equipment. A panel-facilitated dialog culminated in a consensus that further pursuit of standards and references for bioprocess equipment is essential to the industry.

The article then went on to include key points discussed at the meeting in addition to the organization of "the ASME Ad Hoc Standards Writing Committee" and also provided information for application to join the committee.

It was also noted in the meeting minutes that there would be "...three meetings a year for starters with a basic format consisting of one day of subcommittee meetings followed by a second day when the Main Committee would meet."

In that same Para, 90-3-11 of the minutes, the chairman noted that, in anticipation of "*...timely action by the Board on PTCS, it was decided to hold a meeting as soon as possible after the Board takes action.*" That next meeting was scheduled for June 27, 1990, at the ASME offices in New York. On June 4, 1990, the Board on PTCS met and "*Based on the presentation prepared by the Ad Hoc Bioprocessing Equipment Committee and presented by* [Dave Smith] *as Chairman* (Ref. 11)*, the Board on Pressure Technology Codes and Standards (BPTCS), at its 6/4/90 meeting, took action to:...*" and it goes on to say that it recommended to CCS "*...the formation of the Bioprocessing Equipment Main Committee (BPE) with a scope to read: Design, materials, construction, inspection, and testing of vessels, piping, and related accessories such as pumps, valves, and fittings for use in the biopharmaceutical industry.*" Along with that the board on PTCS also approved the officers and members.

The very next day, on June 5, 1990, the Council on Codes and Standards (CCS) met and approved the recommendation of the board on PTCS creating the BPE Main Committee. On June 26, 1990, letters (Ref. 05) reaffirming the membership of officers and members to the newly created BPE Main Committee were sent out. Those letters went to:

David Smith (Chair)—United Engineers and Constructors
William H. (Bill) Cagney (V. Chair)—Biogen, Inc.

Hans Koning-Bastiaan—Genentech, Inc.
Frank J. (Chip) Manning—TEK Supply, Inc.
Richard E. Markovitz—Trinity Industries, Inc.
Theodore Mehalko—Kinetic Systems, Inc.
Joseph Van Houten—Merck & Company, Inc.
Ivy Logsdon—Eli Lilly and Company
Lloyd Peterman—Tri-Clover, Inc.
Mark E. Sheehan—ASME Secretary
Frederick D. Zikas—Parker Hannifin Corp.

At the second meeting, held on June 27, 1990, Lloyd Peterman made meticulous notes for his in-house report (Ref. 12) to Tri-Clover. In it he writes:

Ten of the twelve original committee members were present, and we were immediately informed that this is no longer an ad hoc committee inasmuch as this comm. was accepted by the Board of ASME to become a full-fledged committee which will probably last a lifetime. Our official designation within ASME is BPE-1. Further the original 12 members (writer included) had our terms approved for a five-year period ending June 30, 1995.

Lloyd goes on to write, in those same notes:

One of the visitors, but who also was elected to the permanent committee, Randy Greashman, Director, Bio-Process Research, Merck & Co., in Rahway, NJ, is really pushing frantically for quick results of this comm. He actually stated he wd. prefer a document by the end of 1990; however, the group as a whole, especially those with more experience with these types of meetings, said they would be quite happy with a document by the end of 1991.

I tend to think that the group's naïveté worked in the industry's favor. Had they known what lay ahead, there may have been a very different outcome.

At the following meeting that took place in Dallas, TX, November 28–30, Lloyd writes in his notes (Ref. 13):

"At this meeting, the (4) Task Groups assigned at the previous meeting were scheduled to give a brief report on their activities during the last two months." He goes on to say, "All these (4) Task Groups were voted upon to become full-fledged subcommittees. The writer [Lloyd] was selected as Chairman for the sub-comm. on gaskets and seals.

Contained in the May 1991 issue of the BPEP Newsletter (Ref. 06) was an article on the fledgling BPE Standard. It read, in part:

The ASME BPE Standards Writing Committee continues to expand and formalize its subcommittees.

There are currently 23 members on the main committee divided into the following 6 subcommittees: Dimensions and Tolerances; Material Joining; Gaskets and Seals; Surface Finish; General Requirements; and the newly formed subcommittee, Equipment Design for Sterility and Cleanability.

Later that same year the ASME PR machine went to work soliciting members and notifying interested parties of this new standard. A pamphlet (Ref. 07) produced by Lloyd Peterman and mailed out in late 1991 stated up front that:

"The ASME Committee of Bio-Processing Equipment (BPE) is actively soliciting participation of all interested parties involved in bioprocessing. The BPE committee was formally established and approved by the ASME Board of Pressure Technology Codes and Standards (BPTCS) in June 1990 to create a standard for design, materials, construction, inspection, and testing of vessels, piping, and related accessories such as pumps, valves, and fittings for use in the biopharmaceutical industry." The standards may provide for adaption of other ASME and related national standards, and when so referenced become a part of the standard.

The pamphlet also announced that the BPE Committee had met six times since their "*establishment*" in June 1990, their most recent meeting having been September 12, 1991. It went on to list the officers and subcommittee chairs on the BPE Main Committee at that time. That list included:

David Smith (Chair)—United Engineers
William H. (Bill) Cagney (Vice Chair)—Abbott Biotech
Frank J. (Chip) Manning (Dimensions and Tolerances Chair)—TEK Supply
Randy Cotter (Welding Chair)—Cotter Corp.
Lloyd Peterman (Seals Chair)—Tri-Clover, Inc.
Ted Mehalko (Surface Finishes and Cleanliness Chair)—Kinetic Systems
Casimir A. (Cas) Perkowski (General Requirements Chair)—United Engineers
Nigel Brooks (Equipment Design for Sterility and Cleanability Chair)—Fluor Daniel

After six years of earnest work in establishing the framework of a main committee and six subcommittees, creating a formula for functioning in a harmonized manner and creating content for what would become the international benchmark standard for the bioprocessing and pharmaceutical industry, an initial draft (Ref. 08) of the first ASME BPE Standard was finally pulled together and printed on April 16, 1996. A final proof of the standard was issued on February 14, 1997, and on May 20, 1997, the American National Standards Institute (ANSI) approved the first edition of the ASME BPE Standard. On May 21, 1997, the ANSI Board of Standards Review submitted a letter of approval (Ref. 09) to Ms. Silvana Rodriguez-Bhatti, Manager, Standards Administration, ASME, acknowledging approval of the ASME BPE Standard. The standard was finally printed on October 17, 1997, followed five months later by an ASME press release (Ref. 10) issued on March 14, 1998.

It has since grown and improved with every edition since, which includes a 2000 Addenda, a new edition in 2002, a 2004 Addenda, and subsequent editions in 2005, 2007, 2009, and 2012. In that time period from the inaugural 1997 issue to the 2012 issue, the BPE Standard grew from 108 pages in 1997 to 292 pages in the 2012 issue. Membership reflected in the 1997 issue lists fifty-nine personnel and includes the following:

D. Smith, *Chair*, Raytheon Engineers & Constructors
A. P. Cirillo, *Vice Chair*, Life Sciences International
P. D. Stumpf, *Secretary*, the ASME
D. S. Alderman, Waukesha Cherry-Burrell
W. R. Anton, Advance Fittings
G. A. Attenborough, Fluor Daniel, Inc.
M. A. Atzor, Bayer Corp.
D. D. Baram, Consultant
H. D. Baumann, *Chair SG*, H. D. Baumann Assoc., Ltd.
R. Becker, Tri-Clover, Inc.
E. A. Benway, Cajon Co.
I. Bemberis, W. R. Grace Amicon, Inc.
W. K. Black, Cashco
R. L. Boraski, Robert James Sales
N. R. Brooks, *Chair SD*, Fluor Daniel, Inc.
C. R. Brown, Whitey Co.
W. H. Cagney, IDC
R. D. Campbell, Welding Solutions
G. W. Christianson, Ribi ImmunoChem Research, Inc.
D. C. Coleman, ICOS Corp.
R. A. Cotter, *Chair MJ*, Cotter Corp.
S. D. Dean, Pall Trinity Micro Corp.
J. A. Declark, Trent Tube
P. G. El-Sabaaly, *Chair SF*, Alloy Products Corp.
D. F. Fijas, American Precision Industries
B. E. Fisher, Waukesha Cherry-Burrell
T. J. Gausman, Nupro Co.
K. Gilson, Kinetic Systems
G. C. Grafinger, ITT Standard
J. W. Harrison, Rath Manufacturing Co.
B. K. Henon, AMI
M. A. Hohmann, Eli Lilly & Co.
T. Hoobyar, Asepco
V. L. Horswell, G&H Products, Corp.
S. T. Joy, Jenson Fittings Corp.
M. W. Keller, Amgen, Inc.
A. G. Leach, APV Crepaco, Inc.
J. T. Mahar, Dorr-Oliver
J. R. Maurer, Allegheny Ludlum Steel

F. J. Manning, *Chair DT*, TEK Supply, Inc.
T. Mehalko, TMA
F. Menkel, Bayer Corp.
M. C. Miller, FST Consulting Group
J. P. Netzel, John Crane, Inc.
C. A. Perkowski, Promega Corp.
L. J. Peterman, Tri-Clover, Inc.
E. L. Sandstrom, Pall Trinity Micro Corp.
A. Shekofski, Lederle-Praxis Biologicals
D. P. Sisto, Purity Systems, Inc.
T. Sixsmith, Advanced Ind.
P. Smith, Spirax Sarco
S. R. Swanson, Tri-Clover, Inc.
D. Todhunter, The Seal Source, Inc.
R. E. Trub, Alfa Laval Separations
C. A. Trumbull, Cotter Corp.
W. J. Uridel, Badger Metal Sales
R. T. Warf, Merck & Co.
J. A. Yoakam, Advanced Microfinish, Inc.
R. J. Zinkowski, ITT

In comparison the 2012 issue of the BPE Standard lists a membership of 185, a growth in excess of 300%.

One group of individuals that cannot be left out of the telling of this narrative are the ASME BPE secretaries. These men and women have been and still are instrumental in guiding the leadership and assisting the BPE general membership in so many facets of the functioning of an organization such as the ASME BPE Standard. In the years leading up to the first edition of the BPE Standard in 1997, in which Paul Stumpf served as secretary, there were five others that preceded him. They include:

Mark Sheehan
S. Weinman
G. Eisenberg
J. Gonzalez
Christine Krupinsky

Like Paul, his predecessors served as an irreplaceable and integral part of the BPE Standard Committee helping the leadership as well as the general membership of the BPE Standard with any myriad of things from guiding committees through the procedural spider web of regulatory conformance issues to assisting on balloting protocol and mediating hotel meeting room issues.

Throughout Paul's tenure with the BPE Standard, he has also become a good friend and colleague of the BPE leadership teams whom he has traveled the globe with. Paul is always on call serving as the standard's short-term and long-term memory for each sitting BPE chair, reflecting the attributes of a chief of staff rather than a secretary.

Exceptional Leadership Time and Again

Typically leadership walks a fine line between cajoling and harassment of, in this case, a volunteer membership in attempting to reach a specific set of goals. What seems to occur with the role of chair in the BPE organization is something tantamount to a NASCAR driver sitting perched in a Formula 1 race car; he is certainly someone who has driven fast before, but not like this. With a membership made up largely of self-made type A individuals, whoever takes on the leadership role of such a membership group has to be prepared to meet the challenge or get the pointy end of the spear. From its very beginning the BPE has been fortunate in both the availability and its selection of chairs and vice-chairs.

The hard, timely, and diligent work of the BPE Standard membership, and in particular its leadership, has made it the poster child, or benchmark if you will, for standards development. As Tony Cirillo, past chairman (1999–2005), has said, and I paraphrase, "*The BPE has set the bar for the time it takes to develop a new standard and for its inroads into the international marketplace*."

As you will notice in the text outline that follows, but is more readily discernible in Figure 1, is the overlap of personnel in the vice-chair and chair offices throughout the years. This overlap has provided a great deal of continuity in the passing of responsibility that has not gone unnoticed. The various individuals who have selflessly given of their time and expertise in serving as chair and vice-chair for the BPE Standards Committee include Dave Smith, the first chair of the BPE Standard, being transitioned into that office by the CCS in June of 1990 and continued as chair of BPE until June 1999. William H. (Bill) Cagney served as vice-chair under Dave until 1993 at which time Anthony P. (Tony) Cirillo was elected as vice-chair. He would serve in that capacity until 1999. At that time Tony was then voted in as chair of the BPE Standards Committee. Tony served three consecutive three-year terms and retired from that position in June 2008. Barbara Henon served as vice-chair under Tony from 1999 to 2005 when Rick Zinkowski was elected as vice-chair. Rick would serve as vice-chair under the remainder of Tony's term as chair until 2008. Jay Ankers was duly elected as chair of the BPE Standards Committee in 2008 with Rick Zinkowski continuing to serve as vice-chair. Rick would step down in 2011 to be replaced by Marc Pelletier as vice-chair under Jay.

The nominating committee for the upcoming 2014 elections for the offices of chair and vice-chair of the BPE Standards Committee nominated both Jay and Marc to continue serving in their respective positions. This decision was arrived after all prospective nominees had been interviewed and chose to waive the right to be nominated on the presupposition that Jay and Marc would continue for another term in their respective positions, a huge vote of confidence.

A Never Ending Work in Progress

In the early stages of development, thinking among the fledgling membership at that time quickly coalesced into establishing six primary categories or parts for the standard, SD being the last. These six parts would supposedly cover all of the

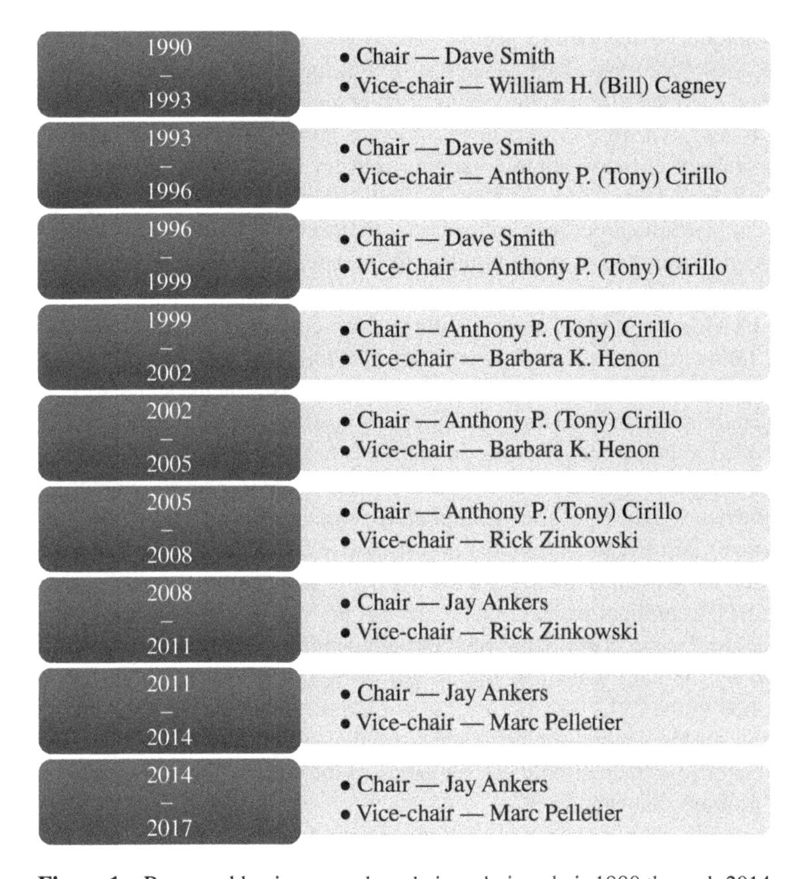

Figure 1 Personnel having served as chair and vice-chair 1990 through 2014

criteria identified as applicable to what the BPE Standard should encompass. The 1997 issue of the BPE Standard reflected that thinking by publishing the standard with the following six parts:

1. General requirements (Part GR)
2. Design relating to sterility and cleanability of equipment (Part SD)
3. Dimension and tolerances (Part DT)
4. Material joining (Part MJ)
5. Surface finishes (Part SF)
6. Seals (Part SG)

As Tony Cirillo, vice-chair at the time, recalls, in Philadelphia at the 1996 meeting, the decision was made to publish the standard the following year in 1997. At the executive meeting, chaired by then vice-chair of the BPE Standard Committee Tony Cirillo, Tony went around the table of subcommittee chairs and asked where each one was with relation to their part being complete.

Each announced that their respective parts were ready, that is, until he got to Nigel Brooks who confessed that he was still working on expanding SD. Tony then stood up and said to everyone, "...*as of right now stop. Everything you're doing, stop. Tweak it. Make it right; because we're going into publication next year.*" There was simply so much low-hanging fruit for Nigel and the SD group; it was like kids in a candy store. And as you will see in the following paragraph, and quite simply throughout the various issues themselves over the years, a line had to be drawn in the sand at some point or the standard would have never been published; it is to this day still expanding.

To provide some basic understanding of the continued expansion of the standard, following its inaugural issue, Part PM for polymers and elastomers was added in 2002. This effort was championed by Ted Hutton the standard's resident nonmetallic guru. In 2009 Part MMOC, spearheaded by the untiring Ken Kimbrel, was added to house the specifics on materials being used in the industry. Also in 2009 Part CR for certification, a segment of the standard that required the strong will and determination of Rich Campbell to forge into place, became a reality. The addition of Part CR was prompted in response to having to gain an upper hand on the issue of nonconforming components being manufactured and distributed as being BPE compliant when in fact there was nothing in place to assure that compliance. And again in that same year, the subcommittee for Part PI was introduced, guided by the hard work and tireless effort of Dan Kleese. Part PI was finally published in the 2012 edition. And by not identifying the hard work and tireless effort of all the general membership also involved in the process does not deny their major role in creating these segments of the standard. Leadership is only as good as those that support it.

Shepherding these various parts of the standard from inception to publication is a long and arduous undertaking, three to five years on average, an undertaking that requires a dedication to its concept, a full understanding of the need for its particular subject matter to be included in the pages of the standard, and a desire to ensure that its content provide the essential elements needed for the industry. This is no small task.

Anecdotal Recollections

Back in 1988 Barbara Henon recalls that she:

> *... received a brochure about a meeting in Chicago that would discuss materials to be used in the newly developing bioprocessing industry.*

She attended the meeting and:

> *... one of the materials being promoted was Al-6XN. The presenter said that Al-6XN required the use of filler metal during welding to prevent loss of corrosion resulting from molybdenum segregation. Since orbital welding with filler metal is not used in biopharmaceutical applications, I arranged to get some of the material and weld*

some autogenously and some using Hastelloy C-22 insert rings. Allegheny Ludlum did corrosion tests on the welds and I wrote a paper with Jack Maurer of Allegheny Ludlum at that time on these corrosion studies [in preparation for] the ASME Winter Annual Meeting in 1989. Randy Cotter was also a presenter at this same venue. I gave presentations at the ASME Bioprocessing Seminars, which at that time were being held at the University of Virginia and, for a couple of years, at organized ASME Bioprocessing Conferences being held at the ASME Winter Annual Meetings.

Barbara goes on to recall that:

I attended early BPE meetings but don't know exactly when I became a member. Randy, Chip Manning, Tony Cirillo were already on the committee at the time I joined. Dave Smith was Chair. The entire committee could go to dinner together after the meeting.

As mentioned in the opening dialogue, one of the more pressing issues that the BPE chose to address early on was that of welding issues and more specifically orbital welding. This was and still is a crucial aspect in the construction of hygienic piping systems.

Arguably one of the biggest issues pertaining to orbital welding was the max. 0.030% sulfur content in 316L stainless steel stipulated by ASTM. Resolution of this issue took place just prior to 1995, but in knowing its beginnings we actually have to go further back to 1984.

Barbara Henon worked for AMI at the time. She recalls that back in 1984:

...my boss, Lou Reivydas, handed me a paper by Fihey and Simoneau at Ontario Hydro describing the effects of sulfur on welds of 304 stainless steel. Later, in 1985 I did some research on the sulfur content of heats of tubing material supplied to me by AMI customers. Following the research I issued a memo to those same customers advising them to specify heats of 316 or 304 with a sulfur content range of between 0.005 and 0.017 wt%. This did not go over well with some suppliers or steelmakers.

When I joined MJ in the early 1990's I presented the findings on the sulfur data to them. With help from Tony and Randy, we were then able to propose this change to Chip who was chairing DT at the time who in-turn proposed it to ASTM.

Barbara goes on to say that:

... this really paved the way for achieving repeatable orbital welds in the biopharmaceutical industry. While an individual can't change an industry, a group working together can make big positive changes.

Under the leadership of Frank J. (Chip) Manning as chair of the DT subcommittee and material experts such as Jack Declark, with Trent Tube at the time and also a member of DT, the DT subcommittee hit the ground running and began to impact the industry soon after its subcommittee was organized, years before publication of the 1997 BPE Standard.

In order for the orbital welding of 316L stainless steel tubing to be used to its utmost benefit, the sulfur content between two butt weld components has to be within a range much more refined than the 0.030% max. allowed under ASTM standards as written at that time. Barbara Henon, as mentioned in the foregoing paragraphs, had handed a resolution to the DT subcommittee. Jack Declark, a member of the DT subcommittee, was also a member of the ASTM A 270 standard. The DT subcommittee realized that the best approach in mitigating the issue was to affect the root cause and add a modifier to the mill specifications. The data that Barbara had handed to DT was the answer.

The DT subcommittee, through Jack Declark, petitioned the ASTM A 270 committee for a supplement that would provide a more explicit option for weldable material to be used in high-purity piping systems. DT was looking for an A 270 option that refined both sulfur content and surface finishes that would be more compatible with the needs of the pharmaceutical industry. This petition and Jack's appeal to the committee convinced ASTM that this supplement was indeed needed. In 1995, three years prior to the actual publication of the first issue of the BPE Standard, ASTM issued a revised A 270 containing the S2 supplement stipulating the option for a much refined 0.005–0.017% sulfur content requirement.

In 2001 Tony Cirillo, chair of the BPE Standard Committee during that period, was contacted by a European tube and fitting manufacturer requesting a meeting with regard to the inclusion of a DIN material specification into the BPE. The meeting took place during the May 2001 BPE meeting at the Hilton Anatole Hotel in Dallas, TX. Uncertain as to what to expect, Tony, in preparation for the meeting, armed himself with Chip Manning and Dr. Rich Campbell.

What this visiting group wanted to discuss was the addition of the DIN equivalent to 316L stainless steel containing a sulfur range of 0.003–0.005% into the BPE Standard. Rich conveyed to the European representatives that the BPE Standard could not recommend a stainless steel chemical composition with a sulfur content in that range. As Rich explained, and I paraphrase, "*Such a low sulfur content would inhibit weld flow and penetration, providing less than acceptable results for the BPE Standard.*" Chip then went on to explain, and I paraphrase, "*We worked extremely hard with ASTM in getting acceptance and inclusion of the S2 supplements into the ASTM A 270 standard. We then worked on the mills to gain acceptance by the manufacturers. We would therefore request that you do the same in order to get DIN to meet the supplemental requirements of BPE.*"

Things were not quite the same after that. Through that meeting the BPE was fortunate enough to gain an addition to its membership that still exists to this day in the form of Dr. Jan Rau, one of the two European representatives at the meeting. Jan has since been a selfless advocate in the giving of his time and dedication to the betterment of the BPE Standard, helping to keep it aligned with the needs of the international community.

Many of the industry-accepted criteria, in the days prior to the BPE Standard, were taken as sacrosanct, but had indeed no substantive basis of origin. These anecdotal criteria were nevertheless carried over into the standard until such time as the industry-accepted values could be researched. One such criterion was the slope issue.

For hygienic requirements, such as those to be included in the BPE Standard, the need to establish clarity for cause and foundational empirical data in creating a baseline of values for slope requirements, and any other requirements for that matter, was and still remains critical to the BPE's goal of getting it right.

It was because of this tenet that Cotter Corp., guided by Randy Cotter back in 1992, elected to sponsor testing that established slope requirement values that are the basis and determinant factor for slope requirements used in the BPE Standard and elsewhere around the globe.

This testing and analysis takes into account the variables in surface tension between mechanically polished and electropolished stainless steel as well as weld joint concavity and convexity. Like the work that Barbara had done with the sulfur content issue, these were early signs that the content within the BPE Standard would be content that would not only change the industry but also do so at a level of scrutiny and consideration that would be hard to match.

In 1999, soon after Tony Cirillo became chair of the BPE Standards Committee, he approached Ernie Benway, who worked for Swagelok at the time, about how many facilities Swagelok had around the world. When Ernie told him about their facilities in Europe and Asia, Tony then went to Paul Stumpf and asked that ASME provide Ernie with as many BPE Standards as he could use in distributing them on his overseas trips. The same was done with others such as Lloyd Peterman who was uniquely invested in Asia. This effort expedited acceptance of the BPE Standard throughout the world.

In a 2003 e-mail between Tony and Lloyd Peterman, Tony is imploring Lloyd, who is now working for United Industries, who by the way became the first recipient of the BPE Certification Program in January of 2013, saying that "*I once again need your help. We need to get moving on promoting the standard internationally*" asking Lloyd if he could put together a list of countries that are using the standard. In response Lloyd provided a list of 19 countries that at the time were using the standard.

Tony recalls that at the Raleigh meeting back in January of 2001, Nigel Brooks cornered him between meetings exclaiming that he "*… simply could not continue to chair SD. It is becoming too much so I will need you to find someone to replace me.*" To understand the angst that was apparently being felt by Nigel, it would help to at least understand the person to a very limited extent. Along those lines Barbara Henon writes:

> *I do remember Nigel Brooks showing slides of mechanical seals that were so complex they rivaled diagrams of Intel chips. He was a very intense individual.*

After his angst-filled conversation with Nigel, Tony's thoughts immediately turned to a recent attendee by the name of Jay Ankers as a potential replacement. Jay had attended only two other BPE meetings that had taken place in May 2000 at Atlantic City followed by a September meeting in San Francisco. In those first two meetings, Jay had made quite an impression on the SD membership and apparently had impressed a few others outside of SD. Word had gotten back to

Tony, and fortunately for the BPE he drew the right conclusion and approached this, fresh off the boat, member who had barely gotten his feet wet in his work with the standard—talk about getting tossed into the deep end of the pool, one full of sharks.

Jay recalls that:

> *...the first meeting I attended was in Atlantic City with Thys Smit, my friend from Fluor Daniel who worked with Nigel Brooks and I in the U.S. and South Africa. Thys and I heard about the BPE from Nigel while at Fluor Daniel. He had invited us to our first meeting. The SD Subcommittee meeting was in a HUGE room with 7-8 people. We went through the WHOLE SD Part with David Baram making editorial comments..., that were actually all very good.*

He remembers Tony talking to him during the 2001 Raleigh meeting about taking over SD. It helps to imagine this new guy being asked to take over a subcommittee with only two meetings under his belt. He has to be thinking this is probably normal...right? How little he knew.

Fortunately for the BPE Jay could not have possibly known what he was getting into and accepted the position of SD chair. In thinking back he recalls that:

When Nigel Brooks handed me his files (box and electronic) he handed me an OLDER version of the SD Part, the one that had the Sprayball Testing from before the update that Tony Cirillo added. After it was published in the 2002 issue and we next met in San Diego Tony called me into a meeting with just him and me to ask, "What the hell happened?!?" That was my first publication as Chair of Part SD and Nigel got me good! We issued an errata revision immediately.

That erratum went out in 2004.

It is difficult to say, but Marc Pelletier was apparently a bad influence on Jay during his early days in the BPE. He caused Jay to become addicted to golf during some of the early meetings in Puerto Rico. You can almost picture Marc "the pusher" Pelletier standing at the first tee, 9 iron in hand pushing his drug of choice on to Jay: ahhh the feel of the soft tropical breeze against sun-tanned skin, the smell of hibiscus in the air as you drive through the lush green, undulating fairways chasing that white dimpled object that just won't go the distance or won't stop hooking to the right, who wouldn't fall victim to that. As Jay recalls:

> *... we used to play the old Bahia Beach course, located just east of San Juan. Before the SD Subcommittee meetings Marc and I could get in several rounds of golf; showing a severe lack of commitment to Tony.*

He goes on to say that:

> *When Tony and Humphrey [Murphy] planned the big launch to Europe, starting in Cork Ireland [in 2001] Marc and I planned a one week golf trip around Ireland with our wives. We rented a van and played a great Irish course each day the whole week, staying in different B&Bs and towns every night. That did not go over well with Tony, especially when we rolled into Cork just before the meeting started on that Tuesday;*

or worse yet, during the meetings when several of us would get up early to go play a scheduled round. Years later in Dublin [June 2004], Tony took us out to the countryside for a meeting. The hotel had a golf course with the first tee near the front door! I remember walking into the Executive meeting with my golf shoes on and grass stains on my pants.

In 2005 Tony would be going into his third three-year term, which would end in 2008. At the May 2005 meeting in Dublin, Tony, along with Paul Stumpf and Jennifer Delda, pulled Jay to the side as he did in Raleigh and asked him, once again, to step up to the plate and allow his name to be nominated for chair of the BPE Standards Committee.

Jay of course said yes and has proven over his past two terms that Tony was right once again. This was confirmed in a statement contained in his letter of nomination from the Nominating Committee for the 2014 elections, which states:

Mr. Ankers has proven, throughout his past two terms in that office, to be a dedicated and passionate leader and advocate of the BPE Standard. He has guided the growth and continued development of the BPE Standard during a time in which it experienced a 215% growth in content from its 2007 edition through its 2012 edition. During that same period the BPE Standard has been acknowledged and accepted into the 2012 edition of the B31.3 Process Piping Code, and has experienced a substantial increase in its acceptance and impact as an internationally accepted industry standard.

During Jay's first two terms, the BPE's expansion and use throughout Asia and South America as well as the involvement from both regions has substantially increased as an offshoot of the international push under Tony's watch and Jay's persistence. As Jay puts it:

There is no question that Lloyd Peterman blazed the trail for the BPE in Asia and SE Asia. Sei Murakami, our BPE Japan delegate, connected ASME to all the engineering codes and standards in Japan better than anyone ever could or, I expect, ever will. Let's just say the ASME BPE was well established across the Pacific when I started working full time in Singapore in 2006.

Folks like Lloyd Peterman, Chuck Chapman with Gemu and Carl [Kettermann] with Rath were writing articles (I am sure there were many others I am forgetting). I had a couple of Large Biopharm Projects in Singapore and I was the Chairman at the time, so it was as easy for me in SE Asia as when Tony was building projects in Ireland to expand the BPE. Those two little islands (with no natural resources) allowed a US standard and US goods to flow naturally.

He goes on to say that:

We set up many of our BPE folks with Singapore reps. who capitalized on the 4–5 large biopharm projects within a 5-block area in Tuas. I think I had over 50 BPE colleagues visit Singapore in three years. We changed sections of the BPE based on Ministry of Manpower requirements (some on the fly as Roche was building ECP-1 and needed "re-closable relief devices that were BPE compliant" The re-write was approved and literally FLOWN to Singapore the next week to show the officials. Not procedurally correct…but hey…it showed ASME's responsiveness.

When we had minor welding issues, Barbara Henon would answer questions that same day. The BPE was used in every step of the process by Kenyon Engineering, our Singapore high-purity contractor and they got the full support of all our BPE suppliers.

Our expansion into China was really the hard work of Lloyd Peterman and Jason Kuo with King Lai. King Lai Fittings and Tubing invited the chair and vice chair to present at China Pharm in 2011. They additionally arranged for some very high profile meetings with China's FDA and the US Consulate. Our BPE and its active China Delegate and active membership from China got the attention of their FDA leaders when we had a sit down meeting with the deputy director of China's FDA and explained that the ASME BPE Standard was developed with China and Japan.

During my first two terms we have appointed the China Delegate and recently the Brazil Delegate. I feel strongly about the fact that they are young, but VERY qualified to fill their Delegate roles. I have also closed the European Subcommittee. Europe was leading our subcommittees, including our main committee and didn't need to be a separate subcommittee.

Probably one of the things I am most proud of with respect to the BPE organization is the number of women engineers, scientists, and business leaders that our standard has attracted and elevated to the highest levels in our committees and subcommittees. I think Barbara is the only woman that I have not nominated to Main committee. She was there before me.

Lead-Up to the BPE Standard

In the years leading up to that first panel discussion held in 1988 at the ASME Winter Annual Meeting were a few dogged individuals beating the drum and curing the ills of the industry at that time. Unbeknownst to them, during the early 1980s these future members would be the vanguard of an organization that was nothing more than a vague concept in their minds' eye, a not yet formed idea just beyond their mental grasp. Most of us in the industry at that time were up to our eyeballs dealing with issues that the BPE Standard would eventually rectify in coming years, too busy in fact to travel around and educate others on what we knew.

But as part of a marketing effort, with a bent toward educating the industry, these future BPE members such as Barbara Henon, Randy Cotter, Pat Banes, and others were running interference, unknowingly in advance of the BPE, by holding seminars in concert with the University of Virginia and at the ASME Winter Annual Meetings, as well as at other venues. The interaction among this small contingent of individuals and the ASME, not to mention the interrelated work these same folks were doing on capital projects throughout the world, would eventually bring them together in the late 1980s at the ASME Winter Annual Meetings to eventually coalesce into the genesis of the BPE Standards Committee.

The essentials of the BPE Standard, in the form of individuals who were expert in their respective field, had been coming together throughout the early 1980s. The frustration we all felt at having to repeatedly work without a net in designing and constructing pharmaceutical and emerging bioprocessing industry facilities was the catalyst. The spark was the interaction between individuals at Genentech and Verax that prompted the panel discussion at the 1988 ASME Winter Annual Meeting.

Beyond 2014

As mentioned in the outset of this narrative, people like Gene Kranz seem to come out of nowhere to be at the right place at the right time. Just as Gene served as NASA Flight Director at the right time in NASA's history, so too have people arrived on the doorstep of the BPE in a timely manner. Key proponents of this standard were there at the very beginning, people such as Bill Cagney, Chip Manning, Randy Cotter, Barbara Henon, Lloyd Peterman, Tony Cirillo, Pat Banes, Dave Delucia, Rick Zinkowski, David Baram, the late Cas Perkowski, and several others no longer associated with the standard.

This was an eclectic group of individuals that arrived from various corners of the industry spectrum, bringing with them their own perspective and knowledge of the pharmaceutical and bioprocessing industries, and it worked. They can each be proud that they came together as a cohesive team to work on a single objective: to improve an industry for the common good. And that is not a simple rhetorical statement.

While I heard similar statements made at motivational talks at a pharmaceutical company years earlier, a statement using the same refrain was repeatedly made and reinforced by Tony in saying to people at BPE meetings:

> *I know you are out there to make money, we all are; it's a necessity. But while you're doing that and while you are involved in these meetings realize one thing, that what you are doing here, your work on the BPE Standard, directly or indirectly impacts not only the well-being, but the essential lives of people around the world. The medications they are required to take in order to live a long and healthy life depend a great deal on what we do here, in these meetings.*

Whether these are indeed noble efforts we are all involved in, whether we are protecting our own interests, or whether we are simply networking remains to be seen for each of us. What I do suspect is that there is a core group of people that are more deeply involved in the sustaining and continuing improvement of this standard than even they may realize. It is those individuals that magically bind the 185 plus individuals that currently make up the BPE Standard into a single cooperative effort that far and away exceeds the earliest expectations of those who sat in on a panel discussion back in 1988 and decided to do something.

In Remembrance

They say that history is written by the victors, or in this case by those that remain. The dates, names, places, and accounts of what occurred in the founding and development of the BPE Standard, as recorded in this narrative, are based on related documentation that has survived by the grace of those who were intent on saving those documents or they were simply laid aside years ago and forgotten about until this project brought them to light again. My thanks to those that did share this information.

Individuals who have come and gone over the twenty-five-year history of the BPE Standard, while not all having been identified within the pages of this narrative, are no less remembered than those who *are* listed herein. The list of names has simply grown too numerous to list in such a casual writing as this.

There are, however, those in the past who remain an integral part of the BPE Standard, having dedicated their time, their intellect, and indeed their heart to a piece of work they indelibly left their mark on. While each of their lives gave us strength in their presence, their passing has imbued us with a staunch realization that they left us with a legacy, one that demands we do not forget them and that we make every attempt to meet the excellence they imprinted on a document we are all a part of:

Lest We Forget

In Memorium

Paul Smith
Theodore (Theo) Wolfe
Casimir (Cas) Perkowski
John Henry
Tom Hoobyar

To those who provided their personal experiences and firsthand information in the writing of this history, I wish to thank you for your recollections, your stories, and your documents, documents which lend credence and substance to the telling of this history.

To the BPE membership in general, documenting the history of the BPE had been parked on the back burner of my brain for at least the past three years prior to this writing. I had approached Chip Manning one morning outside the Starbucks at the Caribe Hilton back in January 2011 hoping we could cut out a time for an interview to discuss his BPE-related experience. Chip was certainly agreeable, but neither of us could find the time during that meeting. I had planned on successive interviews with the others listed on the cover requesting an interview next with Lloyd Peterman in Baltimore at the October 2012 meeting. But again, due to the hectic schedules at these meetings, that interview never took place either. Not until the upcoming 25-year anniversary was announced at the October 2013 meeting in Kansas City did the urgency of this project hit home. And only then was time made to pull this narrative together with the help of those listed on the front cover.

Looking back on the unfolding events as the BPE came to life in the late 1980s and early 1990s, it is not difficult to recognize key moments or key individuals who perpetuated and sustained such an effort. Many would say that work such as creation of the BPE Standard achieves a certain momentum that tends to drive it forward on its own. This sounds a lot like perpetual motion, which is a theoretical impossibility, and so too is the assumption that this is what drives the success of the BPE. Continuous hard work and dedication from all those involved are what sustains the momentum and serves as the driving force necessary to make the BPE Standard the success that it is, a force that is tangible at each and every meeting. It has been an absolute pleasure for me to have had the opportunity to discover and document this history with those directly involved in its making.

Understanding Codes and Standards

In an effort to keep this as brief but as informative as possible, I will keep it relative to the chemical process industry (CPI). The term "CPI" characterizes a broad set of industries that utilize and/or manufacture chemicals in a myriad of ways. This includes such industries as pharmaceutical and bioprocessing along with the conversion of raw materials and intermediates into chemicals and petrochemicals, fats and oils, paints and coatings, food and beverages, the refining of petroleum, and the manufacture of biofuels.

To expand on this subject matter would be to bring into the discussion elements of legislative regulations that can be found under the Code of Federal Regulations (http://www.ecfr.gov/cgi-bin/ECFR?page=browse). Those regulations are codified under the US Code (http://uscode.house.gov/), thus becoming statutory law or code. This would be considered top-down regulation whereby government legislates and passes down such regulations as the Clean Air Act of 1963, the Clean Water Act of 1972, and other such national regulatory precepts and edicts. These are broad sweeping laws that affect the entire country and could only be mandated from the top down.

But there are industry-specific standards and codes, which are created and promulgated from the bottom up, meaning that they are typically generated from the affected industry, defined, vetted through an accredited organization, and published as an industry standard for the use and adoption of end users, states, and municipalities if they so choose. When adopted and enacted by the legislative branch of a government, such industry standards become statutory law (code). If integrated into a performance contract, they become legally binding whether required by regulatory compliance or not.

Aside from one small subtlety, there is essentially no difference between a published code and a published standard. That small, but impactful subtlety refers to the wording and phrasing used to convey a requirement rather than a suggestion, recommendation, or option. The fact that ASME B31.3 is titled "ASME Code for

Pressure Piping" does not mean to imply that, although it is written and titled as a code, any work performed within the United States and within the limits and boundaries of the defined scope of B31.3 has to automatically adhere to its requirements. B31.3 or the BPVC do not become legally enforceable documents until they are adopted by state or local governments or specified as an integral part of a contract.

Case in point: As you can see in Table C1 in the succeeding text, compiled from the National Board Synopsis of Boiler and Pressure Vessel Laws, Rules and Regulations, there are ten states that have adopted only the Boiler Code; thirty-eight states, Puerto Rico, and all of the provinces in Canada have adopted both the Boiler Code and the Pressure Vessel Code. Only one state, Wyoming, has adopted only the Pressure Vessel Code, and only one state, Idaho, has adopted neither code. What it means in stating that they have adopted the Boiler Code and/or the Pressure Vessel Code is this: If they have adopted just the Boiler Code, it means they have adopted Sections I, II, IV, V, VI, and VII of the BPVC. Adopting both the Boiler Code and the Pressure Vessel Code indicates that they have adopted Section VIII of the BPVC in addition to the other sections, point being that these documents only become code upon adoption or by inclusion into a contract.

The 2015 National Board Synopsis Map (Figure 2) graphically reflects, with the exception of indicating adoption of B31.1 and B31.3, the data shown in Table C1.

To date four states including California, Kentucky, Michigan, and Oregon and all of the Canadian provinces have adopted B31.3 as their piping code for CPI-related piping. Adoption of the B31.3 Process Piping Code by a state is typically done indirectly through a state's boiler code. The other 46 states, choosing not to adopt, apparently have not felt the need to require such regulatory dictates and have left that decision to the engineer or end user.

Because steam, steam condensate, water, and other utility piping systems are ancillary to the function of a boiler, the B31.1 Power Piping Code is usually adopted by reference in various state boiler codes. What typically happens, with regard to CPI-type facilities whose piping requirements cover a much broader range of utility and process services than B31.1 can accommodate, is that some state inspection boards feel the need to establish a basic set of rules and regulations for the design, fabrication, and construction of these types of facilities as well. As mentioned in the preceding text, they do this by referencing B31.3 in their boiler code, which therefore makes it a statutory requirement through reference as part of their adopted boiler code. As an example the following is an excerpt from a state boiler code:

R 000.0000 Non-boiler external piping; power boilers; adoption of standards by reference.

Rule 32.

(1) The owner shall ensure that the installation of piping not covered by the ASME boiler and pressure vessel code, section I, 2007 edition, and its 2008a addenda is installed as prescribed by the ASME code for pressure piping, B31.1, 2007 edition, adopted by reference in R 408.4025.

Table C1 – ADOPTION OF THE BPVC BY STATE, PROVINCE, TERRITORY, AND COUNTRY									
UNITED STATES & PUERTO RICO –BASED ON 2015 NATIONAL BOARD DATA									
State	Boiler Code Only	B& PVC	B31.1	B31.3	State	Boiler Code Only	B& PVC	B31.1	B31.3
Alabama		■	■		Nebraska		■	■	
Alaska		■	■		Nevada		■	■	
Arizona		■	■		New Hampshire[3]		■	■	
Arkansas		■			New Jersey		■		
Califormia[1]		■	■	■	New Mexico	■		■	
Colorado		■	■		New York		■		
Connecticut	■				North Carolina		■	■	
Delaware		■	■		North Dakota		■	■	
Florida	■		■		Ohio		■	■	
Georgia		■	■		Oklahoma		■	■	
Hawaii		■	■		Oregon[1]		■	■	■
Idaho					Pennsylvania		■		
Illinois		■	■		Puerto Rico		■		
Indiana		■			Rhode Island		■	■	
Iowa[3]		■	■		South Carolina	■			
Kansas		■	■		South Dakota	■			
Kentucky[3]		■	■	■	Tennessee		■	■	
Louisiana	■				Texas	■			
Maine		■	■		Utah		■	■	
Maryland		■	■		Vermont		■		
Massachusetts		■			Virginia		■		
Michigan	■		■	■	Washington		■		
Minnesota		■	■		West Virginia	■			
Mississippi		■			Wisconsin[5]		■	■	
Missouri		■	■		Wyoming		Note 8		
Montana	■								
CANADIAN PROVINCES and TERRITORIES									
Alberta		■	■	■	Nova Scotia		■		
British Columbia[1]		■	■	■	Nunavut Territory[6]		■	■	■
Manitoba[4]		■	■	■	Ontario[5]		■	■	■
New Brunswick[5]		■	■	■	Prince Edward Island[7]		■	■	■
Newfoundland[1]		■	■	■	Quebec[7]		■	■	■
Labrador[1]		■	■	■	Saskatchewan		■	■	■
Northwest Territories		■	■	■	Yukon Territory[6]		■	■	■

Notes:

1. Also adopted ASME B31.5 and B31.9
2. Also adopted ASME B31.5 Plastic pipe unacceptable unless permitted by specific safety orders.
3. Also adopted ASME B31.9
4. Also adopted ASME B31.4, B31.5, B31.8, and B31.9
5. Also adopted ASME B31.5
6. Also adopted ASME B31.4, B31.5, and B16.5
7. Also adopted ASME B31.2, B31.4, and B31.5
8. Wyoming has adopted only the Pressure Vessel Code Section VIII

Figure 2 2015 National Board synopsis map. *Source*: National Board of Boiler and Pressure Vessel Inspectors (National Board), https://www.nationalboard.org/ © 2016

(2) The owner of a chemical plant or petroleum refinery shall comply with subrule (1) of this rule or shall ensure the installation is installed as prescribed by the ASME code for chemical plants and petroleum refineries, B31.3, 2007 edition.

(3) A licensee under this rule is not required to possess an ASME code symbol stamp, but shall hold a valid installer's license.

(4) The owner shall ensure that the installation of all of the following piping is in accordance with subrule (1) of this rule:

(a) Blowoff piping beyond the second valve out to the safe point of discharge.

(b) Steam piping out to the load.

(c) Feed-water piping from the pump.

(d) Condensate piping

As you can see in the earlier state code excerpt, paragraph (2) gives the option of compliance under B31.1 or B31.3, leaving the final decision to the owner or end user.

Rather than attempt to regulate these requirements, as the four states mentioned earlier have done, the majority of states simply leave the CPI-related piping code selection process to the owner or engineer of record (EOR). EOR refers to an engineering firm who has assumed responsibility, through contractual obligation, of adhering to the fundamental rules of good engineering practice and compliance with all governing codes and regulations in the design and construction of a CPI facility. They will henceforth become the EOR for their scope of work within a facility. If that work is modified by others, then those that engineered the modifications will then assume responsibility for those modified systems, essentially relieving the initial engineer of those obligations. This, of course, is all contractual with final determination of responsibilities made clear in such contracts. Responsibility, I might add, of the fabrication, installation, and testing of piping systems is generally that of the contractor or contractors performing such work.

Having assumed such responsibility they, the EOR, then becomes the responsible party for the design elements of a CPI-type plant or facility. Shifting responsibility of code assignment to the EOR does then have some merit, the thinking being that the entity responsible for the design and engineering of a plant or facility will have skin in the game (liability) and would therefore be more diligent in selecting the most appropriate codes and standards by which to build a facility, allowing regulation in effect to take place from the bottom up.

In those states that have adopted B31.3 as their piping code, it then becomes a requirement by law, and, unless amended otherwise by the state, any work performed in conjunction with a facility characterized as a CPI-type facility has to be done in accordance with B31.3. In the other 46 states, B31.3 has to be adopted by an engineer, contractor, or owner through contract stipulation and/or integration into specifications embedded into a contract, work order, or a facility's SOP, which would then be included by reference.

The Language of Codes and Standards

Statements have been made declaring that when a code or standard has been adopted through legislation or by contract that the engineer would have to comply with everything in the standard or code—not true. This is not true because not everything in the standard or code is a compliance requirement. Using the B31.3 Process Piping Code as further example, the difference between the B31.3 code and the BPE Standard is basically in the wording. Within a code actions are written as either "requirements" or "recommendations" and they have to be declared as such, one way or the other. A "requirement" is written in such a manner as to make an action specific and without variation. This is done by preceding the action with the word "shall." A recommendation is declared by preceding the action with such words as "should" or "may." See example Ex.1 in the succeeding text from B31.3 Para. 301.2.2*(a)*:

> *Provision **shall** be made to safely contain or relieve (see para. 322.6.3) any expected pressure to which the piping may be subjected. Piping not protected by a pressure relieving device, or that can be isolated from a pressure relieving device, **shall** be designed for at least the highest pressure that can be developed.* Ex. 1

In the earlier paragraph the word "shall" in both cases makes this statement an imperative and therefore a requirement. Furthermore, in example Ex. 2 in the succeeding text, B31.3 states the following in Para. 302.2.4(i):

> *The application of pressures exceeding pressure–temperature ratings of valves may under certain conditions cause loss of seat tightness or difficulty of operation. The differential pressure on the valve closure element **should** not exceed the maximum differential pressure rating established by the valve manufacturer. Such applications are the owner's responsibility.* Ex. 2

Stating that the "...differential pressure on the valve closure element ***should*** not exceed..." makes it a suggestion or recommendation, not a requirement. These nuances, between requirements and recommendations in B31.3 and the other B31 codes have been developed and vetted over time to ensure that what needs to be a "requirement" is stated as an imperative. And where an action has some flexibility it is simply a recommendation, a suggestion as to how to accomplish an action.

When a state, province, country, engineer, or end user adopts a standard such as BPE they do so with the assumption that when they write a specification or contract stating that "the engineer or constructor shall comply with the BPE standard" there will be no conflicting information within the standard and there will be clear indications as to what is considered a "requirement" and what is merely a "recommendation" or "suggestion." They are assuming that each "shall" and each "should" will be applied in the proper context. But, unlike the B31 codes, such an assumption may not be entirely valid with regard to the BPE Standard.

As shown in the following example Ex. 3 from BPE, Para. SD-3.4.4*(a)* states:

> *Butt welds **should** be used, if possible, minimizing lap joint welds and eliminating stitch welding.* Ex. 3

When adopted as a code or contractual standard the above statement should not be a recommendation as indicated by the word "should," but should instead be a code requirement written as "shall" with a stated caveat if needed as shown in example Ex. 4:

> *Butt welds* ***shall*** *be used* ***unless the following occurs:***
> *(1)* ***A circumstance in which a butt weld cannot be made and the alternative of which is approved by the owner or owner's engineer.*** Ex. 4

The level of hygienics required in the bioprocessing industry demand the use of butt welds in piping, equipment, and components except for special circumstances, which should require owner approval. From a code standpoint this is not conveyed in the current wording or phrasing. The end user or engineer adopting the BPE standard is relying on and indeed depending on the fact that the content of the standard is not soft on those requirements that are needed to make their system hygienic and cleanable.

The same can be said of Para. SD-3.4.4 *(b)*, as referenced below in example Ex. 5:

> ***Flanges are not recommended, and their use shall be minimized.*** *The bore of weld ...* Ex. 5

As with the above mentioned butt weld requirement the same applies here as well. In referring to example Ex. 5, the soft suggestion in recommending that flange joints should not be used is a statement that is not in line with BPE's characterization of hygienic requirements. With exception of the hygienic clamp joint union, flange joints, along with other types of mechanical joints, are not suitable for hygienic piping systems unless, on a limited basis, there is no other alternative. As a code the statement should be more explicit and definitive as in the following, shown in example Ex. 6:

> ***Flange joints shall not be used unless first approved by the owner or owner's engineer. If approved their use shall be minimized.*** *The bore of weld ...* Ex. 6

The other essential necessary in being considered as a code is in validating and providing supporting data for the quantitative values published within the standard, which is a process the BPE Standard is going through at this time. Publishing values for L/D and clamp joint union pressure ratings without the data to support those values is a luxury reserved for a standard, not a code. From a professional and pragmatic standpoint, such information should also be included in a published standard as well. A code requires such published values to be supported by empirical and/or theoretical data. By performing empirical testing for dead leg and hygienic clamp joint union pressure rating analysis, the BPE Standard is moving in that direction.

The one glaring difference that separates the BPE Standard from B31 codes is the statement in the introduction of the B31 codes, which states: "*The designer is cautioned that the Code is not a design handbook...*" Such a statement does not apply to the BPE Standard because a big part of its benefit for industry lies in defining what constitutes the concept of hygienic design. The design aspect of this standard is integral to the requirements it conveys for component manufacture, system fabrication, and cleanability. This fact makes the BPE Standard a quasi-design handbook.

Serving this dual role should not interfere with the BPE Standard being accepted as a code or merely being adopted as a project or end user standard. Again, it comes down to appropriate content. One example is in design details.

A code does not represent what is not acceptable. Multiple *acceptable* variations of a design element can be provided, but one or more variations on design elements that are *not acceptable* are not a practical thing to do. Representing what is not acceptable is conceivably a never ending proposition. If an engineer or designer wishes to alter what is represented in the BPE Standard as an acceptable design, they either accept responsibility for that noncompliant alteration or they request an interpretation. If it turns out that their alteration is acceptable, it could then become an intent interpretation that actually precipitates a change in the standard. Bottom line is that any details representing nonacceptable design elements would need to be deleted from the BPE Standard in considering it as a code.

Summing Up

1. There are two reasons for which the design, construction, and maintenance[1] of a facility is required to comply with a code:
   - It has been adopted by the jurisdiction within which the facility resides.
   - It is stated in the design basis or specifications, which are made part of a project's contract.
2. If a code is not adopted by a state, province, or country in which a facility resides in or will reside in, then there is no legal or regulatory requirement to comply with the code, only good and practical judgment.
3. A code stipulates the minimum requirements of which a facility is designed, constructed, and maintained. Such documents are written in a manner that makes it explicit in what those requirements are with the basic precept that the document be an accredited set of rules that establish a set of minimum criteria of which a facility can be designed, constructed, and maintained in a manner which is conveyed by the code itself. If a rule is preceded by either the word "shall" or "must," it is imperative that the rule be met. If a rule is preceded by either the word "should"

[1] When referring to maintenance the reference is made to that pertaining to piping modifications or repair.

or "may," it is considered a suggestion or recommendation, but not a requirement. In saying that, it is inferred that not all statements of action in a code are a requirement.

4. A standard is essentially written to provide guidance and methodology within a specific set of parameters, but is typically not written in the same explicit manner as a code. This lends added flexibility to the writing of a standard. However, that is not to say that a standard may, and frequently does, adhere to the same rigorous rules of grammar as a code such as those written for material standards or component standards. A standard can be any document that attempts to standardize on a set of rules and methodology on a specific subject matter. It can be an accredited industry standard or a proprietary corporate standard. And while such a premise does not preclude an accredited industry standard from being adopted as a code or regulatory standard, its validity as such should be reviewed and considered carefully in the explicit nature of its words and phrases.
5. Any firm or legislative body wishing to adopt, as a contractual obligation or government regulation, a standard such as the BPE should consider whether or not they wish to allow some of the current soft spots to remain as is or whether such paragraphs need to be amended to be more explicit.

Creating and Maintaining an American National Standard

The American National Standards Institute (ANSI)

The BPE Standard, as well as any other code or standard created and maintained by ASME, is referred to as an American National Standard (ANS) or, more simply, a National Standard. Those organizations accredited by ANSI to develop such industry standards are referred to as American National Standards Developers or, again more simply put, Standards Developers, also referred to as Accredited Standards Developers.

There are currently over 200 Standards Developers that have created and maintain over 10,000 National Standards. And this is all done in accordance with a well-defined and strict set of rules, guidelines, and regulations created, set forth, and monitored by ANSI.

The ANSI organization was the precipitant culmination resulting from a meeting in 1916 called by the American Institute of Electrical Engineers (AIEE), predecessor, along with the Institute of Radio Engineers (IRE), of the Institute of Electrical and Electronics Engineers (IEEE). This meeting pulled together the ASME, American Society of Civil Engineers (ASCE), American Institute of Mining and Metallurgical Engineers (AIME), and ASTM to discuss the need for a monitoring organization. An autonomous group that would establish and oversee an impartial program and process by which all Standards Developers would adhere to in the development of industry standards and codes. This fledgling program would quickly develop into an internationally recognized accreditation program by which those organizations abiding by the rules and procedures set forth by ANSI would become ANSI accredited.

In the 1880s and 1890s, various professional organizations such as the AIEE and ASME began to emerge throughout the various facets of industry, driven in large part by the catastrophic and oftentimes fatal failures of equipment and machinery in the workplace. There were no standard methodologies at that time much less consensus-approved engineering methods that could establish any degree of assurance that electrical wiring was properly installed or that a boiler was properly designed and built. Everyone from equipment manufactures to electrical wiring contractors were doing what they thought to be right at the time. But without a cohesive set of rules and guidelines to go by or adhere to, it was like the Wild West of industrial growth. And at the turn of the century, growth in industry was rapid and expansive, quite often with deadly consequences.

It came down to the old chicken and egg analogy with regard to what had to happen first in getting manufacturing on the same page with good engineering and manufacturing/fabrication practices. Government, in the form of politicians trying to hold on to their political office, felt the need to do something, anything to stem the all-too frequent injuries and fatalities caused by a lack of regulation. However, without the necessary and proper standards written and ready to put into place, the government could not act. So two things had to happen in order for the safety and well-being of the general public and labor to be realized during this industrial and new technological boom: (i) Engineering standards had to be developed and galvanized by accredited Standards Developers, which would assure their validity, and (ii) government would have to adopt these standards and guidelines as code and then promulgate them as the rule of law.

Established in 1919 as the American Engineering Standards Committee (AESC), ANSI was, as the name implies, a simple committee. It had yet to prove itself as the autonomous and impartial accreditation organization it would eventually become. By 1928, nine years after its creation, ANSI outgrew its committee status and was reorganized and renamed the American Standards Association (ASA). In 1966 ASA was again reorganized and given the name of United States of America Standards Institute (USASI). For the brief time it existed as the USASI, the name was no more than a placeholder for its current name of ANSI, which the organization adopted three short years later in 1969. It still retains that name at the time of this writing.

One aspect of what ANSI does and doesn't do regards the development and publication of codes and standards. It may seem a little confusing, but ANSI does not develop or publish codes and standards. Their job is to instead provide procedural oversight, monitoring, and accreditation for those that do. However, there are information, articles, and Internet links that would make you think otherwise. With such statements as "ANSI Flanges," "ANSI Pumps," "ANSI B31.3," "ANSI B16.34," and others, many designers and engineers are inclined to believe that ANSI is developing and publishing many of these codes and standards. In actuality, stating "ANSI Standard" implies the standard is an accredited National

Standard. This same implication can also apply to flanges, pumps, and myriad other components and equipment.

Some of the confusion, as it relates to ANSI's role in codes and standards, has multiple sources for its cause. The first has to do with the evolution of a product, that product being the function of an organization, which in the case of ANSI is accreditation. In the years leading up to the days in which this book on the *Companion Guide to the BPE Standard* was written, codes and standards were initially published with ANSI being an integral part of the code or standard title. This was done to announce the fact that such codes and standards were indeed accredited by an official organization. It was also done to create familiarity or name recognition with the name ANSI and its predecessor titles. As shown in Figures 3 and 4, the early names of ANSI, such as ASA and USAS, were included in the title to indicate that they were respectfully ASA and USAS accredited standards.

By the 1980s both ANSI and the Standards Developer were indicated in the title. This is represented in such a case as seen in Figure 5 in which ANSI/ASME is represented in the title to indicate an ANSI-accredited ASME code. By the time the 1990s rolled around, the title included only the developer's name, or acronym as it were, as shown in Figure 6. By this time it was felt that ASME and many other Standards Developers were recognized on their own as being ANSI accredited without the need to continue announcing that fact in a standard's title.

Throughout those years in which ANSI, under its various organizational names, was front and center in the title of codes and standards, procedures and specs were written, catalogs were published, and all manner of marketing literature were produced and distributed. In such documents were the words everyone became familiar with in relation to codes and standards, that being ANSI. Not only did ANSI achieve name recognition, they did it so well that the acronym ANSI had essentially become "genericized" or synonymous with various accredited products, like referring to all facial tissue as "Kleenex." ANSI became so synonymous with various standards, and even products, that these documents and products are still quite frequently referred to as ANSI standards, which makes reference to ANSI accreditation, rather than the Standards Developer they were written and published by, or as ANSI Pumps, ANSI Flanges, and so on. These types of references are fine as long as all parties involved in a discussion on such topics as piping, pressure vessels, pumps, and so on are familiar with the implication.

Including ANSI in referencing standards or codes would not be considered a misnomer; a more practical way in which to refer to a standard or code today is by including the name of its developer only, such as in ASME B31.3, ASME BPE, or API RP520. The only time the date of the code or standard needs to be included would be when listing it as a compliant standard in a specific contract, stating that, as an example, "Piping specifications and requirements within the scope of this contract shall comply with the ASME B31.3-2014 Process Piping Code. All piping designated as high-purity fluid service shall additionally comply with the ASME BPE-2014 (Bioprocessing Equipment) Standard."

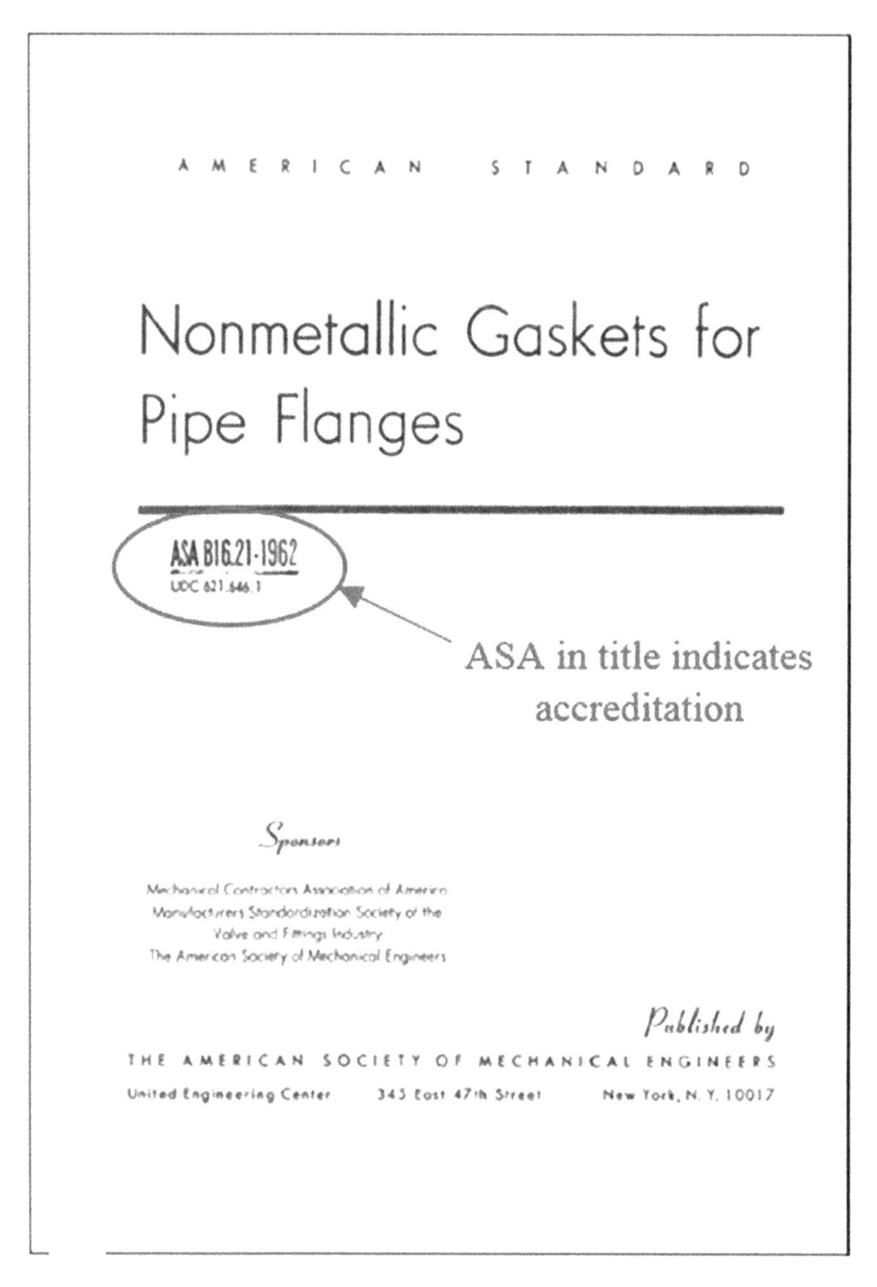

Figure 3 B16.21-1962 cover

Typically, any codes or standards listed in a company's standing portfolio of procedures and specifications, the type of documents that would be retained in a library of such design, engineering, and construction requirements, would contain a statement such as "Unless otherwise stated, the Codes and Standards referred to

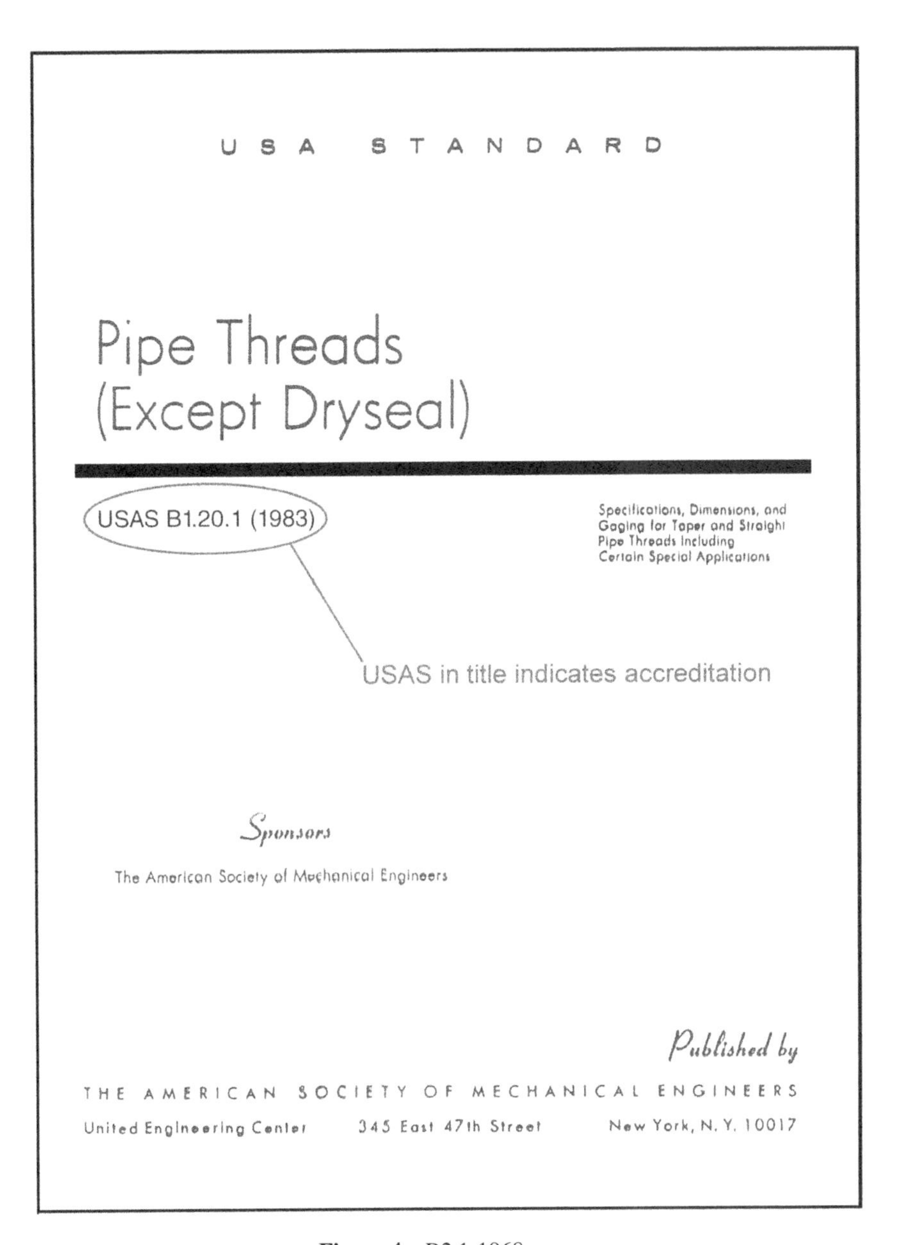

Figure 4 B2.1-1968 cover

herein shall be the most current version of such documents in effect at the time a contract goes into effect." This is due to the fact that such documents are maintained in a company's specification library over a period of years and decades. Rather than having to continuously update and issue such specs every time a standard or code is reissued, it is much more efficient to make the proclamation that

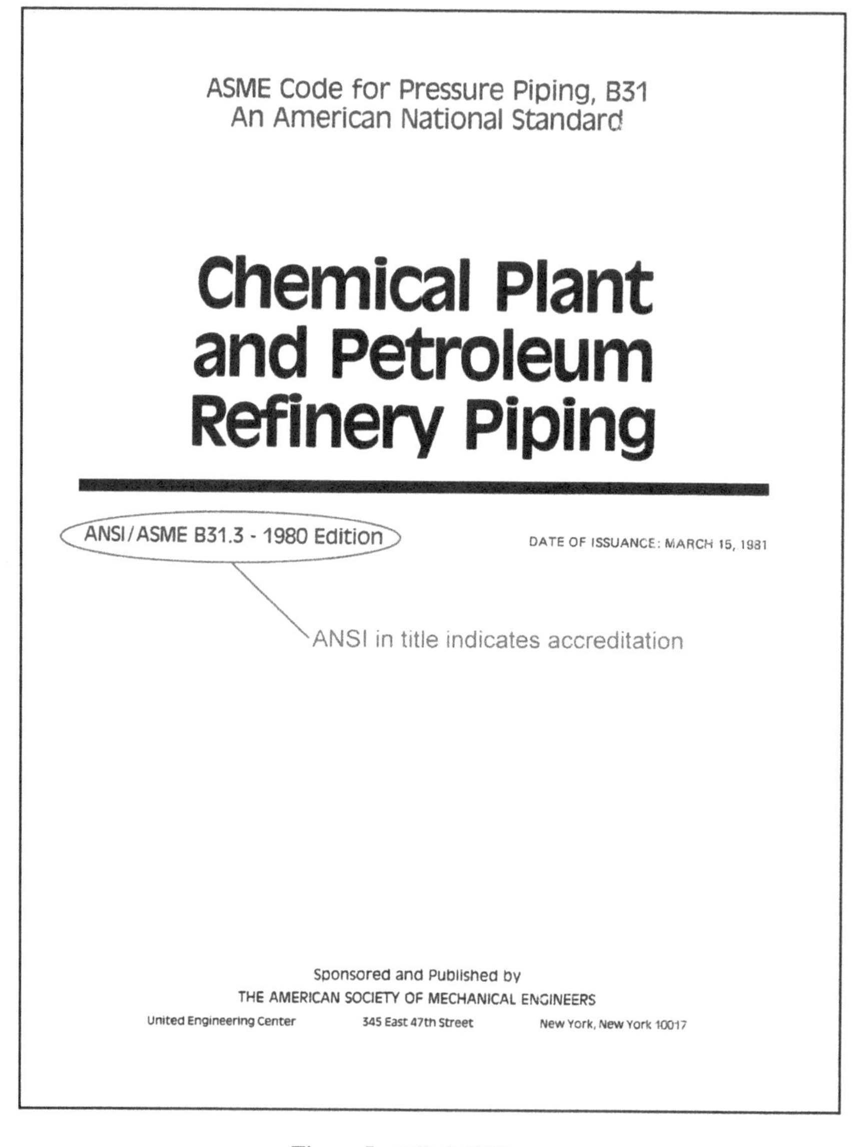

Figure 5 B31.3-1980 cover

"...the most current version of such documents in effect at the time a contract goes into effect." When a project is rolled out, specific issue dates of standards and codes can be added.

Until such time as catalogs referring to ANSI are revised to instead make reference to the appropriate Standards Developer, there will still be a degree of misunderstanding involved.

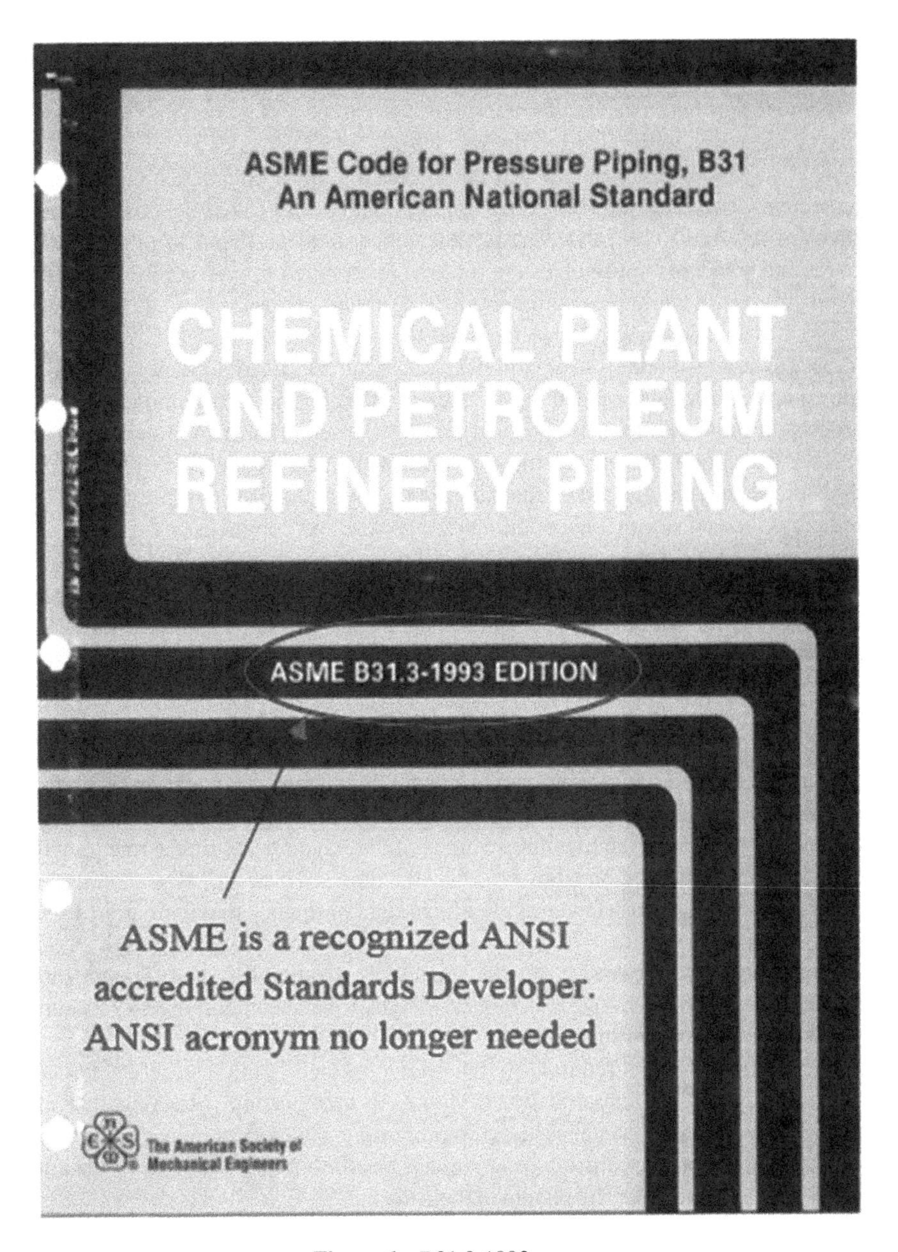

Figure 6 B31.3-1993 cover

The Process of Maintaining an ANSI-Accredited Code or Standard

Accreditation of an ANSI-accredited Standards Developer is the process by which, in the case of ASME, an organization proves, on an ongoing periodic basis, that they are competent to perform the duties and services they have elected and have committed to perform. In the case of codes and standards committees, such as the ASME BPE Committee, there are very strict rules with regard to developing and amending consensus standards as well as the process of responding to requests for interpretation with regard to the standard.

The ASME BPE Standard currently consists of a Standards Committee, Executive Committee, and ten subcommittees that are dedicated to sections of the standard referred to as parts. These same subcommittees found in other ASME codes and standards committees are referred to as subgroups. In addition to these main groups or committees within the BPE Standard, there are also ad hoc groups referred to as task groups. These are small focus groups assigned and dedicated to a specific task as it relates to a particular subcommittee. If resolution of an issue is too complex or requires added information in order for the subcommittee to make a decision on the issue, it is assigned to a task group. That task group will then resolve the issue between meetings, which, in the case of the BPE Standard, is three times per year. Results of that task are then written up as a proposal and submitted to the sponsoring subcommittee for review and comment or for approval. These ad hoc task groups can be made up of one or several people, and they can be resolved before the next meeting, or they may take five or more years to complete.

There are multiple sources that can affect change in a standard such as the BPE. And such sources include those from outside the membership of the standard as well as those within the membership of the standard. These standards, as well as the codes, are public documents and are therefore open to comment and recommendation from the public.

In a merciful attempt at avoiding the crushingly mundane explanation necessary to explain in any rational detail the process by which changes from minute editorial details to the addition of entire sections of the standard are orchestrated, I will default to only touching on a few key essential points. These ASME standards are referred to as consensus standards. This is in light of the fact that their approval at every stage of the process is done by consensus. From task group work to the subcommittee or subgroup, to the Standards Committee, the Supervisory Committee, and finally ANSI, voting is done by a balanced contingent of members at each of these voting plateaus.

Each voting group is required to maintain a balance of interest, referred to as "interest category or classification" among its membership. Meaning that no single interest group, such as those representing component manufactures or material suppliers or end users or any other specific group, is able to ostensibly hijack an issue and steer the results in their favor.

Before closing out on this section, I will leave you with some additional information. The ASME BPE Standards Committee currently meets three times per year in January, the May/June time frame, and the September/October time frame. There is a meeting notice sent out to members about one month prior to a meeting date giving the dates, hotel, discounted hotel rate, and other essential information. This notice is also posted on the BPE Committee web page at https://cstools.asme.org/csconnect/CommitteePages.cfm?Committee=N10120000. On the web page (Ref. Figure 7) under "Meetings" in the left-hand frame, single click on "BPE Meeting Notice" and you will be able to download and print the meeting notice.

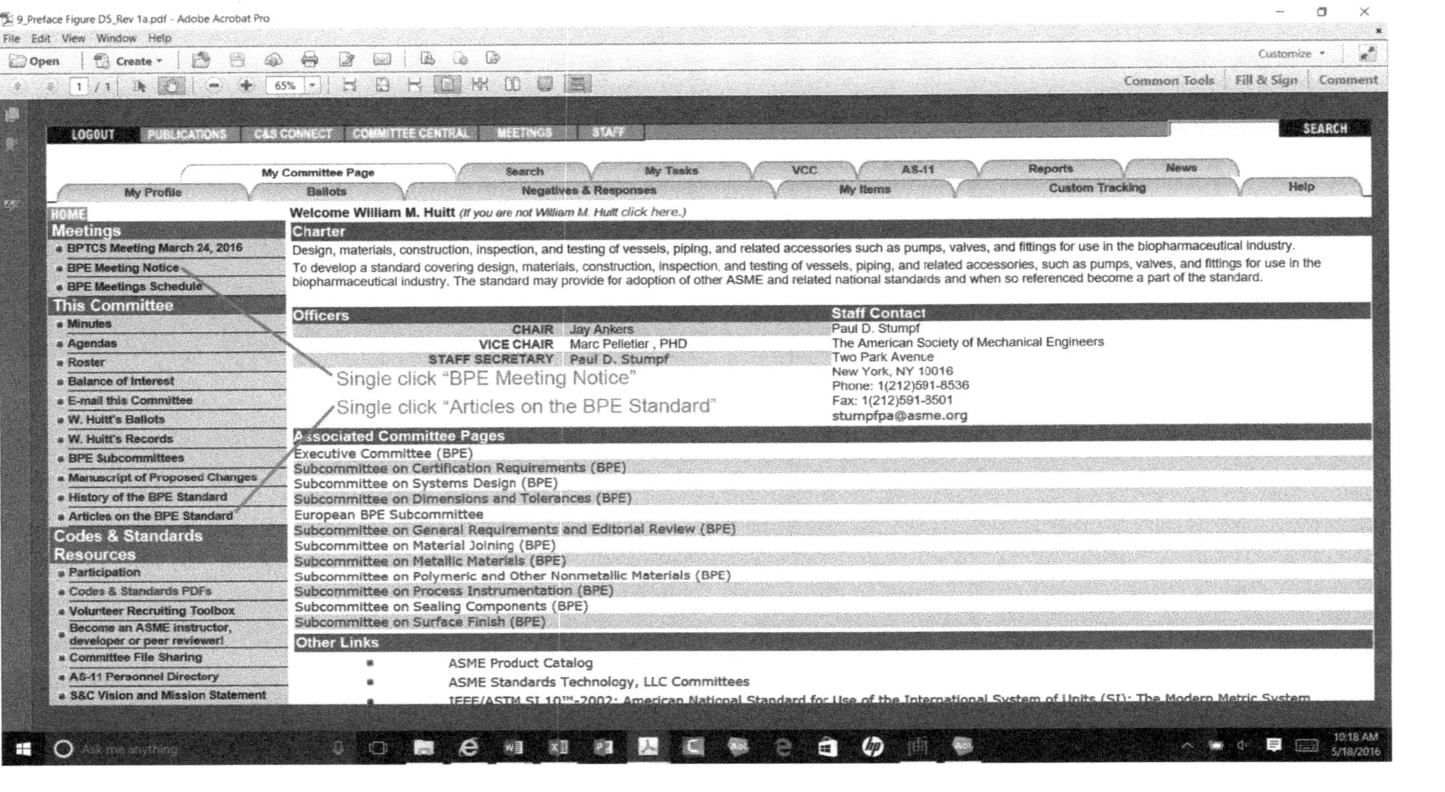

Figure 7 ASME BPE Committee web page

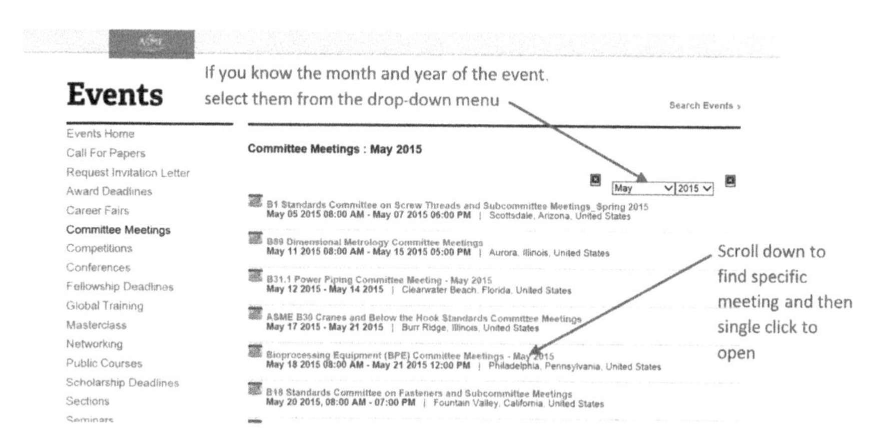

Figure 8 ASME BPE Committee meeting calendar

Meeting information can also be found at http://calendar.asme.org/home.cfm?EventTypeID=4 as shown in Figure 8. Upon reaching the linked page, enter the month and year of the meeting. After it jumps to that time period, scroll down until the specific meeting you are looking for is found then single click on it to open the page for that meeting.

In addition, there are a number of articles published by various members of BPE that can be found on the BPE web page. Referring to Figure 7, under "This Committee" in the left-hand frame, single click on "Articles on the BPE Standard." This will take you to the page link shown in Figure 9 providing a listing of a number of downloadable articles and journal papers written by committee members and published by various means. These articles and papers cover topics that specifically relate to the BPE Standard and to high-purity piping in general. Double-click any of the articles to open and download as a .pdf file.

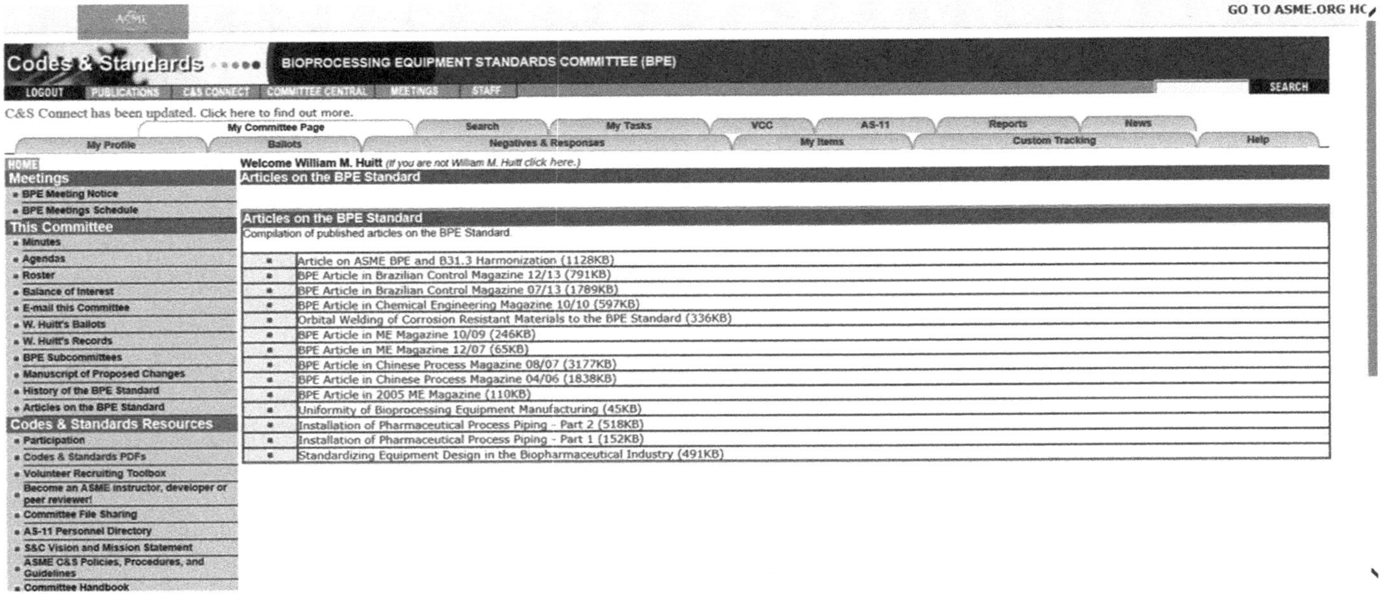

GO TO ASME.ORG HO

Codes & Standards

BIOPROCESSING EQUIPMENT STANDARDS COMMITTEE (BPE)

LOGOUT | PUBLICATIONS | C&S CONNECT | COMMITTEE CENTRAL | MEETINGS | STAFF | SEARCH

C&S Connect has been updated. Click here to find out more.

My Committee Page | Search | My Tasks | VCC | AS-11 | Reports | News

My Profile | Ballots | Negatives & Responses | My Items | Custom Tracking | Help

HOME

Meetings
- BPE Meeting Notice
- BPE Meetings Schedule

This Committee
- Minutes
- Agendas
- Roster
- Balance of Interest
- E-mail this Committee
- W. Huitt's Ballots
- W. Huitt's Records
- BPE Subcommittees
- Manuscript of Proposed Changes
- History of the BPE Standard
- Articles on the BPE Standard

Codes & Standards Resources
- Participation
- Codes & Standards PDFs
- Volunteer Recruiting Toolbox
- Become an ASME instructor, developer or peer reviewer!
- Committee File Sharing
- AS-11 Personnel Directory
- S&C Vision and Mission Statement
- ASME C&S Policies, Procedures, and Guidelines
- Committee Handbook

Welcome William M. Huitt *(If you are not William M. Huitt click here.)*

Articles on the BPE Standard

Articles on the BPE Standard

Compilation of published articles on the BPE Standard

- Article on ASME BPE and B31.3 Harmonization (1128KB)
- BPE Article in Brazilian Control Magazine 12/13 (791KB)
- BPE Article in Brazilian Control Magazine 07/13 (1789KB)
- BPE Article in Chemical Engineering Magazine 10/10 (597KB)
- Orbital Welding of Corrosion Resistant Materials to the BPE Standard (336KB)
- BPE Article in ME Magazine 10/09 (246KB)
- BPE Article in ME Magazine 12/07 (65KB)
- BPE Article in Chinese Process Magazine 08/07 (3177KB)
- BPE Article in Chinese Process Magazine 04/06 (1838KB)
- BPE Article in 2005 ME Magazine (110KB)
- Uniformity of Bioprocessing Equipment Manufacturing (45KB)
- Installation of Pharmaceutical Process Piping - Part 2 (518KB)
- Installation of Pharmaceutical Process Piping - Part 1 (152KB)
- Standardizing Equipment Design in the Biopharmaceutical Industry (491KB)

Figure 9 Articles currently listed on the BPE Standard

Acknowledgments

Chris Mahler and Mary Grace Stefanchik

A special thanks to Chris Mahler, Manager Business Development, Third-Party Sales with ASME, for initiating the onset of this book. We became acquainted in his marketing work with the Bioprocessing Equipment (BPE) Standard and more specifically his work with the BPE Certification Program. At the New Orleans BPE meeting in May of 2014, Chris approached me with the idea of my writing a companion guide to the BPE Standard, asking if I might be interested in such a project. I acknowledged that I did indeed have an interest in taking on such a project and had been sitting on the idea for some time. Having indicated my interest Chris then introduced me, by way of e-mail, to Mary Grace Stefanchik with ASME Press who was, from that point on, instrumental in guiding me through the ASME Press publication process. I am deeply grateful for the consideration and trust in me shown by Chris and for the help and understanding by Mary Grace. After a partnership between the American Society of Mechanical Engineers (ASME) and Wiley Publishing, Inc. was made apparent, Wiley became the primary editor/producer of this book. It was a wonderful experience working with Ramya, Vijay, and Sandra at Wiley.

Others

A book such as this is never done in a silo. The writing and development of such a project is done with the help of others. Others that over the years I have come to respect and know I can depend on, individuals that selflessly give off their time and knowledge to help others. These are individuals that have perfected the art of giving back as volunteers on multiple professional organizations and by serving relentlessly on multiple committees. My professional and personal life has been made better in knowing each of the following individuals who were true assets and resources in writing this book.

The following is a list of those who contributed firsthand accounts and documents that were so pertinent and essential to the historical accuracy in telling the story of the founding and development of the ASME BPE Standard in a narrative titled *Early History and Development of the ASME BPE Standard.* Those that provided recollections and documents from that period include the following in alphabetical order:

Jay Ankers—M + W Group
William H. (Bill) Cagney—JSG, LLC
Dr. Richard D. Campbell—Bechtel Corp.
Anthony P. (Tony) Cirillo—Cirillo Consulting Services, LLC
Randy Cotter—Cotter Bros.
Dr. Barbara K. Henon—Magnatech, LLC
Frank J. (Chip) Manning—VNE Corp.
Lloyd J. Peterman—United Industries, Inc.
Paul Stumpf—ASME Staff Secretary to the BPE Standard

As original founders and protagonists of the ASME BPE Standard, the earlier listed individuals, along with several others identified in this historical account, are the faces of those who, some 30 years ago, understood the need for such an industry standard and did something about it. At the publication of this book, those individuals, and many others named in the historical narrative found in the preface, are still very much involved these 30 years later. That alone speaks volumes with regard to the character and dedication of those individuals who are so much a part of the founding, development, and continued growth of the ASME BPE Standard.

The following also are people that helped keep my writing concise, accurate, and to the point by taking time out of their busy schedules to review many of the pages of this book, individuals whom I trust to give me their opinion in the most intelligent and forthright manner and without embellishment. It includes such individuals as:

Dr. Barbara K. Henon, a consultant, who I knew through her many fine articles on orbital welding long before I first met her in Cork, Ireland, in 2004. She has proven to be an invaluable part of the work we do on codes and standards as well as a trusted friend. Her deep knowledge and understanding of welding was a key resource for this book.

Mr. Richard Shilling is an expert on international export regulations, administration, and communication. He also has an unabashed love affair with his 1965 356C Porsche, a topic of which he lectures on around the country. Having been involved with orbital welding for many years, Richard was kind enough to perform a review on such subject matter providing insight and well-founded recommendations.

Dr. James D. Fritz, a metallurgist with TMR Stainless, provides the BPE Standard with his depth of knowledge and insight into the world of metallurgy playing a major role in the creation of the subcommittee on metallic materials.

Jim is outspoken and active in subcommittee meetings and always a source of reliable and in-depth information when it comes to metallic materials and a valuable resource for this book as well.

Mr. James D. Vogel, founder and director of the BioProcess Institute, past chair of the subcommittee on sealing components, and current vice-chair of the same subcommittee. Jim is a proponent of and has written many informative and timely articles on the subject of single-use systems. Jim was kind enough to share that knowledge in his review.

Mark Embry, with Asepco, is chair of the General Requirements Subcommittee. Mark has played a huge role in helping to harmonize the various sections of the BPE Standard, which does require constant oversight. From a general requirements standpoint, his responsibilities touch on every segment of the standard. With his understanding of general requirements, Mark provided insightful recommendations in the pages of this book he was kind enough to find time to review.

Ken Kimbrel, V. President Ultraclean Electropolish, Inc., has been a long time member of ASME BPE and spearheaded the creation and adoption of Part MM in the standard of which he is past chair. Unable to rest on his laurels, Ken currently chairs the subcommittee on surface finish. He provided knowledge and experience in his review of the manuscript on the topic of surface treatments and conditions.

John Kalnins, Technical Service Manager at Crane ChemPharma, is chair of the ASME B31.3 subgroup F on nonmetallics and metals lined with nonmetallics. He is also a member of the new ASME Nonmetallic Pressure Piping Systems (NPPS) subcommittee on thermoplastic piping. I have known John for a few decades now and have been impressed time and again with the depth of his knowledge regarding nonmetallic materials. John was kind enough to review and comment on the nonmetallic subject matter.

About the Author

William M. (Bill) Huitt has been involved in industrial piping design, engineering, and construction since 1965. Positions have included design engineer, piping design instructor, project engineer, project supervisor, piping department supervisor, engineering manager, and president of W. M. Huitt Co., a piping consulting firm founded in 1987. His experience covers both the engineering and construction fields and crosses industry lines to include petroleum refining, chemical, petrochemical, pharmaceutical, pulp and paper, nuclear power, biofuel, and coal gasification.

Mr. Huitt has written numerous procedures, guidelines, papers, and magazine articles on the topic of pipe design and engineering. He is a member of various industry-related organizations including the American Society of Mechanical Engineers (ASME) where he is a member of the B31.3 section committee, B31.3 Subgroup H on High-Purity Piping, and three ASME BPE subcommittees and several task groups, as well as the ASME Board on Conformity Assessment for BPE Certification where he serves as vice-chair. He was a member of the American Petroleum Institute (API) Task Group for the development of RP-2611; he additionally serves on two corporate specification review boards and was on the Advisory Board for ChemInnovations 2010, 2011, and 2012, a multi-industry Conference and Exposition.

Mr. Huitt authored the training program and provides training to ASME auditors for auditing BPE-compliant fitting manufacturers for BPE Certification. Bill has presented at the Annual ASME Consultants Meeting and at the Annual Inspectors Meeting for the National Board of Boiler and Pressure Vessel Inspectors. He has also presented at the ISPE Annual Meeting as part of the educational track.

1

Introduction, Scope, and General Requirements of the BPE

1.1 Introduction

While the scope of the BPE Standard defines the physical and technical boundaries of what it covers, its introduction provides an overview of its intent. It also touches on the philosophical aspect of the Standard when content relevant to the needs of a user is not specifically addressed in the Standard. Meaning that, based on the relevant requirements and recommendations contained in the Standard, it is advisable to interpolate from those guidelines and requirements what would be appropriate for a particular situation not specifically covered in the Standard.

The BPE Standard, as explained in its introduction, applies to components in a processing system that come in contact with product, product intermediates, and raw material fluids. What is not mentioned, but is included by inference, is process fluids. And to be clear, what is also covered by the BPE Standard are utility services such as those that handle compendial waters and steam. Compendial waters as well as steam are those utilities that meet the requirements of the pharmacopoeias for utility fluid services. Fluid services include Water for Injection (WFI), purified water, and clean steam.

In the case of pharmaceutical manufacturing, a compendial is essentially a specification that complies with the US Pharmacopoeia (USP). There are three main pharmacopoeia organizations around the world: USP, European Pharmacopoeia (Ph. Eur.), and the Japanese Pharmacopoeia (JP). These organizations harmonize in creating drug formulary (specifications) as well as specifications for water used in the manufacture of drugs and drug products.

Bioprocessing Piping and Equipment Design: A Companion Guide for the ASME BPE Standard,
First Edition. William M. (Bill) Huitt.

The BPE Standard is not a stand-alone standard, but instead works in association with the Boiler and Pressure Vessel Code (BPVC) Section VIII and the B31.3 Process Piping Code. From a pressure vessel standpoint, the primary requirements for any pressure vessel will default to Section VIII requirements. The high-purity (HP) BPE aspects and requirements of a pressure vessel will overlay those Section VIII requirements. That is to say, Section VIII provides the safety and integrity requirements of a pressure vessel, while BPE provides the HP cleanability requirements in addition to that.

The same holds true for BPE's association with the B31.3 Process Piping Code. B31.3 provides the safety and integrity requirements of a piping system while BPE provides the HP cleanability requirements in addition to that. The reader of the BPE Standard will find many references to Section VIII and B31.3 throughout the Standard. There are liaison members of the BPE Standard that are also members of the BPVC and B31.3. This indeed is essential in harmonizing these codes and standards as they change and evolve over the years. Liaison efforts between the various codes and standards help provide coordination and harmonization between such codes and standards. Section 7.4.3 goes into greater detail with regard to the ways in which BPE and B31.3 work together.

1.2 Scope of the ASME BPE Standard

The ASME BPE Standard (hereinafter referred to as "Standard," "BPE," or "BPE Standard" depending on context) is the pseudo accepted international standard for system design, component standardization, and equipment design for the bioprocessing and pharmaceutical industries as well as other industries that require clean-in-place (CIP) or steam (sanitize)-in-place protocols. Essentially this is any process or segment of a process in which living organisms are used to facilitate the manufacturer of a product. Whether the end product is related to pharmaceutical, food, biofuel, or any other end product whose manufacturing process at some point contains living organisms, it does not matter. The need to preserve the appropriate cleanliness of a process system and its ability to prevent cross-contamination, external contamination, and leachable contamination from wetted parts is essential.

While the Standard is specifically designated to apply only to new systems, it is acceptable to apply it to in-service or existing systems. During the period of time this book was being written, modification to the wording of the Standard was voted on and approved to make this fact abundantly clear in the Standard. This is discussed further in Chapter 4 of this book. Before applying the Standard to existing systems, it is recommended that the existing system's Fitness for Service (FFS) be assessed. Where applicable, this means wall thickness examinations, fatigue assessment, corrosion under insulation examination, and much more depending on the expanse of the system, its years in service, and its operating conditions while in service. All of this being of chief concern with regard to a system's integrity with respect to its intended continued service. Such analysis should be performed by personnel experienced in FFS analysis.

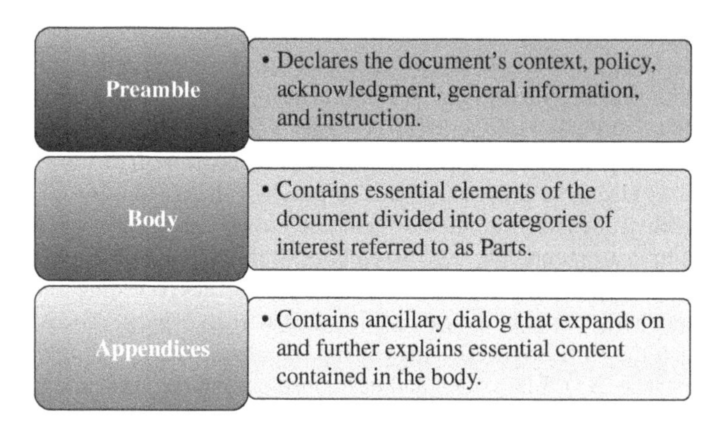

Figure 1.2.1 Main segments of the BPE standard

With regard to existing systems that have been abandoned in place, from a purity standpoint, it is suggested that while such systems may prove to be structurally sound, they may pose a contamination risk that is simply too problematic to undertake in attempting to remediate for hygienic use. If, on the other hand, after being thoroughly examined with consideration given to verifiable cleanability, an acceptable remediation plan established, and a proper protocol written, an existing system may be deemed acceptable for possible use in HP processing. Otherwise such systems should be dismantled, left as is, or simply not considered for the purpose of a HP process or utility service.

In referring to Figure 1.2.1, the BPE Standard consists of three main segments: Preamble, Body, and Appendices, which is typical of most ASME codes and standards. And, I might add, some of the terminology used here is not necessarily what ASME would use in describing the makeup and compilation of their codes and standards. It is instead my way of compartmentalizing and describing how these code and standard books are assembled. To continue, The Preamble portion of the book describes the document's context, points out ASME policy, acknowledges the membership, provides general information, gives a brief timeline history, and provides instruction. The Body of the Standard contains essential elements of the document divided into categories of interest referred to as Parts. This is where the requirements, recommendation, and guidance of the Standard reside. The Appendices is where ancillary dialog is added to expand on and further explain essential content contained in the body.

With regard to the Body of the Standard, the 2014 edition is currently divided into 10 sections referred to as Parts, as follows:

- **Part GR**—General Requirements: This section sets the tone and defines the scope and intent of the Standard. It defines terms that are specific to the bioprocessing industry and other terms that may have originated elsewhere and have been adopted by the BPE standard and given a definition that better relates to its intended use. It also provides a listing of documentation that is essential in meeting Food and Drug Association (FDA) compliance requirements. These are documents that would serve other industries well in proving verifiable and traceable evidence of material and workmanship.

- **Part SD**—Systems Design: Part SD creates a forum for lessons learned in the bioprocessing industry and also establishes standardized methodologies for achieving cleanable process systems. It provides discussion on how to design cleanability and sterility into a system and covers specific design issues with regard to instrumentation, hose assemblies, filtration, and other equipment. In addition to hydrostatic testing, this section also touches on testing fundamentals for spray balls, drainability, cleanability, and sterility. This section is one place in which the BPE Standard steps beyond the main focus of the B31.3 format. For instance, B31.3 is written and developed around the cornerstone of safety and system integrity, and the BPE, while also integrating safety and integrity, is focused mainly on providing acceptable criteria for system design. Since its inception, the SD subcommittee (SC) has taken on the task of researching accepted industry design practices that are currently being used in the bioprocessing industry. This, in an effort to validate, and, where necessary, rectify those largely unsubstantiated design practices and criteria while developing new and beneficial design criteria for adoption into the BPE Standard.
- **Part DT**—Dimensions and Tolerances for Process Components: Part DT has created standard dimensions for HP fittings. Prior to the availability of the BPE Standard and Part DT, there were no industry standard dimensions on fittings and valves and no common set of manufacturing tolerances. This meant that components from one manufacturer to the next were not necessarily interchangeable. This was a situation that had long presented a nightmare for many designers, forcing a situation in which all fittings had to be purchased from the same manufacturer to ensure compatibility. By working with the fitting manufacturers, they were able to create a standardized set of fittings. In addition they added a much needed option for the sulfur content of ASTM A270 stainless steel to support the use of autogenous orbital welding. This will be discussed in depth in Chapter 4 of this book.
- **Part MJ**—Materials Joining: For both metallic and nonmetallic material Part MJ touches on all aspects of the welding and bonding of pressure vessels, tanks, tubing, and fittings. It provides guidance on acceptable requirements related to material selection, inspection, examination, and testing. It also discusses joining processes and procedures, weld joint design and preparation, weld acceptance criteria, procedure and performance qualification, and documentation requirements. Several tables list weld acceptance criteria, and detailed graphics illustrate acceptable and unacceptable welds.
- **Part SF**—Process Contact Surface Finishes: A crucial element in the ability to attain and maintain a clean system is in the quality of the finish on the product contact surface. Whether in the bioprocessing industry or other sectors in which at least a segment of the processing scheme involves bioprocessing (such as biofuel production), the cleanability of the product contact surface is crucial. In addition to Part SF providing the methods by which surface finishes are classified, it also spells out the acceptance criteria for compliance.

- **Part SG**—Sealing Components: This part covers equipment seals and provides a classification scheme that describes the required integrity of a seal under specific service conditions. Seals are segregated into two groups: static and dynamic. The static seal is the type that would be used in a hygienic clamp union. A dynamic seal is the type that would be used to seal two surfaces, one of which is a nonstatic surface such as a rotating ball in a ball valve.
- **Part PM**—Polymeric and Other Nonmetallic Materials: Added to the Standard in 2002, this section covers criteria related to polymers in the form of thermoplastics, thermosetting resins, and elastomers as well as other nonmetallic forms. It touches on design considerations, joining methods, interior product contact surfaces, and materials of construction.
- **Part CR**—Certification: This part was first included in the 2009 publication of the standard and gives users a way to ensure that the tubing and fittings they purchase are compliant with ASME BPE Standard requirements. This is achieved through a well-defined and implemented certification program for compliance with the BPE Standard by those manufactures, fabricators, and service providers that qualify. The certification process is a multifaceted program based on an in-depth quality management system (QMS) program that is defined in Part CR. Specifically, the program requires that the applicant for certification create a QMS manual, as defined in the BPE Standard, which is expected to mirror the quality program actually being used in their production process. Among many other requirements, the manual should reflect a company's organizational hierarchy, inspection protocols, materials handling procedures (from receiving through manufacturing and shipping), procedure for segregation of materials, inspection personnel qualifications, reject resolution, and documentation needs.
- **Part MM**—Metallic Materials: This section was first published in the 2009 issue of the BPE Standard. Its incorporation into the standard was driven by the growing importance of alternative materials of construction beyond Type 316L stainless steel. The main objective of this section is to help system designers and facility owners improve system quality and sustainability and to improve compatibility with fluids that are too aggressive for Type 316L stainless steel. Adding Part MM allows the standard to elaborate and expand its information on metallic materials in a centralized and comprehensive way. This section offers a definitive but ever-changing listing of acceptable materials in their various forms and provides further information on pitting resistance equivalent number (PREN) rankings, corrosion test references for alloys, discussion points on superaustenitics, duplex stainless steels, nickel alloys, ferrite content restrictions, and much more.
- **Part PI**—Process Instrumentation: First included in the 2012 publication of the BPE Standard, this much needed section of the Standard establishes standard requirements for instrumentation as it applies to bioprocessing and other HP process requirements. It touches on minimum requirements for such instrument items as transmitters, analyzers, controllers, recorders, transducers, final control elements, signal converting or conditioning devices, and computing devices. It also discusses electrical devices such as annunciators, switches, and pushbuttons.

As mentioned, the sum of the aforementioned parts make up what is referred to as the body of the BPE Standard. Following are what could be considered extensions of the Standard in the form of a mandatory appendices and a nonmandatory appendices.

- Mandatory Appendices: It is editorial policy to be concise, to the point, and without elaboration in the body of any standard or code when stating a requirement or recommendation. Such documents are written and published under the accreditation guidelines of American National Standards Institute (ANSI), which do not permit, under those rules, educational-type dialog within the body of an accredited standard or code. However, there are instances in which it is felt that further explanation and guidance on various compliance statements within a standard or code are needed. Unable to properly make such elaborations within the body of a standard or code such guidance is provided in the mandatory appendices. The mandatory appendices in the BPE Standard are therefore an extension of the body of the Standard and allows for elaboration, clarification of subject matter, and guidance on *required* elements contained in the Standard itself—content that would otherwise not be permitted. Content of the mandatory appendices shall be complied with the same as that found in the body of the Standard unless otherwise amended by specification.
- Nonmandatory Appendices: Much like the mandatory appendices, the nonmandatory appendices are an extension of the Standard itself. Wherein, guidance and extended information on various topics pertaining to the Standard can be provided. Information contained in the nonmandatory appendices are, as the name suggests, not mandatory. It can be treated as information for guidance only or it can be adopted as needed in requiring compliance for particular activities. In the event that it is adopted, it then becomes mandatory.

The Preamble of the Standard along with its Foreword, Statement of Policy, Committee Roster, Summary of Changes, the Body, and the Appendices constitute all of the elements necessary in an industry standard. Like all other codes and standards, the BPE will also continue to add, modify, and remove content as it keeps pace with industry and technologies in finding new and better ways to improve how we design and build safer and more productive pharmaceutical manufacturing facilities.

1.3 Intent of the BPE Standard

While the actual content of the Standard is clear and concise in stating the requirements necessary in creating a piping system conducive to FDA regulations and cleanability, there are nuances and interpretations that can allow the designer to expand on what is written. These nuances and interpretations are similar to performing interpolation between data points. For example, if you are given what the pressure rating value of a mechanical joint is at 300°F and its pressure rating value at 400°F, you can determine the pressure rating value for that same joint at

360°F through interpolation. Understanding and applying the logical intent of a standard such as the BPE are much the same way. This can be referred to simply as applying the "philosophical intent" of the Standard.

The same logic used in what is written in the Standard can be expanded into areas that may be of a proprietary nature to a company and not covered specifically in the Standard. Such needs can be addressed in a manner that follows the philosophical intent of the Standard. Such interpolations of the Standard will be touched on as we move through this book.

1.4 ASME B31.3 Chapter X

At a 2005 meeting of the ASME B31.3 Process Piping Code committee, a presentation was made to its section committee pointing out the need to adopt a chapter on Ultrahigh-purity (UHP) piping. UHP is a term used to define a level of cleanness more associated with the semiconductor industry than any other industry. The initial vanguard in this effort to develop a new chapter in the piping code consisted mainly of personnel from the semiconductor industry, thus the use of the term UHP in describing the chapter.

Members of the ASME BPE Standard, who are also members of the B31.3 committee, learned of the work being done on this new chapter in B31.3, referred to as Chapter X, and saw an opportunity to better integrate the BPE Standard with the piping code it most closely relates to. Two members of the fledgling B31.3 Chapter X UHP piping section were subsequently invited to the October 2007 BPE meeting held in Philadelphia. The visiting gentlemen were introduced and gave a presentation to the BPE Standards Committee with regard to the new B31.3 Chapter X.

At the time BPE members became involved in the development of Chapter X, it was titled UHP Piping and the chapter prefixes were "U." It was pointed out that if the new chapter were to include more than just content on the semiconductor industry, it would have to change its title since the term "UHP" pertained almost singularly to the semiconductor industry. It was then agreed to title it simply "Chapter X High Purity Piping."

After another 5 years of writing, rewriting, balloting, and more balloting, Chapter X was finally approved and then published in the 2012 issue of ASME B31.3. This effort accomplished a number of things but two in particular. First, it influenced and impacted the safety aspect of the BPE with regard to piping and equipment in industries which heretofore operated to a large extent under their guidance. Safety, based on integrity of design, is the hallmark of B31 codes. And to be integrated into that process is to inherit that philosophy and methodology.

Secondly, references made from the BPE Standard to B31.3 prior to the addition of Chapter X did not harmonize as they should. The addition of Chapter X provided improved continuity between BPE and B31.3 that did not previously exist. The content of Chapter X helped B31.3 interface better with both the BPE and the SEMI International Standards and in so doing melded the essentials of design for HP and UHP piping with that of the safety and integrity requirements of B31.3.

1.5 Terms and Definitions

Many, if not the majority of words and terminology, have variations in their meaning. Codes, standards, and legal documents have to generally be very explicit in what it is they mean when using certain terms. In declaring something to be a requirement in codes and standards, there has to be specific meaning in the term or terms that are used. This is to prevent the intended meaning from becoming lost or misconstrued due to possible nuances in the term's general use and definition. In other words there can be no variation in what is meant when making statements that direct the user of a code or standard to do something or in describing a requirement. The essence of each word takes on a heightened degree of import when safety, integrity, and even personal health are at stake, as is the case when writing content for industry codes and standards.

A "term of art" therefore describes a term that has been adopted or contrived with a very specific definition that has been applied to a term for a special or particular meaning in context with, or more appropriate for its intended application. The generally accepted definition for a term that already exists may need to be nuanced in order to more accurately define its specific purpose in a code or standard. Codes, standards, and legal documents frequently have to provide such specific definitions for some terminology in order to make a term's meaning explicitly accurate as to its intent within the context the term is used in these types of documents.

In cases in which a code or standard is being referenced in material specifications, a design basis, or in a procedural document, it is advisable to understand the definitions of the terminology defined within those codes or standards. Terms that are intended to apply in a particular manner, in the context of the code or standard, are specifically defined in that document to avoid having their intent misconstrued or misinterpreted.

As an example we will consider the term "hygienic," defined in the Merriam–Webster dictionary as follows:

Hygienic
a: of or relating to hygiene
b: having or showing good hygiene
c: of, relating to, or conducive to health or hygiene

Such definitions, as previously listed, while describing the general use of the term, do not convey the explicit description for which the term is used in the BPE Standard. The BPE Standard has therefore adopted the term (making it a "term of art") for use under a definition that better fits its specific use and application within the Standard; a definition that more directly and explicitly relates to its intended use in the Standard as follows:

Hygienic: of or pertaining to equipment and piping systems that by design, materials of construction, and operation provide for the maintenance of cleanliness so that products produced by these systems will not adversely affect human or animal health.

And there are, of course, terms particular to an industry or to a standard that are contrived, created, or simply find their way into the vernacular of an industry. Over time such terminology becomes adopted to define certain aspects related specifically to an industry or content within a standard. Some of these terms are adopted by an accredited standard and legitimized in the process. With regard to the BPE Standard and the bioprocessing industry that it serves, there are such terminology as:

Autogenous weld: A weld made by fusion of the base material without the addition of filler.

Closed head: For orbital GTAW, a welding head that encapsulates the entire circumference of the tube/pipe during welding and that contains the shielding gas.

Compression set: Permanent deformation of rubber after subscription in compression for a period of time, as typically determined by ASTM D395.

Dead leg: An area of entrapment in a vessel or piping run that could lead to contamination of the product.

Clean-in-place (CIP): Internally cleaning a piece of equipment without relocation or disassembly. The equipment is cleaned but not necessarily sterilized. The cleaning is normally done by acid, caustic, or a combination of both, with WFI rinse.

Open head: For orbital GTAW, a welding head that is open to the atmosphere external to the tube/pipe being welded and that does not enclose the shielding gas, which is still provided through the torch.

Passivation: Removal of exogenous iron or iron from the surface of stainless steels and higher alloys by means of a chemical dissolution, most typically by a treatment with an acid solution that will remove the surface contamination and enhance the formation of the passive layer.

Rouge: A general term used to describe a variety of discolorations in HP stainless steel biopharmaceutical systems. It is composed of metallic (primarily iron) oxides and/or hydroxides.

Under CFR Title 21 of the FDA, there also exists definitions for terminology used in context with FDA regulations, such as:

Active ingredient: Any component that is intended to furnish pharmacological activity or other direct effect in the diagnosis, cure, mitigation, treatment, or prevention of disease or to affect the structure or any function of the body of man or other animals. The term includes those components that may undergo chemical change in the manufacture of the drug product and be present in the drug product in a modified form intended to furnish the specified activity or effect.

Components: Any ingredient intended for use in the manufacture of a drug product, including those that may not appear in such drug product.

In-process material: Any material fabricated, compounded, blended, or derived by chemical reaction that is produced for, and used in, the preparation of the drug product.

Such terms as those listed previously and their definitions can be found in the ASME BPE Standard or under CFR Title 21 of the FDA. And there are other "terms of art" in which a word or phrase in a published work, such as this book, has a meaning that may be more specific or slightly different from that which might otherwise be inferred by definition from another source—a meaning specifically defined in order to be made more explicit in its use. In such cases those terms or phrases are defined as to their intended meaning for the context in which they are used. The definitions that follow are for such terms used in this book and are defined here to more appropriately relate to their specific use herein, as follows:

Biopharmaceutical: Pharmaceutical products manufactured using bioprocessing.
Bioprocessing: Any chemical process in which living cells are utilized.
End user: A company or named person or persons designated as having overall ownership and/or responsibility for the manufacture of an end product or raw material.
Piping and tubing: These two terms are synonymous within the context of this discussion.
Piping system: All pipe/tube, fittings, inline components, equipment, instrumentation, insulation, and supports that make up a processing system.
Process solution (aka process): Any chemical or other additive solution that is combined with other chemicals or solutions to become an integral part of a finished product.
Process contact surface: Any surface (component, equipment, instrument, single-use component) that comes in direct contact with a process solution including the surfaces of ancillary systems handling fluids that come in contact with the process system on a secondary basis such as CIP.
Product solution (aka product): The final solution that makes up a finished product even though additional steps in the finishing process may still be required (e.g., encapsulation, crystallization, etc.).
Product contact surface: Any surface (component, equipment, instrument, single-use component) that comes in direct contact with a product solution.
Wetted (aka wetted surface): The surface of any part of a single component or equipment item that comes in contact with the product or process at any time during operation of a system.

Allow me to explain also the capitalization used in this book with regard to the terms "standard" and "code." Where either the term "standard" or "code" refers to a specific standard or code as in "BPE Standard" or "B31.3 Code" those terms will be capitalized. Or when making an implied reference to a specific standard or code, such as when Standard is in reference to BPE or when Code is in reference to B31.3, the terms will be capitalized. When making a general statement about standards or codes, the terms will not be capitalized.

1.6 Quality Assurance

Ground rules first. The term "components," as used in CFR Title 21, refers to (chemical) ingredients used in the manufacture of an in-process material, drug product, food, or cosmetic product. It does not refer to fitting components such as tubing, tees, or elbows. So in referring to documentation requirements related to "components" such as those stated in CFR Title 21 Section 211.180 subparagraph (b) as follows:

> (b) Records shall be maintained for all components, drug product containers, closures, and labeling for at least 1 year after the expiration date…

It does not pertain to tubing and fittings. The question then arises as to if the FDA does not state a requirement for documentation and traceability for process system material that comes in contact with process or product fluids, then why the need for all of the documentation requirements found in the BPE?

The reason for requiring documentation and traceability for such process or product contact material is a statement made under CFR Title 21 Part 211 Section 211.65 subsection (a), which reads:

> (a) Equipment shall be constructed so that surfaces that contact components (i.e., chemical ingredients or process fluid), in-process materials, or drug products shall not be reactive, additive, or absorptive so as to alter the safety, identity, strength, quality, or purity of the drug product beyond the official or other established requirements.

Within the hidden implication of the previous statement that reads, *…surfaces that contact components, in-process materials, or drug products shall not be reactive, additive, or absorptive…* lies the need to prove to the on-site FDA inspector the fact that, "*…surfaces that contact components, in-process materials, or drug products…*" are indeed not, "*…reactive, additive, or absorptive…*" Though such a latent requirement as the need for documented proof and traceability for components and equipment that come in contact with a process is not explicitly stated, but is instead implied, does not diminish the impact of the statement. In clarifying such regulatory compliance requirements, Standards Developers such as ASME, API, NFPA, and many others, peel back the layers of regulatory dialog, parse the rhetoric, and assess nuances in those regulations, such as the earlier statement, and then define those latent implications to the user in a manner that will make such underlying requirements much more apparent.

Within the FDA organization is the Office of Regulatory Affairs (ORA). This is the department that carries the responsibility of inspecting facilities that manufacture in-process material, drug product, food, or cosmetic product and whose job is to enforce regulatory requirements that such facilities are obliged to follow.

FDA inspectors, working through the ORA to inspect new and operating drug manufacturing facilities, are guided by three resources: the regulations set forth under CFR Title 21, its counterpart USC (US Code) Title 21 Chapter IX, and their FDA training guidelines. Training for FDA inspectors plays a big role in how they perceive, not only the rule of law under the USC but also to understand the nuances and implications of the unwritten variants within the many complex guidelines and laws the manufacturer is obliged to follow under CFR Title 21.

One of the difficulties lies in the fact that many regulations governing the manufacture of food, drugs, and cosmetics are, in many cases, intentionally vague. This is due in large measure to two basic facts:

- Much of the manufacturing in the drug, food, and cosmetic industries is proprietary and specialized. It would be impossible to write detailed requirements that would apply to all the varied manufacturing schemes without constraining or interfering with innovation and without inhibiting new concepts in design and manufacturing.
- The criteria that the inspector must base their field analysis on is relative by nature and is subject to a subset of nuances that would be impossible to capture in words, making broad statements in the CFR a necessity.

It helps to somewhat understand the perception of the FDA inspector in carrying out their fundamental responsibility of ensuring that the regulations relative to the product being manufactured are being met. Not their mind-set but rather their training portfolio. There are a set of "Inspection Guides" and "Inspection Technical Guides" used in the FDA inspectors' training that are available in Appendices B through J found in the back of this book. What is found in this book can also be found at www.fda.gov/iceci/inspections. At this site you will also find access to other guides used by FDA inspectors, such as the following:

- Field Management Directives: These are mainly for internal management.
- IOM: Investigations Operations Manual.
- Guide to International Inspections and Travel.
- Medical Device GMP Reference Information.
- QS Regulation/Design Controls.

Point being, that FDA regulations are written in a manner that is somewhat tangential to formal regulations in that the inspectors' guidelines allow, indeed it prescribes for them the means to make on-site judgment calls with respect to interpreting the meaning of a regulation based on their firsthand assessment. And this is where the BPE Standard comes to the aid of the designer, engineer, and end user. The BPE Standard compiles and assimilates information such as that which is contained in, not only the regulations themselves but also in such guides and training documents as those mentioned previously.

So when a regulatory statement is made to the effect that, as mentioned earlier, *...surfaces that contact components, in-process materials, or drug products shall* *not be reactive, additive, or absorptive...* the Standard expands on such a statement by determining, from that, just what it is the FDA will require of the end user in showing proof that they are in compliance. That is why documentation and traceability are such a significant requirement in the BPE Standard. All material used in the construction of drug manufacturing that comes in contact with the process or product will require a paper trail that is not insignificant. Meaning, verifiable material documentation, from that of gaskets and seals to modular assemblies and from weld filler material to instrument probes, will be required and will need to be traceable to their manufactured source.

1.6.1 Documentation

Using Section 1.6 as a premise, this section will describe the type of documentation necessary in providing evidentiary proof that the material expected to be in contact with process and product fluids are acceptable, are in accordance with FDA dictates, and are fabricated, assembled, and installed in accordance with the ASME BPE standard.

1.6.1.1 Trust but Verify

Whether you are a raw material supplier, a component manufacturer, a fabricator, a service provider, or a distributer, you are a link in a supply chain. And as such, you, as well as the engineer or end user, should be continually checking upstream in that supply chain to ensure that consistently compliant material, components, and services are delivered to your doorstep. This may involve periodic audits of a provider's QMS manual or even an on-site audit. The QMS and other facets of quality assurance (QA) will be discussed in greater detail in Chapter 6.

1.6.1.2 Source Verification of Material, Product, and Services

Documentation is the foundation of HP piping systems. From a manufacturer's QMS to an installing contractor's turnover package and everything in between, proper documentation provides assurance, traceability, and accountability for material, components, and services procurement of anyone in the supply chain leading up to an installed process system. Such documentation provides the following:

- Assurance, as to procurement and delivery of proper construction material and that the work was done in accordance with specifications.
- Traceability, in that all material that comes in contact with the process or product is traceable back to its original Material Test Report (MTR) or Certificate of Compliance (C of C) for its original chemical composition.
- Accountability, in that all personnel responsible for the welding, assembly, installation, and testing of process or product contact systems are identified and on record.

It is not enough to verify and document that the welding of two sections of tubing was done properly and in compliance with a governing code or standard. The chemical properties and mechanical properties of the material from which each section of tubing was formed has to be traceable back to the mill from which it was originally melted.

A batch of melted material and chemical additives is referred to as a heat of material. Each batch, or heat of material, also referred to simply as a "heat," has its own characteristics, or fingerprint, if you will. These characteristics are based on the heat's chemical composition and its mechanical properties. Each heat of material is assigned a specific heat number and will carry that number with it throughout its recorded life cycle and product evolution.

As it is separated into various product forms, the heat number is subsequently marked on each of its separate forms of material as it moves from mill to market. The document that contains the material's chemical and mechanical properties, plus other information as identified in the bulleted list in the following, is referred to as a Material Test Report, Mill Test Report, Certification Report, Certified Material Test Report, or simply a Test Report.

Figure 1.6.1, is a simple diagram that represents the path taken from mill to market by tube and tube fitting products, such as the tubing itself and various fittings made from tube such as elbows, tees, laterals, etc. Welded tubing is manufactured from sheet material. The sheet material formed at the mill is rolled on to spools. These spools are then either slit into skelp (strips of metal) at the mill or sent to a tubing manufacturer in bulk coils to be slit as needed. The bulk coils of steel, prior to entering the production line, are, as mentioned, slit into skelp and then formed and welded longitudinally into tubing. Some of that tubing will be used by fitting manufacturers and formed into various fittings such as elbows, tees, and laterals.

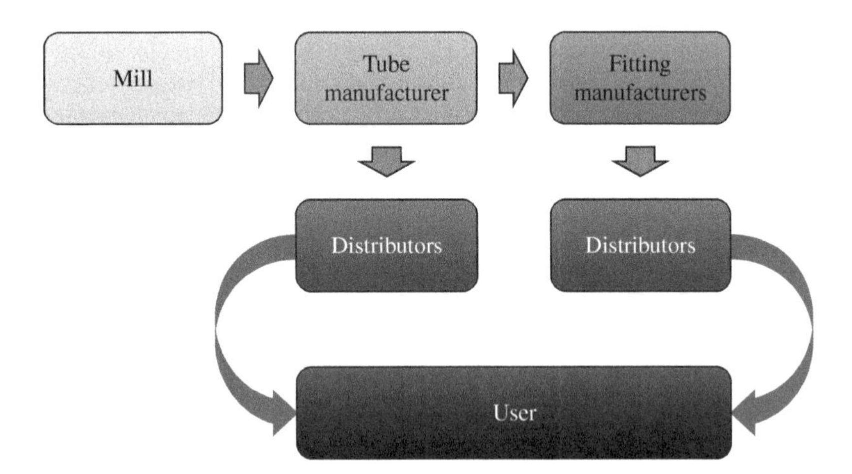

Figure 1.6.1 Mill to market of tube and fitting products

As the material moves along its path from mill to market to become an end product and to then be distributed to users of the product in constructing HP facilities, it requires one more thing—its MTR. The MTR has to be requested by the purchaser and will contain a requested laundry list of information applicable to the material. As an example, ASTM A450 states that,

> "When specified in the purchase order or contract, the producer or supplier shall furnish a Certified Test Report certifying that the material was manufactured, sampled, tested, and inspected in accordance with the Specification, including year date, the Supplementary Requirements, and any other requirements designated in the purchase order or contract, and that the results met the requirements of that Specification, the Supplementary Requirements, and the other requirements."

It then goes on to state that,

> "In addition, the Certified Test Report shall include the following information and test results, when applicable:
>
> - Heat number,
> - Heat analysis,
> - Product analysis, when specified,
> - Tensile properties,
> - Width of the gage length when, when longitudinal strip tension test specimens are used,
> - Flattening test acceptable,
> - Reverse flattening test acceptable,
> - Flaring test acceptable,
> - Flange test acceptable,
> - Hardness test values,
> - Hydrostatic test pressure, Non-destructive Electric Test method,
> - Impact test results, and other test results or information required to be reported by the product specification."

The MTR originates at the mill or foundry from which the material is produced. Referring to the mill or foundry as the producer, this is the source of the original heat or batch of material, and they, the producer, will assign an original heat number to each batch or heat of steel. That heat number is then marked on the various forms of the steel from its initial billets, bars, and sheets to its final product in the supply chain.

As the product is modified by downstream tubing and fitting, manufacturers' information contained on the original MTR may get transposed on to the various manufacturers' personalized MTR forms, forms that better suit the needs of each manufacturer. This does not give the product manufacture a license to modify the chemical properties or the original heat number. And depending on the type of forming that is done in the manufacture of a component, the mechanical properties may undergo a transition. A transition that may alter the mechanical properties of a material such as its yield and tensile strength.

Because of this possibility, it is not a requirement to include the mechanical properties in the MTR of fittings (Ref. BPE Para. MM-6.3). If included they must comply with the specifications of the raw material from which they are manufactured.

1.6.1.3 Turnover Package

As an installing contractor, who is very likely the fabrication contractor, begins completing the installation of piping systems, a turnover process begins. This is the phase of a project in which the mechanical contractor turns over all required documentation related to each system that the contractor is responsible for. Typically the documentation package, referred to as a turnover package, is handed over to the owner or the owner's representative by systems, which is an organized way of handling what could be, and typically is, thousands of documents. It is between the owner and contractor to agree on the most efficient and organized way in which the handover of documentation is done.

If the turnover procedure is set up for the installing contractor to hand over the turnover documentation at the time each system or package of systems is completed, then the turnover packages can serve as notification to the owner that a system is completed and ready for their review. Once the package is logged in, an assigned inspection group can then go about verifying that all of the required documentation has been included, based on a predetermined set of criteria. This in turn triggers a walk-down of the installed system to verify its accuracy with the documentation and to verify its completion.

Within each turnover package, there are certain documents that are to be accrued by the contractor in the performance of their work and then transferred to the owner in a manner dictated by the owner or agreed to between the owner and contractor. This list of documentation is considered to be a minimum requirement under the BPE Standard but can be expanded on. That list includes:

- For materials
  - MTR
  - C of C
  - Material Examination Log
- For welding
  - Welding Procedure Specification (WPS)
  - Procedure Qualification Record (PQR)
  - Welder Performance Qualification (WPQ)
  - Welding Operator Performance Qualification (WOPQ)
  - Examiner Qualifications approved by owner or owner's representative
  - Inspector Qualifications approved by owner or owner's representative
- For weld documentation
  - Weld map
  - Weld log

  - Weld examination and inspection log
  - Coupon log
- For testing and examination
  - Passivation report
  - Spray ball testing
  - Pressure testing (actually a "leak test")
  - Final slope check documentation
  - Calibration verification documentation
  - Purge gas certifications
  - Signature log
  - Number of welds—both manual and automatic
  - Number of welds inspected expressed as a percentage (%)
  - Heat numbers of components must be identified, documented, and fully traceable to the installed system

An additional item, not a BPE requirement for the turnover package but a document that can nonetheless be added to the previous list, is a simple but effective method used to identify and number test circuits used in leak testing the installed piping systems. This will be explained along with a scheme for the leak testing process in Appendix A. The aforementioned list of documentation will be discussed and explained in the appropriate chapters throughout this book.

1.7 An Essential Understanding of Codes and Standards

An understanding by those that write and develop codes and standards, an understanding that goes largely unacknowledged and unstated within codes and standards, is that codes and standards provide the **essential minimum requirements** necessary to achieve their intended goal. And that initiative is to provide guidance and establish requirements needed in order to achieve a safe working and operating environment within the scope and confines of that code or standard's responsibility. In other words the Standards Developer creates a line in the sand when it comes to writing and developing codes and standards for industry. A line that on one side provides prescriptive requirements and guidance to the user and on the other side refrains from becoming burdensome with excessive and overly conservative requirements. What the engineer should understand in working with codes and standards is that those minimum requirements should be treated as benchmark values and not as a crutch. Meaning that each circumstance should be considered on its own merit, and if good engineering judgment leads to more conservative values, then those values should be determined and applied.

Engineers should have awareness in the fact that ASME and other such Standards Developers walk a fine line in their efforts to create and establish guidance and requirements that create safety in the workplace while at the same

Case in Point

In an incident that occurred at a Texas refinery on February 16, 2007, a leak from a ruptured liquid propane pipeline in a propane deasphalting (PDA) unit caused an explosion that ripped off a nozzle on a PDA extractor column causing ignited propane to erupt from the now-opened nozzle on the column at a velocity sufficient to create a jet fire (Figure 1.7.1).

The blowtorch-like flame discharged toward a main pipe rack approximately 77 ft away. As the temperature of the nonfireproof structural steel of the pipe rack reached its plastic range and began to collapse in on itself, the piping in the rack, which contained additional flammable liquids, collapsed along with it (Figures 1.7.2 and 1.7.3).

Due to the loss of support and the effect of the heat, the pipes in the pipe rack, unable to support its own weight, began to sag. The allowable bending load eventually being exceeded from the force of its unsupported weight, causing the rack piping to rupture spilling its flammable contents into the already catastrophic fire. The contents of the ruptured piping, adding more fuel to the fire, caused the flames to erupt into giant fireballs and thick black smoke.

While the engineer was certainly in compliance with the governing code, with regard to fire proofing, a thorough risk analysis may have determined that the 50 ft avoidance perimeter stated in the standard might not have been sufficient for such an installation. Proprietary circumstances, therefore, make it the imperative responsibility of the engineer or the owner to make risk assessments based on specific design conditions; conditions that may require good engineering practice to push design beyond the minimum requirements of an industry code or standard when such analysis dictates.

Figure 1.7.1 From plant surveillance camera 90 s after ignition. Courtesy: U.S. Chemical Safety and Hazard Investigation Board

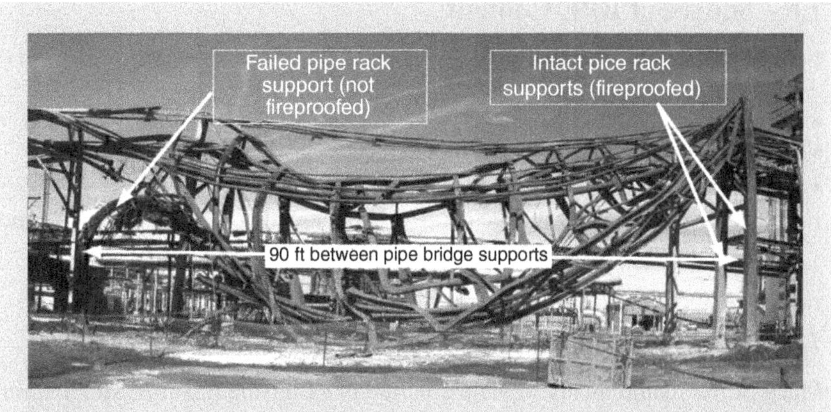

Figure 1.7.2 View of PDA unit pipe rack. Courtesy: U.S. Chemical Safety and Hazard Investigation Board

Figure 1.7.3 Aerial view of PDA unit with PDA extractor columns in the upper right. Courtesy: U.S. Chemical Safety and Hazard Investigation Board

time trying not to step over that line of infringement, getting into the area of overregulation, and undue influence. Knowing this should embolden the engineer and designer to look beyond these basic requirements and recommendations when necessary to determine whether particular design conditions fall within the conditional parameters set forth by the standard or whether conditions are such that more conservative values or a more conservative approach should be considered.

1.8 Source of BPE Content

Content of the BPE Standard is developed from three primary sources. That being:

- Government regulations
- Generally accepted principals and practices of the industry
- Research and testing done by BPE membership
  - R&T is prompted by the recognition of the membership that standardization or clarification is needed to support and/or inform the industry.

1.8.1 Government Regulations

Much of the requirements stipulated by the FDA for the bioprocessing industry and the pharmaceutical industry in general, along with industry itself, create a context within which the BPE Standard is developed and maintained. Government regulations (top-down directives) are assessed and analyzed by the various BPE SCs to determine what the designer, constructor, and facility owner will need to know and do in order to comply with these many regulations.

As an example, under Title 21 of the FDA, Part 211—Current Good Manufacturing Practice for Finished Pharmaceuticals, Subpart D Section 211.65—Equipment Construction, it states:

(a) Equipment shall be constructed so that surfaces that contact components, in-process materials, or drug products shall not be reactive, additive, or absorptive so as to alter the safety, identity, strength, quality, or purity of the drug product beyond the official or other established requirements.
(b) Any substances required for operation, such as lubricants or coolants, shall not come into contact with components, drug product containers, closures, in-process materials, or drug products so as to alter the safety, identity, strength, quality, or purity of the drug product beyond the official or other established requirements.

In essence the statement in subparagraph (a) is telling the manufacturer of pharmaceutical products that the material of construction of the equipment and components that make up a process system, such as tubing, fittings, valves, seals, pressure vessels, etc.—those items that come in contact with the process or product—shall not alter it in any way that would be counter to its processing design nor can any lubricants or sealants, under paragraph (b), applied to processing equipment, come in contact with the process or product.

The various SCs within the BPE Standard take such government regulations and interpret them in a manner that, when added to the Standard, provides the necessary guidance to the designer and constructor on what needs to be done in order to meet those requirements.

The core of design within the BPE Standard resides mainly within Part SD. And this is not to diminish the content of the other parts of the Standard.

On the contrary, each of the ten parts that make up the Standard are integral with one another. Each providing essential information necessary in designing and constructing piping systems that meet the HP demands of FDA regulations in conjunction with the safety and integrity requirements of the ASME B31.3 Process Piping Code.

1.8.2 Generally Accepted Principals and Practices of the Industry

Each segment of industry, over time, develops an ever-evolving set of metrics that include a range of materials that seem to be more compatible with respect to service conditions within that industry; design methods developed that have, over time, proven to provide the best results; quantitative values that seem to achieve the needed result; and so on. These metrics are principles and practices developed by designers, engineers, constructors, and product manufactures that, over time, become integral and essential in achieving safety and efficiency in the workplace while meeting regulatory and design requirements. These can be referred to as "generally accepted principals and practices of the industry." The bioprocessing industry is no different.

Much of what constitutes safety and design practices in the BPE Standard, or in the B31.3 Piping Code, comes from this huge resource—that of the workplace. From the engineer working on a new concept to the designer taking a new approach at making something work or from a fabricator working a design into a physical reality to a plant operator making it function properly, each of these workplace functions develop their own set of methods, equations, and principals in carrying out their job responsibility. From these efforts, principals and practices evolve into quasi-industry standardization. In the absence of any standardization, these are the fundamentals that the workplace relies on.

The problem with such industry-derived principals and practices is the fact that in most cases there is no record of what their basis for acceptance is or how such practices were arrived at. When such principles and practices are adopted and accepted into a standard or code, they are then assessed, evaluated, and clarified as to their validity and application and whether or not they should be adopted as a recommendation or requirement in the code or standard.

1.8.3 Research and Testing Done by the BPE Membership

In consideration for adopting and refining various generally accepted principals and practices of the industry, a code or standard will vet the premise of the value or action by testing its premise under a controlled and documented procedure. Much of this testing and research is either done wholly by the membership or is coordinated by the membership in using third-party testing contractors.

When referring to "membership," I am making reference to an SC Task Group (TG) with, on average, five members whose assignment is to resolve issues that require at-length discussions or research and testing to accomplish the task.

When the issue has been resolved in TG, a proposal is written and submitted to the SC for review and comment and/or balloting for approval to be entered into the Standard.

Not all proposals are based on generally accepted principals and practices of the industry. Some are simply a case of the BPE membership understanding the need for standardization and clarification of various subject matter such as those listed in the following, which are currently being worked on:

- Maximum acceptable dead leg
- Standardized clamp joint pressure ratings
- Hold-up volume
- Lyophilization system
- Chromatography
- Filtration

Topics such as those listed previously, and many more I might add, are discussed as to the merits of including them in the BPE Standard. Like the other ASME codes and standards, the BPE goes to great lengths to avoid including superfluous information in the Standard. If a topic is covered elsewhere by another Standards Developer in a sufficient manner, there is no need to expound on that same topic in the BPE Standard. If such a topic needs to be touched on in the BPE Standard for any reason whatsoever, it simply references the document that already addresses that topic. This prevents conflicts that might otherwise occur over time when a topic contained in one document is somewhat duplicated or paraphrased in another.

During the time this book was being written, research, analysis, and testing had been recently completed for determining maximum dead leg requirements and for establishing hygienic clamp joint pressure ratings. Both are topics of considerable interest to the bioprocessing industry. Up until this point in time, maximum dead leg values were theoretical, and standardized pressure ratings for the hygienic clamp joint assembly were proprietary to each manufacture and inconsistent throughout the industry. Results of the testing and research for both of these topics are expected to appear in an upcoming issue of the BPE Standard.

1.9 ASME B31.3 Process Piping Code Chapter X

Two members of the B31.3 TG, on what would eventually become Chapter X of that Code, were invited to attend the 2007 meeting of the BPE in Philadelphia. They graciously accepted, and at that meeting they gave a presentation to the main committee as to what this TG was working on. As a side note, there is a great deal of harmonization between codes and standards committees, which is due in large part to the fact that so many members are on multiple committees within organizations such ASME (BPVC, B31 Committees, B16 Committees, etc.) as well as interorganizational committees (ASME, API, CGA, NFPA, etc.) in much the same manner.

This creates an integrated network of communication between code and standard committees and their affiliate organizations in arresting and resolving any conflicts that might otherwise exist in the pages of these codes and standards.

Work on this new chapter of the B31.3 Process Piping Code was initially proposed to the B31.3 committee in 2004. Once approved, its initial writing was influenced to a large extent by a TG composed mostly of individuals representing the semiconductor industry. Consequently the terminology, design and testing methodology, and quantitative values were based mainly on that of the semiconductor industry. The initial draft of Chapter X was therefore based largely on the requirements and guidelines of the semiconductor industry rather than HP piping in general. It was not written broad enough to encompass other industries that also required HP piping systems, such as the pharmaceutical, bioprocessing, food and dairy, and biofuel industries. The requirements of the semiconductor industry are far more stringent but for altogether different reasons than those of the bioprocessing industry. To make this proposed Chapter X work, it had to have a broader scope. Focus of the future Chapter X had to therefore change from an UHP philosophy, which relates mainly to the semiconductor industry, to that of a simple HP philosophy in an effort to broaden its scope of use.

The reason that B31.3 Chapter X High Purity had to exist at all is due to the fact that B31.3 is a construction and safety code for pressure piping, whereas the BPE Standard is chiefly about design and cleanability. The safety component within the BPE Standard defaults, by reference, to B31.3. The problem, prior to the 2010 publication of B31.3, was that no content existed within B31.3 acknowledging the BPE Standard, or the mechanisms used in HP piping, such as orbital welding, weld coupons, hygienic clamp joints, and other such BPE-related topics. The advent of Chapter X thereafter would be closing the loop on HP piping.

1.9.1 B31.3 Chapter X as Supplement to the Base Code

In B31.3 the base code is considered to be the content found in Chapters I through VI. These chapters are essentially written for metallic piping intended for fluid services that can be categorized under B31.3 as normal and Category D fluid services. These are the basic essential elements with regard to designing, constructing, and installing steel piping within the scope of what is considered normal and Category D fluid services. Any requirements beyond those essentials, such as requirements for nonmetallic piping, high-pressure piping, toxic or hazardous fluids, etc. are considered supplemental to those base requirements.

The supplemental requirements for nonmetallic piping and piping lined with nonmetallic materials can be found in Chapter VII. Nonmetals were initially introduced to the Code in its 1976 publication but were not given their own chapter until the 1980 publication. The paragraphs in Chapter VII are numbered with respect to the paragraphs in the base code with the added prefix A.

Supplemental requirements associated with handling toxic fluids, defined by ASME B31.3 as Category M fluid services, can be found in Chapter VIII.

This chapter was first added to the Code in its 1976 publication. The chapter establishes more stringent requirements for toxic fluid services and was also developed to supplement the base code. The paragraphs in Chapter VIII are numbered with respect to the paragraphs in the base code with the added prefix M.

Chapter IX, added in the 1984 publication, provides supplemental requirements for operations involving high-pressure fluids. The paragraphs in Chapter IX are numbered with respect to the paragraphs in the base code with the added prefix K.

The most recent addition to those supplemental chapters is Chapter X High Purity Piping. This new chapter was first included in the 2010 issue of the ASME B31.3 Code. The 2010 issue was actually published in March 2011. As in Chapters VII, VIII, and IX, Chapter X is supplemental to the base code, so that the respective base code paragraphs included in Chapter X carry the added prefix U to identify them with the HP requirements in Chapter X.

1.9.2 Harmonization of the BPE Standard and B31.3 Chapter X

ASME B31.3 Chapter X was born out of the harmonization efforts that interconnect the various Standards Developer organizations as well as the committees within those organizations. Members of the BPE Standard worked with members of the B31.3 Chapter X subgroup to help in drafting Chapter X until its approval and publication. During this process members of BPE also became members of B31.3 and Chapter X. One such member, Dr. Barbara K. Henon, took on the responsibility of serving as the first liaison between the two committees in providing a liaison report at each of the code and standard meetings to keep both the B31.3 Code and BPE Standard committees updated as to the changes that were taking place in each of those documents. I have since taken over that roll as of 2014 and continue to do so at this writing.

The liaison approach has kept both the Code and the Standard well in tune and abreast of one another with respect to HP piping system requirements. And over the foreseeable decades, well into the future, this same intercommunication is expected to be maintained in much the same manner as it is today.

2

Materials

2.1 Scope of this Chapter

This chapter will cover materials, both metallic and nonmetallic, that are typically, and some not so typically, used for pipe, tubing, fittings, seals, and equipment that are used in high-purity applications such as bioprocessing. In this chapter we will identify the select few materials used for high-purity applications and touch on the reasons for their being selected for such applications.

2.2 Materials of Construction

It is a requisite that materials of construction, those intended for use in pharmaceutical bioprocessing systems, "…shall not be reactive, additive, or absorptive so as to alter the safety, identity, strength, quality, or purity of the drug product beyond the official or other established requirements," as stated in CFR Title 21 Sec. 211.65(a). It is therefore incumbent upon the designer/engineer to select materials that will not only withstand the mechanical forces acting upon a pressurized piping system but also select a material that is also highly compatible with the process fluids as well.

Such materials of construction, as those used in pharmaceutical bioprocessing systems, may either be metallic or nonmetallic in polymeric form. I make the distinction of polymeric form in the case of nonmetallic materials due to the broad spectrum of nonmetallic materials that exist in both life and the industrial

Bioprocessing Piping and Equipment Design: A Companion Guide for the ASME BPE Standard, First Edition. William M. (Bill) Huitt.

marketplace, but are not remotely related to material that the pipe, fittings, or equipment would be manufactured from for bioprocessing applications.

In the 1997 inaugural issue of the BPE Standard, there was only passing mention of nonmetallic piping. And in fact, there were only a handful of metallic materials that were suggested at that time, materials such as 316, 316L, AL-6XN®, and so on, and oh, by the way, "...elastomers/fluoroelastomers that are in compliance with FDA regulations" was the extent to which BPE mentioned nonmetallic piping.

Since that time the nonmetallic material market has grown to stake its claim on a large part of the piping, equipment, and ancillary components within the pharmaceutical industry. With significant inroads into nonleachables, durable pressure containing components, and single-use products, nonmetallics have earned their place alongside the metallic stainless steels and other metal alloys for use in the pharmaceutical industry. Both types of material will be discussed in this book.

2.3 Metallic Materials

Metallic materials selected for use in bioprocessing and in pharmaceutical systems in general are selected for service based on the following precepts:

- Compatibility with the fluid service
- Able to operate within the pressure/temperature design range of the contained fluid service
- Able to withstand the CIP or SIP regimen
- Will not release any part of its chemistry makeup into the fluid stream through corrosion or erosion
- Will not alter the safety, identity, strength, quality, or purity of the drug product in any way

In the process of selecting a material, after having considered the listed precepts of doing so, is the material's cost, availability, and the ability to acquire the specified surface finish. As the earlier listed material requirements are checked off in narrowing down the choices of acceptable material, the engineer will be left with a wide range of acceptable material with an equally wide cost range.

In making a final material selection from those materials identified as acceptable, there are two additional considerations, as mentioned, that can help to further refine that list: cost and availability. Typically these two considerations are connected at the hip. Meaning that the more available an item is, the less costly it will be in comparison to other, less available items. It's a root cause factor in economics in which supply and demand have a big impact on product cost.

Upon developing a short list of possible materials by following the earlier five material selection precepts, materials under consideration for use in a bioprocessing system can then be pared down to a material that is readily available while also

Table 2.3.1 Acceptable wrought stainless steels

UNS	SAE	EN
Austenitic stainless steels		
S30400	304	1.4301
S30403	304L	1.4307
		1.4306
S31600	316	1.4401
S31603	316L	1.4404
		1.4435
Superaustenitic stainless steels		
S31703	317L	1.4438
N08904	904L	1.4539
N08367	(AL-6XN[a])	—
S31254	—	1.4547
N08926	—	1.4529
Duplex stainless steel		
S32205	—	1.4462

Designated materials listed as the same line item may not be equivalent, but are comparable.
[a]Not an SAE identifier. AL-6XN is a registered trademark of ATI Properties, Inc., parent company of ATI Allegheny Ludlum.

Table 2.3.2 Acceptable wrought nickel alloys

UNS	Trade name	EN
N06625	Alloy 625	2.4856
N10276	Alloy C276	2.4819
N06022	Alloy 22	2.4602

Designated materials listed as the same line item may not be equivalent, but are comparable.

having the least installed cost. The BPE Standard, in Part MM, provides a laundry list of acceptable materials in their various product forms that can be utilized in the material selection process. These include material as shown in Tables 2.3.1, 2.3.2, 2.3.3, and 2.3.4.

2.3.1 Understanding ASTM Material Designations

At this juncture, before going further into materials, there should be a basic understanding of the earlier listed material standards as well as the composition of these materials and why they, in particular, are selected for bioprocessing fluid services and pharmaceutical piping systems in general.

Table 2.3.3 Acceptable cast stainless steel and nickel alloys

UNS	ACI	EN	Approximate wrought equivalent	
			UNS	EN
Austenitic stainless steels				
J92600	CF8	1.4308	S30400	1.4301
J92500	CF3	1.4309	S30403	1.4307 1.4306
J92900	CF8M	1.4409	S31603	1.4401
J92800	CF3M	1.4409	S31603	1.4404 1.4435
Superaustenitic stainless steels				
J92999	CG3M	1.4412	S31703	1.4438
J94651	CN3MN	—	N08367	—
J93254	CK3MCUN	1.4557	S31254	1.4547
Duplex stainless steel				
J92205	CD3MN	1.4470	S32205	1.4462
Nickel-based alloys				
N26625	CW6MC	—	N06625	2.4856
N30002	CW12MW	—	N10276	2.4819
N26455	CW2M	—	N10276	2.4610 2.4819
N30107	CW6M	—	N10276	2.4819
N26002	CX2MW	—	N26022	2.4602

Designated materials listed as the same line item may not be equivalent, but are comparable.

Table 2.3.4 Acceptable wrought copper

UNS	EN
C10200	—
C12000	—
C12200	CW024A

In doing so the first thing we will touch on is the material designations beginning with ASTM. Within this book and its subject matter, we will be focused on ferrous and nonmetallic polymeric material as it relates to pipe, tubing, and other components, including equipment. In this section we are concentrating on ferrous material. We will also touch on nonferrous metals only as a comparison in putting ferrous materials into context.

Simply put, ferrous alloys contain iron as the primary alloy addition; nonferrous alloys contain very little or no iron. Ferrous alloys include such generic types as carbon steel, stainless steel, mild steel, cast iron, and wrought iron. Nonferrous

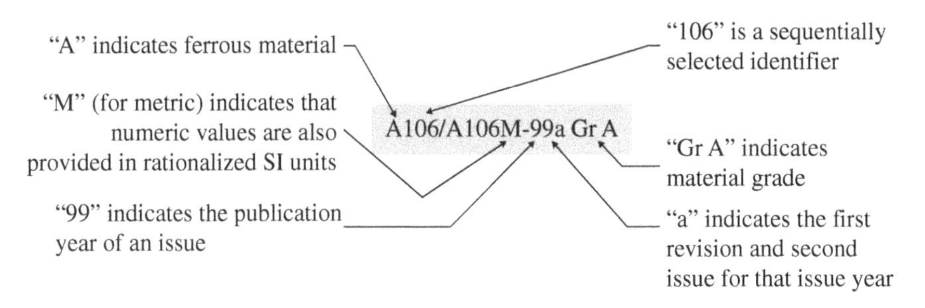

Figure 2.3.1 ASTM material standard designator description for A106

alloys include such generic types as aluminum, brass, copper, nickel, tin, lead, and zinc and additionally include what is referred to as precious metals such as gold and silver. Ferrous alloys are widely used for their strength, durability, and high-temperature applications when alloyed with proper constituent chemicals. Nonferrous alloys are widely used for their corrosion resistance (CR), their comparative light weight in many cases, and their formability.

With that general comparison of ferrous and nonferrous alloys provided, we will now concentrate more on ferrous alloys and more particularly on stainless steels. The prefix that ASTM applies to ferrous materials is "A." The prefix that ASTM applies to nonferrous materials is "B." As an example we will consider the carbon steel standard ASTM A106/A106M-99a Gr A Seamless Carbon Steel Pipe for High-Temperature Service.

As can be seen in Figure 2.3.1, the "A" indicates that the material specified is considered ferrous. The "106" that follows is a sequentially but arbitrarily selected number that is specific to this material standard. The standard number in this case is repeated after the forward slash, but includes the suffix "M" indicating that the numeric values given in the standard are also given in SI units. Following that is the year of issue, which in this case is 1999. The "a" following the date indicates that within the publication period of this issue, the standard was revised and reissued for the first time. For standards in which multiple grades exist, the specifier, when calling out A106 material, will need to specify a particular grade of material. The "Gr A" specifies the required grade of material.

Some material standards will contain multiple grades, such as those found in A106, indicating a variance in chemical composition, heat treatment, and/or mechanical strength. An increase in tensile strength and yield strength will be indicated by an ascending alphabetical grade suffix. Whereas, in the referenced A106 standard called out in Figure 2.3.1, Gr A indicates a min. tensile strength of 48 000 psi and a min. yield strength of 30 000 psi. Gr B on the other hand indicates a min. tensile strength of 60 000 psi and a min. yield strength of 35 000 psi. And finally, Gr C indicates a min. tensile strength of 70 000 psi and a min. yield strength of 40 000 psi.

A stainless steel standard designator, similar to that of A106 carbon steel, will have a somewhat different representation, as shown in the example of A270 in Figure 2.3.2, in that the grade indicator is represented in a different manner.

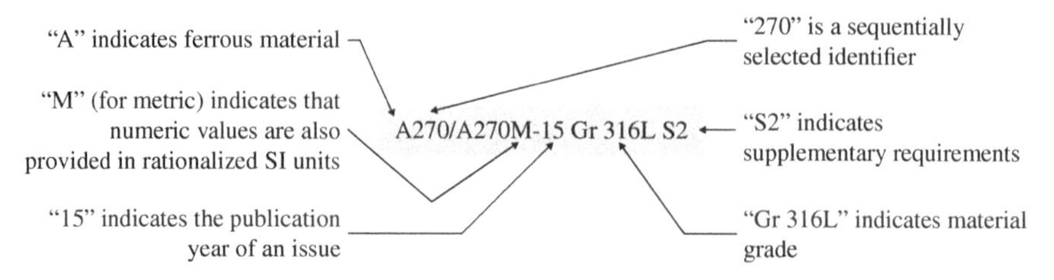

Figure 2.3.2 ASTM material standard designator description for A270

While the first part of the designation is similar to the grade identifier itself, in this case, 316L is significantly different.

The reason for this differently written grade designator is that ASTM adopted the numbering system from the Society of Automotive Engineers (SAE) for stainless steels such as 304, 304L, 316, and 316L. In an ASTM material standard, such as A270 or A269, multiple grades of material within that standard, in addition to its respective UNS number, are represented by either an SAE number (316L), alloy designator (2205), or its UNS number (S31603). And again, this change in grade identifiers is an indication of a variance in chemical composition, heat treatment, and/or mechanical strength, plus, with the stainless steels, a variation in CR.

Similar to Figure 2.3.1, Figure 2.3.2 represents a standard callout for stainless steel, wherein the "A" indicates that the material specified is considered ferrous. The "270" that follows is a sequentially selected number that is specific to this material standard. The standard number in this case is repeated after the forward slash, but includes the suffix "M" indicating that the numeric values given in the standard are also given in SI units. Following that is the year of issue, which in this case is 2015. This standard, having no suffix letter after the year, indicates that its content has not been revised from its original publication. For standards in which multiple grades exist, the specifying engineer will need to specify a particular grade of material. In this case "Gr 316L" is the specified grade of material. In addition, some standards will have supplemental requirements added to the back end of the standard. These are additional but optional requirements that the specifier can add to the material. In this case the specifier indicates the S2 supplemental requirement.

As alluded to in the preceding text, an ASTM material standard will quite frequently cover multiple variations of a material. These variations can occur due to slight modifications to the chemistry of the material, variations in the manufacturing process, or different heat treatment processes. These various subsets of material are differentiated by segregating them into identifiable groups using such terms as grade, type, and class. Generally speaking these terms are used in ASTM standards under fairly loose guidelines that are broadly described as follows:

- Grade: Assigned to identify multiple compositions of material within a single standard.
- Type: Refers to the generic material type such as ferritic, austenitic, and so on, or it may identify the various manufacturing processes such as furnace butt

weld (F), electric resistance welded (E), and so on. This is not to be confused with what many take as an abbreviation for "type" in reading TP 316L. This will be discussed in more detail shortly.

- Class: Further division among the grades and types of material, class is another level (subset) or division within a material standard.

To expand further on ASTM grades, types, and classes, examples are provided in the form of snapshots taken from ASTM standards. Figure 2.3.3 shows Table 1 as copied from ASTM A270-03 Standard Specification for Seamless and Welded Austenitic and Ferritic/Austenitic Stainless Steel Sanitary Tubing. The "grade" of material, identified as marker "A," indicates those SAE designations listed across the top line of ASTM Table 1 as "grades" of material. The SAE designations make up four of the ten listed materials. Note that each SAE designation has a prefix of TP, as in "TP 316L" identified with the marker "B." The TP is not an abbreviation of "type," but is instead the ASTM designation for tube (T) and pipe (P), meaning that this material is applicable to both product forms.

In comparison, refer to Figure 2.3.4, which shows a portion of Table 1 from ASTM A213-03b Standard Specification for Seamless Ferritic and Austenitic Alloy-Steel Boiler, Superheater, and Heat-Exchanger Tubes. The prefix "T" under the "grade" column indicates that the T2, T5, T5b, and so on are applicable in tube form only. Typically, where material is applicable in pipe product form, there will be no prefix or there may be a prefix indicating special or variable conditions that differentiates a material grade.

In referencing Figure 2.3.5, which is a portion of Table 2 contained in ASTM A336/A336M-07 Standard Specification for Alloy Steel Forgings for Pressure and High-Temperature Parts, you will notice that the grade designation for material applicable to forging contains the prefix "F," as in F1, F11, F12, and so on. Were they material grades pertaining to cast material they would have the prefix "C."

In referring to Tables 2.3.1, 2.3.2, 2.3.3, and 2.3.4, the acronyms used in those tables are derived from the organizations that develop and maintain the material, identifying numbers listed in the columns under their respective acronyms. Those acronyms include the following:

- **Alloy Casting Institute (ACI)**
  The Alloy Casting Institute (ACI) has changed its name to Steel Founders' Society of America (SFSA), but still functions much as it has since its founding in 1902. The organization assigns designation numbers to cast materials and liaisons with ASTM, NACE, ISO, and the US government.
- **European Standards (EN)**
  These are standards that have been ratified by one of three recognized authorities otherwise known as European Standardization Organizations. Those organizations include the European Committee for Standardization (CEN), European Committee for Electrotechnical Standardization (CENELEC),

Table 1 Chemical requirements

Element	Grade	TP 304	TP 304L	...	TP 316	TP 316L	...	...	...	...	...
	UNS Designation[A]	S30400	S30403	S31254	S31600	S31603	N08926	N08367	S31803	S32205	S32750
				Composition, %							
Carbon, max		0.08	0.035[B]	0.020	0.08	0.035[B]	0.020	0.030	0.030	0.030	0.030
Manganese, max		2.00	2.00	1.00	2.00	2.00	2.00	2.00	2.00	2.00	1.20
Phosphorus, max		0.045	0.045	0.030	0.045	0.045	0.030	0.040	0.030	0.030	0.035
Sulfur, max		0.030	0.030	0.010	0.030	0.030	0.010	0.030	0.020	0.020	0.020
Silicon, max		1.00	1.00	0.80	1.00	1.00	0.50	1.00	1.00	1.00	0.80
Nickel		8.0–11.0	8.0–12.0	17.5–18.5	10.0–14.0	10.0–14.0	24.0–26.0	23.5–25.5	4.5–6.5	4.5–6.5	6.0–8.0
Chromium		18.0–20.0	18.0–20.0	19.5–20.5	16.0–18.0	16.0–18.0	19.0–21.0	20.0–22.0	21.0–23.0	22.0–23.0	24.0–26.0
Molybdenum		...	...	6.0–6.5	2.00–3.00	2.00–3.00	6.0–7.0	6.0–7.0	2.5–3.5	3.0–3.5	3.0–5.0
Nitrogen[C]		...	...	0.18–0.22	...	...	0.15–0.25	0.18–0.25	0.08–0.20	0.14–0.20	0.24–0.32
Copper		...	...	0.50–1.00	...	...	0.50–1.5	0.75 max	...	...	0.50 max

[A] New designation established in accordance with Practice E 527 and SAE J 1086.
[B] For small diameter or thin walls or both, where many drawing passes are required, a carbon maximum of 0.040 % is necessary in grades TP304L and TP316L. Small outside diameter tubes are defined as those less than 0.500 in. (12.7 mm) in outside diameter and light wall tubes as those less than 0.049 in. (1.24 mm) in average wall thickness (0.044 in, (1.12 mm) in minimum wall thickness).
[C] The method of analysis for nitrogen shall be a matter of agreement between the purchaser and manufacturet.

A B C

Figure 2.3.3 Table 1 of ASTM A270-03

Table 1 Chemical requirements for ferritic steel

Grade	Composition, %									Other Elements
	Carbon	Manganese	Phosphorus, max	Sulfur, max	Silicon	Chromium	Molybdenum	Titanium	Vanadium, min	
T2^	0.10–0.20	0.30–0.61	0.025	0.025	0.10–0.30	0.50–0.81	0.44–0.65	...	...	
T5	0.15 max	0.30–0.60	0.025	0.025	0.50 max	4.00–6.00	0.45–0.65	...	...	
T5b	0.15 max	0.30–0.60	0.025	0.025	1.00–2.00	4.00–6.00	0.45–0.65	...	...	
T5c	0.12 max	0.30–0.60	0.025	0.025	0.50 max	4.00–6.00	0.45–0.65	...	...	
T9	0.15 max	0.30–0.60	0.025	0.025	0.25–1.00	8.00–10.00	0.90–1.10	...	...	
T11	0.05 min–0.15 max	0.30–0.60	0.025	0.025	0.50–1.00	1.00–1.50	0.44–0.65	...	...	
T12^	0.05 min–0.15 max	0.30–0.61	0.025	0.025	0.50–max	0.80–1.25	0.44–0.65	...	...	
T17	0.15–0.25	0.30–0.61	0.025	0.025	0.15–0.35	0.80–1.25	...	...	0.15	
T21	0.05 min–0.15 max	0.30–0.60	0.025	0.025	0.50 max	2.65–3.35	0.80–1.06	...	...	
T22	0.05 min–0.15 max	0.30–0.60	0.025	0.025	0.50 max	1.90–2.60	0.87–1.13	...	...	
T23	0.04–0.10	0.10–0.60	0.030	0.010	0.50 max	1.90–2.60	0.05–0.30	...	0.20–0.30	W 1.45–1.75 Cb 0.02–0.08 B 0.0005–0.006 N 0.030 max

Figure 2.3.4 Table 1 of ASTM A213-A213M-03b

Table 2 Chemical requirements

	Composition, %							
	Grade							
Element	F1	F11, Classes 2 and 3	F11, Class 1	F12	F5[A]	F5A[A]	F9	F6
Carbon	0.20–0.30	0.10–0.20	0.05–0.15	0.10–0.20	0.15 max	0.25 max	0.15 max	0.12 max
Manganese	0.60–0.80	0.30–0.80	0.30–0.60	0.30–0.80	0.30–0.60	0.60 max	0.30–0.60	1.00 max
Phosphorus, max	0.025	0.025	0.025	0.025	0.025	0.025	0.025	0.025
Sulfur, max	0.025	0.025	0.025	0.025	0.025	0.025	0.025	0.025
Silicon	0.20–0.35	0.50–1.00	0.50–1.00	0.10–0.60	0.50 max	0.50 max	0.50–1.00	1.00 max
Nickel	…	…	…	…	0.50 max	0.50 max	…	0.50 max
Chromium	…	1.00–1.50	1.00–1.50	0.80–1.10	4.0–6.0	4.0–6.0	8.0–10.0	11.5–13.5
Molybdenum	0.40–0.60	0.45–0.65	0.44–0.65	0.45–0.65	0.45–0.65	0.45–0.65	0.90–1.10	…

Figure 2.3.5 Table 2 of ASTM A336A-336M-03b alloy steel forgings

and European Telecommunications Standards Institute (ETSI). Of course the standards organization most related to ASME BPE is the CEN.

- **SAE (SAE International)**
  Formerly known as the Society of Automotive Engineers, SAE International and the American Iron and Steel Institute (AISI) worked in conjunction with one another, beginning in the 1930s, in developing and maintaining an alloy steel numbering system. In 1995 the AISI stepped aside leaving the SAE to continue its maintenance of the program. The 200 series, 300 series, and 400 series stainless steel alloys are all a small part of that numbering system. SAE alloy numbers such as 304, 304L, 316, 316L, 317, 321, and so on all translate into various grade designations when included in such ASTM material standards as A269, A270, A312, and so on.
- **Unified numbering system**
  The assignment of numbers or designations of steel product formulations and forms has historically been done by standards developers such as ASTM International, the American Welding Society (AWS), and the SAE, the US government, and trade associations such as the Aluminum Association (AA), the AISI, and the Copper Development Association (CDA).

However, without a designated central organization, early on, to control and coordinate the process of assigning numbers to the various steel product forms, duplication occurred quite frequently in which the same product form would be assigned a number by the SAE and another number by the AISI. Or the same number would be assigned to one product by the SAE and to an altogether different product by the AISI.

By 1967 these problems became untenable prompting discussions between ASTM and the SAE. In May 1969 the US Army became aware of these discussions and very interested in the fact that there might possibly be a solution at hand for developing a universally accepted numbering system for steel product forms, so much so that they issued a contract to the SAE to conduct a "Feasibility Study of a Unified Numbering System for Metals and Alloys." This turned out to be an effort sponsored and researched jointly by ASTM and the SAE.

The study was completed in 1971 informing the US Army that a universal numbering system was indeed feasible. On the heels of that announcement, in April 1972 ASTM and the SAE assembled an advisory board to further develop and refine the proposed numbering system. In 1974 the advisory board completed its work with the publication of "SAE/ASTM Recommended Practice for Numbering Metals and Alloys." It was the culmination of work that began some 7 years earlier.

The UNS numbering system consists of 18 designation groups with each group designation consisting of a single-letter prefix followed by five numbers as shown in Table 2.3.5. The current 12th edition of the UNS contains entries for more than 5 600 metals and alloys, 4 100 cross-referenced specifications, and 15 350 trade names.

Table 2.3.5 UNS metal group designations

UNS series	Metal group
Axxxxx	Aluminum and aluminum alloys
Cxxxxx	Copper and copper alloys
Dxxxxx	Steels—designated by mechanical property
Exxxxx	Rare earth and rare earthlike alloys
Fxxxxx	Cast irons
Gxxxxx	AISI and SAE carbon and alloy steels
Hxxxxx	AISI H-steels
Jxxxxx	Cast steels
Kxxxxx	Miscellaneous steels and ferrous alloys
Lxxxxx	Low-melting metals and alloys
Mxxxxx	Miscellaneous nonferrous metals and alloys
Nxxxxx	Nickel and nickel alloys
Pxxxxx	Precious metals and alloys
Rxxxxx	Reactive and refractory metals and alloys
Sxxxxx	Stainless steels, valve steels, superalloys
Txxxxx	Tool steels
Wxxxxx	Welding filler metals
Zxxxxx	Zinc and zinc alloys

2.3.2 Stainless Steel

In the selection of material applicable for use in constructing bioprocessing and other similar type systems, it is relevant to understand, at the very least, some of the basic essentials of this very narrow group of materials. The first imperative in the selection of these materials, as stated in the first paragraph of this chapter, is that any material that comes in contact with a drug product or process cannot be "…reactive, additive, or absorptive so as to alter the safety, identity, strength, quality, or purity of the drug product…."

In explaining these basic essentials, we will need to get into the microscopic weeds of the material to see where these essentials lie. That said, we will first discuss the austenitic stainless steels in the wrought form of 316L. What makes this the workhorse material in not only the wrought form of tubing and fittings but also forgings and castings used for valves and other components and equipment. In clarifying the reason this material is used so frequently, we will first consider why it was dubbed "stainless steel" in 1913 by cutlery manager Ernest Stuart while working with one of the material's inventors, Harry Brearley. Others, during the late 1800s into the early 1900s, are also credited with developing similar formulations of what would ultimately be categorized as "stainless steel."

The phenomenon that makes stainless steel stainless is the natural inclination for the material's surface to self-passivate in an oxygen-enriched environment as a result of the chromium metal, as a percentage, contained in the material's chemistry composition. Passivation is the natural occurring process by which the surface of stainless steel develops a self-generating passive chromium-rich oxide surface layer, a layer on the order of 10–50 Å thick that protects the predominantly iron constituent of the base material from corrosion.

The passivation process, while it does occur rather organically on its own in an oxygen-containing environment, does not always evolve within the needed time frame or to a point sufficient to blanket the surface, inner or outer, of tubing, fittings, and equipment, to protect that metal surface from an attacking process fluid or external environment. In order to instigate and promote the passivation process in a timely and controlled manner and ensure a well-blanketed passivation layer, the entire length of the inside diameter of the process system will require a passivation procedure. There will be further discussion on passivation in Section 2.8. For now we will look into the microscopic world of steel alloys.

2.3.3 *The World of Crystallography*

Based on heat treatment and chemistry, there are three basic categories of stainless steel: ferritic, austenitic, and martensitic. These terms also make reference to the microscopic crystal structure of the material in which the ferrite crystal is referred to as body-centered cubic (bcc), the austenite crystal is face-centered cubic (fcc), and the martensite crystal is distorted tetragonal. These basic elements of steel can be defined in the following manner:

- Ferrite: a form of pure iron with bcc crystal structure or lattice occurring in low carbon steel
- Austenite: A nonmagnetic solid solution of iron with another alloying element and having face-centered crystal structure or lattice
- Martensite: A crystal structure, typically steel or another mineral, formed as a result of distortion in transformation and having body-centered tetragonal (bct) crystal structure or lattice

In the world of crystallography, there are 14 types of crystal unit cell structures within seven so-called lattice systems referred to as cubic (3 unit cell structures), tetragonal (2 unit cell structures), hexagonal (1 unit cell structure), rhombohedral (1 unit cell structure), triclinic (1 unit cell structure), monoclinic (2 unit cell structures), and orthorhombic (4 unit cell structures). Of the 14 unit cell structures, there are three that have been selected for discussion with respect to the metallurgical structures of stainless steel. Two of them are the cubic crystal lattice, those being the bcc and the fcc as shown in Figure 2.3.6a and b and as wire frame graphics in Figure 2.3.7a and b. The other is the bct crystal lattice which is represented as a wire frame graphic in Figure 2.3.7c.

You will notice in Figure 2.3.7a, b, and c there are black dots that represent atoms within the cell lattice. In the bcc, for instance, there are atoms at each of the cube's eight corners and one in the center of the cube. The fcc has an atom at each of its eight corners and one in each of the cube's six sides or faces. The bct, like

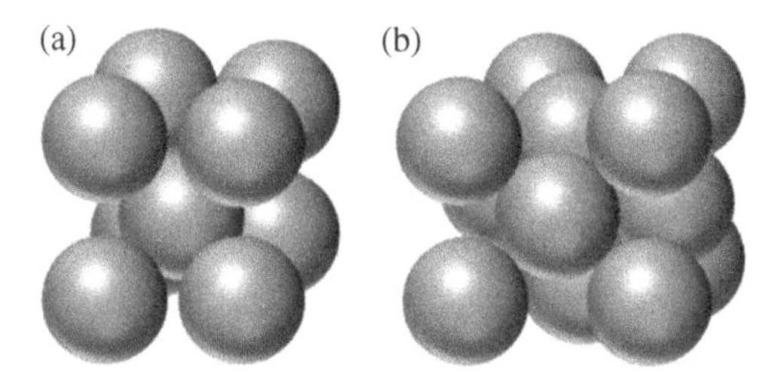

Figure 2.3.6 Graphic representation of crystallographic structures: (a) body-centered cubic and (b) face-centered cubic

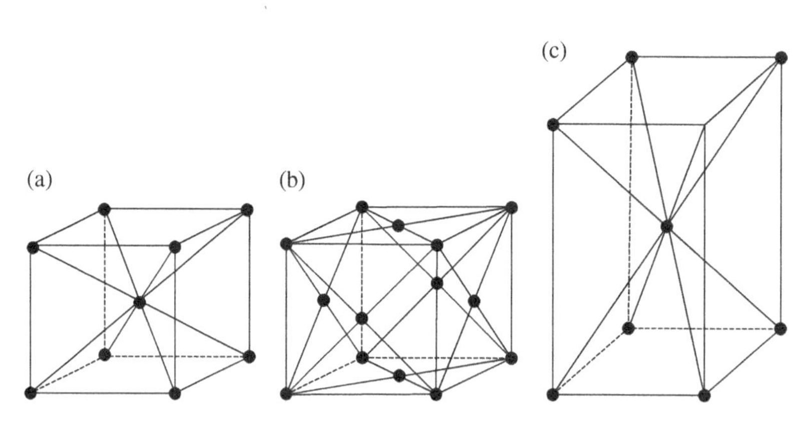

Figure 2.3.7 Wire-frame representation of crystallographic structures: (a) body-centered cubic, (b) face-centered cubic, and (c) body-centered tetragonal

the bcc, has atoms at each of its eight corners and one in the center of the tetragonal. Its shape is basically like that of a cube that has been elongated or distorted in one direction.

What does this actually mean with regard to selecting stainless steel? Keep in mind that each of the three types of unit cells described in the preceding text is representative of the types of microscopic cells that coalesce into a matrix of material to make up the composition of steel, stainless steel in this case. Each of these cells is connected by their specific lattice configuration, which in the case of alpha or delta ferrite (Ref. Table 2.3.6) are the bcc-type unit cells. One bcc unit cell will attach to another by way of each of the eight atoms located at each of the eight corners of the cube. Each corner atom will be common to the corner of eight bcc-type unit cells, meaning that there is 1/8 of an atom attributed to each corner of a bcc lattice, and therefore the eight corners equal to one whole atom. Adding that to the one atom in the center gives a net total of two atoms for the bcc structure.

In the case of the austenite fcc crystal structure, it, like the bcc structure, contains an atom at each of its eight corners, but it also has an atom at the center

Table 2.3.6 Metallurgical construct of stainless steel (at room temperature)

Metallurgical structure	Crystal type	Phase	Atoms
Ferritic	Body-centered cubic	Alpha or delta ferrite	9
Austenitic	Face-centered cubic	Gamma iron	14
Martensitic	Distorted tetragonal	Quenched or tempered martensite	10–13

Table 2.3.7 Unit cell packing factor

Crystal type	Number of atoms/unit cell	Coordination number	Packing factor
Body-centered cubic (bcc)	2	8	0.68
Face-centered cubic (fcc)	4	12	0.74
Body-centered tetragonal (bct)	2	8	0.70
Hexagonal close-packed (hcp)	6	12	0.74
Simple cubic (sc)	1	6	0.52

of each of its six faces (Figure 2.3.7b). The four corners add up to one atom, and the six face-centered atoms are each shared by a mating face allowing one half of an atom at each of the six faces which equals three atoms per unit cell. The three face atoms plus the combined one atom for the corners equal four atoms for each fcc unit cell.

The bct unit cell is similar to the bcc unit cell in that it also has a net number of two atoms. One of the essential data points in knowing the aforementioned is in understanding that the packing factor or atomic packing factor can now be understood. The packing factor is the volume of atoms in a cell per the total volume of a cell. The more closely and tightly packed the atoms are, the higher the factor value. Table 2.3.7 provides examples of some of these packing factor values, as follows:

As can be seen in Table 2.3.7, the fcc arrangement of atoms has a higher packing value than does the bcc arrangement of atoms. The bcc (i.e., ferrite crystal) arrangement of atoms does not permit the atoms to pack as tightly as the fcc (i.e., austenite crystal) arrangement of atoms. The fcc crystalline structure of austenitic material leads to improved ductility, allowing the material to be malleable and deform under load before reaching its fracture point, whereas materials such as alpha iron and tungsten, consisting of the bcc crystalline structure, are not as closely packed (Ref. Table 2.3.7) and therefore not as ductile a material as that of the fcc crystalline structure. But, while it is not as malleable as the austenite material, such material has greater strength.

In order to change the steel crystalline structure from bcc to fcc, the steel has to go through a phase change induced by heat treatment in which the material changes from ferritic to austenitic. At its base level iron, combined with 6.67% chromium, forms iron carbide or cementite. Therefore, at room temperature, this basic alloy contains cementite and ferrite. When this basic steel is heated to a temperature in excess of 725°C (1340°F), the cementite dissolves into the matrix and a secondary austenite phase is formed. To this various alloying constituents are added to the chemistry of the steel to achieve various degrees of CR, weldability, strength, and ductility. For 316L and a few other stainless steels, those alloying constituents include those shown in Table 2.3.8.

The benefits of austenitic stainless steels that drive its selection and considerable use in bioprocessing industry applications as well as food and dairy applications are as follows:

- CR
- Durability
- Ductility
- Cleanability
- Nonadditive to process stream
- Economically priced in comparison to other acceptable alloys

In the previous paragraphs it was described how and why the austenite phase of stainless steel is achieved. But there are other forms of stainless steels, as indicated in Table 2.3.8, with such descriptive names as "superaustenitic" and "duplex," and two groups of material not even indicated in Table 2.3.8, referred to as "super duplex" and "lean duplex" stainless steels. These materials can be described as follows:

- Superaustenitic stainless steels (e.g., AL-6XN [N08367]), while these stainless steels may not exhibit an enhanced resistance to chloride pitting and crevice corrosion, the high (>6%) molybdenum content (referred to as a 6 moly material) along with the addition of nitrogen and a higher nickel content enhances the steel's resistance to stress corrosion cracking beyond that of the 300 series steels. The higher alloy content of the superaustenitic steels does tend to make these types of steels more costly than that of the 300 series stainless steels. The number 6, in 6 moly, is more commonly used in the industry than the written six.
- By increasing the amount of chromium and molybdenum in the chemical composition of stainless steel, while also reducing the amount of nickel, an approximate equal balance of ferrite and austenite phases can be achieved, thereby earning this form of the metal its name of duplex stainless steel.
- Super duplex is a duplex stainless steel with a pitting resistance equivalent number (PREN) ≥ 40, where $PREN = \%Cr + 3.3 \times \%Mo + 16 \times \%N$. If tungsten (W) is part of the chemistry, it is taken into account by the equation: PREN = %

Table 2.3.8 ASTM chemical composition of some austenitic stainless steels

UNS	EN	Alloy	C	Mn[a]	Si[a]	Cr	Ni	Mo	N[a]	Cu
Austenitic										
S30400	…	304	0.080	2.00	0.75	18.00–20.00	8.00–11.00	…	0.10	…
…	1.4301		0.070	2.00	1.00	17.50–19.50	8.00–10.50	…	0.11	…
S30403	…	304L	0.035	2.00	0.75	18.00–20.00	8.00–20.00	…	0.10	…
…	1.4307		0.030	2.00	1.00	17.50–19.50	10.00–12.00	…	0.11	…
…	1.4306		0.030	2.00	1.00	18.00–20.00	10.00–13.00	…	0.11	…
S31600	…	316	0.080	2.00	0.75	16.00–18.00	11.00–14.00	2.00–3.00	0.10	…
…	1.4401		0.070	2.00	1.00	16.50–18.50	10.00–13.00	2.00–2.50	0.11	…
S31603	…	316L	0.035	2.00	0.75	16.00–18.00	10.00–15.00	2.00–3.00	0.10	…
…	1.4404		0.030	2.00	1.00	16.50–18.50	10.00–13.00	2.00–2.50	0.11	…
…	1.4435		0.030	2.00	1.00	17.00–19.00	12.50–15.00	2.50–3.00	0.11	…
Superaustenitic stainless steels										
S31703	…	317L	0.030	2.00	0.75	18.00–20.00	11.00–15.00	3.0–4.0	0.10	…
…	1.4438		0.030	2.00	0.75	17.50–19.50	13.00–17.00	3.0–4.0	0.11	…
N0S904	…	904L	0.020	2.00	1.00	19.00–23.00	23.00–28.00	4.00–5.00	0.10	1.0–2.0
…	1.4539		0.020	2.00	0.70	19.00–21.00	24.00–26.00	4.00–5.00	0.15	1.2–2.0
N08367	…	AL-6XN	0.030	2.00	1.00	20.00–22.00	23.00–25.00	6.00–7.00	0.18–0.25	0.75
S31254	…	254	0.020	2.00	0.80	19.50–20.50	17.50–18.50	6.0–6.5	0.18–0.22	0.5–1.0
…	1.4547		0.020	2.00	0.70	19.50–20.50	17.50–18.50	6.0–7.0	0.18–0.22	0.5–1.0
N08926	…	926	0.020	2.00	0.50	19.00–21.00	24.00–26.00	6.0–7.0	0.15–0.25	0.5–1.5
…	1.4529		0.020	1.00	0.50	19.00–21.00	24.00–26.00	6.0–7.0	0.15–0.25	0.5–1.5
Duplex stainless steels										
S32205	…	2205	0.030	2.00	1.00	22.00–23.00	4.5–65	3.0–3.5	0.14–0.20	…
…	1.4462		0.030	2.00	1.00	21.00–23.00	4.5–6.5	2.5–3.5	0.10–0.22	…

Sulfur is limited to 0.030 max. and phosphorous to 0.045.

Only the primary chemical constituents are listed in this table. For a complete listing refer to the appropriate material specification.

[a] Maximum amount given.

Cr + 3.3 × (%Mo + 0.5 × %W) + 16 × %N. Usually super duplex grades have a chromium content of 25% or more.

- Lean duplex stainless steel was developed in an effort to provide the benefits of a duplex at less cost. This was accomplished by reducing the amount of molybdenum and nickel while increasing nitrogen and manganese to stabilize the austenite phase. Reducing these alloys (making the material "lean" in other words) reduced the cost of the material, but at the same time retained the strength and CR so beneficial in duplex stainless steels.

2.3.4 Pitting Resistance Equivalent Number (PREn)

PREN or PREn, mentioned in the preceding text, is a value derived from the equations given earlier and repeated here as PREN = %Cr + 3.3 × %Mo + 16 × %N, or when tungsten (W) is part of the material's chemistry, PREN = %Cr + 3.3 × (% Mo + 0.5 × %W) + 16 × %N. These equations base the potential for CR on the alloy makeup or chemistry of the material. These equations attribute weighted values to certain alloys within the chemistry of the stainless steel. But, there is also CR testing based on temperature that can be performed under ASTM G48 Standard Test Methods for Pitting and Crevice Corrosion Resistance of Stainless Steels and Related Alloys by Use of Ferric Chloride Solution. There are six test methods described in G48 that are as follows:

- **Method A**—Ferric chloride pitting test
- **Method B**—Ferric chloride crevice test
- **Method C**—Critical pitting temperature (CPT) test for nickel-base and chromium-bearing alloys
- **Method D**—Critical crevice temperature (CCT) test for nickel-base and chromium-bearing alloys
- **Method E**—CPT test for stainless steels
- **Method F**—CCT test for stainless steels

Under the G48 tests the CPT or the CCT of stainless steel alloys can be determined. However, it must be understood that the CPT and CCT measured with the G48 test methods relate specifically to the ferric chloride test solution and cannot be used to predict pitting and crevice corrosion in other environments. However, the relative pitting and crevice CR determined by G48 testing can help in selecting an appropriate alloy for specific service conditions. The higher the PREN, the more corrosion resistant the material is.

In doing a side-by-side comparison of some of the materials mentioned in the preceding text, it will help sort out or determine what material might be better suited or more appropriate over another for a specific service application by comparing some of the key considerations of various acceptable materials. Table 2.3.9 compares three primary aspects or decision points of acceptable materials for consideration. Those key considerations are CR, strength (tensile, yield, and allowable stress), and cost.

Table 2.3.9 Three-point material comparison

Alloy designation			CR[a]	Strength of material							Cost[b]
UNS	EN	Alloy	PREX[c]	Min. tensile	Min. yield	ASME allowable stress (KSI) °F/°C					Factor
				KSI	KSI	100/0	200/93	300/149	400/204	500/260	
Austenitic											
S30400	…	304	18	70	25	16.7	16.7	16.7	15.7	14.8	1.00
…	1.4301										
S30403	…	304L	18	70	25	16.7	16.7	16.7	15.7	14.8	1.00
…	1.4307										
…	1.4306										
S31600	…	316	22.5	70	25	16.7	16.7	16.7	15.7	14.8	1.25
…	1.4401										
S31603	…	316L	22.5	70	25	16.7	16.7	16.7	15.7	14.8	1.25
…	1.4404										
…	1.4435										
Superaustenitic stainless steels											
S31703	…	317L	28	75	30	20.0	20.0	20.0	18.9	17.7	1.66
…	1.4438										
N08904	…	904L	32	71	31	20.7	20.7	20.4	18.7	17.1	2.72
…	1.4539										

(*Continued*)

Table 2.3.9 (Continued)

Alloy designation			CR[a]	Strength of material							Cost[b]
UNS	EN	Alloy	PREX[c]	Min. tensile	Min. yield	ASME allowable stress (KSI) °F/°C					Factor
				KSI	KSI	100/0	200/93	300/149	400/204	500/260	
N08367	…	AL-6XN	43	94	43	30.0	30.0	29.9	28.6	27.7	3.46
S31254	…	254	41	98	45	30.0	30.0	29.5	27.5	25.8	1.85
…	1.4547										
N08926	…	926	40.5	94	40	26.7	26.7	26.7	24.7	23.3	3.50
…	1.4529										
Duplex stainless steels											
S32205	…	2205	34	95	65	31.7	31.7	30.6	29.4	28.7	1.16
…	1.4462										

[a] Corrosion resistance (CR) value as pitting resistance equivalent number (PREN).
[b] Cost values are given in relation to the base cost of 304 stainless steel with 304 given a factor of 1.0.
[c] PREN values are of wrought product form based on data from the Nickel Development Institute.

2.3.5 Alloying Constituents in Austenitic Stainless Steel

The following are various alloying constituents and their effects and attributes on alloy steel composition:

- **Delta (δ) ferrite**: The high-temperature form of iron, delta ferrite, is formed when a low carbon concentration of an iron-carbon alloy is cooled from its liquid state before making the transformation to austenite. In highly alloyed steels, such as stainless steel, delta ferrite can be retained at room temperature. A small amount of delta ferrite is held within the microstructure of austenitic stainless steel to help prevent hot cracking of the material in the weld during fabrication. However, too much delta ferrite in low nickel austenitic stainless steels can have an adverse effect on a material's CR. With an increase in martensite comes a decrease in ductility and susceptibility to hydrogen embrittlement. Too high a level will predispose any surface delta ferrite to be dissolved during electropolishing leaving a dull white surface. A high level of delta ferrite can also make the austenitic stainless steel more susceptible to corrosion, corrosion which it might otherwise withstand.
- **Carbon (C)**: Carbon, at the proper levels, adds strength to austenitic materials. But carbon, with the ability to readily diffuse through the cell matrix structure and take up residence at the grain boundaries of a material, also has, at levels arguably ≥0.08%, a tendency to be adverse to the material's composition. What can occur is precipitation of chromium carbide at the grain boundaries in the heat-affected zone (HAZ) during welding causing depletion of the corrosion protection of the chromium alloy in the HAZ. This is a process referred to as "sensitization" and makes the material susceptible to intergranular attack (IGA) or intergranular corrosion (IGC). With carbon content less than or equal to 0.030% instead of at the 0.070–0.080% level, the low carbon (i.e., 304L and 316L) stainless steels essentially prevent sensitization from occurring during normalizing processes.
- **Manganese (Mn)**: This metal not only combines with sulfur to form manganese sulfides, which can be detrimental to pitting resistance, but also promotes the formation of austenite and the solubility of nitrogen.
- **Silicon (Si)**: Silicon enhances the fluidity of the molten steel and assists in deoxidation after argon oxygen decarburization (AOD) refining. A high content of silicon can diminish the ductile properties of the steel.
- **Chromium (Cr)**: With a chromium content greater than or equal to 10.5%, the reaction that takes place with that chromium, in contact with oxygen in the environment, is what makes stainless steel stainless. This anomaly creates a chromium-rich oxide on the surface of stainless steel making it corrosion resistant to many forms of corrosion. Chromium also enhances scale resistance and tensile strength and helps the material resist erosion.
- **Sulfur (S)**: The addition of sulfur enhances weld penetration, but too much sulfur creates inclusions, while low levels of sulfur make good weld penetration difficult. With autogenous automatic orbital welding used extensively in the

bioprocessing industry, another sulfur-related issue came to bare welding. With manual welding the welder can manipulate the weld to maintain good, complete, and centered penetration at the joint. What the welder sometimes has to overcome in making an acceptable weld is the variation in sulfur content between the two joining pieces. With orbital welding there is no manipulation. Once the machine is programmed and set to weld, it will follow the circumferential butt joint without deviation. If the sulfur content between the two heats of material varies sufficiently enough, incomplete penetration could cause the weld to be rejected. There will be more on this in Chapter 5.

- **Nickel (Ni)**: Besides stabilizing the austenite structure of austenitic stainless steels, nickel also increases the material's strength at elevated temperatures and enhances ductility and CR in certain environments such as reducing acids.
- **Molybdenum (Mo)**: In addition to also increasing strength at elevated temperatures and CR, molybdenum provides an increase in creep resistance. Molybdenum also helps inhibit any tendency the austenitic stainless steel should have toward pitting.
- **Nitrogen (N)**: Nitrogen is an austenite stabilizer that enhances the resistance to chloride pitting and crevice corrosion. It will also slow the kinetics of the formation of undesirable intermetallic precipitates such a sigma and chi. As part of the composition of "austenitic" and "duplex" stainless steels, the addition of nitrogen will increase the resistance to localized pitting attack. Low carbon "austenitic" grades (i.e., 304L, 316L) containing less than 0.03% carbon are recommended when welding is required. The lower carbon minimizes the risk of sensitization. However, the low carbon levels have a tendency to reduce the yield strength of the material. By including nitrogen in the material's chemical composition, it raises the yield strength levels back to the same level as standard or straight grades of austenitic stainless steels.
- **Copper (Cu)**: Copper is added in small amounts to some austenitic stainless steels, particularly those that fall under the category of "superaustenitic," to improve CR for reducing acid environments such as sulfuric acid service.

2.3.6 Dual Certified Stainless Steels

Indicating dual certification of a stainless steel implies that the material either falls within the overlapping chemical and mechanical properties of two different material standards or those of two different grades within the same material standard. In order to determine if a material can qualify for dual certification, and be stamped as dual certified, its material composition, mechanical properties, and other elements, such as manufacturing tolerances, have to fall within, what I will refer to as, the manufactured sweet spot. This is the overlapping area in which chemical element requirements, mechanical properties, or dimensional tolerances between two or more materials can align themselves.

To explain this we will first look at dual certified ASTM A269/A270 Gr 316L stainless steel tubing. That identifier indicates that the tubing is dual certified under two different ASTM material standards, A269 and A270, as grade 316L stainless steel. In comparing the two material standards—and the comparison has to be made between ASTM standards or between EN standards and not between ASTM and EN standards—we look for the following:

- Chemical composition for Gr 316L is identical for both standards.
- Mechanical properties for Gr 316L are identical for both standards.
- Manufacturing tolerances and surface finish of the finished product are different:
  - Meeting the manufacturing tolerances of A270 also meets those of A269.
  - Meeting the surface finish requirements of A270 exceeds those of A269.

According to the earlier bulleted points, with chemical composition and mechanical property requirements identical between A269 and A270, what needs to occur for the material to qualify for dual certification is that both the manufacturing tolerances and surface finish need to align with the requirements of A270. Compliance with the S2 supplement within A270 also lies within the requirements of A269.

Using UNS S31600/S31603 (316/316L) as an example for a material qualifying for dual grade certification, the differences in the chemical composition between the two grades of stainless steel lie in the carbon content, which limits carbon to a max. of 0.08% for 316 and a max. of 0.035% for 316L, and in the nickel content range of 11.0–14.0% for the 316 and 10.0% to 15.0% for the 316L. If a lot of stainless steel tubing having a carbon content less than or equal to 0.035% and a nickel content between 11.0 and 14.0% also meets the higher mechanical properties of the straight 316 stainless steel, it therefore meets the requirements of both grades of stainless steels allowing it to be dual certified for both.

2.3.7 So Why 316L Stainless Steel?

Providing the information on metallic materials to this point has led us to revisit the opening statement in Section 2.3.1 in which was written:

> *Metallic materials selected for use in bioprocessing and in pharmaceutical systems in general are selected for service based on the following precepts:*
>
> - *Compatibility with the fluid service*
> - *Able to operate within the pressure/temperature design range of the contained fluid service*
> - *Able to withstand the CIP or SIP regimen*
> - *Will not leach any part of its chemistry make up into the fluid stream through corrosion or erosion*
> - *Will not alter the safety, identity, strength, quality, or purity of the drug product in any way*

The reason that 316L stainless steel has proven to be the workhorse material for use in high-purity bioprocessing applications is due to the fact that it meets all of the bulleted criteria listed in the preceding text and then some. The following are clarifications to those criteria with respect to 316L stainless steel:

- Compatibility with the fluid service
  - Because of its chromium-rich passive layer of protection, 316L stainless steel is compatible with a broad range of chemicals and processes typically found in the pharmaceutical industry.
- Able to operate within the pressure/temperature design range of the contained fluid service
  - The limiting factor with regard to design pressures and temperatures of a process system will be the mechanical joint, which in this case would be the hygienic clamp joint.
- Able to withstand the CIP or SIP regimen
  - Chemically cleaning or sanitizing a system at 15 psig at 250°F (121°C) saturated steam is well within the compatibility range of 316L.
- Will not leach any part of its chemistry makeup into the fluid stream through corrosion or erosion
  - The passive chromium oxide layer that develops on 316L establishes and maintains a barrier that protects the base metal of the stainless steel while at the same time preventing any interaction with the process fluid that would initiate leaching of that base metal into the process fluid.
- Will not alter the safety, identity, strength, quality, or purity of the drug product in any way
  - By 316L not being additive to the process fluid stream as pointed out in the previous bullet point, it meets this final criteria as well.

In addition to meeting the earlier bulleted criteria, 316L, because of its fcc crystalline structure, is highly ductile and relatively easy to weld. The option of requesting the S2 supplement under ASTM A270, which specifies a much more narrow sulfur range, allows the metal to be welded autogenously with an automatic orbital welder. This provides the advantage of acceptable welds on a repeatable basis when setting up and performing production welds. For more in-depth discussion on the topic of sulfur content, and welding in general, refer to Chapter 4.

In meeting the earlier criteria, 316L does so with a low comparative cost as compared to other acceptable alloys. It is also a very available material in a wide range of product forms from tubing to fittings made from the tubing as well as forgings and castings for valves as well as sheet product for manufacturing pressure vessels.

Where added corrosion allowance, strength, and other enhanced chemical and mechanical attributes need to be met in a service application, there is a wide selection of material, beyond that of 316L stainless steel, to choose from in the form of steel alloys as listed in the previous tables and nonmetallic materials as discussed next.

2.4 Nonmetallic Materials

Nonmetallic is a term, when taken at face value, that covers basically anything not metallic. This makes the term uniquely ambiguous. If not metallic, as nonmetallic implies, it could essentially be anything from water to wood and from bamboo to diamond. But, in all practicality, Part PM in the BPE Standard has drilled down on what it considers to be "nonmetallic" materials that are applicable to high-purity applications, materials that fit within the framework of what could possibly qualify for use in a bioprocessing service application.

2.4.1 What Are Nonmetallic Materials?

Part PM, in an effort to describe the niche of nonmetallic materials acceptable to BPE, has divided this class of materials into three primary categories that include polymeric material, solid single-phase material, and solid multiphase material. These are materials selected for their acceptability of use in the form of tubing, fittings, instrumentation, equipment, and seals that come in contact with a bioprocessing fluid.

Tables 2.4.1, 2.4.2, 2.4.3, 2.4.4, 2.4.5, and 2.4.6 are compilations of Tables PM-2.1.1-1, PM-2.1.2-1, and PM-2.1.3-1 found in the BPE Standard. In Table 2.4.1 you will find the three primary categories of nonmetallic material, polymeric, solid single phase, and solid multiphase, considered currently acceptable for use in

Table 2.4.1 Nonmetallic material categories

Polymeric		Solid single phase		Solid multiphase
Thermoplastic	Thermoset	Amorphous	Crystalline	Composites
General	Elastomers	Glass	Sintered materials	Reaction bonded
Polyolefins	Rigid	Borosilicate	Aluminum oxide	Silicon carbide
Fluoropolymers		Soda lime	Silicon carbide	Silicon nitride
Elastomers			Silicon nitride	Siliconized carbon graphite
			Tungsten carbide	Resin-impregnated carbon graphite
			Zirconium dioxide	
			Cemented materials	
			Tungsten carbide with alloyed binder	
			Tungsten carbide with nickel binder	
			Tungsten carbide with cobalt binder	

Table 2.4.2 Polymeric thermoplastic materials

General	Polyolefins	Fluoropolymers	Elastomers
The following types of thermoplastic are typically used to manufacture fittings, connectors, filter housings, piping, rigid tubing, column tubes, and filter media	The following types of thermoplastic are typically used to manufacture fittings, connectors, piping, rigid tubing, filter media, capsules, and bags	This form of thermoplastic is typically used to manufacture fittings, piping, tubing, flexible hose, filter media, capsules, diaphragms, pumps, vessel liners	This form of thermoplastic is typically used to manufacture tubing and bags
Polyester (PET)	Polypropylene (PP)	Fluorinated ethylene propylene (FEP)	Blends of EPDM with polypropylene
Polyamide (nylon)	Ultra-low-density polyethylene (ULDPE)	Perfluoroalkoxy (PFA)	Styrene-isobutylene-styrene block polymers
Polycarbonate	Low-density polyethylene (LDPE)	Polytetrafluoroethylene (PTFE)	Copolymers of ethylene and octane
Polysulfones	High-density polyethylene (HDPE)	Ethylene tetrafluoroethylene (ETFE)	Ethylene-vinyl acetate (EVA) copolymer
Polyether ether ketone (PEEK)	Ultra-high-molecular-weight polyethylene (UHMW)	Polyvinylidene fluoride (PVDF)	

Table 2.4.3 Polymeric thermoset materials

Elastomers	Rigid
The following types of thermoplastic are typically used to manufacture tubing, seals, gaskets, diaphragms, and hoses	This form of thermoplastic is typically used to manufacture tanks and pipe
Ethylene propylene diene (EPDM)	Fiber-reinforced polymer (FRP/GRP) composites
Ethylene propylene rubber (EPR)	
Silicon (VMQ)	
Fluoroelastomers (FKM)	
Perfluoroelastomer (FFKM)	

Table 2.4.4 Solid single-phase amorphous materials

Glass
The following types of glass are typically used to manufacture sight glasses, vessel lights, optical sensors, and glass electrodes.
Borosilicate
Soda lime

Table 2.4.5 Solid single-phase crystalline materials

Sintered	Cemented
The following types of crystalline materials are typically used to manufacture mechanical seals, bearings, and process sensors	This form of crystalline materials is typically used to manufacture mechanical seals and bearings
Aluminum oxide	Tungsten carbide with alloyed binders
Silicon carbonate	Tungsten carbide with nickel binder
Silicon nitride	Tungsten carbide with cobalt binder
Tungsten carbide	
Zirconium dioxide	

Table 2.4.6 Solid multiphase composite materials

Reaction bonded	Siliconized carbon graphite	Resin impregnated carbon graphite
The following types of composites are typically used to manufacture mechanical seals	The following type of composite is typically used to manufacture mechanical seals	This following type of composite is typically used to manufacture mechanical seals
Silicon carbide	Multiphase mixture of crystalline silicon carbide, carbon, and graphite	Multiphase mixture of carbon, graphite, organic resin, and potential inorganic nonmetallic additives
Silicon nitride		

high-purity applications. Also in this table you will find a further segregation of these three primary categories of material into five additional groups of material referred to as thermoplastics, thermosets, amorphous, crystalline, and composites. Further still are the types of material listed under each of those groups.

In Table 2.4.2 are listed the thermoplastic types of material identified as general, polyolefins, fluoropolymers, and elastomers. Under each of these types of material are the various forms of materials referred to as general, polyolefins, fluoropolymers, and elastomers. Also described under each of the forms of material are some of the end products typically manufactured with these types of material. You will find the same information in Table 2.4.3 (thermoset materials), Table 2.4.4 (amorphous materials), Table 2.4.5 (crystalline materials), and Table 2.4.6 (composite materials).

There currently is no definition in the BPE Standard or elsewhere for "nonmetallic material" that puts it in context with BPE requirements. Until there is one, and based on what we have discussed so far on this topic, we will provide a definition here, one that pertains only to the discussions within these pages and has not gone through the rigors of debate, consensus, and approval by any recognized and accredited committee. That definition reads:

Nonmetallic material: a product form containing no appreciable or discernible amount of a metallic substance taking the form of a polymeric, solid single-phase, or solid multiphase material.

2.4.2 Extractables and Leachables

Metallic materials have the potential to corrode or erode, shedding metallic particulate into a fluid stream. Such particulate matter can contaminate the process stream to the point of nonconformance or adulteration. However, the passive chromium layer of 316L stainless steel, when sufficiently generated, prevents this from occurring. With nonmetallic materials there is a similar concern, not with particulate matter, but with chemical constituents leaching into the fluid stream, referred to as "leaching." In determining what chemical constituents have a propensity to separate from a base material matrix and leach into a contact fluid, testing for "extractables" is performed on targeted material. The FDA therefore defines "extractables" and "leachables" as:

- *Extractables*: compounds that can be extracted from the container closure system (CCS) when in the presence of a solvent
- *Leachables*: compounds that leach into the drug product formulation from the container closure as a result of direct contact with the formulation

While these definitions are of course accurate, they speak more to the packaging of a finished drug product. The term "CCS" refers to packaging of a finished drug product. Whether it is bulk storage or shipping container, or a vial or box container sitting on the pharmacist's shelf, it is a CCS. To define this in a more generic but pointed manner as it applies to bioprocessing manufacturing systems, with pipe, fittings, instruments, and equipment, we will define it in a slightly different way, a way that better suites and is more appropriate for the manufacture of a drug product, as follows:

- *Extractables*: chemical constituents, organic or inorganic, that are prone to be released or extracted from a base material matrix by and into a solvent while in contact with that solvent under controlled test conditions. It is a means by which potential leachables in a base material matrix can be identified and quantified.
- *Leachables*: chemical constituents, organic or inorganic, that have a tendency to be separated and released from their base material matrix by a contact fluid allowing the released constituents to migrate from the base material to and into the contact fluid.

Such leachables, should they find their way into a drug product fluid, would become a trace drug anomaly within the matrices of that drug product. This could quite possibly render that product adulterated and out of spec. Because of this potential it is required that polymeric tubing, fittings, instrumentation, and equipment that come in contact with the process fluid of a food or drug product have documentation certifying that its material of construction has been tested in accordance with USP <87> (or ISO 10993 Part 5) and USP <88> Class VI (or ISO 10993 Parts 6, 10, and 11) and found to be acceptable for use as a material coming in contact with a food or drug product.

USP <87> is a cytotoxicity test performed *in vitro* (in a glass lab vessel such as a petri dish) with mammalian cell cultures to determine the cell's reaction to a selected polymer. ISO 10993 Part 5 is a similar *in vitro* reactivity test for cytotoxicity. USP <88> Class VI is a cytotoxicity test performed *in vivo* (inside a host organ) to determine the host's reaction to a selected polymer. Class I through VI testing under USP <88> requires progressively more test variations as the class numbers increase with Class VI requiring the most at nine different tests. ISO 10993 Parts 6, 10, and 11 are similar *in vivo* reactivity tests for cytotoxicity. Part 6 is titled *Tests for local effects after implantation*, Part 10 is titled *Tests for irritation and delayed type hypersensitivity*, and Part 11 is titled *Tests for systemic toxicity*. Cytotoxicity is essentially the degree to which a sample material, such as a polymer, is destructive to targeted test cells or the host subject.

In order to verify compliance with the necessary regulatory requirements each nonmetallic component intended for a process contact application will require documented verification in the form of a Certificate of Compliance (C of C). Table PM-2.2.1-1 in the standard, as referenced in Figure 2.4.1, provides an excellent summary and at-a-glance breakdown of the information required in a C of C for various nonmetallic product forms. The C of C should be requested from the manufacturer or supplier for each of the items listed and should include, at a minimum, the information indicated in Table PM-2.2.1-1.

	Applications									
Label/mark	Polymeric seals	Hygienic union seals	Diaphragms	Hoses	Tubing	Single-use components and assemblies	Filters	Columns	Other nonmetallics	Steam-to/through connections
Compliance to ASME BPE	X	X	X	X	X	X	X	X	X	X
Manufacturer's name	X	X	X	X	X	X	X	X	X	X
Manufacturer's contact information	X	X	X	X	X	X	X	X	X	X
Part number	X	X	X	X	X	X	X	X	X	X
Lot number or unique identifier or serial number	X	X	X	X	X	X	X	X	X	X
Material(s) of construction (process contact)	X	X	X	X	X	...	...	X	X	X
Compound number or unique identifier	X	X	X	...	X	...	...	...	X	...
Cure date or date of manufacture	X	X	X	...	X	X	X	...	...	X
USP <87> or ISO 10993-5	X	X	X	X	X	X	X	X	X	X
USP <88> or ISO 10993-6, –10, –11	X	X	X	X	X	X	X	X	...	X
Intrusion category (SG-4.2)	...	X	...	...	...	...	...	...	...	...

Figure 2.4.1 Content required on the certificate of compliance. *Source*: https://www.asme.org/products/codes-standards/bpe-2014-bioprocessing-equipment. © The American Society of Mechanical Engineers

2.4.3 Single-Use Systems and Components

Single-use systems (SUS), as it applies to pharmaceutical bioprocessing, are processing systems made up of single-use, nonmetallic components such as bags, connectors, instruments, and even mixers and other equipment items. The idea stems from the use of flexible polyvinyl chloride (PVC) bags that have been in use in the healthcare industry for decades as containers for blood, blood plasma, IV drips, and other similar uses.

Single-use components have evolved over recent years from the 50 mL to 10 L flexible PVC containers to now include materials such as single- and multilayer multimaterial designs using ethylene vinyl acetate (EVA) and ethylene vinyl alcohol (EVOH) typically with polyethylene (PE) as the inner contact layer. These materials allow single use to now be applied to mixers, fermenters, and bioreactors. Gusseted designs in concert with various laminated material combinations permit single-use containers to reach storage volumes of 10 000 L.

What the single-use or disposable device concept hopes to promote is a shift in current thinking, thinking that perhaps changes the concept, indeed the precept, of designing an industrial bioprocessing system from the basis of a 20-year life cycle in which equipment and ancillary components such as piping are specified on the basis of being in service for 20 years to that of a single batch cycle.

Such a paradigm shift in conceptual engineering for industrial-type bioprocessing manufacturing has created a new set of metrics to be considered when conceptualizing the manufacture of a drug product. Such consideration will typically take place during scale-up from clinical to industrial production or when duplicating another operating manufacturing process facility.

In the conceptual stage of facility design, the engineer cannot necessarily default any longer to piping and equipment constructed of stainless steels and other metallic alloys based on a presumptive 20-year operating life cycle. With the advent and growth of single-use disposable processing components, these legacy-type operating system designs are no longer the only game in town. The engineer now, at least from a theoretical standpoint, has to determine whether or not a proposed process is conducive to, or in any way merits, a facility based on a single-use processing philosophy or platform, in whole or in part.

Some of the basic decision point metrics to be considered are listed as follows:

- Extent of single-use disposable components
  - At the conceptual stage of a project, determine the extent of viability. Is the entire process system conducive to a single-use platform? Or should only strategic segments of the process system be considered for single-use components?
- Initial labor cost
  - Looking at only those segments of the process considered viable for the single-use platform, what is the estimated labor cost difference to commission and start up those process segments designed around single-use components and equipment when compared to that of reusable metallic or nonmetallic components and equipment?
- Long-term operational labor cost
  - Looking at only those segments of the process considered viable for the single-use platform, what is the estimated ongoing labor cost difference for the ongoing operation of those process segments designed around single-use components and equipment when compared to that of reusable metallic or nonmetallic components and equipment?
- Initial material cost
  - Looking at only those segments of the process considered viable for the single-use platform, what is the estimated initial material cost difference between those process segments designed around single-use components and equipment when compared to that of reusable metallic or nonmetallic components and equipment?
- Long-term material cost
  - Looking at only those segments of the process considered viable for the single-use platform, what is the estimated ongoing material cost difference between those process segments designed around single-use components and equipment when compared to that of reusable metallic or nonmetallic components and equipment?
- Campaign turnaround efficiency
  - What is the estimated difference in turnaround cost, labor, and material between a processing system based on the single-use platform and that of reusable metallic or nonmetallic components and equipment?
  - What is the estimated turnaround time difference between process campaigns of an SUS and those of a system of reusable metallic or nonmetallic components and equipment?

- Cross contamination
  - Is there a concern with cross contamination with either the SUS or the reusable system?
- Product holdup
  - How much residual (holdup) product is expected to remain in the single-use disposable components compared to that of the reusable components and equipment?
- CIP/SIP
  - What is the added cost in the design, fabrication, installation, and ongoing maintenance of CIP and SIP systems with regard to system utilizing reusable metallic or nonmetallic components and equipment?
- Manufacturer's quality management system (QMS)
  - What impact would a process system designed around a single-use platform have on the content of a facility's QMS.
- Documentation
  - What impact would a process system designed around a single-use platform have on the documentation requirements of the operating system?

Polyvinyl alcohol (PVA)-lined braided hose was first introduced at the New Jersey manufacturing facility of the Resistoflex company in 1936. After WWII they began working with a new product referred to as polytetrafluoroethylene (PTFE) and by 1953 were manufacturing PTFE-lined hose. By 1956 Resistoflex was manufacturing PTFE-lined pipe. In those early and initial steps, some 80 years ago, of cross-linking these newly discovered nonmetallic polymeric products with pressurized metallic fluid handling components such as hose and pipe, industry has evolved to a point in which entire systems consist of polymeric nonmetallic components and equipment, both solid and lined steel pipe.

As these plastics and elastomers continue to evolve, we seem to be becoming more and more a disposable society. While technology improves on the front end with providing more and more useful products in new product forms and materials, we have to be mindful of the back end. On the back end of such product growth are the remnants of these products after they have completed their life cycle. What we should be doing in this particular case, when referring to single-use components and equipment, is to modify our thinking. Instead of referring to these items as disposables, perhaps we should refer to them as reclaimables.

The steel industry took this issue in hand years ago, with regard to the amount of scrap metal being discarded, by incorporating scrap metal into their steelmaking processes. The basic oxygen furnace (BOF) process, which manufactures 40% of the steel produced in the United States, uses 25–35% scrap steel in making new steel. The electric arc furnace (EAF) process, which manufactures the remaining 60% of the steel produced in the United States, essentially uses 100% scrap steel to make new steel, adding alloying chemicals to produce the various steels needed for the marketplace.

Worldwide, scrap metal has been treated for decades as a commodity. According to the United Nations Commodity Trade Statistics Database, the amount of scrap metal traded on the world market in 1990 amounted to 9.3 million tons. By 2011 it had increased to around 106 million tons. In 2011 alone, the Bureau of International Recycling recorded that the total amount of scrap metal traded on the world market increased 7.6% to 570 million tons. This is no cottage or niche industry, but an actual industry of its own.

Recycling on such a large scale has a significant impact on airborne emissions, energy, water resources, mineral resources, and cost. If the polymer industry follows a similar closed-loop philosophy of counterbalancing the back-end recycling with front-end production, then single-use reclaimable components and equipment should certainly have no problem finding their place in industrial bioprocessing system applications.

2.5 Surface Finish

Due to the impact that microscopic elements such as bacteria and particulates can potentially have on a pharmaceutical process fluid, the contact surface finish of the tubing, fittings, and other inline components such as valves, filters, and so on is of major concern. As described in Section 7.3, microscopic gouges and pits in process contact surfaces can be a costly issue with high-purity bioprocessing systems and so too can the onset of rouging, which will be discussed in Section 2.6.

While it may be highly possible to prevent foreign particulates from entering a sealed process system, it is virtually impossible to eliminate the possibility of transient bacteria from gaining access to the internals of a piping system[1] (Ref. Section 7.3.1). But there are steps that can be taken to at least minimize such microbial contamination in protecting a process fluid. One such way is to create a process contact surface finish that is not conducive to allowing bacteria to find a place of residence and set up house.

As seen in Figure 2.5.1, bacteria in a fluid stream have the ability, as planktonic organisms being forced through a piping system with the fluid flow, to find microscopic crevices, pits, or a recess in the wall of a pipeline where they can gain a foothold and begin setting up a colony. The pipe wall seen in Figure 2.5.1 is the mechanically polished surface of 316L stainless steel tubing.

Biofilms, as those pictured in the preceding text in Figures 2.5.1 and 2.5.2 scanning electron microscopy (SEM) micrographs, can consist of many different species of bacteria and archaea (single-celled organisms) living within a matrix of excreted polymeric compounds. This matrix, the plastic looking goo seen in Figure 2.5.2, is the structure that holds the colony together and helps them expand. It protects the cells within it and facilitates communication among the cells through chemical and physical signals. Some biofilms have been found to contain water channels that help distribute nutrients and signaling molecules.

If left unabated these cells will grow at an exponential rate. Under perfect growth conditions, a bacterial cell divides into two daughter cells once every 20min. This means that a single cell and its descendants will grow exponentially to more than two million cells in 8 hours.

Figure 2.5.1 Biofilm at 2000×. *Source*: Reproduced with permission of Frank Riedewald (2004)

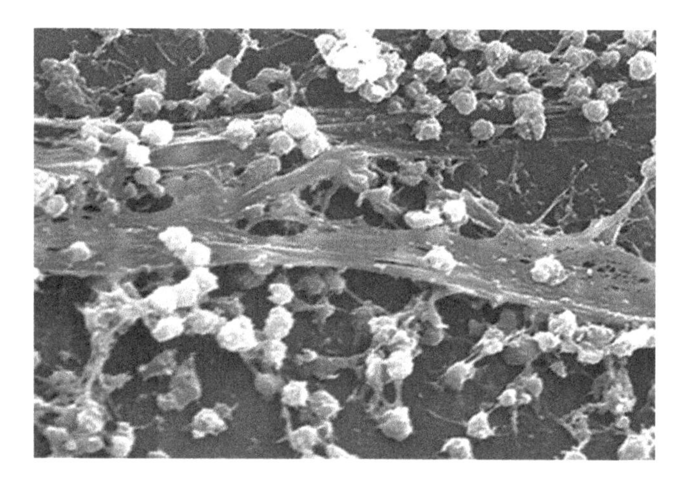

Figure 2.5.2 Staphylococcus biofilm at 2363×

At the CDC in Atlanta, researchers took plastic pipe sections and filled them with water contaminated with two strains of bacteria. After allowing the bacteria to incubate for 8 weeks, the scientists emptied out the infested water and doused the pipes with germ-killing chemicals, including chlorine, for 7 days. They then refilled the pipes with sterile water and periodically sampled the "clean" water. The team reported that both strains survived in the chemically treated pipes and reestablished colonies there.

It therefore stands to reason that in specifying material of construction for hygienic systems, we take every precaution to specify a process contact material that will create an environment within the piping and equipment that makes it painfully difficult for bacteria to exist. And in specifying such material its surface finish should comply with BPE Table SF-2.2-1 as shown in Figure 2.5.3. Should the material have an electropolished surface, it should then also comply with Table SF-2.2-2 as represented in Figure 2.5.4. Polymeric materials would comply with BPE Table SF-3.3-1 as shown in Figure 2.5.5.

Table SF-2.3-2 (Figure 2.5.5) is essentially a list of unacceptable consequences that occur occasionally as a result of time, temperature, and chemical complexities involved in the process of electropolishing. Such are the lengths that need to be gone through in order to reduce the risk of losing product to bacterial adulteration.

Surface finishes are quantified by the values of Ra (referred to as center line avg. (CLA) by ISO or as arithmetic average (AA), RMS, and ISO grade numbers and described in terms of roughness, waviness, and lay. These terms are defined in ASME B46.1 "Surface Texture (Roughness, Waviness, and Lay)" as follows:

- *Lay*: the predominant direction of the surface pattern, ordinarily determined by the production method used.
- *Root mean square (RMS) roughness, Rq*: the RMS average of the profile height deviations taken within the evaluation length and measured from the mean line.
- *Roughness*: the finer spaced irregularities of the surface texture that usually result from the inherent action of the production process or material condition.
- *Roughness average, 9 Ra*: the arithmetic average of the absolute values of the profile height deviations recorded within the evaluation length and measured from the mean line.
- *Surface texture*: the composite of certain deviations that are typical of the real surface. It includes roughness and waviness.
- *Waviness*: the more widely spaced component of the surface texture. Waviness may be caused by such factors as machine or workpiece deflections, vibration, and chatter. Roughness may be considered as superimposed on a wavy surface.

Represented in Table 2.5.1 in the succeeding text are comparisons of the earlier Ra, RMS, ISO grade number (N), and CLA values. Prior to these very specific quantifying values, surface finishes were identified in very general terms referred to as "polish numbers," that is, #4, #8, and so on, and "grit numbers," that is, 80, 180, 240, and so on. The polish numbers were very broad subjective values that could mean what the engineer or inspector wanted them to mean. The grit numbers were based on sandpaper grit. The grit value is based on the number of grit particles per square inch on sandpaper. Therefore, the higher the grit number, the finer the grit and the smoother the surface being sanded or polished with a grit compound.

The intent behind creating a smooth and near flawless process contact surface is twofold: (i) to eliminate places in which planktonic organisms moving through a piping system can get a foothold to begin building a colony that becomes biofilm and (ii) to enhance the cleanability of the system by improving drainability and by mitigating the possibility of cross contamination on a microscopic level.

Table SF-2.2-1 Acceptance criteria for metallic process contact surface finishes

Anomaly or Indication	Acceptance criteria
Pits	If diameter <0.020 in. (0.51 mm) and bottom is shiny [Notes (1) and (2)]. Pits <0.003 in. (0.08 mm) diameter are irrelevant and acceptable.
Cluster of pits	No more than 4 pits per each 0.5 in. (13 mm) × 0.5 in. (13 mm) inspection window. The cumulative total diameter of all relevant pits shall not exceed 0.040 in. (1.02 mm)
Dents	None accepted [Note (3)].
Finishing marks	If R_a max. is met.
Welds	Welds used in the as-welded condition shall meet the requirements of MJ-8. Welds finished afterwelding shall be flush with the base metal, and concavity and convexity shall meet the requirements of MJ-8. Such finishing shall meet the R_a requirements of Table SF-2.4-1.
Nicks	None accepted.
Scratches	For tubing, if cumulative length is <12.0 in. (305 mm) per 20 ft (6.1 m) tube length or prorated and if depth is <0.003 in. (0.08 mm). For fittings, valves, and other process components, if cumulative length is <0.25 in. (6.4 mm), depth <0.003 in. (0.08 mm), and R_a max. is met For vessels, if length <0.50 in. (13 mm) at 0.003 in. (0.08 mm) depth and if <3 per inspection window [Note (4)].
Surface cracks	None accepted.
Surface inclusions	If R_a max. is met.
Surface residuals	None accepted. visual inspection
Surface roughness (R_a)	See Table SF-2.4-1.
Weld slag	For tubing, up to 3 per 20 ft (6.1 m) length or prorated, if <75% of the width of the weld bead. For fittings, valves, vessels, and other process components, none accepted (as welded shall meet the requirements of MJ-8 and Table MJ-8.4-1).
Porosity	None open to the surface.
Buffing	None accepted.

GENERAL NOTE: This table covers surface finishes that are mechanically polished or any other finishing method that meets the R_a max.

NOTES:

(1) Black bottom pit of any depth is not acceptable.

(2) Pits in super-austenitic and nickel alloys may exceed this value. Acceptance criteria for pit size shall be established by agreement between owner/user and supplier. All other pit criteria remain the same.

(3) For vessels, dents in the area covered by and resulting from welding dimple heat transfer jackets are acceptable.

(4) An inspection window is defined as an area 4 in. × 4 in. (100 mm × 100 mm).

Figure 2.5.3 Acceptance criteria for metallic process contact surface finishes. *Source*: https://www.asme.org/products/codes-standards/bpe-2014-bioprocessing-equipment. © The American Society of Mechanical Engineers

Table SF-2.2-2 Additional Acceptance criteria for electropolished metallic process contact surface finishes

Anomaly or indication	Acceptance criteria
Blistering	None accepted
Buffing	None accepted
Cloudiness	None accepted
End grain effect	Acceptable if R_a max. is met
Fixture marks	Acceptable if electropolished
Haze	None accepted
Orange peel	Acceptable if R_a max. is met
Stringer indication	Acceptable if R_a max. is met
Weld whitening	Acceptable if R_a max. is met
Variance in luster	Acceptable if R_a max. is met

Figure 2.5.4 Additional acceptance criteria for EP metallic process contact surface finishes. *Source*: https://www.asme.org/products/codes-standards/bpe-2014-bioprocessing-equipment. © The American Society of Mechanical Engineers

Table SF-3.3-1 Acceptance criteria for polymeric process contact surface finishes

Anomaly or indication	Acceptance criteria
Scratches	For rigid tubing/piping, if cumulative length is <12.0 in. (305 mm) per 20 ft (6.1 m) tube/pipe length or prorated and if depth <0.003 in. (0.08 mm). For other process components, surface finish must be agreed upon by supplier and owner/user.
Surface cracks	None accepted
Surface inclusions	None accepted
Surface roughness, R_a	See Table SF-3.4-1.

GENERAL NOTE: All process contact surface finishes shall be defined by the owner/user and supplier using the criteria described in SF-1, purpose and scope.

Figure 2.5.5 Acceptance criteria for polymeric process contact surface finishes

Table 2.5.1 Surface roughness values

Ra value		ISO grade numbers Roughness N	RMS value		Center line avg. (CLA) (μin.)
Microinches (μin.)	Micrometers (μm)		Microinches (μin.)	Micrometers (μm)	
0.25	0.006		0.2	0.006	0.25
0.5	0.0125		0.5	0.012	0.5
1	0.025	N1	1.1	0.028	1
2	0.05	N2	2.2	0.055	2
4	0.1	N3	4.4	0.11	4
8	0.2	N4	8.8	0.22	8
16	0.4	N5	17.6	0.44	16
32	0.8	N6	32.5	0.81	32
63	1.6	N7	64.3	1.61	63
125	3.2	N8	137.5	3.44	125
250	6.3	N9	275	6.88	250
500	12.5	N10	550	13.75	500
1000	25	N11	1100	33	1000
2000	50	N12	2200	55	2000

CLA, center line average in microinches; N, new ISO (grade) scale numbers; Ra, roughness average in micrometers or microinches; RMS, root mean square in microinches.

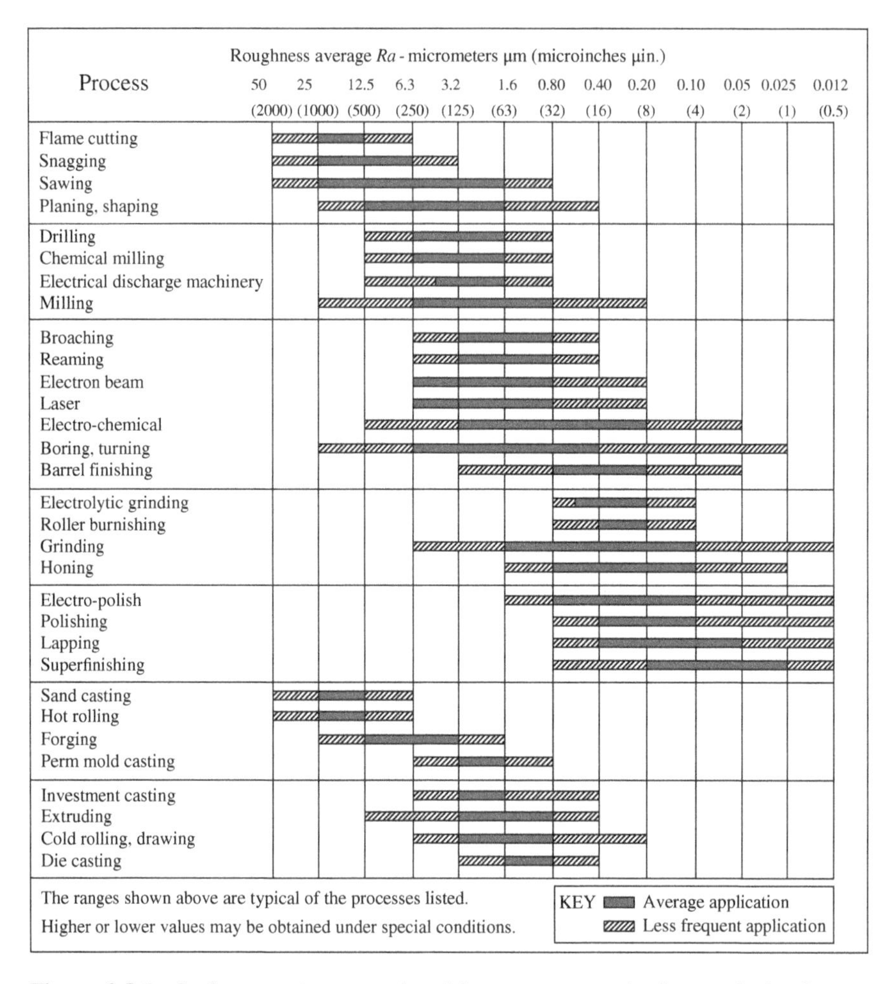

Figure 2.5.6 Surface roughness produced by common production methods. *Source*: https://www.asme.org/. © The American Society of Mechanical Engineers

Designing a system with an electropolished process contact surface that is drainable helps to both stem the onset of biofilm and assist in its eradication when it does develop.

Figure 2.5.6, taken from ASME B46.1, represents a table of surface roughness finish ranges that are achieved through various production methods. The roughness values are given in both micrometers (μm) and in microinches (μin.). It's apparent to see that the electropolishing method used in achieving the Ra finishes needed in the bioprocessing industry is indeed the most appropriate means available.

2.6 Rouge

In specifying material of construction for hygienic piping systems, we typically default to the proven standby 316L stainless, as described and reinforced time and again in this and other chapters of this book and beyond this book in countless articles, white papers, presentations, and company specifications. If 316L is so great, why then does this anomaly, referred to as *rouge*, persist in being a thorn in the side of operations and maintenance personnel by appearing internally on high-purity piping systems on an incidental basis or on an ongoing basis? In some systems periodic remediation becomes an integral part of the operating procedure, part of doing business if you will.

The BPE Standard defines rouge in the following manner:

- *Rouge*: a general term used to describe a variety of discolorations in high-purity stainless steel biopharmaceutical systems. It is composed of metallic (primarily iron) oxides and/or hydroxides. Three types of rouge have been categorized:
  - *Class I rouge*: a rouge that is predominantly particulate in nature. It tends to migrate downstream from its origination point. It is generally orange to red-orange in color. These particles can be wiped off a surface and are evident on a wipe. Surface composition of the stainless steel under the rouge remains unchanged (Ref. Figure 2.5.1).
  - *Class II rouge*: a localized form of active corrosion. It occurs in a spectrum of colors (orange, red, blue, purple, gray, black). It can be the result of chloride or other halide attacks on the surface of the stainless steel (Ref. Figure 2.5.2).
  - *Class III rouge*: a surface oxidation condition occurring in high-temperature environments such as pure steam systems. The system's color transitions to gold, to blue, and to various shades of black, as the layer thickens. This surface oxidation initiates as a stable layer and is rarely particulate in nature. It is an extremely stable form of magnetite (iron sesquioxide, Fe_3O_4) (Ref. Figure 2.5.3).

Codes and standards, with respect to piping and equipment, typically do not cover issues that relate to systems that have been in operation for a period of time. Like the ASME B31 piping codes, the BPE Standard is directed specifically at new piping systems, as stated in paragraph GR-2 where it reads, "This Standard is intended to apply to new fabrication and construction." However, the issue with the occurrence and reoccurrence of rouge in high-purity piping systems is so prevalent that its subject matter has been addressed in the BPE Standard.

The subject of rouge is covered extensively in Nonmandatory Appendix D of the BPE Standard, but in order to direct the reader and user of the standard to Appendix D, mention of the subject had to be made in the body of the standard. This grammatical link to Appendix D from the body of the standard was made with paragraph SF-2.8 where it touches on the subject matter of rouge and refers the reader to Appendix D for an in-depth discussion on the topic.

Within Appendix D are five tables, D-2-1 and D-2-2, D-3.1-1 and D-3.2-1, and D-4.1-1, that embody prevention, analysis, classification, and remediation of rouge in providing concise and worthwhile information on the following:

- "Considerations that Affect the Amount of Rouge Formation during the Fabrication of a System" (D-2-1)
  - Table D-2-1 provides point-by-point methodologies in attempting to prevent the onset of rouging as affected by fabrication.
- "Considerations that Affect the Amount of Rouge Formation during the Operation of a System" (D-2-2)
  - Table D-2-2 provides point-by-point methodologies in attempting to prevent the onset of rouging as caused during system operation.
- "Process Fluid Analyses for the Identification of Mobile Constituents of Rouge" (D-3.1-1)
  - Table D-3.1-1 suggests four analytic protocols and methodologies that might be used in detecting, quantifying, and analyzing particulate matter that could be the early onset of rouging.
- "Solid Surface Analyses for the Identification of Surface Layers Composition" (D-3.2-1)
  - Table D-3.2-1 provides five methodologies that can be used in detecting, quantifying, and analyzing tubing or component wall surface conditions for the presence of rouging.
- "Rouge Remediation Processes Summary" (D-4.1-1)
  - Table D-4.1-1 provides a list of various rouge remediation or derouging processes based on rouge classification.

But what is rouge, where does it come from, and why can't it be prevented? The answer to that lies with equal parts astute testing and analysis, conjecture, and alchemy. As mentioned in Section 2.3.2 and to a further extent in Section 2.7, the passivated chromium-rich surface that forms on stainless steel is the mechanism by which stainless steel is protected from corrosion. If this protective barrier is compromised in any way, the base metal, iron, becomes exposed and vulnerable to the contact liquid.

To answer the three-part question earlier of "what is rouge, where does it come from, and why can't it be prevented," we can, in first answering the "what" question, look to the definition of rouge provided in the BPE Standard as stated in the preceding text. This tells us that it is composed of "…metallic (primarily iron) oxides and/or hydroxides." But, to learn the specifics of what rouge is, we have to dig a little deeper and look at the three classifications assigned to the various types of rouge that appear in operating systems under varying conditions.

2.6.1 Class I Rouge

In a research study done on rouge by Mr. Tom Hanks, P.E., CRB Consulting Engineers, Inc., Carlsbad, CA, he writes that "The incipient stages of [Class I] are examples of hydrated oxides (limonite) that are easily cleaned deposits and the

Figure 2.6.1 Class I and II rouge in pump casing. *Source*: Tom Hanks, CRB Consulting Engineers, Inc., presentation to ASME BPE (*See plate section for color representation of this figure*)

passive layer of the stainless steel is uncompromised. The rouge deposits may exhibit the identical composition as that of their [upstream] source. Rouge concentration is typically denser near the source and diminishes with distance. Color of the rouge can also vary with distance from source; generally beginning as orange or red 'hematite' developing into other valence forms such as grey or black 'magnetite'. The hematite form of iron oxide is the most common state for ferrous oxides found in most rouge events and requires oxygen and water for formation. The formation of hematite and magnetite from the lowest iron oxidation states appears to form over time of exposure and temperature."

The Class I rouge collects on the interior surfaces of a piping system as a result of the electrostatic attraction of the pipe, components, and equipment walls.

Figure 2.6.1 shows a pump casing with a thorough coating of rouge, as hematite, that ranges from a Class II to a Class I rouge. The Class II is represented by the yellow-hued high-velocity impingement region located at the center of the impeller. From there it spreads outward toward the high-velocity region of the outer circumference of the impeller vanes as the color transitions to a rusty red and a Class I rouge on both the impeller vanes and the casing itself.

2.6.2 Class II Rouge

Figure 2.6.2 shows another pump casing, this one covered in Class II rouge. With regard to Class II rouge, Mr. Hanks notes that this type of rouge "Derives from iron oxides that originates [from a source elsewhere] within the system.

Figure 2.6.2 Class II rust-colored rouge in pump casing. *Source*: Tom Hanks, CRB Consulting Engineers, Inc., presentation to ASME BPE (*See plate section for color representation of this figure*)

The sources are typically breaches of the passive layer or un-passivated welds, etc. Class II rouge is corrosion by halides, usually chloride, forming at the breaches of the passive layer. Stainless steel corrosion cells are in the surfaces of the system and must be removed mechanically or by acid etching." Class II is similar in chemistry to that of Class I with the exception being that the Class II particulate matter that makes up the rouge is imbedded into the base material, whereas Class I is affixed to the surface of the stainless steel.

Figure 2.6.3 is another form of Class II hematite rouge, this one appearing as a purplish tint. Class II rouge is developed in two stages: the first of which is the dissolution of the passive chromium oxide layer exposing the base metal to the environment and making it susceptible to corrosion. The corrosive environment then begins to attack the base metal.

2.6.3 Class III Rouge

Figure 2.6.4 shows a Class III rouge. Mr. Hanks notes in his report that Class III rouge "...is a more complex formation of iron II and iron III oxides that occur in higher temperature (steam) systems. It may be orange or red in its incipient stage but turns gray or black with exposure to elevated temperature (the temperature driving the reactions that forms the more complex crystal structure). It is commonly referred to as 'Magnetite'. It may appear powdery or at higher temperature

Figure 2.6.3 Class II purple-colored rouge in pump casing. *Source*: Tom Hanks, CRB Consulting Engineers, Inc., presentation to ASME BPE (*See plate section for color representation of this figure*)

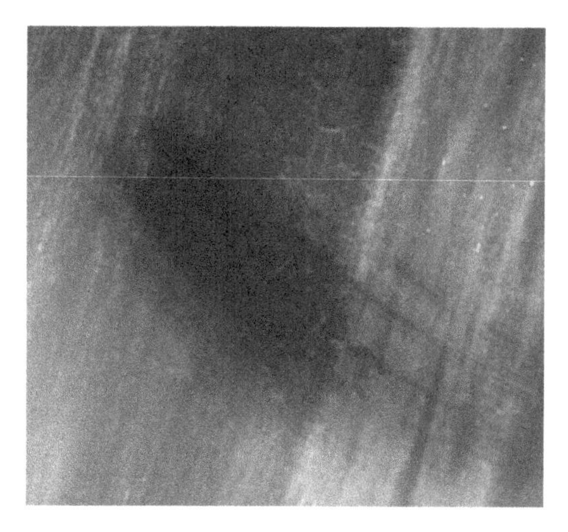

Figure 2.6.4 Class III rouge. *Source*: Tom Hanks, CRB Consulting Engineers, Inc., presentation to ASME BPE (*See plate section for color representation of this figure*)

have a glazed appearance. The developed form is iron sesquioxide and is not only stable but limits the diffusion of oxygen and other ions to or from the passive layer. It can be removed only by mechanical means or strong acid (i.e. derouging by oxalic acid)."

2.6.4 Background on Rouge

Before continuing I would like to give credit where credit is due. Back in 1999 a paper was written on "Rouging of Stainless Steel in WFI and High Purity Piping Systems." This paper was written by John C. Tverberg, who, at the time, was working for Trent Tube, and James A. Ledden, who, at the time, was working for Pharmacia/Upjohn. The paper was presented in October 1999 at an International Research Symposium, "Preparing for Changing Paradigms in High Purity Water," held in San Francisco, CA.

The report and presentation was the culmination of a 2-year study that began in late 1997, taking advantage of the dismantling of an operating unit that included piping from WFI and high-purity water systems in a Pharmacia/Upjohn facility in Kalamazoo, MI. The system had been in service for 15 years and, having shown signs of extensive rouging, was a prime candidate for the study of this heretofore elusive and perplexing anomaly.

Contained in that 1999 report and presentation was the phrase "In these rouging examples we are faced with what appears to be **three different mechanisms for formation**." That phrase and the descriptions of those three "mechanisms" established the premise for the three classifications used to segregate and identify the three basic forms of rouge.

Classifications described in the preceding text, adopted by industry standards; repeated time and again in articles, books, and papers; and referred to in presentations around the world, have been a huge asset to high purity-related industries. The dismantling of the Pharmacia/Upjohn operating unit was an opportunity that seldom comes along. Its benefits, from the carefully orchestrated research performed by Messer's Tverberg and Ledden and their support team, not to mention the cooperation from Pharmacia/Upjohn, continue to pay dividends for industry these 17 years later and will continue to pay dividends far into the future.

Here then are the ways the three classifications were first described by Messer's Tverberg and Ledden in their 1999 research paper:

Class 1: Rouge from external sources

a. Forms:
   (i) Metal particles that are oxidized in WFI. These particles are generated from external surfaces by erosion or cavitation.
   (ii) Oxides from foreign sources such as carbon steel bolts, nuts or tie rods.
b. The particles and/or oxides are held on to the tubing surface by electrostatic attraction.
c. The stainless steel surface under the oxide is unaffected.
d. Composition of the oxide is not that of corroding stainless steel:
   (i) Particles and/or oxides appear to have a composition that matches that of the stainless steel from which it originates, for example, a pump impeller.

Class 2: Rouge from in situ oxidation of the stainless steel surface where rouge is located.

a. Results from low Cr/Fe ratio (<1) and appears related to mechanical polished and non-chemically passivated surfaces.

b. The surface under the rouge appears “active” with pits, crevice corrosion, etc.
c. Chlorides may play a major role, but this must be proven with additional tests.

Class 3: Black oxide rouge
a. Originates from high temperature steam service.
b. Appears to exist in two forms:
   (i) Glossy black over electropolished surfaces. These are very stable.
   (ii) Powdery black. These are the high temperature analog of Class 2 rouge. This rouge will rub off.
c. Rouge cannot be removed with simple cleaning, must be chemically removed

The earlier outline description of the three classes of rouge is presented here in the manner in which it was presented in 1999. Their descriptive analogy still holds up to this day.

2.6.5 Source of Rouge

To answer the “where does it come from” question, we can again refer back to the BPE definition and declare that there are two places from which rouge typically originates, that being from a corrosion source upstream from its deposited location (Class I) or from the site where it is actually detected (Classes II and III).

For rouge originating from upstream sources, it will typically come from areas in which the passive layer has been breached. This could be from internal surface areas at weld HAZ having a heat tint color indicative of a weld area being exposed to an unacceptable amount of oxygen during welding. Or it could be surface areas in high-velocity zones of a system. These high-velocity zones could be at pump impellers where not only velocity has an effect but also pump cavitation if it exists. Or it could be in the seat area of valves that are throttled to almost closed, creating a sonic flow rate, or in the case of clean steam, superheated steam for a brief distance downstream. Each of these situations can be further described as follows:

2.6.5.1 Unacceptable Heat Tint in HAZ

When welding on stainless steel, it is essential that a backing gas be applied, typically argon or an argon mixture, to displace the air containing oxygen surrounding the weld area. When oxygen is present at the HAZ, depending on the amount by percentage, the internal surface of the tubing can oxidize leaving a straw to deep blue to purple tint on the inside wall of the tubing. The oxidation that forms in the HAZ of the weld is rich in chromium oxide; however the heat tint is not an effective barrier for preventing corrosion. The chromium present in the heat tint comes from the surface of the alloy leaving a shallow region on the metal’s surface that is depleting it of chromium. To restore the CR, both the heat tint and the underlying chromium depleted region must be removed.

The thickness of the oxide layer that develops in the HAZ is largely dependent upon the percentage of oxygen that is present in the area surrounding the HAZ. Refer to acceptable heat tint color in BPE Fig. MJ-8.4-2 and MJ-8.4-3. The variation in heat tint caused by increases in oxygen content can be seen in AWS D18.1 Exhibit C. The thickness of this oxide layer, made determinant by the discoloration, will dictate how much material will have to be removed by grinding and/or acid etching before passivation of the base metal can occur.

Should this unwanted oxide layer not be fully removed, it will make it difficult, if not impossible, for the passive layer to form, thus leaving the base material unprotected from corrosion providing a localized source for dissolved iron that can flow downstream to settle out as iron oxide, typically at an impingement area within the piping system.

2.6.5.2 High-Velocity Erosion

High-velocity erosion of austenitic stainless steels on various forms of construction material becomes a real concern where liquid velocities can erode surface material, including the passive protective surface of stainless steel. Once the passive layer is worn away, you then have the combined attack of the ongoing erosion plus that of corrosion, which is now able to attack the now vulnerable base metal.

These high-velocity zones can be found at the tip of impeller blades or flutes, in the seat area of throttling valves, and at the impingement areas of flow directional changes.

2.7 Electropolishing

The BPE Standard defines electropolishing in the following manner:

Electropolishing: a controlled electrochemical process utilizing acid electrolyte, direct current (DC), anode, and cathode to smooth the surface by removal of metal.

To elaborate, electropolishing is the process by which an electrolytic solution in conjunction with DC electricity, time, and temperature combines under controlled conditions to create anodic dissolution of microscopic surface materials. This is done by connecting the workpiece to be finished to the anodic (+) side of a DC power source and connecting an inert piece of metal (stainless steel or lead) to the cathodic (−) side of the same DC power source. This can be achieved by immersing the workpieces in an electrolytic bath, as represented in Figure 2.7.1, or in situ, with installed piping and/or equipment.

Because of their size and varied physical characteristics, there are various methods needed in electropolishing vessels, pipe, and components. Vessels, for

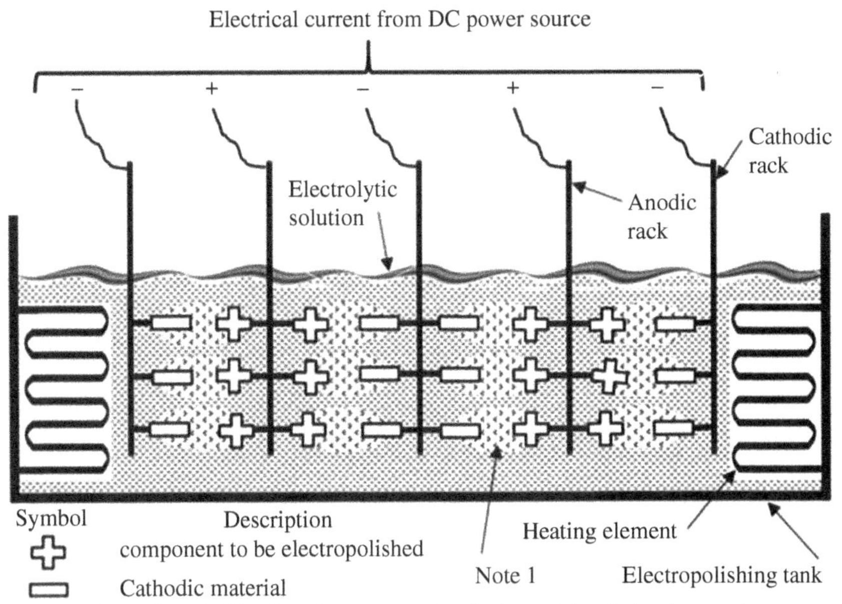

Note 1: Microparticles shedding from the component (anode +) material and being drawn to the target (cathode–) material.

Figure 2.7.1 Schematic of electropolishing bath

instance, can be used as their own container for the electropolish solution. If they happen to be too large to be completely filled with the electropolish solution, they can be laid horizontally, partially filled, and then rotated about a cathode as the solution coats the inside of the vessel.

Pipe, on the other hand, can be placed on a multipipe electropolishing rack that, once connected to the electrolytic solution source, will have the electropolishing solution pumped through each section of pipe with either a stationary or movable cathode running through the inside of each pipe. A similar type procedure can be used on installed piping systems. If the site does not have a waste treatment facility capable of handling the dilute acid with trivalent chromium and nickel ions contained in the rinse solution, it will need to be collected and hauled to a licensed waste treatment facility.

But what we will use to describe the electropolishing process is the electropolishing vat. The principals used here apply to the electropolishing process in general. Referring to Figure 2.6.1, the entire fixture with the anodic rack, containing the workpieces to be finished along with the cathodic rack, containing the inert material, is immersed into an electrolytic bath of an acidic chemical solution, typically consisting of phosphoric and sulfuric acids. Upon submersing the material into the chemical bath, time, temperature, and electrical current density become essential elements of the process. Each of these elements, the anodic and cathodic materials, current density, composition of the electrolytic

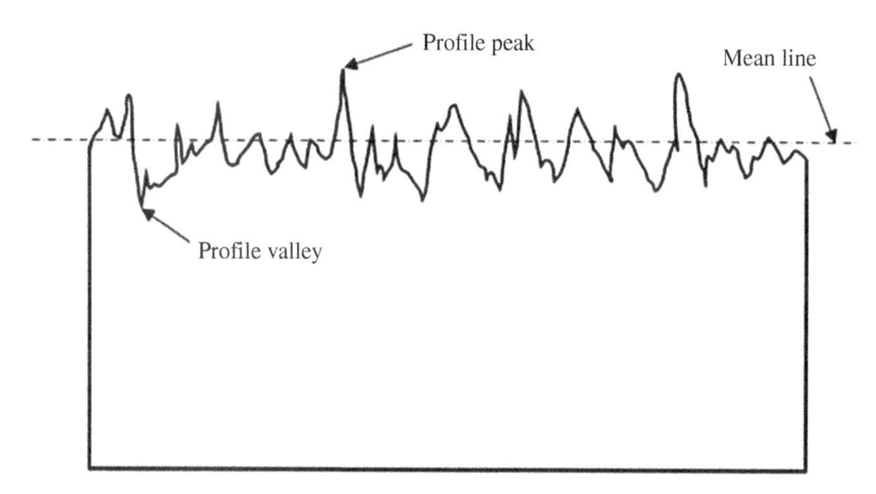

Figure 2.7.2 Profile of pre-electropolished 316L stainless steel

solution, time, and temperature all have to be proportional to each other in this process or the pseudo-alchemy does not work.

When these materials are immersed into the electrolytic solution, the acids begin to break down the exposed higher peaks of the jagged surface, those shown in Figure 2.7.2, through oxidation and dissolution. The peaks on the surface are leveled resulting from when the polishing conditions are set up creating a higher anodic current density at the peaks versus the valleys. This breaking down and dissolution of the microscopic material at the workpiece is enhanced and controlled by the electrical current, controlled in the sense that the oxide metal ions are drawn away from the workpiece by the electrochemical current and pulled toward the cathodic material where it collects.

As the electropolishing process continues to loosen and draw off particulate material from the anodic component, the peaks and valleys of the initial jagged surface of the material will begin to dissolve into a smooth, rolling surface minus the crags and crevices found in the initial form of the material, as depicted in Figure 2.7.3. The finisher has to time exposure of the material in the electrolytic soup in conjunction with the temperature of the solution as well as the density of the electrical current.

Figure 2.7.4 shows the magnified surface of a 316L stainless steel surface in its mechanically polished Ra 20 condition. The striations on the surface are the result of granular polishing grit being forced across the surface of the stainless steel under pressure. While this process, depending on grit size, does improve the visual appearance of the stainless steel, it is insufficient for the needs and requirements of high-purity applications. Each of those gouges and overlaps of surface material visible under the magnification of SEM is a potential housing community for bacteria and microscopic particulate matter.

Figure 2.7.5 is a post-electropolished 316L stainless steel surface with the same magnification as that of Figure 2.7.4 but without the apparent striations seen

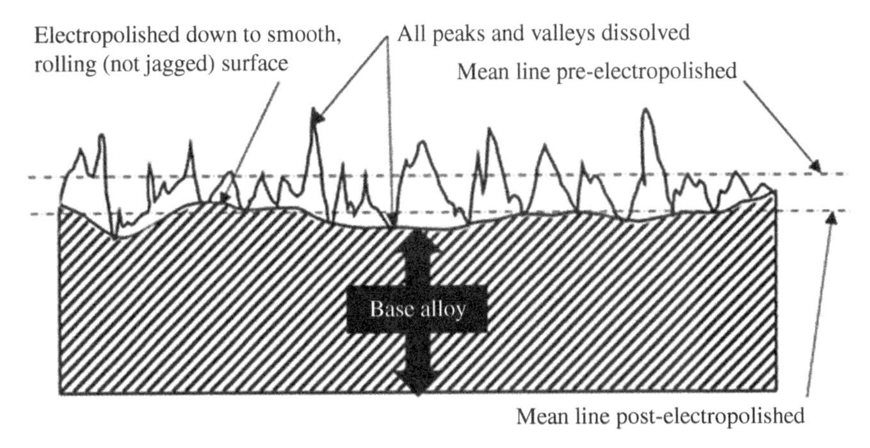

Figure 2.7.3 Profile of post-electropolished 316L stainless steel

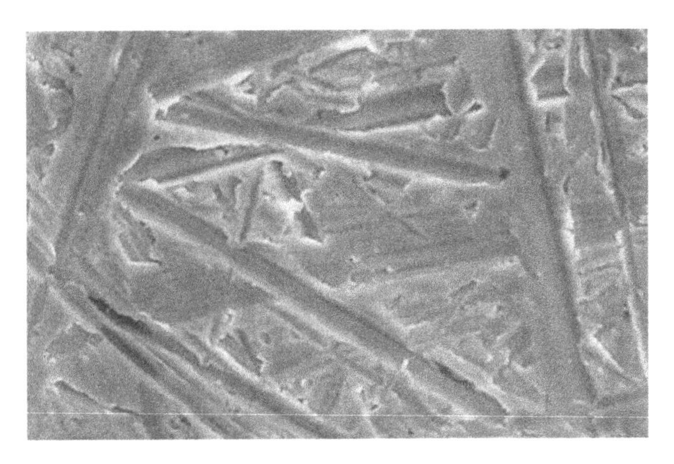

Figure 2.7.4 316L stainless steel mechanically polished Ra 20. *Source*: http://ultracleanep.com/. © 2015 Ultraclean Electropolish, Inc.

Figure 2.7.5 316L stainless steel electropolished Ra 7. *Source*: http://ultracleanep.com/. © 2015 Ultraclean Electropolish, Inc.

across the face of the surface shown in Figure 2.7.4. That surface in Figure 2.7.5, at Ra 7, lacks any identifiable marks or features. The Ra 7 surface roughness value is based on the high points and low points of the smooth rolling electropolished surface rather than the erratic peaks and valleys of the mill or mechanically finished surface.

2.7.1 *Irregularities or Flaws in Electropolishing*

With the electropolishing process there are so many essential variables, each one dependent on the others, that have to be controlled against a backdrop of material discrepancies, human error, timing, and handling that any number of rejectable results can occur at any time. Here are a few:

Frosting
Frosting, as seen in Figure 2.7.6, can be caused by "over EP" resulting in an etched surface. Over EP (long cycles, high current densities) can be attributed to the molybdenum constituent in 316L stainless steel, which helps reduce pitting in service. Molybdenum polishes out much faster than other alloying constituents causing an etched surface, which in turn causes frosting.

Cloudiness
The descriptive term cloudiness, as detected on an electropolished surface, as seen in Figure 2.7.7, has the appearance of a milky white or light discoloration. Such cloudiness can typically be found covering portions of an electropolished surface and not the entire electropolished surface area. This can be the result of insufficient exposure to the EP process or poor rinsing practices afterward.

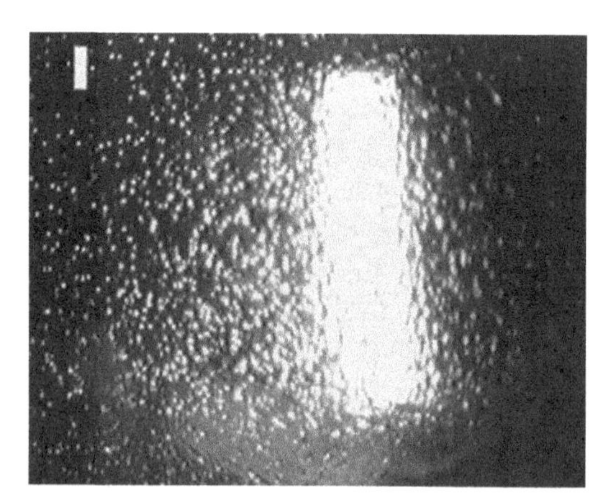

Figure 2.7.6 Example of frosting of electropolished 316L (*See plate section for color representation of this figure*)

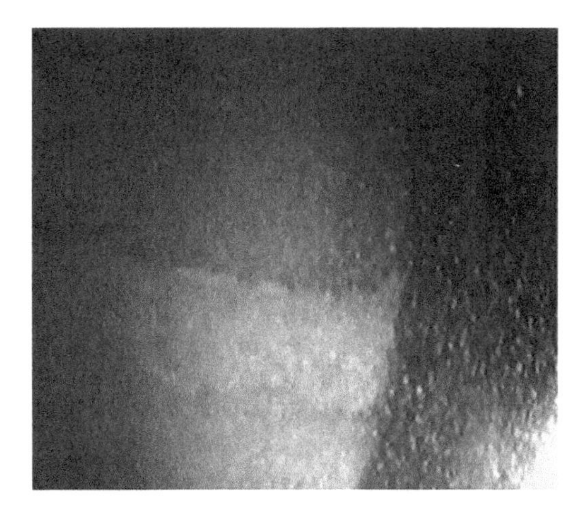

Figure 2.7.7 Example of cloudiness of electropolished 316L (*See plate section for color representation of this figure*)

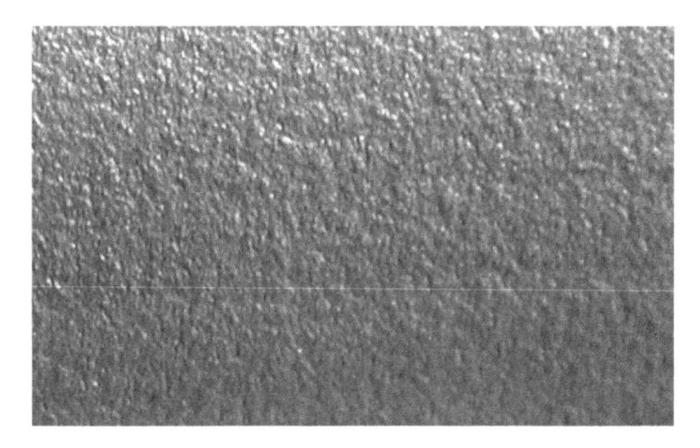

Figure 2.7.8 Example of orange peel 316L on stainless steel resulting from electropolishing

Orange Peel

The orange peel surface flaw, as seen in Figure 2.7.8, on an electropolished surface can be the result of an unanticipated and unknown material chemistry composition and not necessarily the result of an improperly executed EP process. The effect has the ripple appearance of an orange peel, thus the name.

Beyond the graphic examples of those given in the preceding text, there are also electropolishing flaws described as various "luster" anomalies that appear on the surface as dull, grainy, or just generally patchy discolorations, as well as flaws described as "shadows," "pitting," "streaks," and "stains." Such flaws can be attributed to contamination from exposure to industrial process materials, industrial pollutants, lubricants, nonmetallic inclusions, surface decarburization during annealing, and other extraneous contaminants found both in the material itself and in an industrial manufacturing environment.

2.8 Passivation

The previous discussion on electropolishing can quite literally carry over into this discussion on passivation. The electropolishing process, in accordance with ASTM B912, *Standard Specification for Passivation of Stainless Steels Using Electropolishing*, can provide a quality passive surface. The quality of the passive surface layer is predicated on the formulation of the electropolish solution and the conditions under which the procedure is performed.

Passivation is defined in the BPE Standard as the:

> removal of exogenous iron or iron from the surface of stainless steels and higher alloys by means of a chemical dissolution, most typically by a treatment with an acid solution that will remove the surface contamination and enhance the formation of the passive layer.

Rather than defining the term "passivation" itself, the earlier definition, in actuality, better describes a chemically induced cleaning and passivating process. Passivation, when not chemically induced, is a naturally occurring phenomenon that takes place in an oxygen-containing environment in which the oxides or nitrides of the prevalent metal alloying constituent form an inert surface layer or "passive" protective layer also referred to as a "native oxide layer" on its own. However, what this "naturally" occurring passivation process does not do is clean away surface impurities that may inhibit a thorough and resolute forming of this passive layer making comprehensive corrosion protection suspect and possibly problematic.

Contaminants on the surface of metallic material can prevent or compromise the thorough and comprehensive passivation of an alloy's surface. In the case of 316L stainless steel, chromium oxides may be inhibited from forming at the material's surface due to surface contaminants preventing oxygen from sufficiently reaching the material's surface to initiate and perpetuate the natural passivation process. Or there may be free iron present at the surface as a result of work area contamination or contact with unalloyed steel tools, equipment, or other components. Free iron particulates, upon being moistened, will form iron oxides (rust), which, when left unabated, can cause pitting to the base metal.

Free iron can be detected by testing in accordance with testing described under ASTM A380, *Standard Practice for Cleaning, Descaling, and Passivation of Stainless Steel Parts, Equipment, and Systems*. Part of the chemically induced passivation process is intended to remove free iron, along with other contaminants, from the passivated surface.

Achieving comprehensive passivation under controlled conditions is a two-step process. To begin with there is a set method in achieving proper passivation. Each finisher has their own procedures and chemical formulations specific to the needs of each particular alloy and component form. Therefore the following describes the passivation process in general terms and not as an exact science.

Table E-5-1 Test matrix for evaluation of cleaned and/or passivated surfaces

Type of test	Test description	Pros	Cons
1. Gross Inspection of Cleaned and/or Passivated Parts per ASTM A380/A967 (Pass/Fail)			
Visual Examination [CT (test for cleanliness), RT (test for the presence of rouge)]	Bench or field test. Visual examination is the direct or indirect visual inspection of, in this case, a passivated metallic surface.	Can be performed with minimal preparation and equipment. Good general appearance review.	Not quantitative. Subjective interpretation of findings.
Wipe Test ASTM A380 (CT, RT)	Bench or field test. This test consists of rubbing a test surface with a clean, lint-free, white cotton cloth, commercial paper product, or filter paper moistened with high-purity solvent.	Useful for testing surfaces that cannot be readily accessed for direct visual examination. Removable surface contamination can be easily identified and compared.	Not quantitative. Difficult to inspect hard to reach areas of large tube diameters. There is also a risk of leaving errant fibers behind from the wipe or plug. Can be detrimental to electropolished surfaces.
Residual Pattern Test ASTM A380 (CT)	Bench or field test. After finish-cleaning, dry the cleaned surface per ASTM A380, The presence of stains or water spots indicates the presence of contaminants.	A simple test with rapid results.	Not quantitative. Not very sensitive.
Water-Break Test ASTM F22	The water-break test is performed by withdrawing the surface to be tested, in a vertical position, from a container overflowing with water. The interpretation of the test is based upon the pattern of wetting.	General cleanliness of surface is easily determined. Useful in detecting hydrophobic contamination.	Not quantitative. This test identifies the presence of retained oils and greases. The test is not applicable on all surfaces including, but not limited to, electropolished surfaces.
A 380 Water-wetting and Drying; ASTM A967 Water Immersion Practice A [PT (test for passivation)]	Bench or field test. Immersed in, or flushed with distilled water then air dried. Repeated for a minimum of 12 times. A modified version of this test requires a solution of 3 to 7% salt water, with a final rinse prior to inspection, using DI quality water or better.	Staining is evidence of free iron, which is detected through visual examination. Identifies possible pitting corrosion sites or imbedded iron.	Not quantitative.
High Humidity Test ASTM A380 & ASTM A967 Practice B (PT)	Bench test. Sample coupon is immersed or swabbed with acetone or methyl alcohol then dried in an inert atmosphere. The coupon is then subjected to 97% humidity at 100°F for 24 hrs or more.	Staining is evidence of free iron, which is detected through visual examination.	Not quantitative. Not used for installed tubing. Sample coupons can be used, but does not prove complete coverage. Lengthy test. Containment cabinet required.
Salt Spray test ASTM A 967 Practice C (PT)	Bench or field test. This test is conducted in accordance with ASTM B117 subjecting the test area to a 5% salt solution for a minimum of 2 hours.	Rust or staining attributable to the presence of free iron particles imbedded in the surface will become noticeable upon visual examination of the metal surface.	Not quantitative. Longer term testing is required to test for passive film quality or corrosion resistance. However, exposures over about 24 hours may show light staining resulting from differences in micro finish texture.

Figure 2.8.1 (a–d) Listing of various testing methods for cleanliness and passivation. *Source*: https://www.asme.org/products/codes-standards/bpe-2014-bioprocessing-equipment. © The American Society of Mechanical Engineers

2. Precision Inspection of Cleaned and/or Passivated Parts under ASTM A 380/A 967 (Pass/Fail)			
Solvent Ring Test ASTM A 380 (CT)	Bench test. Place a single drop of high-purity solvent on the surface to be evaluated, stir briefly then transfer to a clean quartz microscope slide and allow the drop to evaporate. If foreign material has been dissolved by the solvent, a distinct ring will be formed on the outer edge of the drop as it evaporates.	Good test for organic contamination on the test surface.	Not quantitative.
Black Light inspection ASTM A 380 (CT)	Bench test. This test requires the absence of white light and a flood type ultra-violet light.	Suitable for detecting certain oil films and other transparent films that are not detectable under white light. Good test for organic contamination on surface.	Not quantitative. Not practical when testing for passivation
Atomizer Test ASTM A 380 (CT)	Bench test. This test is conducted in accordance with ASTM F 21 using DI quality water or better. A variation of the water-break test, this test uses an atomized spray, rather than a simple spray or dip to wet the surface.	Test for presence of hydrophobic films. This test is more sensitive than the Water-Break Test.	Not quantitative. Requires direct visual examination.
Ferroxyl Test For Free Iron ASTM A 380/Potassium Ferricyanide-Nitric Acid ASTM A-967 Practice E (PT)	Bench or field test, Apply a freshly prepared solution of DI water or better, nitric acid, and potassium ferricyanide to the coupon using an atomizer having no iron or steel parts. After 15 seconds a blue stain is evidence or surface iron. Remove solution from the surface as soon as possible after testing, per ASTM A 380 or A 967.	Identification of free iron contamination on surface. Very sensitive test.	Not quantitative. This is a very sensitive test and must be performed by personnel familiar with its limitations. Either a sacrificial coupon is used for this test or the test area is cleaned as described in the respective ASTM practice and/or specification. Safety and disposal issues exist with the test chemical. Easy to get a false-positive result.
Copper Sulfate Test ASTM A 380/ASTM A 967 Practice D (PT)	Bench test. Prepare a 250-cm^3 solution consisting of 1 cm^3 of sulfuric acid (s.g.1.84), 4 g copper sulfate, and the balance in DI water or better. Apply this to a sacrificial coupon using a swab. Keep the surface to be tested wet for a period of 6 minutes with additional applications as needed.	Identification of free iron contamination on the test surface. Is effective in detecting smeared iron deposits.	Not quantitative. Imbedded Iron is detected, but difficult to detect small discrete iron particles.

Figure 2.8.1 (*Continued*)

The components are first cleaned with a degreaser or by pickling to remove any surface residue as a result of machining or fabrication. The pickling process alone will cause a passive layer to form precluding the need to passivate afterward. And it should also be noted that applying a pickling solution to an electropolished surface will very likely etch that surface. The cleaning is then followed by a deionized (DI), or better, water rinse(s). After the components are cleaned and rinsed, the surface to be passivated is then thoroughly exposed to a solution consisting of either nitric, phosphoric, or citric acid. Contact of any of these acids with a metal surface triggers the onset propagation of, in the case of austenitic stainless steel such as 316L, the chromium oxide passive layer. This too is followed by a thorough DI water rinse.

3. Electrochemical Field and Bench Tests			
Cyclic Polarization measurements	This technique uses cyclic polarization measurements similar to the ASTM G61 test method to measure the critical pitting potential (CPP). The more noble (more positive) the CPP, more passive the stainless steel surface. Similar results may be obtained with the ASTM G150 test which measures critical pitting temperature (CPT).	This test method provides a direct measurement of the corrosion resistance of a stainless steel surface. The measured CPP provides a quantitative measurement of the level of passivation. The test equipment is relatively inexpensive.	The method requires a potentiostat and corrosion software package to make the measurements . To ensure reliable results, operators should be trained in electrochemical test techniques.
Electrochemical Pen (ec-pen) (PT)	The result is based on pre-set values. Being the size and shape of a writing instrument, the ec-pen makes electrolytic contact when placed on the test surface. Capillary action causes electrolyte to flow from the reservoir to the surface through a porous polymer body while preventing the electrolyte from leaking out of the pen. There is a stable electrode inside the pen mechanism. By simply positioning the ec-pen on the sample surface electrolytic contact is established and electrochemical characterization is possible. The measured area is typically 1.5 mm^2.	Easy to handle. Short sample preparation time, real-time results, and the possibility to run experiments on virtually any size object with various surface geometries. The ec-pen is a portable instrument for the measurement of corrosion potential suitable for field use.	This test does not quantify the passive layer, but instead provides a pass-fail indication of passivity. The local test area needs to be cleaned and re-passivated after testing.
Koslow Test Kit 2026 (PT)	Similar to the ec-pen, in that it measures the corrosion potential of the metal surface, the Koslow 2026 consists of a meter, a probe and an inter-connecting cable. An electrical charge is first applied to the test piece after which a moist pad is placed on the surface of the same test piece. The probe is pressed into the moist pad to complete the circuit. Within a couple of seconds the cell voltage result appears on the digital meter.	Measures corrosion potential at the surface.	User sensitive

Figure 2.8.1 *(Continued)*

Because the chemical properties of the material within the HAZ of welded joints may become altered due to the heat generated by the welding process, it is recommended that the passivation process be performed in situ on all installed welded piping systems. However, if the oxidation indicated by the heat tint is too thick (a darker blue hue of the heat tint), the chemicals used for passivation may not be strong enough to remove oxidation formed in the HAZ of welds, thereby prohibiting passivation from taking place at such locations. If there is a need to remove heat tint, a determination made by inspection, it is highly recommended that this be done prior to final passivation of the system.

4. Surface Chemical Analysis Tests			
Auger Electron Spectroscopy (AES) (PT, RT)	Secondary and Auger electrons, in the targeted area of the test coupon, are bombarded with a primary electron beam, which is used as en excitation source. Photoelectrons are subsequently ejected from the outer orbital of atoms in the target material. The ejected photoelectrons are then detected by means of electron spectroscopy. The method by which the ejected photo-electrons are detected and analyzed is AES. This test is useful for surface analysis from 2 angstroms to a depth greater than 100 Angstroms.	Provides quantitative analysis. Using e scanning primary beam, secondary electron images yield information related to surface topography. Auger electrons, when analyzed as a function of energy, are used to identify the elements present. Elemental composition of the surface to a depth of 2–20 Angstroms is determined and can be used in depth profiling applications.	The specimen chamber must be maintained at Ultra High Vacuum (UHV). The specimen must be electrically conductive. Instrument is not readily available. Expertise is needed for data interpretation.
Electron Spectroscopy for Chemical Analysis (ESCA) aka, X-ray Photoelectron Spectroscopy (XPS) (PT, RT)	Using X-ray as an excitation source, photoelectrons are ejected from the inner-shell orbital of an atom from the target material. The ejected photoelectrons are then detected by means of XPS. The method by which the ejected photoelectrons are then detected and analyzed is ESCA (or XPS). Useful for surface analysis to a depth of 10–100 Angstroms.	Provides quantitative analysis in measuring the following: 1 . Elemental composition of the surface (10–100 Angstroms usually) 2. Empirical formula of pure materials 3. Elements that contaminate a surface 4. Chemical or electronic state of each element in the surface 5. Uniformity of elemental composition across the top of the surface (aka. line profiling or mapping) 6. Uniformity or elemental composition as a function of ion beam etching (aka, depth profiling)	The specimen chamber must be maintained at Ultra High Vacuum (UHV). Instrument is not readily available. Expertise is needed for data interpretation.
GO-OES (Glow-Discharge Optical Emission Spectroscopy) (PT, RT)	GO-OES uniformly sputters material from the sample surface by applying a controlled voltage, current and argon pressure. Photomultiplier tube detectors are used to identify the specific concentrations of various elements based on the wavelength and intensity of the light emitted by the excited electrons in each element when they return to the ground state.	The GO-OES method is particularly useful for rapid, quantitative, depth profiling of thick and thin-film structures and coatings.	Relatively expensive. Instrument not widely available.

Figure 2.8.1 (*Continued*)

There are a number of tests that can be performed on surfaces that have been passivated in order to determine the quality of the passivation. The BPE, within Appendix E of that standard, not only describes passivation procedures but also provides a listing of the various tests applicable to validating a passivated surface. Figure 2.8.1a–d represents those various tests along with the pros and cons of each.

3

Process Components

3.1 Process Components

The term "process components" is a very general term that, by definition, covers "...piping [tubing], fittings, gaskets, vessels, valves, pumps, filter housings, and instruments." Throughout the world there are myriad "sanitary" mechanical joint-type fittings used in the food, dairy, and beverage industries as well as those used in the pharmaceutical and bioprocessing industry. However, the BPE compliant clamp joint assembly is the only mechanical type joint assembly that the BPE Standard recognizes. Chapter 4 discusses the hygienic clamp joint assembly in great detail.

In this chapter we will touch on not only the hygienic clamp joint fittings but also welded fittings as well as valves, seals/gaskets, and instruments.

3.2 Pressure Ratings

3.2.1 Pressure Ratings of Welded Components

Before launching into discussions on the various components, we will touch on the issue of component pressure ratings. When a code or standard refers to pressure ratings, it is making reference to mechanical-type joints, not a butt welded joint. A statement such as, "The allowable pressure ratings for fittings designed in accordance with this Standard may be calculated as for straight seamless pipe of equivalent material," as found in B16.9 factory-made wrought butt welding fittings, can be somewhat confusing in its implication that a pressure rating exists; they do not exist, at least not in standardized form. What can be

Bioprocessing Piping and Equipment Design: A Companion Guide for the ASME BPE Standard,
First Edition. William M. (Bill) Huitt.

determined, by calculation, is the maximum allowable design pressure P_{MAX} of welded pipe and fittings. The results of which depend on the type of material, how the pipe or fittings were manufactured, its wall thickness, and what type of weld is made in attaching the fitting to the pipe.

A manufactured weld joint, such as a longitudinal weld joint performed in the manufacture of welded pipe or tubing, is assigned a safety value as a factor in calculating the strength or the maximum allowable design pressure of that pipe and any welded fitting of the same material welded to it.

B31.3, table 302.3.4, shown here in Figure 3.2.1, provides a joint quality factor, E_j, for the various manufactured longitudinal welds used in the manufacture of welded pipe. B31.3 table A-1B, shown here, in part, as Figure 3.2.2, lists the joint quality factor, E_j, per material specification. As an example, referring to Figure 3.2.1, a straight furnace butt weld has a joint quality factor, E_j, of 0.06. Whereas a single electric fusion butt weld that is 100% radiographed has a joint quality factor, E_j, of 1.00. Referring now to the outlined ASTM A269 material specification listed in Figure 3.2.2, it can be seen that there are three different tube manufacturing choices for determining the joint quality factor, E_j, for the A269 material.

Welds performed in attaching a fitting to pipe or another fitting have a similar safety factor that is applied in calculating the strength of the welded pipe and fitting in combination with the joint quality factor, E_j. These weld joint strength reduction factors can be found in B31.3, table 302.3.5, shown here in Figure 3.2.3. In referring to Figure 3.2.3, the outlined area identifies austenitic stainless steel welded autogenously and with filler material. As an example an autogenous weld made in joining a section of tubing and a fitting, both of A269 Gr 316L stainless steel, has a weld joint strength reduction factor, W of 1, as shown in the table. Meaning it retains the full maximum allowable design pressure of the pipe.

Before going further, it must be noted that under ASME B31.3, paragraph 322.6.3, the distinction is made between the two terms *maximum allowable pressure* and *vessel* as they are used in Section VIII of the BPVC and their respective counterpart terms, *design pressure* and *piping system*, as used for piping systems in B31.3. This is mentioned so that in reading maximum allowable design pressure, it is understood that it is related to piping systems and not pressure vessels as used in this context.

The two factors, E_j and W, apply in the following manner in determining the maximum allowable design pressure for pipe and weld fittings. In referencing B31.3, paragraph V304, an equation is shown here as Equation 3.2.1 for determining the maximum allowable design pressure, P_{MAX}, for straight pipe under internal pressure:

$$P_{MAX} = \frac{2\left(\bar{T} - c - \text{mill tol.}\right)SEW}{D - 2\left(\bar{T} - c - \text{mill tol.}\right)Y} \tag{3.2.1}$$

where:

c = sum of mechanical allowances (thread or groove depth) plus corrosion and erosion allowances

D = nominal pipe outside diameter

E = quality factor from B31.3 table A-1B

P = internal design gage pressure
P_{MAX} = maximum allowable design pressure (*referred to in B31.3 as maximum allowable gage pressure*)
S = stress value for material from B31.3 table A-1
$\bar{T}$ = nominal wall thickness of pipe
t = pressure design thickness, as calculated in accordance with B31.3 paragraph 304.1.2 for internal pressure or as determined in accordance with paragraph 304.1.3 external pressure
W = weld joint strength reduction factor in accordance with B31.3 paragraph
Y = coefficient from B31.3 table 304.1.1, valid for $t < D/6$ and for materials shown.

Table 302.3.4 Longitudinal weld joint quality factor, E_j

No.	Type of Joint		Type of Seam	Examination	Factor, E_j
1	Furnace butt weld, continuous weld		Straight	As required by listed specification	0.60 [Note (1)]
2	Electric resistance weld		Straight or spiral (helical seam)	As required by listed specification	0.85 [Note (1)]
3	Electric fusion weld				
	(a) Single butt weld		Straight or spiral (helical seam)	As required by listed specification or this code	0.80
	(with or without filter metal)			Additionally spot radiographed in accordance with para. 341.5.1	0.90
				Additionally 100% radiographed in accordance with para. 344.5.1 and Table 341.3.2	1.00
	(b) Double butt weld		Straight or spiral (helical seam) [except as provided in 4 below]	As required by listed specification or this code	0.85
	(with or without filter metal)			Additionally spot radiographed in accordance with para. 341.5.1	0.90
				Additionally 100% radiographed in accordance with para. 344.5.1 and Table 341.3.2	1.00
4	Specific specification				
	API 5L	Submerged arc weld (SAW) Gas metal arc weld (GMAW) Combined GMAW, SAW	Straight with one or two seams Spiral (helical seam)	As required by specification	0.95
				Additionally 100% radiographed in accordance with para. 344.5.1 and Table 341.3.2	1.00

NOTE:
(1) It is not permitted to increase the joint quality factor by additional examination for joint 1 or 2.

Figure 3.2.1 Longitudinal weld joint strength reduction or quality factor by weld type. *Source*: https://www.asme.org/. © The American Society of Mechanical Engineers

Table A-1B Basic Quality Factors for Longitudinal Weld Joints in Pipes, Tubes, and Fittings, E_j (Cont'd)
These quality factors are determined in accordance with para. 302.3.4(a). See also para. 302.3.4(b) and Table 302.3.4 for increased quality factors applicable in special cases. Specifications, except API, are ASTM.

Spec. No.	Class (or Type)	Description	E_j [Note (2)]	Appendix A Notes
Stainless Steel				
A182	...	Forgings and fittings	1.00	...
A249	...	Electric fusion welded tube, single butt seam	0.80	...
A268	...	Seamless tube	1.00	...
		Electric fusion welded tube, double butt seam	0.85	...
		Electric fusion welded tube, single butt seam	0.80	...
A269	...	Seamless tube	1.00	...
		Electric fusion welded tube, double butt seam	0.85	...
		Electric fusion welded tube, single butt seam	0.80	...
A312	...	Seamless tube	1.00	...
		Electric fusion welded tube, double butt seam	0.85	...
		Electric fusion welded tube, single butt seam	0.80	...
		Electric fusion welded, 100% radiographed	1.00	(46)
A358	1, 3, 4	Electric fusion welded pipe, 100% radiographed	1.00	...
	5	Electric fusion welded pipe, spot radiographed	0.90	...
	2	Electric fusion welded pipe, double butt seam	0.85	...
A376	...	Seamless pipe	1.00	...
A403	...	Seamless fittings	1.00	...
		Welded fitting, 100% radiographed	1.00	(16)
		Welded fitting, double butt seam	0.85	...
		Welded fitting, single butt seam	0.80	...
A409	...	Electric fusion welded pipe, double butt seam	0.85	...
		Electric fusion welded pipe, single butt seam	0.80	...
A487	...	Steel castings	0.80	(9)(40)

Figure 3.2.2 Longitudinal weld joint strength reduction or quality factor by material. *Source*: https://www.asme.org/products/codes-standards/bpe-2014-bioprocessing-equipment.

Example

Using ASTM A269 Gr 316L seamless stainless tubing, 2″ OD with a 0.065″ WT as an example, the key elements of this tubing would be:

$c = 0.008$
$D = 2''$
$E = 1$ (from B31.3 table A-1B, Figure 3.2.2)
$S = 16\,700$ (*at ambient conditions*)
$\bar{T} = 0.065''$
$W = 1$ (from B31.3 table 302.3.5, Figure 3.2.3)
$Y = 0.04$

Equation 3.2.1 is as follows:

$$P_{\text{MAX}} = \frac{2\left(\bar{T} - c - \text{mill tol.}\right)SEW}{D - 2\left(\bar{T} - c - \text{mill tol.}\right)Y} \tag{3.2.1}$$

Replace the equation nomenclature with the real values:

$$P_{\text{MAX}} = \frac{2(0.065 - 0.008)16\ 700 \times 1 \times 1}{2.0 - 2(0.065 - 0.008)0.4}$$

Table 302.3.5 Weld Joint Strength Reduction Factor, W

Steel Group	Component Temperature, T_i, °C (°F)														
	427 (800)	454 (850)	482 (900)	510 (950)	538 (1,000)	566 (1,050)	593 (1,100)	621 (1,150)	649 (1,200)	677 (1,250)	704 (1,300)	732 (1,350)	760 (1,400)	788 (1,450)	816 (1,500)
CrMo [Notes (1)–(3)]	1	0.95	0.91	0.86	0.82	0.77	0.73	0.68	0.64	...	...	...	...	...	...
CSEF (N + T) [Notes (3)–(5)]	...	...	...	1	0.95	0.91	0.86	0.82	0.72	...	...	...	...	...	...
CSEF [Notes (3) and (4)] (Subcritical PWHT)	...	...	1	0.5	0.5	0.5	0.5	0.5	0.5	...	...	...	...	...	...
Autogenous welds in austenitic stainless grade 3xx and N088xx and N066xx nickel alloys [Note (6)]	...	...	...	1	1	1	1	1	1	1	1	1	1	1	1
Austenitic stainless grade 3xx and N088xx nickel alloys [Notes (7) and (8)]	...	...	...	1	0.95	0.91	0.86	0.82	0.77	0.73	0.68	0.64	0.59	0.55	0.5
Other materials [Note (9)]	...	...	...	...	...	...	...	...	...	...	...	...	...	...	...

GENERAL NOTES:

(a) Weld joint strength reduction factors at temperatures above the upper temperature limit listed in Appendix A for the base metal or outside of the applicable range in Table 302.3.5 are the responsibility of the designer. At temperatures below those where weld joint strength reduction factors are tabulated, a value of 1.0 shall be used for the factor W where required; however, the additional rules of this Table and Notes do not apply.

(b) T_{cr} = temperature 25°C (50°F) below the temperature identifying the start of time-dependent properties listed under "NOTES – TIME-DEPENDENT PROPERTIES" (Txx) in the Notes to Table 1A and 1B of the BPV Code Section II, Part D for the base metals joined by welding. For materials not listed in the BPV Code Section II, Part D, T_a shall be the temperature where the creep rate or stress rupture criteria in paras. 302.3.2(d)(4), (5), and (6) governs the basic allowable stress value of the metals joined by welding. When the base metals differ, the lower value of T_{cr} shall be used for the weld joint.

(c) T_i = temperature, °C (°F), of the component for the coincident operating pressure–temperature condition, i, under consideration.

(d) CAUTIONARY NOTE: There are many factors that may affect the life of a welded joint at elevated temperature and all of those factors cannot be addressed in a table of weld strength reduction factors. For example, fabrication issues such as the deviation from a true circular form in pipe (e.g., "peaking" at longitudinal weld seams) or offset at the weld joint can cause an increase in stress that may result in reduced service life and control of these deviations is recommended.

(e) The weld joint strength reduction factor, W, may be determined using linear interpolation for intermediate temperature values.

NOTES:

(1) The Cr–Mo Steels include: ½Cr–½Mo, lCr–½Mo, 1¼Cr–½Mo– Si, 2¼Cr–1Mo, 3Cr– lMo, 5Cr–½Mo, 9Cr–1Mo. Longitudinal and spiral (helical seam) welds shall be normalized, normalized and tempered, or subjected to proper subcritical postweld heat treatment (PWHT) for the alloy. Required examination is in accordance with para. 341.4.4 or 305.2.4.

(2) Longitudinal and spiral (helical seam) seam fusion welded construction is not permitted for C–½Mo steel above 850°F.

(3) The required carbon content of the weld filler metal shall be ≥0.05 C wt. %. See para. 341.4.4(b) for examination requirements. Basicity index of SAW flux ≥1.0.

(4) The CSEF (Creep Strength Enhanced Ferritic) steels include grades 91, 92, 911, 122, and 23.

(5) N + T = Normalizing + Tempering PWHT.

(6) Autogenous welds without filler metal in austenitic stainless steel (grade 3xx) and austenitic nickel alloys UNS Nos. N066xx and NO88xx. A solution anneal after welding is required for use of the factors in the Table. See para. 341.4.3(b) for examination requirements.

(7) Alternatively, the 100,000 hr Stress Rupture Factors listed in ASME Section III, Division l, Subsection NH, Tables 1–14.10 A-xx, B-xx, and C-xx may be used as the weld joint strength reduction factor for the materials and welding consumables specified.

(8) Certain heats of the austenitic stainless steels, particularly for those grades whose creep strength is enhanced by the precipitation of temper-resistant carbides and carbonitrides, can suffer from an embrittlement condition in the weld heat affected zone that can lead to premature failure of welded components operating at elevated temperatures. A solution annealing heat treatment of the weld area mitigates this susceptibility.

(9) For carbon steel, W = 1.0 for all temperatures. For materials other than carbon steel, CrMo, CSEF, and the austenitic alloys listed in Table 302.3.5, W shall be as follows: For $T_i \leq T_{cr}$, $W = 1.0$. For $T_{cr} < T_i \leq 1.500$°F, $W = 1 - 0.000909(T_i - T_{cr})$. If T_i exceeds the upper temperature for which an allowable stress value is listed in Appendix A for the base metal, the value for W is the responsibility of the designer.

Figure 3.2.3 Weld joint strength reduction factor. *Source*: https://www.asme.org/products/codes-standards/bpe-2014-bioprocessing-equipment. © The American Society of Mechanical Engineers

Run the calculation:

$$P_{MAX} = \frac{1903.8}{1.9544} = 974.1 \text{ say } 974 \text{ psig}$$

The maximum allowable internal design pressure, P_{MAX}, under ambient conditions for two A269 Gr 316L seamless tubing components autogenously welded together is therefore 974 psig.

Should the design temperature increase the longitudinally or circumferentially welded piping material to the point of reducing the allowable stress value (S_a), as found in B31.3, table A-1, it will alter the maximum allowable design pressure, P_{MAX}, of the pipe, tubing, or fitting, downgrading it to a lesser value.

As stated throughout the BPE Standard, the butt weld is the preferred joint and more specifically the autogenous orbital welded joint. This method of assembling a high-purity piping system, limiting the number of mechanical joints, promotes the use of drainability, reduces the possibility of liquid holdup, and provides a high-integrity piping system. Manual welds, *welding in which the entire welding operation is performed and controlled by hand*, shall be approved by the owner and identified in the weld log and on the weld map.

3.2.2 Pressure Ratings and Other Fundamentals of Hygienic Clamp Joint Unions

While the butt welded system is preferred, the BPE hygienic clamp joint union or assembly is still a necessity when a break in the pipeline is needed. Such joints or connections may be needed for connecting to equipment, where in-line components require periodic removal, where a section of tubing requires periodic removal for cleaning or possible removal access for equipment.

The hygienic clamp joint union is discussed in this book in detail. That discussion can be found in Chapter 4, under Section 4.5. What we will touch on in this chapter are some of the primary essentials of the clamp joint assembly itself as well as installation fundamentals.

3.2.2.1 Clamp Joint Assembly Components

The entire hygienic clamp joint assembly is composed of four separate and independent although interrelated components: two ferrules, a ferrule gasket, and a clamp. These individual components are assembled by a secondary influence, the installer. Proper protocols should be in place and control exercised in the material selection and application of each element in order to attain a joint that has acceptable and proven leak tightness.

Listed metallic materials acceptable for use in the hygienic clamp joint assembly can be found in MM-4.2, MM-4.3, MM-4.4, MM-4.5, and MM-4.6. For unlisted or unknown materials refer to MM-3.3 and MM-3.4, respectively.

3.2.2.2 User Accountability

A key factor governing the integrity of a hygienic clamp joint assembly is in the selection of its material and its installation. The BPE Standard cites duties and responsibilities that are to be assumed by those that select the material of construction for the assembly components and for the proper installation of those components. The integrity of any mechanical type joint, like the clamp joint assembly, depends greatly on the proper material being selected and the quality of the workmanship of the installation.

3.2.2.3 Material Selection

It is the responsibility of the user to apply sound engineering judgment and experience in selecting the proper hygienic clamp joint assembly material for each of its components with regard to the fluid service and conditions they are intended for. It is not within the scope of the BPE Standard to provide guidelines or requirements with regard to how that process should be accomplished.

3.2.2.4 Clamp Size and Type

The BPE Standard covers a clamp assembly size range of 1/4″ through 6″. The clamp types are intentionally generic by design, allowing each manufacturer to put their own conceptual twist on clamp design so long as they adhere to the ferrule fit-up requirements of DT-9.3-1 with the added requirement to withstand a minimum rated leak test pressure of 1.5× a system design pressure.

While there are no specific dimensions given in the BPE Standard for hygienic clamps, there are a set of minimum requirements with which manufacturers are required to comply with under DT-9.4. These guidelines coupled with the specific manufacturing dimensional requirements of hygienic ferrules, as referenced in tables DT-7-1 Type A and B, are the basis by which hygienic clamps are to be designed and manufactured. In addition to the dimensional requirements stated previously, the clamp is also required to meet or exceed the respective pressure/temperature ratings in accordance with BPE paragraph DT-2 and table DT-2.1.

3.2.2.5 Clamp Joint Assembly Gasket Material

Selection of a proper gasket material for the hygienic clamp joint assembly should be based on good engineering practice and experience. It is the user's responsibility to determine the limitations of a gasket material, not only with respect to its compatibility with the fluid but also with consideration given to the design pressure/temperature conditions of the fluid system.

3.2.2.6 Clamp Joint Assembly Temperature Considerations

The use of hygienic clamp joints at either high or low temperatures shall take into consideration the risk of joint leakage due to thermal-related forces and moments developed in the connected piping or equipment. Such considerations shall take into account any secondary line service conditions such as SIP or CIP procedures in which temperatures may be in excess of normal operating conditions.

The designer is cautioned that the hygienic clamp joint assembly is not engineered to withstand excessive bending or torsional loads. For piping systems that will undergo thermal expansion and/or contraction, added consideration must be given in the form of flexibility analysis and the proper placement of anchors and guides in preventing such thermal loads from compromising hygienic clamp joints in the system.

3.2.2.7 Pressure Ratings of Hygienic Clamp Joints

Pressure/temperature ratings for hygienic clamp joints are currently proprietary in that each manufacturer has established pressure rating values for their clamps on an individual basis. And while some of these pressure rating values may be relatively similar from one manufacturer to another, this is accidental. There is no standardized set of pressure rating values for the hygienic clamp joint assembly in the industry. It is therefore necessary, at the present time, to depend on the manufacturer to provide such assurances.

3.2.2.8 Mixed Material Hygienic Clamp Joints

If two connecting ferrules in a hygienic clamp joint are of different materials, the rating of the joint assembly at any temperature, while not stated, shall be rated in accordance with the recommendation of the manufacturer of the lowest rated component.

3.2.2.9 Basis for Pressure Ratings

The pressure used in determining a minimum joint rating is the design pressure. The system design pressure is a predetermined value based on system operating factors that allow for the most extreme coincident of pressure and temperature in conjunction with other incidental loads that subject the piping system to conditions in excess of the normal operating/working pressure of the contained fluid.

The design temperature for a corresponding design pressure is the expected temperature of the pressure-containing shell of the component. In general, this temperature is the same as that of the contained fluid at the design pressure. Use of a design temperature other than that of the contained fluid at the design pressure is the responsibility of the user, subject to the requirements of applicable codes and regulations. For any temperature below −20°F (−29°C), the rating shall be no greater than the rating shown for ambient conditions.

3.2.2.10 Systems

The pressure rating of a piping system is not predicated on the pressure rating of the hygienic clamp joint alone but is instead based, inversely, on a system's design conditions. Once those values of pressure, temperature, and errant or intentional transient dynamic forces have been established, the pressure rating of the weakest component in that system shall exceed any combination of those design elements and shall be able to withstand a hydrostatic test pressure of 1.5 times the systems design pressure. The pressure rating of the weakest component in a system has to therefore exceed those design conditions.

3.2.2.11 Assembled Components

Assembled components, such as valves and in-line instruments, may very well have a component pressure rating different than that of the hygienic clamp joint assembly itself. Should the pressure rating of the assembled component be less than that of the hygienic clamp joint assembly, the pressure rating of the system will be governed by the pressure rating of that lowest rated component.

3.3 Hygienic Clamp and Automatic Tube Weld Fittings

The BPE Standard has 31 fittings segregated into 5 different functional groups, as follows:

1. Elbows and bends
2. Tees and crosses
3. Reducers
4. Ferrules
5. Caps

These fittings are represented and dimensioned according to their function and end connection type. Their functional design types are listed previously, but their end connection type can be:

- Fully automatic weld ends
- Fully hygienic clamp ends
- A combination of automatic weld ends and hygienic clamp ends

When bends, such as those for 90 and 45° elbows, are made, the size of the bend is typically based on or related to the OD of the pipe or tubing. When doing

Table 3.3.1 Fitting bend radius

Size (in.)	A	T	Bend radius (R_b)[a]	R_b/OD[b]
1/4	2.625	1.500	1.125	5.000
3/8	2.625	1.500	1.125	3.000
1/2	3.000	1.500	1.500	3.000
3/4	3.000	1.500	1.500	1.500
1	3.000	1.500	1.500	3.000
1 1/2	3.750	1.500	2.250	1.500
2	4.750	1.500	3.250	1.625
2 1/2	5.500	1.500	4.000	1.600
3	6.250	1.750	4.500	1.500
4	8.000	2.000	6.000	1.500
6	11.500	2.500	9.000	1.500

[a] Bend radius (R_b) = $A - T$.
[b] R_b/OD of the fitting equals the bend radius as it relates to the OD of the fitting, wherein the 1/4″ elbow has a bend radius that is 5× the fitting diameter referred to as a 5*D* bend.

this, the bend radius (R_b) is based on the centerline of the tubing having a bend radius value referred to as *XD* where X = radius factor and D = nominal OD of the tubing or pipe. As an example, in forming a 2″ 90° elbow, it would have, based on BPE dimensions, a 1.625*D* bend. This would translate into a bend radius of 1.625 × 2″ (fitting size) = 3.250″ (radius). In comparison, all long radius 90° elbow bends under ASME B16.9 "Factory Made Wrought Buttweld Fittings" have 1.5*D* bends.

Table 3.3.1 provides a listing of the bend radii (R_b) and the OD multiplier (R_b/OD) for each tubing size based on the dimensional information in the BPE Standard.

Following are the various standardized BPE fittings (Ref. Figures 3.3.1, 3.3.2, 3.3.3, 3.3.4, 3.3.5, 3.3.6, 3.3.7, 3.3.8, 3.3.9, 3.3.10, 3.3.11, 3.3.12, 3.3.13, 3.3.14, 3.3.15, 3.3.16, 3.3.17, 3.3.18, 3.3.19, 3.3.20, 3.3.21, 3.3.22, 3.3.23, 3.3.24, and 3.3.25). Dimensions for these fittings can be found in Part DT of the Standard and are provided in Appendix L of this book for quick reference. In accordance with BPE and the previous bullet points, the following represented BPE fittings are divided into the five bulleted groups as follows:

- Elbows and bends
- Tees and crosses
- Reducers
- Ferrules
- Caps

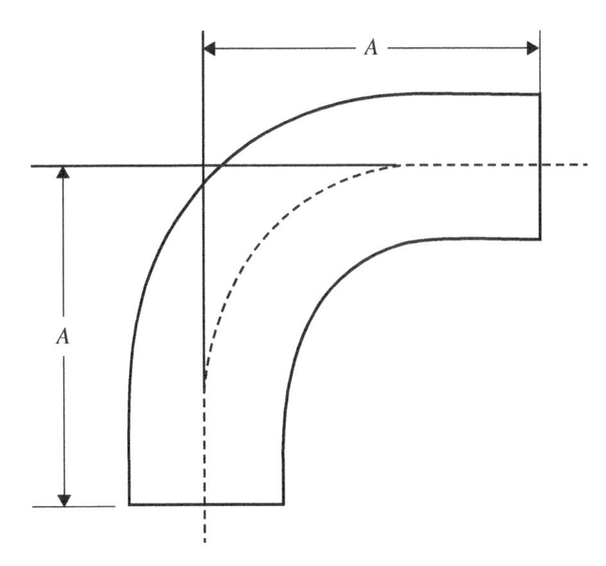

Figure 3.3.1 Automatic tube weld 90° elbow. *Source*: https://www.asme.org/products/codes-standards/bpe-2014-bioprocessing-equipment. © The American Society of Mechanical Engineers

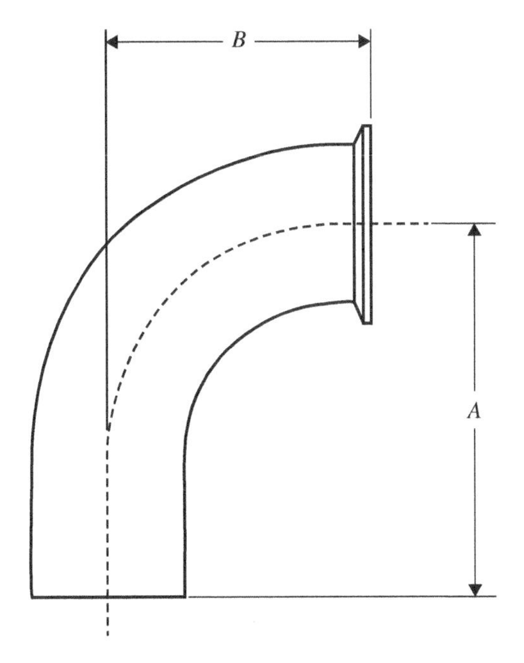

Figure 3.3.2 Automatic tube weld × hygienic clamp joint, 90° elbow. *Source*: https://www.asme.org/products/codes-standards/bpe-2014-bioprocessing-equipment. © The American Society of Mechanical Engineers

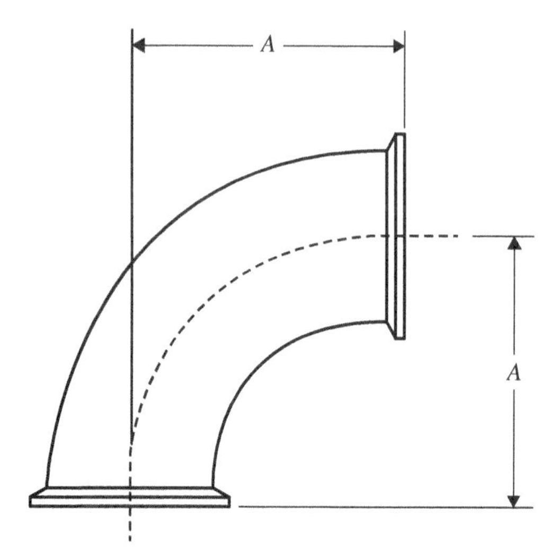

Figure 3.3.3 Hygienic clamp joint 90° elbow. *Source*: https://www.asme.org/products/codes-standards/bpe-2014-bioprocessing-equipment. © The American Society of Mechanical Engineers

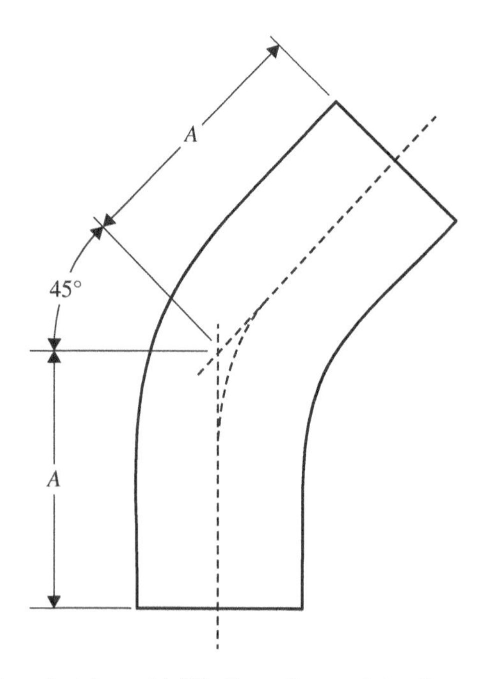

Figure 3.3.4 Automatic tube weld 45° elbow. *Source*: https://www.asme.org/products/codes-standards/bpe-2014-bioprocessing-equipment. © The American Society of Mechanical Engineers

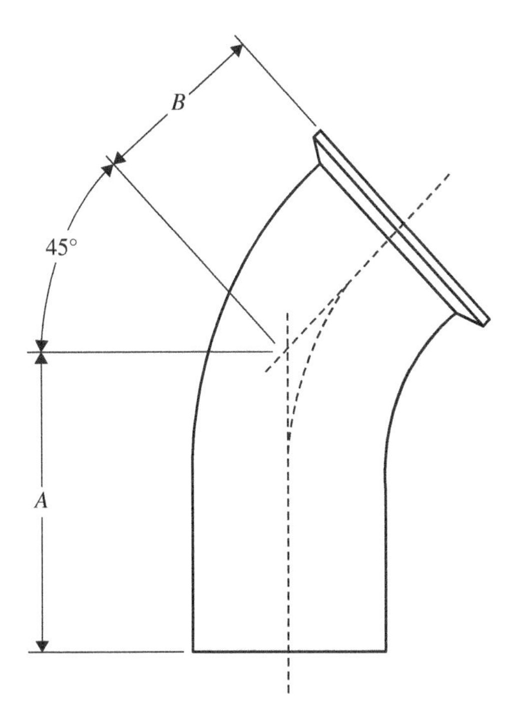

Figure 3.3.5 Automatic tube weld×hygienic clamp joint, 45° elbow. *Source*: https://www.asme.org/products/codes-standards/bpe-2014-bioprocessing-equipment. © The American Society of Mechanical Engineers

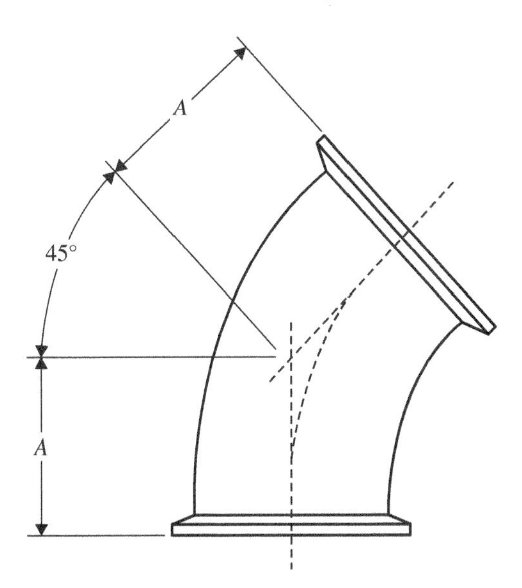

Figure 3.3.6 Hygienic clamp joint 45° elbow. *Source*: https://www.asme.org/products/codes-standards/bpe-2014-bioprocessing-equipment. © The American Society of Mechanical Engineers

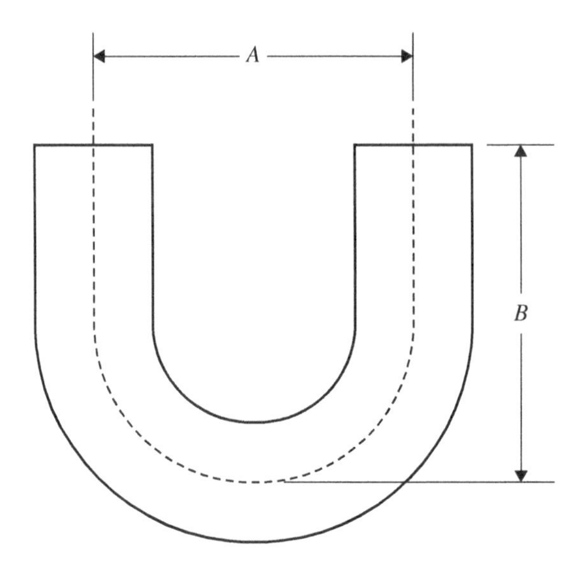

Figure 3.3.7 Automatic tube weld 180° return bend. *Source*: https://www.asme.org/products/codes-standards/bpe-2014-bioprocessing-equipment. © The American Society of Mechanical Engineers

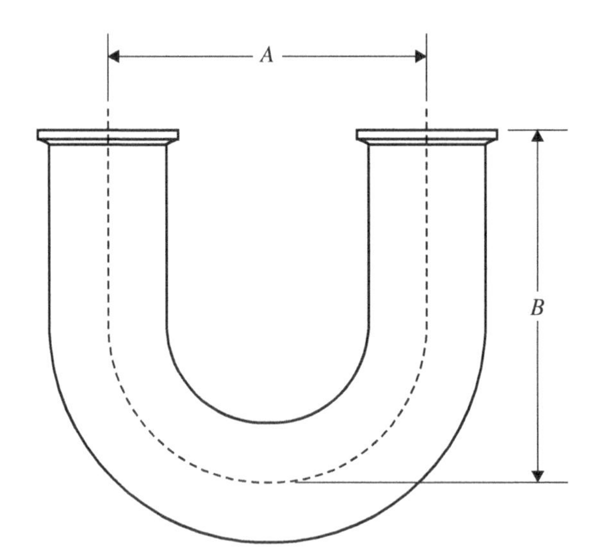

Figure 3.3.8 Hygienic clamp joint 180° return bend. *Source*: https://www.asme.org/products/codes-standards/bpe-2014-bioprocessing-equipment. © The American Society of Mechanical Engineers

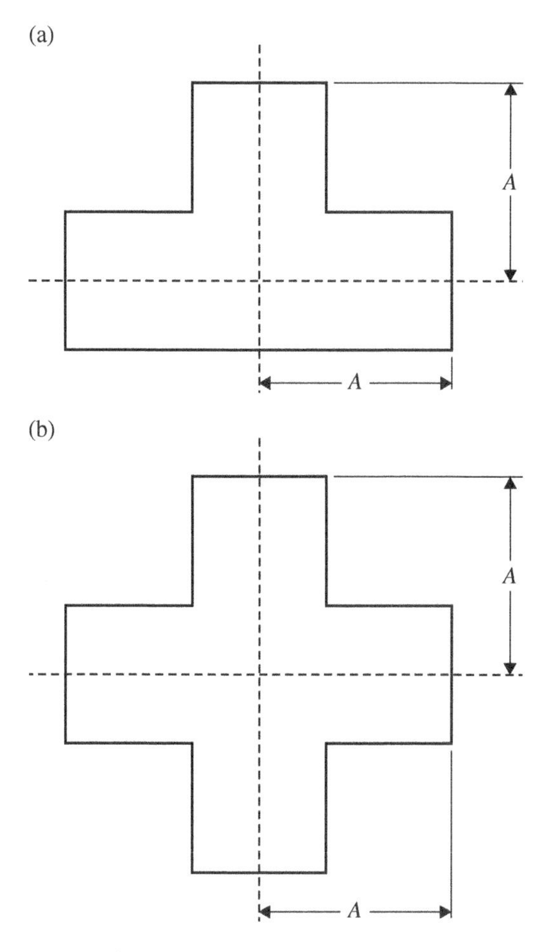

Figure 3.3.9 (a) Automatic tube weld straight tee and (b) automatic tube weld cross. *Source*: https://www.asme.org/products/codes-standards/bpe-2014-bioprocessing-equipment. © The American Society of Mechanical Engineers

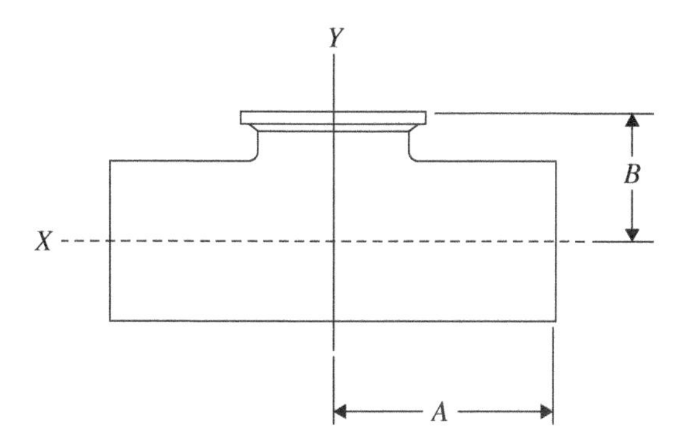

Figure 3.3.10 Automatic tube weld short outlet hygienic clamp joint tee. *Source*: https://www.asme.org/products/codes-standards/bpe-2014-bioprocessing-equipment. © The American Society of Mechanical Engineers

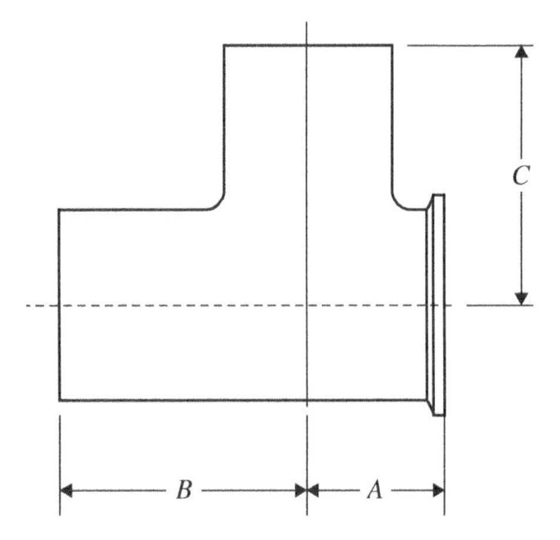

Figure 3.3.11 Hygienic clamp joint short outlet run tee. *Source*: https://www.asme.org/products/codes-standards/bpe-2014-bioprocessing-equipment. © The American Society of Mechanical Engineers

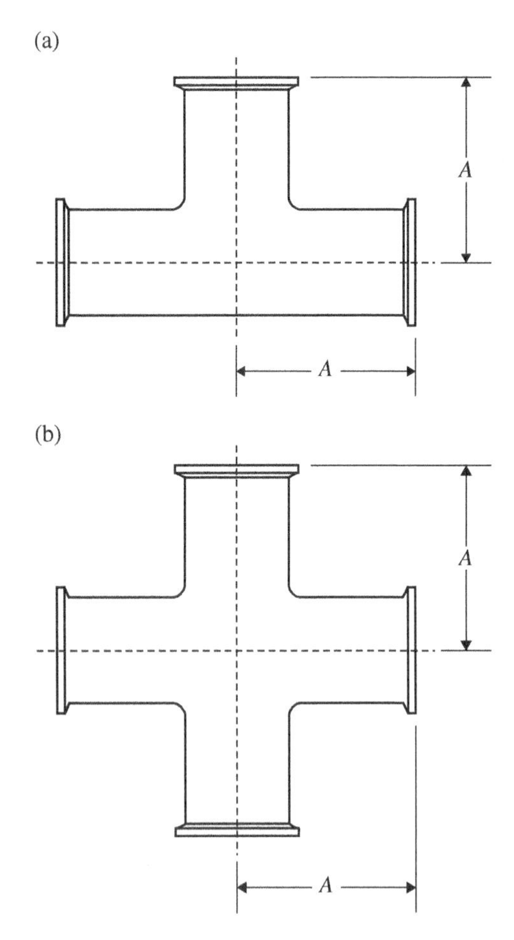

Figure 3.3.12 (a) Hygienic clamp joint straight tee and (b) hygienic clamp joint cross. *Source*: https://www.asme.org/products/codes-standards/bpe-2014-bioprocessing-equipment. © The American Society of Mechanical Engineers

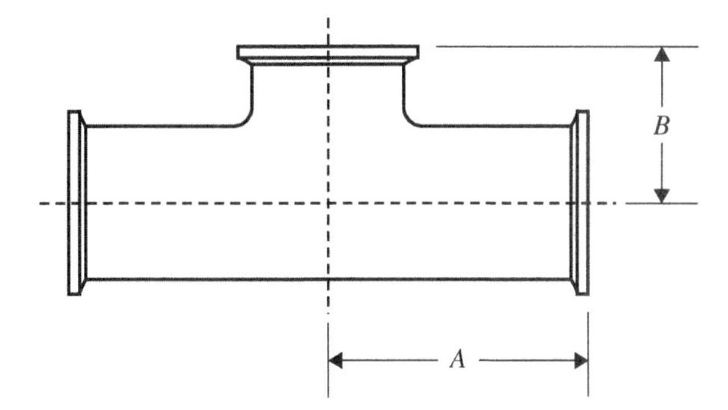

Figure 3.3.13 Hygienic clamp joint short outlet tee. *Source*: https://www.asme.org/products/codes-standards/bpe-2014-bioprocessing-equipment. © The American Society of Mechanical Engineers

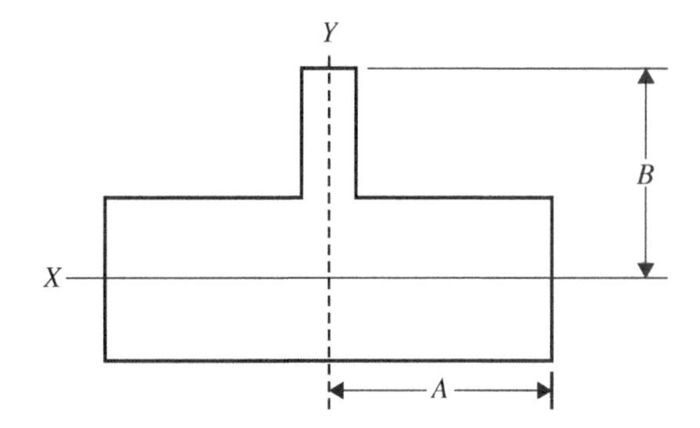

Figure 3.3.14 Automatic tube weld reducing tee. *Source*: https://www.asme.org/products/codes-standards/bpe-2014-bioprocessing-equipment. © The American Society of Mechanical Engineers

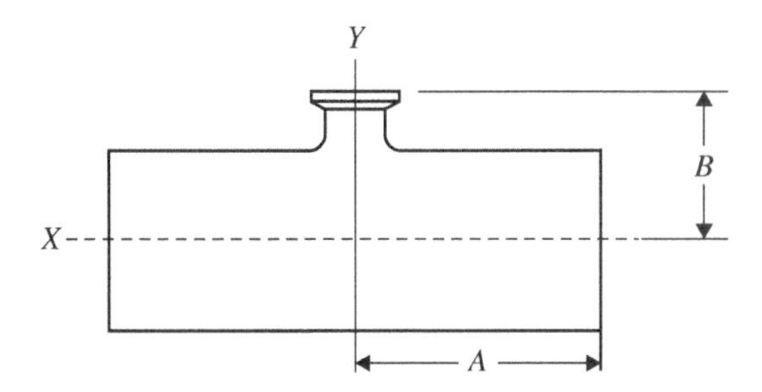

Figure 3.3.15 Automatic tube weld × short outlet hygienic clamp joint reducing tee. *Source*: https://www.asme.org/products/codes-standards/bpe-2014-bioprocessing-equipment. © The American Society of Mechanical Engineers

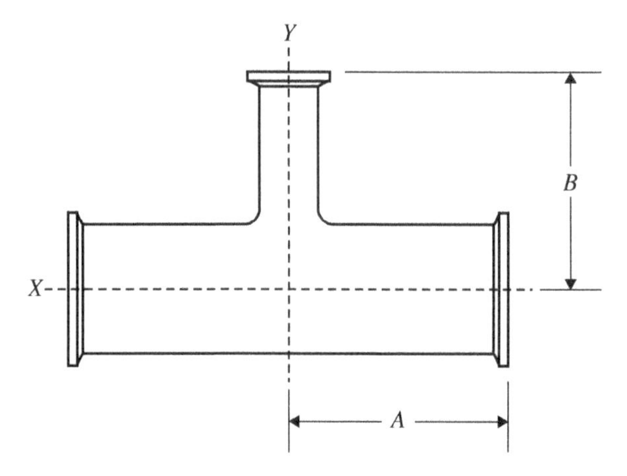

Figure 3.3.16 Hygienic clamp joint reducing tee. *Source*: https://www.asme.org/products/codes-standards/bpe-2014-bioprocessing-equipment. © The American Society of Mechanical Engineers

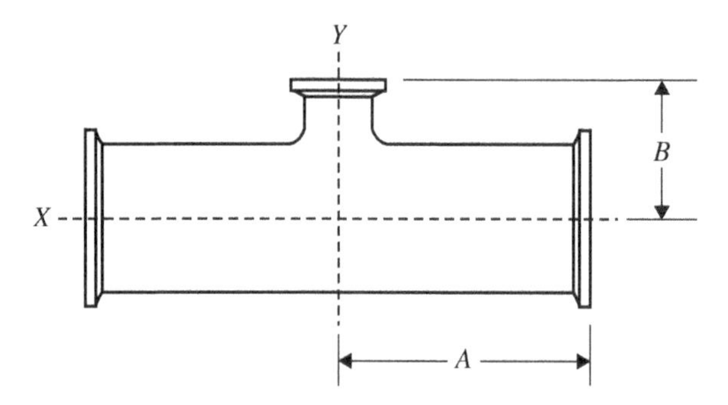

Figure 3.3.17 Hygienic clamp joint short outlet reducing tee. *Source*: https://www.asme.org/products/codes-standards/bpe-2014-bioprocessing-equipment. © The American Society of Mechanical Engineers

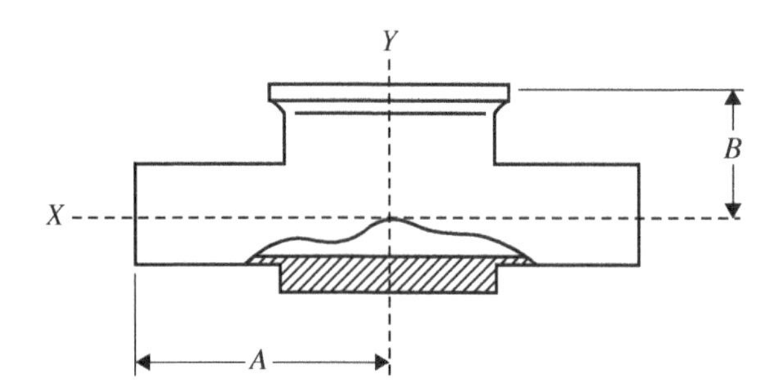

Figure 3.3.18 Automatic tube weld instrument tee. *Source*: https://www.asme.org/products/codes-standards/bpe-2014-bioprocessing-equipment. © The American Society of Mechanical Engineers

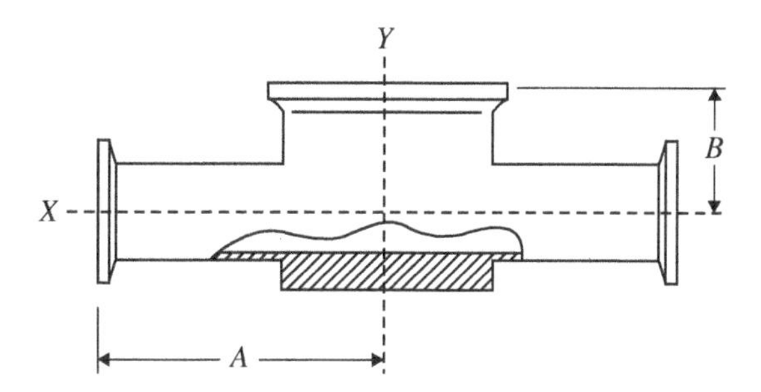

Figure 3.3.19 Hygienic clamp joint instrument tee. *Source*: https://www.asme.org/products/codes-standards/bpe-2014-bioprocessing-equipment. © The American Society of Mechanical Engineers

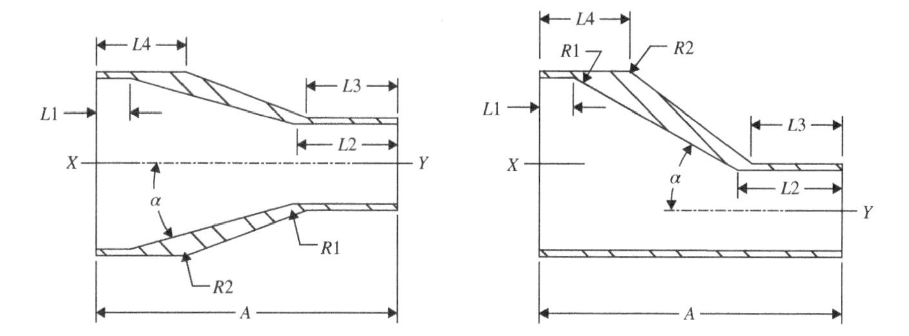

Figure 3.3.20 Automatic tube weld concentric and eccentric. *Source*: https://www.asme.org/products/codes-standards/bpe-2014-bioprocessing-equipment. © The American Society of Mechanical Engineers

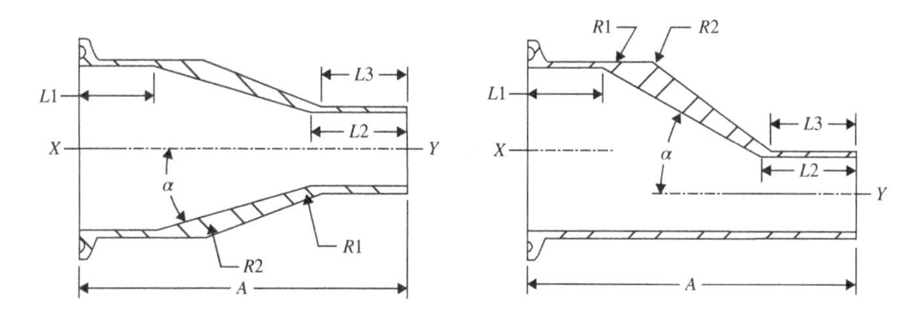

Figure 3.3.21 Hygienic clamp joint × tube weld concentric and eccentric reducers. *Source*: https://www.asme.org/products/codes-standards/bpe-2014-bioprocessing-equipment. © The American Society of Mechanical Engineers

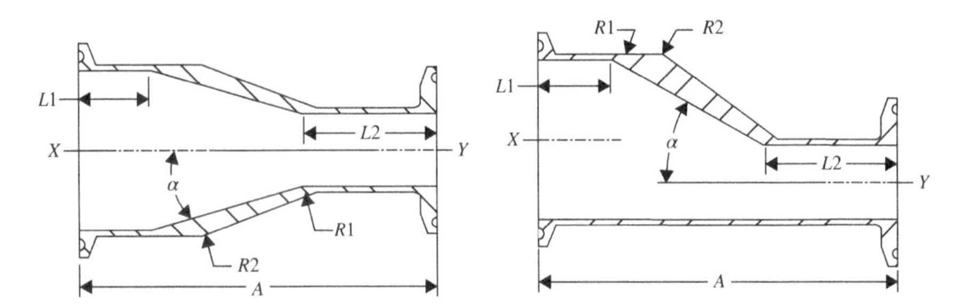

Figure 3.3.22 Hygienic clamp joint concentric and eccentric reducers. *Source*: https://www.asme.org/products/codes-standards/bpe-2014-bioprocessing-equipment. © The American Society of Mechanical Engineers

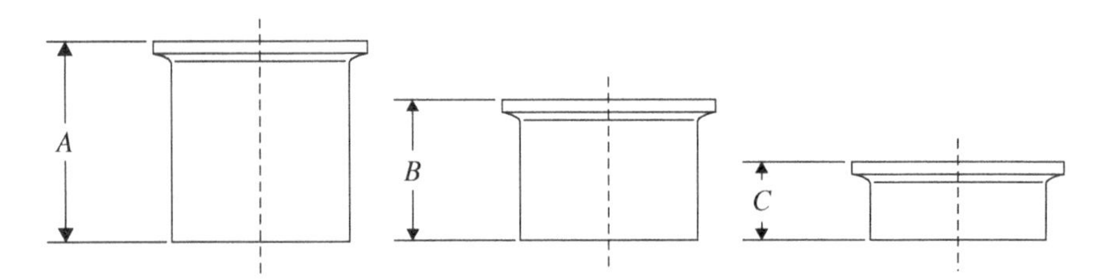

Figure 3.3.23 Automatic tube weld ferrule three standardized lengths. *Source*: https://www.asme.org/products/codes-standards/bpe-2014-bioprocessing-equipment. © The American Society of Mechanical Engineers

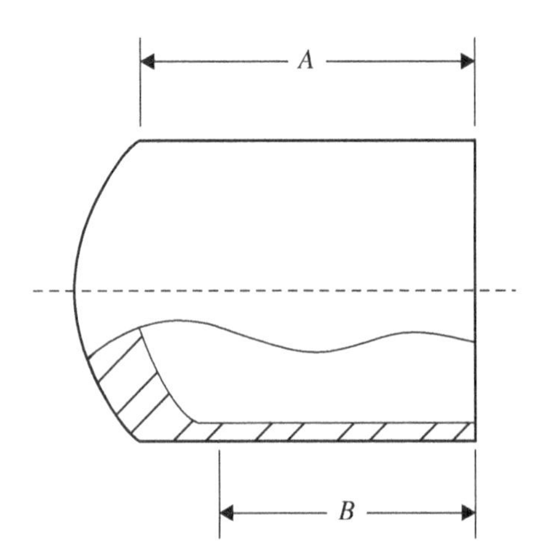

Figure 3.3.24 Automatic tube weld cap. *Source*: https://www.asme.org/products/codes-standards/bpe-2014-bioprocessing-equipment. © The American Society of Mechanical Engineers

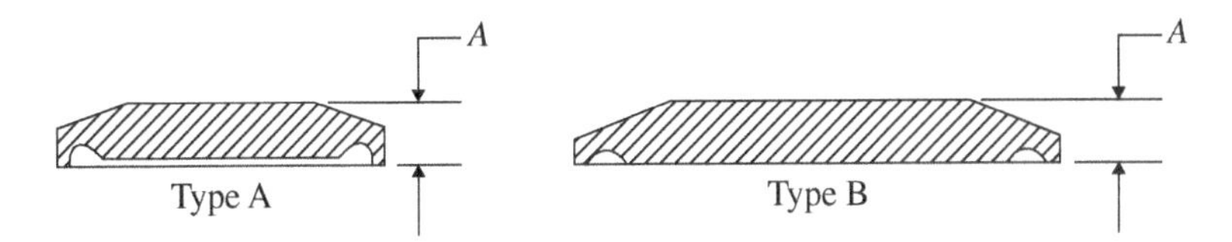

Figure 3.3.25 Hygienic clamp joint solid end caps. *Source*: https://www.asme.org/products/codes-standards/bpe-2014-bioprocessing-equipment. © The American Society of Mechanical Engineers

Dimensions for these fittings can be found in Part DT of the BPE Standard. Appendix L, of this book, also contains BPE compliant fitting dimensions as well as dimensions for fittings not listed in the BPE Standard. For nonstandard fittings, those not listed in the BPE Standard, dimension shall comply with the intent of the standard in accordance with paragraph DT-4.2. Dimensions for such nonstandard fittings can also be found in Appendix L.

3.4 Sanitary Valves

A few quick facts about valves used in bioprocessing systems:

- In accordance with BPE paragraph SD-3.1.2.1(c), valves, like other components that make up a piping system, should have weld ends.
- Hygienic clamp end valves should be used only if disassembly of the piping system might be necessary.
- In determining the distance L, in the L/D dead leg equation, it shall be from the ID of the pipe wall to the seal point of the valve (paragraph SD-3.1.2.2).
- In accordance with BPE, paragraph SD-4.2.3(b) "Ball valves are an acceptable industry standard for isolation purposes on continuous steam service.
- Three-piece body ball valves should be used instead of single-body designs for both cleanability and maintainability. The bore of the ball valve assembly shall match the inside diameter of the tube (see fig. SG-2.3.1.3-1)."
- Ball valves may be considered for use in applications other than product contact applications. Such applications should be considered and agreed to by the owner/user.
- Valves that mount on equipment, including sample valves, shall be of hygienic design (SD-3.4.2(f)).
- Zero-static diaphragm valves are recommended for low-point drains.
- Check valves used in product or process contact applications shall be of hygienic design and able to withstand a CIP procedure.
- Check valve design shall minimize crevices in its assembly and function.
- Check valve design and placement shall minimize liquid holdup.
- Spring-loaded check valves should be avoided for use in process or product contact applications.

- Pressure relief valves shall be of hygienic design on both the upstream and downstream sides of the valve seat.
- Pressure relief valves in process contact applications shall be designed for CIP service as well.
- Pressure relief valves in process contact applications and CIP/SIP service shall be designed with an override allowing the cleaning or sanitization fluid to flow through the valve.
- *All weld end connections for valves shall have a minimum unobstructed weld end length equal to or greater than the minimum control portion as per DT-7.* (paragraph DT-8)
- Two-way weir type diaphragm valves shall be permanently marked on both sides of the body to indicate optimum orientation to achieve full drainage of the system it is installed in. (SG-3.3.2.3(b)(2))

In accordance with BPE paragraph DT-10.1, valves shall be visually inspected for the following criteria, as a minimum. Valves contained in sealed packaging need not be removed from that packaging for inspection, provided the following can be verified without doing so:

Manufacturer's name, logo, or trademark
Alloy/material type
Description, including size and configuration
Heat number/code
Process contact surface finish (SF) designation (only one SF designation allowed)
Reference to ASME BPE

ASME BPE Certificate of Authorization holders shall mark the reference to this standard by applying their ASME certification mark with BPE designator. Refer to fig. CR-1-1.

Non-ASME BPE Certificate of Authorization holders shall only mark "BPE."

Pressure rating for valves

No damage or other noncompliances

3.5 Seals

Part SG provides all the information pertaining to the sealing of piping systems and equipment. This applies to internal seals such as those used in valve seats to shut off flow to those seals used to contain fluids and prevent them from escaping to the environment.

Seals can be divided into two general groups, static and dynamic. A static seal, referred to as a gasket, is characterized in BPE paragraph SG-2.2.1 as a seal with the "…absence of relative motion between sealing surfaces, or between the sealing surface and a mating surface, after initial installation." Whereas a dynamic seal is characterized by movement between the seal surface and a mating surface, after initial installation.

Examples of static seals include the gaskets used in the BPE hygienic clamp union and the DIN 11864 hygienic clamp union. Examples of dynamic seals include those used at the valve seat where the flow of fluid within the pipeline is blocked or controlled. And there is also the valve seals, referred to as packing, that contain the fluid and prevent it from leaking to the atmosphere by way of the valve stem.

In explaining the function of various types of seals, Part SG describes, under paragraph SG-2.3, the operation of:

- Valves
  - Five different diaphragm valves
  - Ball valve
  - Tank bottom ball valve
  - Rising stem single valve
  - Double seat mix proof valve
  - Needle valve
  - Butterfly valve
  - Thermostatic steam trap valve
  - Back pressure control valve
  - Pinch valve
  - Check valve
  - Plug valve
- Mechanical seals
  - Single mechanical seal for pumps
  - Single mechanical seal for top-mounted agitators
  - Dual pressurized mechanical seal
  - Dual unpressurized mechanical seal
- Reciprocating seal
- Oscillating seal

In conjunction with the mechanical seals, Part SG, under paragraph SG-2.3.2.4, has provided descriptions and graphics for the various flush plans first standardized by the American Petroleum Institute (API) under API standard 682 pumps—*shaft sealing systems for centrifugal and rotary pumps*. These seal flushing schemes have been adopted by API 610 and 676 as well as ASME B73.

As an example, flush plan BPE52, Ref. Figure 3.5.1, shows the sealing fluid being pulled off of the discharge of the pump with pressure controlled by a manual valve in the line running to the mechanical pump seal. The sealing fluid is connected to the bottom of the pump seal between the inboard seal (process contact) and the outboard seal (atmospheric contact). Venting and flushing will exit the top of the mechanical seal.

Seals, whether hygienic clamp joint union gaskets, valve diaphragms, or pump seals, need to meet the demands of not only the intended design service conditions and chemical compatibility requirements but also those conditions extraneous to the primary fluid service itself.

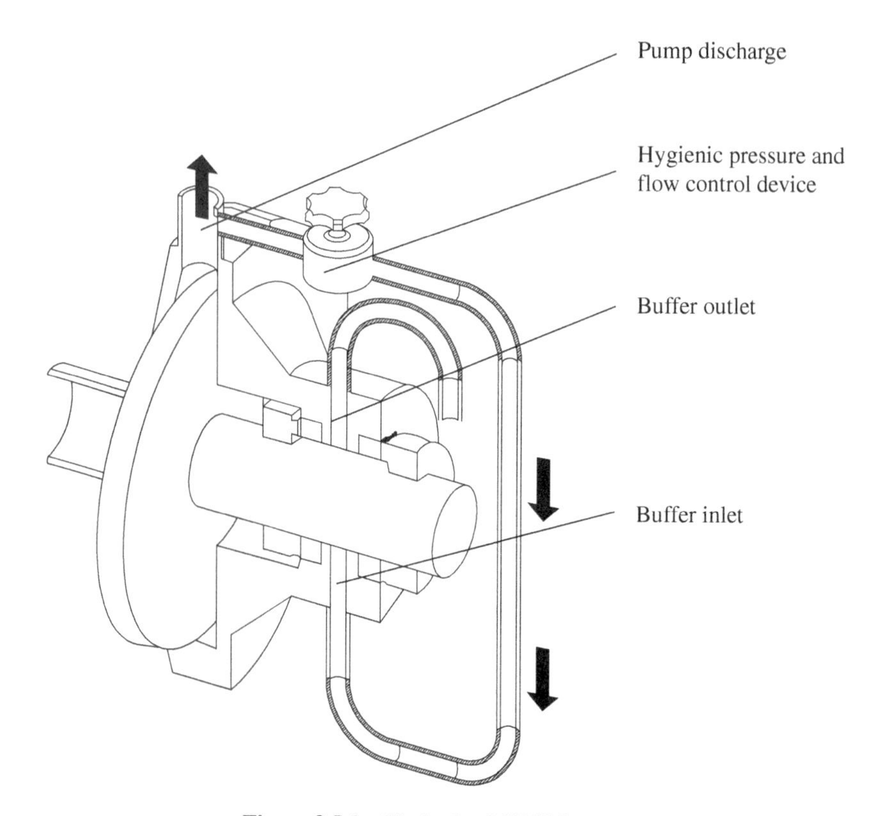

Figure 3.5.1 Flush plan BPE52 for pump

Those extraneous service requirements may include passivation, CIP, SIP, and possible derouging. Each of these added requirements, in specifying seals, will need to be considered when appropriate. In situ passivation is quite frequently done as a final step after sections of tubing have been welded into place. This process is used to promote passivation in the HAZ of welds that were performed in the process of installing the sections of tubing and possibly even those of the prefabricated sections themselves.

The initial DI water flush in the passivation process will be followed by a cleaning and degreasing solution. Following that will be a circulation of the passivation solution that could be a solution of nitric acid, phosphoric acid, citric acid, or other such proprietary solutions. Knowing the chemistry and the temperatures at which a passivation process will be performed should be added to the requirements of a prospective seals material.

This same analogy applies to a piping system having to undergo planned CIP or SIP procedures. The gaskets and seals throughout the system should be given the same considerations with regard to being compatible with the chemicals and temperatures of the CIP solution and with the sanitization temperature of the SIP process.

Derouging, a remediation process, is a consideration that is typically not faced until this anomaly is discovered, either by intent or by accident. Much of what we do in design is done, in part, to preclude the onset of rouge from occurring. But in spite of what is done to preempt rouge from developing in a metallic system, it all too frequently manages to find a way into a system. The question then becomes, should allowances be made in specifying sealing material for a piping system in the off-chance that rouge may form at some point in the future?

As stated, in part, in paragraph GR-1, "The ASME Bioprocessing Equipment Standard was developed to aid in the design and construction of new fluid processing equipment...." That statement is implicit in the fact that the BPE Standard, like the B31 Piping Codes, is predicated on the fact that all of the tubing, fittings, and equipment are new. The Standard does permit the use and application of content with respect to in-service applications, making the point that there are other considerations that need to be taken into account.

Having said that, the only reason the topic of rouge has been made a part of the Standard is due in large part to its ongoing reoccurrence. It is a systemic problem that, even though it does relate to in-service systems, needs to be addressed on behalf of the industry. The problem that occurs when attempting to allow the outside change of derouging impact the material specifications of seals and gasket is that there is no way of knowing what to design for.

Rouge, referring Section 2.3, develops into three different forms, identified as Category I, Category II, and Category III. Categories I and II are basically, without getting into the chemistry, surface particulate matter that has accumulated on the material's surface. These are relatively easy to remove with light acids and solutions. Category III, on the other hand, is an imbedded contaminant that requires much harsher chemical solutions to remove. These chemicals, in remediating the rouge, can actually etch the surface of the tubing or equipment in the process.

Trying to plan for such an occurrence in specifying seals to be compatible with chemicals you may or may not need, for a situation that may or may not occur, does not appear to be economically prudent engineering.

3.6 Instruments

Part PI addresses the high-purity demands of instrumentation in explaining specific design and installation attributes required in achieving compliance for hygienic service applications. To quote BPE paragraph PI-2, "Process instrumentation [Part PI] includes primary elements, transmitters, analyzers, controllers, recorders, transducers, final control elements, signal converting or conditioning devices, computing devices, and electrical devices such as annunciators, switches, and pushbuttons. The term does not apply to parts (receiver bellows or a resistor) that are internal components of an instrument."

The six instrument types currently discussed by Part PI include the following:

1. Flow meters
2. Level instruments

3. Pressure instruments
4. Temperature instruments
5. Analytical instruments
6. Optical devices

3.6.1 Coriolis Flow Meter

The Coriolis type flow meter is the only flow meter type that is discussed in the BPE Standard. It is acceptable with either a straight tube design or as a bent tube design. In either case the meter tubes are required to be flooded at all times with no chance of air pockets.

The meter's installed slope orientation will largely depend on whether it is a straight tube design or a bent tube design. With a bent tube design, the drainable installation will require that the meter be installed in the vertical or near-vertical installation with the direction of flow upward. The straight tube design allows for installation in a sloped horizontal pipeline. SF of the meter tube should match that of the piping it is installed in.

3.6.2 Radar Level Instruments

Radar level instruments, also referred to as "...noncontact radar, free space radar, and through-air radar level instruments" are the only level instrument discussed in the BPE Standard. It is recommended, in accordance with paragraph PI-5.1.3.3, that these level instruments be mounted in the vertical on the top head of a vessel. There should be no obstructions between the antennae and the liquid surface being detected. It is also recommended that the nozzle for mounting the instrument vessel head be 1/3–2/3 of the vessel radius distance from the center of the vessel outward.

3.6.3 Pressure Instruments

Part PI has no content currently on pressure instruments. They are considering this a placeholder for future intended content.

3.6.4 Temperature Instruments

There are two types of temperature elements covered in Part PI. They are resistance temperature detectors (RTDs) and thermocouples. The RTD is the preferred type, but the thermocouple is acceptable upon approval by the owner/user. It is acceptable to use both types with or without a thermowell.

3.6.5 Analytical Instruments

Compendial waters such as purified and WFI are analyzed partially qualified based on their conductivity. This is done with conductivity instruments such as those shown in Figure 3.6.1.

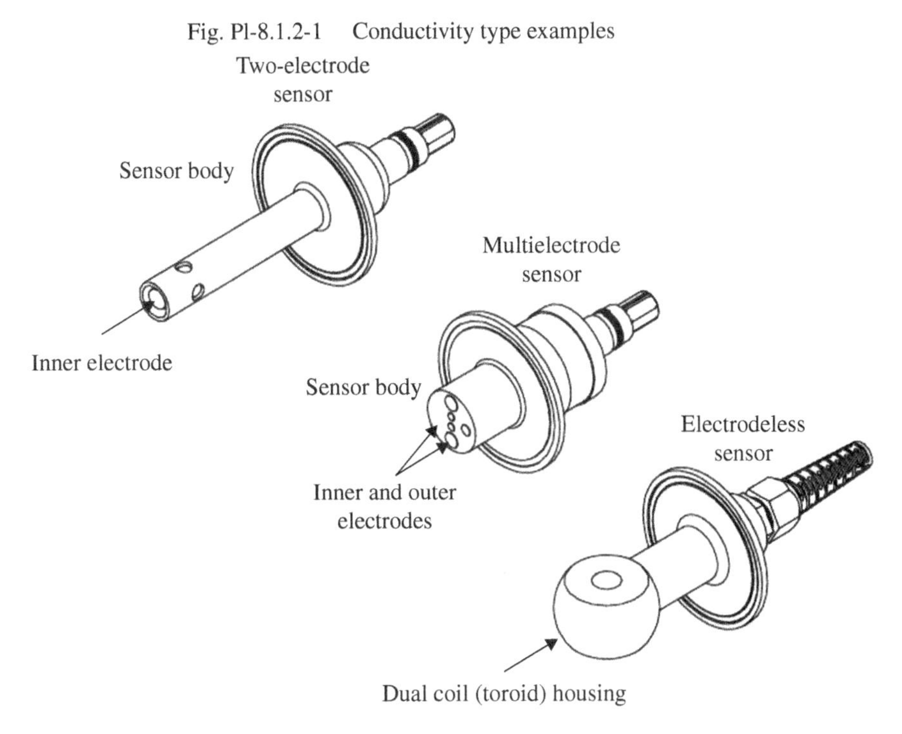

Figure 3.6.1 Conductivity probe examples

It is preferred that the direction of flow be into the probe. And while there is flexibility in the relationship between the probe and direction of flow, the one essential is that the probe remains submerged at all times. When installing into a horizontal pipeline, do not install the probe into the top of the pipe run.

3.6.6 Optical Devices

As stated in BPE paragraph PI-9.1.1, "Optical devices are used to measure various process characteristics including color, turbidity, concentration, percent solids, optical density, particle and cell size/shape, cell density, and cell viability. They are used in filtration, chromatography, fermentation, and water systems. They provide critical control information and process safeguards." In other words, such instrumentation and devices are used to detect the visual and some mass characteristics of a fluid. This includes sight glasses for visual assessment.

4

Fabrication, Assembly, and Installation

4.1 Scope and Introduction to this Chapter

4.1.1 Scope

This chapter discusses the various elements involved in fabricating, assembling, and installing process piping systems for the bioprocessing industry. The discussion on fabrication will cover the various aspects of welding, both metallic and non-metallic components together to create an assembly, touching particularly on the use of autogenous orbital welding for metallic piping and the various joining processes for nonmetallic materials. Also discussed is the hygienic clamp joint and why it is so necessary to adhere to torque values in tightening clamp bolts. The final topic of discussion is system installation, where it all comes together.

4.1.2 Introduction

Before launching into discussions on fabrication, assembly, and installation, it is essential that we establish some ground rules with regard to where the BPE Standard, and indeed the piping codes, stand on the issue of retrofits, modifications, and repairs to existing operating systems.

Such questions about where the BPE Standard, or even where a piping code stands on such issues arise time and again when projects involve tying into an operating process system, modifying an operating process system, or when making incidental repairs to an operating process system. The questions relate to

Bioprocessing Piping and Equipment Design: A Companion Guide for the ASME BPE Standard, First Edition. William M. (Bill) Huitt.

working under the requirements of the current standard, in this case the 2014 issue, while tying into, modifying, or making repairs to a piping system that was designed, fabricated, installed, and tested under an earlier issue of the same standard and code.

BPE paragraph GR-2, in the 2014 edition, states, in part, that:

> This Standard is intended to apply to new fabrication and construction. It is not intended to apply to existing, in-service equipment. If the provisions of this Standard are optionally applied by an owner/user to existing, in-service equipment, other considerations may be necessary. For installations between new construction and an existing, in-service system, the boundaries and requirements must be agreed to among the owner/user, engineer, installation contractor, and inspection contractor.

It explicitly states in the second sentence that "It is not intended to apply to existing, in-service equipment" but then goes on to state that "If the provisions of this Standard are optionally applied by an owner/user to existing, in-service equipment, other considerations may be necessary." These statements are somewhat contradictory to one another and confuse the basic precept or intent. The statement does however coincide with B31.3 paragraph 300(c)(2) which states that:

> This Code is not intended to apply to the operation, examination, inspection, testing, maintenance, or repair of piping that has been placed in service. The provisions of this Code may optionally be applied for those purposes, although other considerations may also be necessary.

The previous B31.3 statement is not contradictory within itself like the BPE statement is as it goes on to allow its optional use for existing operating systems.

To correct this contradiction, a modification to paragraph GR-2 has been approved for the 2016 publication of the BPE Standard, in which the GR-2 paragraph will now read:

> This Standard is intended to apply to new fabrication and construction. If the provisions of this Standard are optionally applied by an owner/user to existing, in-service equipment, other considerations may be necessary. For installations between new construction and an existing, in-service system, **such as a retrofit, modification, or repair**, the boundaries and requirements must be agreed to among the owner/user, engineer, installation contractor, and inspection contractor.

Figure 4.1.1 depicts and makes clear, in a simplified manner, the various analogies when considering retrofits, modifications, or repairs and how they apply under the scope of the BPE Standard or the B31.3 Piping Code. The following therefore describes those analogies shown in Figure 4.1.1 as follows:

Symbols: ⟨TI⟩ = tie-in point or interface connection between new and existing piping.

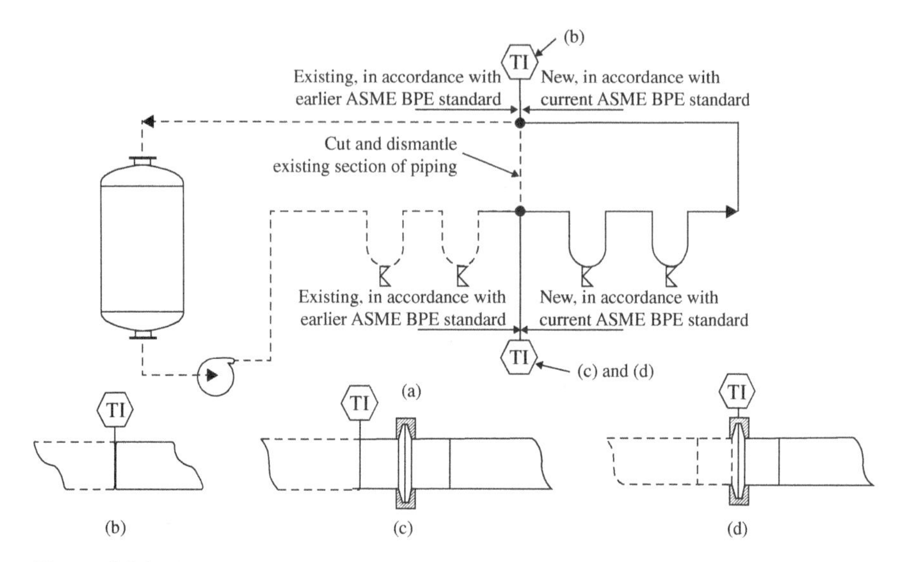

Figure 4.1.1 Rules for process system retrofit modification and repair. (a) Schematic of closed loop piping system, (b) Buttweld existing tube × new tube connection, (c) Buttweld existing tube × new ferrule connection, and (d) existing clamp × new clamp connection

1. When modifying, repairing, or otherwise tying into an existing piping system, the following shall apply:
   a. At the tie-in interface point at which a connection (Ref. Figure 4.1.1b, c, or d) is made in connecting new piping to an existing system, the following is required:
      (i) A new weld, as shown previously in Figure 4.1.1b and c, shall comply with all applicable requirements as stated in the current BPE Standard at the time the project specifications were written and agreed to or, in the case of a repair, at the time the work order was written.
      (ii) When attaching to an existing ferrule, as shown in Figure 4.1.1d, whether on tubing or as an integral part of a component or equipment item, it shall be inspected and approved if it is without damage or distortion.
   b. Including the tie-in interface point, a leak test shall be performed on both the entire portion of a newly installed piping system and that portion of the existing system extending from the tie-in interface to a point at which the existing section of piping can be blocked in a manner that will withstand the leak test pressure.

In performing such modifications or repairs, the untouched portion of the existing system, providing it was initially designed, fabricated, and installed in accordance with an earlier version of the BPE Standard, may retain its initial compliance rating up to the point at which an interfacing joint weld is made, an interfacing connection is made at an existing ferrule, or a clamp joint assembly is installed, creating a tie-in point to connect or integrate new tubing, components, and/or equipment to an existing system.

Examination, inspection, and testing, beginning at the interfacing joint weld or clamp joint and encompassing all new piping, components, and equipment, shall be performed in accordance with the latest edition of the BPE Standard published at the time the system's design was agreed to or, in the case of a repair, at the time work was commenced. Leak testing, as mentioned previously, will extend beyond the tie-in point into the existing piping to a point at which the existing piping can be isolated for the test.

4.2 Fabrication

The topic of piping fabrication can take on many varying aspects. This includes such topics as shop and field fabrication, isometric drawings, weld maps, fabrication drawings, pipe bending, welding, orbital welding, manual welding, TIG welds, MIG welds, autogenous welds, and so much more. In getting started the first thing we will discuss are fabrication drawings.

4.2.1 Fabrication Drawings and Spool Pieces

In order to provide detailed information and instructions to the fabricator, a fabrication drawing or isometric (iso) with associated bill of material is needed. This is a perspective type of drawing drawn at a 30° attitude. It is not a true perspective but gives the illusion of being perspective due to the 30° attitude of the line work. Unlike an orthographic plan view that stacks layers of piping on top of one another, this skewed rendering of the piping system lays open all of the piping and components to be easily viewed by the designer and fabricator. Being able to view all of the runs of the piping, its dimensions, all components, wall or floor penetrations, and its location within a facility is absolutely necessary in order to fabricate, install, and assess the systems compliance requirements for drainability and point-of-use (POU) requirements. The isometric drawings, as you will see, are duplicated and modified for use as weld maps and slope maps as well.

Referring to Figure 4.2.1, each black dot on the pipeline indicates a circumferential butt weld. Some of those butt welds are identified as FW, FFW, and "FCW." Those acronyms are symbols for field weld, field fit weld, and field closure weld, respectively. These indications are what can be referred to as an industry-accepted practice; they are not from an industry standard. These indications typically apply to shop fabricated or shop fab piping only. I say "typically apply" because there are no requirements dictating when such indications can and cannot be applied. Welds indicated with a weld symbol (-•-) but not specifically identified are shop welds or welds made at the fabrication shop.

The difference lies in that fact that the welds identified as the various types of FWs are used to inform both the shop fabricator and the field fabricator wherein their work and responsibilities lie. The shop will make all welds not having an

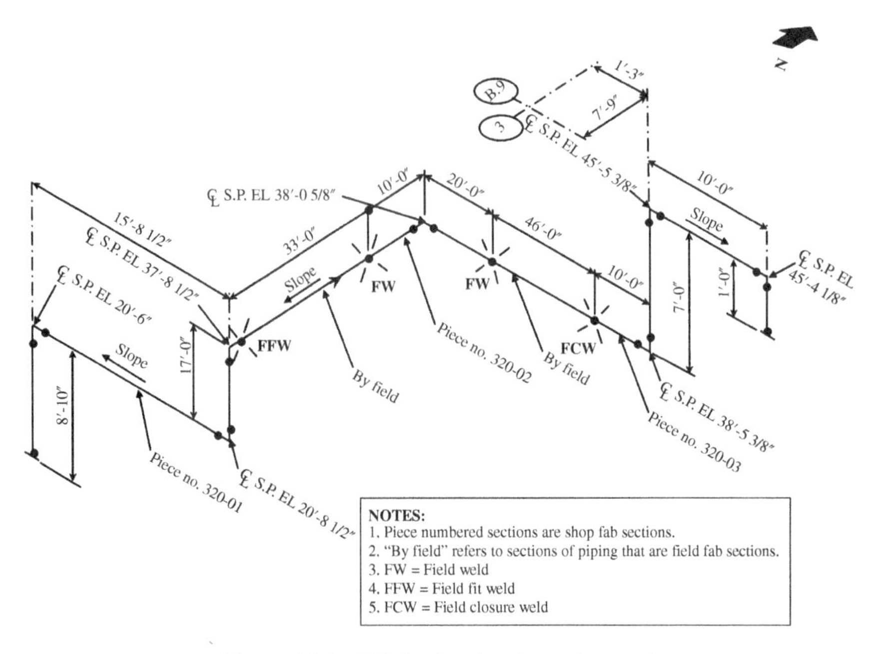

Figure 4.2.1 Fabrication drawing or isometric

indication of an FW and will perform the necessary end prep of both sections of pipe or tubing to be welded together when the symbol "FW" is indicated. If the symbol "FFW" is indicated, the shop will prepare only the end of one section of pipe or tube to be welded. The other end of the other section to be welded will not be prepared for welding, and the unprepared end will be cut 6″ to 9 ft longer than the dimension indicated on the drawing. Descriptions of these various FWs are therefore as follows:

- **FW** = field weld: Identification of welds that are to be made in the field, typically welded in place.
- **FFW** = field fit weld: Identification of a weld joint in which allowances are given in the form of an additional 6–9″ added to the indicated dimensional length of one of the two sections of pipe or tube being joined in order to allow for any compensation that may need to be made due to tolerance buildup as a result of equipment fabrication tolerance, connecting nozzle location tolerance, in-line component tolerance, etc.
- **FCW** = field closure weld: Identification of the final weld to be made of an installed piping system in which, like the FFW, one of the two piping segments to be joined will extend 6–9″ beyond its indicated dimensional length. In determining the location of a closure weld, the design needs to take into consideration the fact that the only means of inspecting a closure weld is by borescopic examination. Meaning that the weld should be within a range of 25–30 ft of a borescope access point in order for the camera lens to reach the otherwise inaccessible internal weld.

You will also notice an indication of the centerline (℄) set point (SP) elevation (EL) at each elbow or turn in the pipe runs. These are interim markers for the installer, indicating that the piping is sloped. The arrow parallel to the pipe run ← Slope indicates that there is a slope in the line and which direction the slope runs in. There is also a set of locator dimensions that locate the pipeline, off of a building column in this case. EL of the pipeline describes what EL or floor the installer will be working at. The location from a column, equipment item, or other building structure describes where in a facility the pipeline is located.

Before continuing, the following are the definitions of two terminologies that will be used:

- *Piping system*: A system made up of piping, equipment, controls, instruments, and ancillary items associated and identified with a particular segment of a process system.
- *Pipeline or piping run*: A run of pipe that includes the piping, equipment, controls, instruments, and ancillary items associated with the pipe run that begins and ends at major equipment items.

Also identified in Figure 4.2.1 are spool pieces such as Piece No. 320-1. When fabricating piping systems in a shop, there has to be consideration given to transporting the fabrication to the site. Because most complete piping runs would be in configurations too large to be shipped and too cumbersome and convoluted to be installed, they need to be segmented. Segmented into smaller assemblies that can be shipped without being damaged and making them easier to install.

A designer, in creating an isometric fabrication drawing, will make a determination as to the most logical approach at segmenting the fabrication of a pipeline into spool pieces by strategically locating FWs. Considerations assumed by the designer will include:

- Configuration of the fabricated spool for shipping.
  - The spool piece should lay as flat as possible to minimize the risk of being distorted.
- Configuration of the fabricated spool for installation.
  - Attempting to weave a fabricated segment of piping through previously installed pipe, cable tray, duct, conduit, supports, and building supports is made more difficult by a long and convoluted section of pipe fabrication.
- Strategic placement of FW.
  - To create shippable spool pieces that do not complicate the installation.
- Strategic placement of FFW.
  - When an FFW is indicated, it will inform the fabricator to extend the length of pipe or tubing beyond the indicated weld point by a predetermined length of approximately 6–9″ that will be trimmed and end prepped as necessary in the field.

  - Such flexibility, enabling field modifications, allows the installer to accommodate a stack-up of installation tolerances.
- Strategic placement of FCW.
  - To allow for possible borescopic internal weld examination of the final weld to be made in the piping assembly.
  - Note that an FCW may also be an FFW.
- Consideration for floor and wall penetrations.
  - Strategically locate an FW at the floor and wall penetrations to ensure ease of installing the pipe through a penetration. Try to avoid having directional changes immediately above and below a floor penetration or on both sides of a wall penetration.

In order to identify the shop fab spool pieces when they arrive on-site, spool piece numbers will be tagged or written on each spool. The spool piece number typically consists of key elements of the isometric drawing number followed by a sequence number. On large projects with multiple floors and areas, it helps to include this type of location information in with the spool piece number. This allows the field to stage the fabricated spools in close proximity to the areas where they will be installed.

It must be kept in mind that the ends of the spool pieces should be capped with temporary plastic caps to protect the pipe ends that have been prepared for welding as well as serving to keep out dirt and debris. The end caps are not only essential for shipping and handling to the job site and while at the job site but are also beneficial during fabrication.

The term "spool," as in "spool piece," comes from welders viewing single plane fabrication drawings they were working from. The piping on the drawing would sometimes consist of a section of pipe with flanges welded at each end (Ref. Figure 4.2.2). Such a drawing gave the appearance of a spool and thus the name.

Some fabricators would, and still do, make fabrication drawings in such a manner, particularly the packaged assembly manufacturers in which there is a great deal of repetitive fabrication. But the fabrication shops that fabricate pipe for on-site installations, and even skid or module-type fabrication, by far utilize the isometric-type drawing. The term "spool," as mentioned previously, evolved from referring to the drawing to eventually referring to the pipe that was being fabricated from the drawing. Consequently the fabricated sections of pipe were

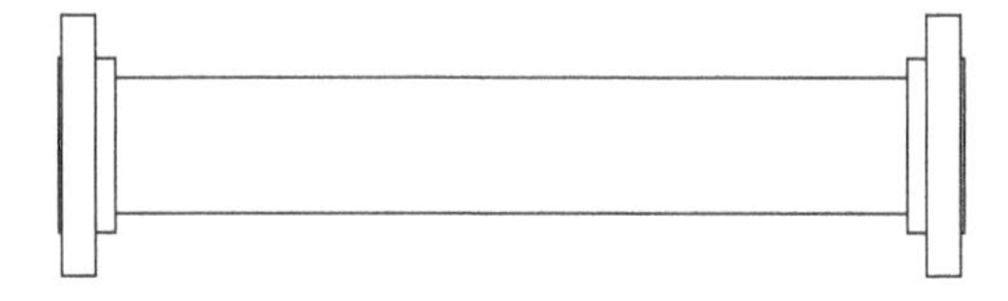

Figure 4.2.2 Spool drawing

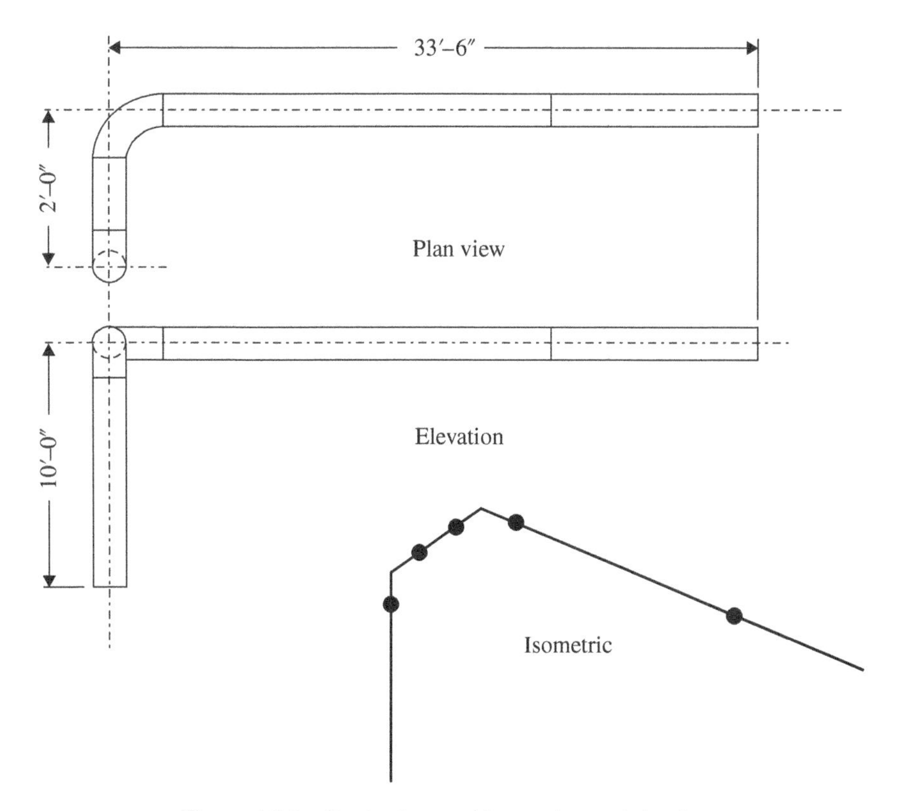

Figure 4.2.3 Single plane and isometric spool drawings

referred to as "spool pieces." With the evolution of isometric drawings, these too were referred to as "spool drawings," and sections of the piping delineated on the drawing for ease of shipping and field installation were identified as "spool pieces."

Figure 4.2.3 shows the same spool piece both in a single plane drawing showing two views of the spool and an isometric view showing a quasi-3-dimensional view of the same spool. Figure 4.2.4 represents the entire pipe run that the aforementioned pipe spool belongs with. The section or spool shown in Figure 4.2.3 is represented in Figure 4.2.4 as spool piece number PC102-2.

Spool piece PC102-2 fits in the pipe run between an FW and an FFW. The piping of both PC102-1 and PC102-2 will be cut to length and have their ends prepped (prepared) for the FW to be made at the FW location. At the other end of PC102-2, where PC102-2 and PC102-3 connect, PC102-2 will be extended beyond the dimension of 33 ft to 6″ by approximately 6–9″ to allow for fit-up adjustment during installation. This FFW would also serve as the FCW or the final weld made in the installation of this piping run. Isometric fabrication drawings will be discussed further as this chapter progresses.

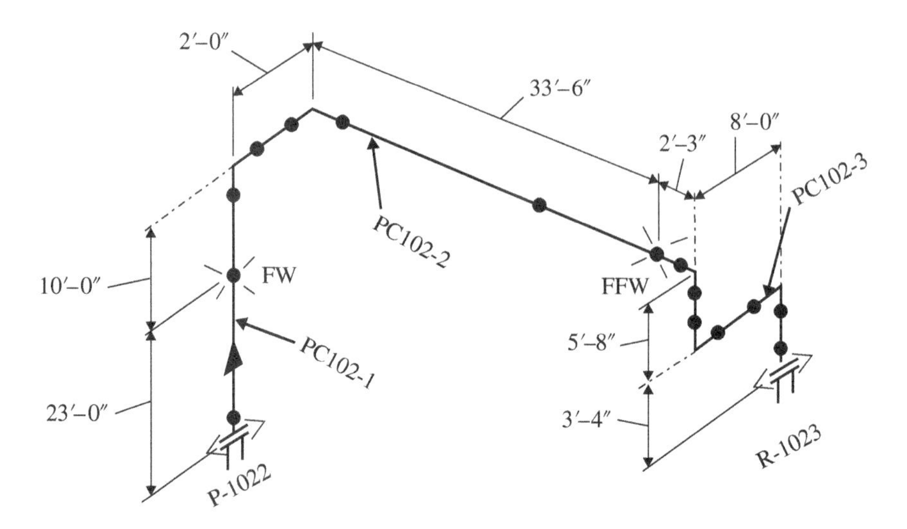

Figure 4.2.4 Simple isometric fabrication drawing

4.3 Fabrication of Metallic Tubing

4.3.1 Welding Documentation and Retention

Aside from a welder or welding operator's expertise and know-how, fabrication of piping intended for high-purity applications is more complex, time consuming, and more costly than other industrial-type pipe fabrication. The complexity, additional time, and cost are not due to the pipe material used in high-purity applications nor is it the results of the type of welding. The added complexity, time, and cost is due in large measure to documentation requirements, the added care in handling the material used in high-purity pipe fabrication, and the necessary examination/inspection requirements.

In Table 4.3.1 is a list of documentation required in validating a high-purity piping system operating under FDA regulatory requirements. The list of required documentation has been compiled from a listing under BPE paragraph GR-5.2.

As you can see by the list of required documentation, not only those documents related to welding, but for everything else as well, the amount of documentation is formidable. For even a modest sized project, and I would estimate modest to be an order of magnitude of around $300 MM US, it would demand the need for thousands of pages of documentation. Such an integral, meaningful, and voluminous part of any project demands that a control procedure for such documentation be articulated early on, written down procedurally, and put into place at the onset of any project. Trying to capture or pull together such information after the fact would not only be extremely time consuming and costly but it could actually jeopardize a project.

Table 4.3.1 Required documentation by category

- Materials documentation
  - Material Test Reports of all process contact metallic material
  - Certificate of compliance for the following
    - All process contact nonmetallic material
    - Filler material
    - Consumable inserts
    - Backing, purge, shield welding gases
  - Material examination logs
- Welding and weld-related inspection and examination qualifications
  - Welding procedure specifications/parameters (WPS/P) in accordance with ASME BPVC Section IX
  - Procedure qualification records (PQRs) in accordance with ASME BPVC Section IX
  - Welder performance qualifications (WPQs) in accordance with ASME BPVC Section IX
  - Welding operator performance qualifications (WOPQs) in accordance with ASME BPVC Section IX
  - Examiner qualifications
  - Documentation of approval of the aforementioned by the owner/user's representative prior to welding
  - Inspector qualifications
  - Document signed and dated by owner/user verifying approval of inspector qualifications
- Weld
  - Weld maps (isometric)
  - Weld logs
  - Weld examination and inspection logs
  - Coupon logs
- Testing and examination
  - Passivation reports
  - Spray device coverage testing)
  - Leak testing
  - Final slope check documentation (isometric)
  - Calibration verification documentation
  - Purge gas certifications (refer also to the aforementioned "materials documentation")
  - Signature logs
  - Number of welds—both manual and automatic
  - Number of welds inspected expressed as a percentage (%)
  - Heat numbers of components that must be identified, documented, and fully traceable to the installed system
  - Surface finish C of Cs
  - NDE (nondestructive examination) reports
- System/equipment
  - Standard operating and maintenance procedures and manuals
  - Installation procedures
  - Piping and instrumentation diagrams
  - Detail mechanical drawings and layouts
  - Technical specification sheets of components and instrumentation
  - Original equipment manufacturer's data
  - Manufacturer's data and test reports
  - Any additional documentation not listed previously that is specifically required by the owner/user

Of that laundry list of project-related documentation, the documents pertaining to welding activities include the following:

Welding and weld-related inspection and examination qualifications
Welding procedure specifications/parameters (WPS/P) in accordance with ASME BPVC Section IX
Procedure qualification records (PQRs) in accordance with ASME BPVC Section IX
Welder performance qualifications (WPQs) in accordance with ASME BPVC Section IX
Welding operator performance qualifications (WOPQs) in accordance with ASME BPVC Section IX
Examiner qualifications
Documentation of approval of the aforementioned by the owner/user's representative prior to welding
Inspector qualifications
Document signed and dated by owner/user verifying approval of inspector qualifications
Weld
Weld maps (isometric)
Weld logs
Weld examination and inspection logs
Coupon logs

As it pertains to piping and tubing, the previous list of documentation, in accordance with BPE paragraph GR-5.5.2(a), which defers to ASME B31.3 paragraph 346.2 and 346.3, shall be "…retained for at least 5 years after the record is generated for the project…." In other words, since the code does not state this, the owner and contractor must agree as to who will retain the documents. But whoever is elected to retain the documents, they have to do so for a minimum of 5 years after the documents have been approved for use. This requirement applies to the piping only and does not carry over to pressure vessels and tanks.

Weld documentation for pressure vessels and tanks, in accordance with BPE paragraph GR-5.5.2(b), defers to the BPVC Section VIII. Under Section VIII, Mandatory Appendix 10, paragraph 10–13 is a list of documents that are required to be retained by the "manufacturer or assembler" for a period of not less than 3 years. That list of documents includes the following:

- Manufacturer's partial data reports
- Manufacturing drawings
- Design calculations, including any applicable proof test reports
- Material Test Reports (MTR) and/or material certifications
- Pressure parts documentation and certifications
- Welding procedure specifications (WPS) and PQRs
- Welders qualification records for each welder who welded on the vessel

- RT and UT reports
- Repair procedure and records
- Process control sheets
- Heat treatment records and test results
- Postweld heat treatment records
- Nonconformance and dispositions
- Hydrostatic test records

Although the amount of documentation, as mentioned earlier, is formidable, it is necessary in bringing certainty, validity, and consistency to how we design, build, and operate drug manufacturing facilities.

4.3.2 Welding for Piping Systems

In discussing metallic fabrication, it must be noted that when handling and fabricating stainless steel, the fabricating facility should be conducive to the segregation of tools and equipment made of plain steel. Coming in contact with the equipment used to handle piping throughout a fabrication shop, moving tools back and forth between carbon steel and stainless steel, and working on carbon steel in close proximity to stainless steel can all lead to contamination of stainless steel.

Airborne free iron from grinding and cutting operations that comes in contact with stainless steel, if it becomes moistened or otherwise worked into the stainless steel, can initiate blooms of rust in the stainless steel material. This surface rust can, if left unabated, evolve into pitting. If the contamination is not removed prior to fabrication or installation, it will continue the corrosion process.

The distance placed between stainless steel fabrication and the fabrication of plain carbon steels is largely dependent upon airflow throughout a combined shop facility. If, for example, loading dock doors are left open for shop ventilation on a hot summer day, the air currents are blowing unabated through a shop area where carbon steel fabrication is taking place. The air then blows through a large open access way and beyond the walls that separate the carbon steel work area from the stainless steel work area. The free iron dust that becomes airborne from grinding, cutting, or brushing operations in the carbon steel shop area can travel on air currents into the adjoining stainless steel work area. Such criteria, when it comes to fabricator qualifications, are further detailed in Chapter 6.

In accordance with BPE paragraph SD-3.1.2.1(c), the primary method for connecting tubing, fittings, valves, and other in-line components for high-purity service is by welding. This is the de facto default methodology of assembling a high-purity tubing system. Mechanical joints, which are essentially any method of assembling tubing other than by welding, are used for strategically placed breakout points in order to dismantle sections of tubing for cleaning, access for equipment removal, and other maintenance purposes.

The BPE Standard has a very limited list of acceptable welding processes that are acceptable for use in high-purity process contact services. Under BPE paragraph MJ-4 Welding Processes and Procedures, BPE identifies two categories of

Table 4.3.2 Acceptable welding processes for high-purity applications

Welding process	Abbreviations
Arc welding	
Gas tungsten arc welding	(GTAW) also known as TIG[a] welding.
Gas metal arc welding	(GMAW) also known as MIG[a] welding.
Shielded metal arc welding	(SMAW) also known as manual metal arc welding (MMAW).
Flux-cored arc welding	(FCAW)
Submerged arc welding	(SAW)
Plasma arc welding	(PAW)
High-energy beam welding	
Electron beam	(EBW)
Laser beam	(LBW)

[a]Prior to WWII the terms TIG and MIG were acronyms for tungsten inert gas and metal inert gas, respectively. This was due to inert gas being used as a shielding gas. During the WWII buildup of warships and land vehicles, a more efficient and faster method of welding was needed and researched. In discovering the benefits of mixing volatile gases with inert gases, they found it necessary to change the name of TIG and MIG by replacing "inert gas" to just "gas" as in gas tungsten arc and gas metal arc welding.

welding processes, which includes paragraph MJ-4.1 Welds Finished After Welding and paragraph MJ-4.2 Welds Used in the As-Welded Condition.

For "Welds Finished After Welding," BPE states that the welding processes used "…shall be limited to the arc or high energy beam (electron beam and laser beam) processes…." For "Welds Used in the As-Welded Condition," BPE states that the welding process used "…shall be limited to the inert-gas arc processes (such as gas tungsten arc welding and plasma arc welding) or the high energy beam processes (such as electron beam or laser beam welding)…."

Those two paragraphs, MJ-4.1 and MJ-4.2, identify the following welding processes as being acceptable for use on high-purity service applications, as listed in Table 4.3.2. But every effort should be made to use an automatic or machine process.

4.3.2.1 Orbital GTA Welding

Based largely on the advent of autogenous orbital welding, a weld can be repeatedly performed, when properly programmed, that will consistently meet the fundamental requirements as defined in tables MJ-8.3-1 and MJ-8.4-1, as well as fig. MJ-8.4-1. But the usefulness and efficiency of the orbital welding program were made that way with the help of the initial BPE committee members and ASTM.

Prior to the early BPE membership composing what would eventually become the BPE Standard, as recently as the mid-1990s, engineers designing a pharmaceutical-grade facility had to specify, and fabricators had to match, heat numbers on tubing and components that were to be welded together by the automatic orbital welding method. This hunt for such specific material was a time-consuming and costly undertaking to specify, locate, procure, and verify material for tubing and components with matching heat numbers. The heat number itself is a tracking number for metallic material that allows the material to be traced back to its original chemical composition at the time of its formulation in the mill.

Documentation, in the form of MTRs, is one part of the rigid documentation records required when building a pharmaceutical manufacturing facility. The MTR provides certification of the chemical composition of the material along with the heat number to validate that data. The reason for going to all the trouble of matching up heat numbers was prompted by the need to match up compatible sulfur content within the chemistry of the material. That information, along with a material's mechanical properties, is included in the MTR.

As declared in a 1982 study and resulting paper by J.L. Fihey and R. Simoneau, the amount of sulfur content in 304 and 316 type stainless steels can have a drastic effect on the weldability of the material. The paper they wrote described the sulfur-related phenomenon of weld pool shift that can occur when welding a flat plate of 304 or 316 stainless steel having a very low percentage of sulfur to a plate that contains a higher percentage of sulfur.

At this same time, in 1981/1982, C.R. Heiple and J.R. Roper were performing studies on surface tension gradients and the effects on the molten weld pool fluid flow patterns and fusion zone geometry. The results of their findings were published in April 1982 in the Welding Research Supplement to the *Weld Journal*. Their study, coincidentally, explained the sulfur anomaly written about by Fihey and Simoneau that same year.

What Heiple and Roper discovered in their research was that sulfur content has an effect on the surface tension of the molten weld pool, thereby affecting the fluid flow of the molten weld pool, and does so in the following manner. It was suggested by Heiple and Roper that fluid flow of the molten weld pool is influenced by surface tension referred to as Marangoni convection. Sulfur affects the weld pool shape by altering the surface tension, which in turn alters the fluid flow of the molten weld pool.

Referring to Figure 4.3.1 two different flow patterns, (a) and (b), are represented. Referring to detail (a) of Figure 4.3.1, when the sulfur content is about 0.007 wt% or less, the temperature coefficient of surface tension becomes negative. This creates areas of lower surface tension under the weld pool. This, in turn, causes the fluid flow at the surface of the weld pool to flow from the center under the arc outward. When this occurs less heat is contributed to weld penetration, and the result is a high width-to-depth (W/D) ratio creating a wide weld bead with respect to depth of penetration.

With a sulfur content of about 0.008 wt% or greater, the temperature coefficient of surface tension becomes positive. Referring to detail (b) in Figure 4.3.1, the heated area under the arc has the highest surface tension. This surface tension gradient causes the fluid in the weld pool to flow from the outer fringes of the cooler, lower surface tension region of the weld pool toward the center. This directs the heat of the arc downward, increasing penetration. This results in a high (D/W) ratio creating a narrow weld bead with respect to the depth of penetration.

This anomaly was a well-known fact in the industry during the 1980s and 1990s, but its root cause, at the time of the two publications, was not understood. Then, in 1984 Dr. Barbara Henon learned of the paper written by Fihey and

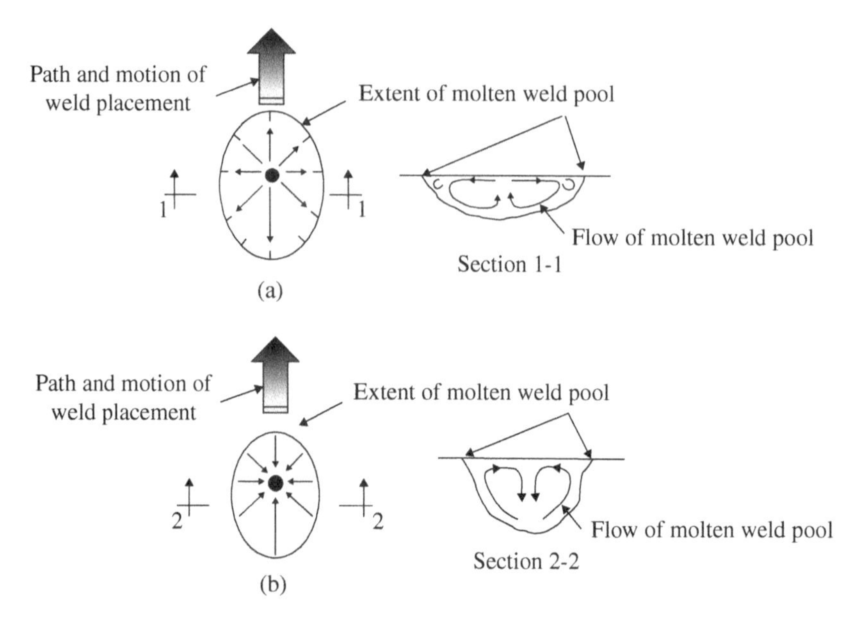

Figure 4.3.1 Effect of sulfur on fluid flow of molten weld pool. (a) Flow direction of a weld pool with negative temperature coefficient of surface tension and (b) flow direction of a weld pool with positive temperature coefficient of surface tension

Simoneau and wanted to see if their work would translate to tube welds. The goal was to determine whether or not this might be the cause of weld pool shift occurring now and then with tube welds made by orbital welding machines.

The problem, as Barbara discovered, did turn out to be the aforementioned sulfur anomalies. In working toward a resolution to this problem of weld pool shift, she had drilled down in her research to develop what would become the resolute fix for the industry, but that industry-wide fix would not come until years later. Once the determination was made that a solution was at hand, a letter was sent to Arc Machine Inc. (AMI), who she worked for at the time, customers explaining the findings of the study with the recommendation that they order 304 and 316 stainless steels with a sulfur range of 0.005–0.017 wt%. Those privy to this information had to then go to the pains of finding components manufactured from heats of material having a sulfur content within the range specified by Dr. Henon, and this was no small feat.

Upon joining the fledgling BPE Standards group in the early 1990s, Barbara presented her findings to Mr. Tony Cirrilo and Mr. Randy Cotter, describing the sulfur content issue and a proposed range of sulfur that would accommodate orbital welding. This in turn was presented to the DT subcommittee, chaired at the time by Mr. Chip Manning. Chip, along with DT subcommittee member Mr. Jack Declark, who was also a member of ASTM, approached ASTM with the proposal of adding a supplemental requirement that would include a sulfur range of 0.005–0.017 wt% in material standard ASTM A270 for UNS S31600 (316) and S31603 (316L) materials. This would allow a material specification to be written that would specify, up front, the needed sulfur content for weld fittings essentially eliminating the need of having to hunt down compatible material.

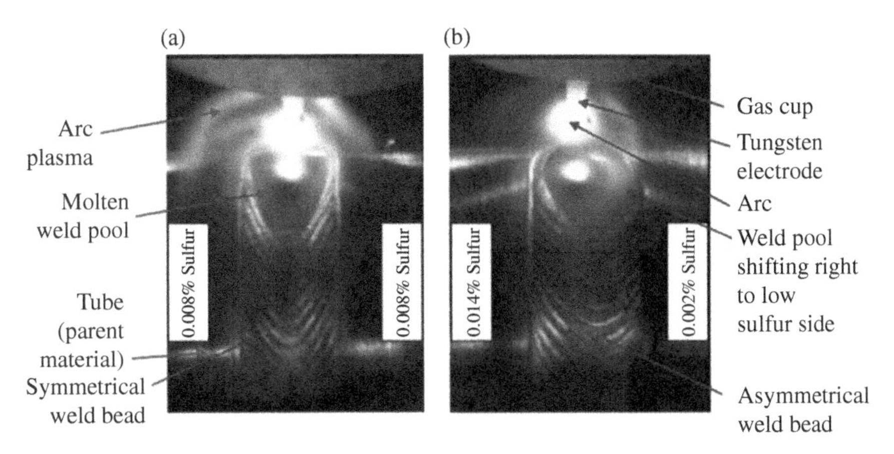

Figure 4.3.2 Effect of sulfur on welding 316L stainless tubing. (a) Symmetrical weld joint of two tubes having similar sulfur content. (b) Asymmetrical weld joint of two tubes having dissimilar sulfur content. *Source*: http://www.arcmachines.com/. © 2016 Arc Machines, Inc. (*See plate section for color representation of this figure*)

Having provided a graphic representation in Figure 4.3.1 describing what occurs in welding with varying percentages of sulfur contained in 304 and 316 type stainless steels, Figure 4.3.2 shows what occurs in an actual weld. The two photos, (a) and (b), in Figure 4.3.2 are actually frame shots extracted from a video taken during Dr. Henon's research into the source of the weld issues that were plaguing industry at that time. The tubing, and there are two different sets of tubing depicted in the two photos, was mounted on a lathe with a manual gas tungsten arc welding (GTAW) torch used to run the weld program. The AMI engineering staff used a special Arc Machines' vision system that filtered out most of the arc light in order to make the weld pool visible.

Photo (a) depicts a gas tungsten arc (GTA) weld being performed in joining two sections of tubing both containing 0.008 wt% sulfur. As is apparent, the weld bead is well centered on the joint, meaning that weld depth and width are equal about the joint interface. Photo (b) depicts a GTA weld being performed in joining two sections of tubing, one containing 0.014 wt% sulfur and the other containing 0.002 wt% sulfur. The tungsten electrode is still centered over the joint interface as it is in photo (a), but the weld bead is not.

The effect that the different percentages of sulfur has on welding material, as shown in detail (b) of Figure 4.3.2, is very apparent. The higher sulfur content of the material on the left promotes a narrow weld bead with good penetration, while the lower sulfur content of the material on the right creates a wider weld bead with less penetration. The issue, prior to the papers written by J.L. Fihey and R. Simoneau and C.R. Heiple and J.R. Roper capped by Barbara's discovery and subsequent acceptance as a supplemental requirement in ASTM A270, was frequent and costly weld rejections. In the first issue of the BPE Standard, published in 1997, the sulfur range of 0.005–0.017 wt% for 316L stainless steel could be found in table 3 of Part DT. In the 2014 publication of the Standard, it can be found in paragraph MM-5.1.1.

4.3.2.2 Welding with GTA Orbital Welding

GTA orbital welding is a preprogrammed automated GTAW process by which a tungsten electrode is rotated about the joint of two fixed tubing components in performing an autogenous (fusion) weld or it can be a weld performed with the use of added filler metal or consumable insert. Filler material is typically applied when added strength is needed or when an over-alloying filler metal is used to mitigate dilution of the alloying elements within the parent material.

The autogenous method of orbital welding is intended for welding OD tubing from 1/8″ (3 mm) to 4″ (101 mm) and to 6″ (152 mm) with added complexity and additional training needed for the welding operator. This is due to OD variation (ovality) tolerances that actually double in allowable tolerance from the 4″ (101 mm) tube OD size to the 6″ (152 mm) tube OD size. The 4″ (101 mm) tube size has a ±0.015″ (±0.38 mm) ovality tolerance, while the 6″ (152 mm) tube size has a 0.030″ (0.76 mm) ovality tolerance. The following will explain why the larger tube sizes and their tolerances can be an issue with orbital welding.

In orbital welding the tubing and/or fitting(s) to be welded are mounted in an enclosed weld head fixture. These orbital welding units typically have a clamshell-type design in which the top portion of the weld head assembly opens up in order to set and align the components to be welded (Ref. Figure 4.3.3). While the tubing and/or fitting components remain fixed in the weld head, the nonconsumable tungsten electrode, mounted in a rotor, is made to autorotate about the joint creating an arc that will melt and fuse the material together. Connected to the weld head are the power supply, computer control cable, purge gas supply, and filler wire feed when needed.

The end tip of the tungsten electrode is set at a fixed distance from the workpiece joint. That distance is referred to as the "arc gap." The orbital weld program

Figure 4.3.3 Weld head of orbital welding machine. *Source*: http://www.arcmachines.com/. © 2016 Arc Machines, Inc.

maintains a constant electric current level to the weld machine, while the arc gap determines the arc voltage. Based on wall thickness and also controlled by the machine's program, that gap will be about 0.050″ (1.27 mm) for a 4″ (101 mm) OD tube or fitting with a 0.083″ (2.11 mm) wall thickness and about 0.070″ (1.8 mm) for a 6″ (152 mm) OD tube or fitting with a 0.109″ (2.77 mm) wall thickness.

The amount of voltage supplied to the tungsten electrode, in conjunction with the rotational speed of the tungsten electrode as it moves around the perimeter of the work piece, is programmed to provide enough heat to melt the thickness of the tube or fitting wall but not so little heat as to not fully penetrate the material. Should the arc gap vary due to inconsistent ovality of the tubing or fitting as the tungsten electrode rotates around the work piece, the arc voltage, and thus the heat of the arc, will fluctuate accordingly. This could lead to incomplete penetration at various points around the circumference of the joint.

Figure 4.3.4 shows ASME BPE Fig. MJ-8.4-1 representing what an acceptable weld profile (a) should look like along with deviation tolerances (b), (c), (d), and (f) from that acceptable weld as well as an example of an unacceptable weld

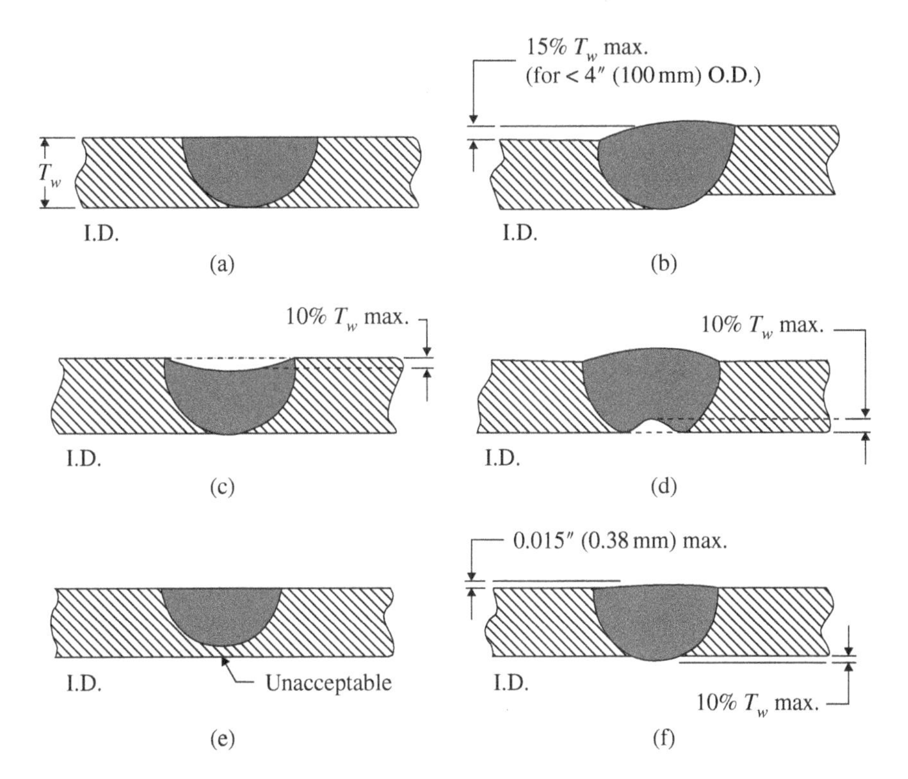

Figure 4.3.4 Tube weld tolerances. (a) Acceptable, (b) misalignment (mismatch), (c) OD concavity, (d) I.D. concavity (suckback), (e) lack of penetration, and (f) convexity. *Source*: https://www.asme.org/products/codes-standards/bpe-2014-bioprocessing-equipment. © The American Society of Mechanical Engineers

(e) with lack of penetration. These are the joint parameters within which the orbital welding machine has to work.

Such welding machines, as the orbital welder, are operated by welding operators. And like a welder the welding operator is required to be qualified in order to produce welds on an orbital welder. And although their task is more of a technical nature in setting up and programming the welding machine, their skill set and qualifications with regard to the orbital welding process are still required.

Taking a few listed items found in Table 4.3.1 on welding qualification documentation, the following describes the various documents and qualifications required for welding, welders, and welding operators:

- WPS: In accordance with Sect. IX, "A WPS is a written qualified welding procedure prepared to provide direction for making production welds to Code requirements. The WPS or other documents may be used to provide direction to the welder or welding operator to assure compliance with the Code requirements."

  The completed WPS shall include and describe all of the essential, nonessential, and, when required, supplementary essential variables for each welding process used in the WPS. It shall also reference the supporting PQR described in Section IX, paragraph QW-200.2. Other information that may be considered helpful in making a code weldment may also be added to the WPS.
- PQRs: "The PQR is a record of variables recorded during the welding of the test coupons. It also contains the test results of the tested specimens. Recorded variables normally fall within a small range of the actual variables that will be used in production welding."
- WPQs: The WPQ is the documentation verifying and attesting to a welder's ability to execute a weld in accordance with a WPS.
- WOPQs: The WOPQ is the documentation verifying and attesting to a welding operator's ability to execute a weld in accordance with a WPS.

4.4 Fabrication of Nonmetallic Piping and Tubing

4.4.1 Fabrication of Polymeric Components

BPE paragraph PM-2.1 states that polymeric "Materials should be compatible with the stated processing conditions, cleaning solutions (where appropriate), and sterilizing conditions (where appropriate), etc., as specified by the owner/user." Integral to such acceptance criteria is the joining method of polymeric materials. As mentioned in BPE paragraph MJ-9.3, the fusion method of welding polymer piping does not require solvents or glues. It uses the base material to form the weld eliminating the possibility of contamination. Paragraph MJ-9.3 lists those fusion joining methods as:

- Butt fusion
- Beadless fusion

- Bead and crevice-free (BCF) fusion
- Infrared (IR) fusion
- Socket fusion

Butt fusion welding (aka contact butt fusion) is a contact fusion process by which two pipes and/or components are aligned and the ends trimmed and faced with a planer tool until they become flush with one another, leaving no visual gaps around the two interfacing pipe and/or component ends. After preparation the two squared ends are first brought into contact with a metal-heating plate, which is placed between the two ends, which are held in place against the plate until the piping material reaches a molten state. Once the ends reach their molten state, they are then pulled away from the hot plates; the hot plates are removed from between the two ends and the two ends forced together and held in that position until the joint cools. The cooled and completed joint will have approximately 90% of the strength of the pipe itself. This type of fusion welding is not acceptable for use in systems that require in-place drainage and bioburden control. It may be used in single-use applications with approval of the end user.

Beadless fusion welding (aka contact beadless fusion) is a modified version of the butt fusion process. The difference lies in the welding process, which is designed and performed to eliminate the inner weld bead at the point of fusion. The process is done as described in the butt fusion process with the exception that the molten material that would normally flow equally to both the inner and outer walls of the pipe is instead forced to flow only to the outside of the joint with the use of an expandable bladder that is inflated on the inside of the pipe at the joint location. After the joint is cooled, the expandable bladder is removed and the external bead can be left as is, or it can be removed as necessary. The beadless fusion weld line is almost undetectable and will have approximately 90% of the strength of the pipe itself. This type of fusion welding is acceptable for use in systems that require in-place drainage and bioburden control.

BCF Fusion is a method of fusion welding that leaves no bead on either the inside or outside of the pipe wall and does not require a contact plate as does beadless fusion welding. The two components to be welded are placed in the welder in the bottom half of a fixture that resembles an orbital welding machine head. On either side of the welding head are fixture clamps that hold the pieces to be welded in place. An inflatable bladder is placed inside the piping and positioned at the joint to be welded. The top half of the welding head is then brought down into place. Inflation of the bladder and fusion heat from the heating elements is computer controlled. The heating element against the OD of the pipe keeps the molten material from forming a bead on the outside, and the bladder applies pressure on the ID of the pipe to keep a bead from forming inside. This type of fusion welding is acceptable for use in systems that require in-place drainage and bioburden control.

IR fusion welding (aka noncontact fusion) is one of the newest forms of welding used in joining thermoplastics. IR welding, like butt fusion welding, is accomplished by a similar fusion process. The difference being that the ends of the pipe or fittings do not come in contact with the heating element. They are instead heated by radiant heat from an IR heating element. Not having to force the ends of the pipe or fitting against a heated plate significantly reduces the weld bead size while also eliminating the possibility of contamination from that source. The computer-controlled welding of the IR welding machines offers more accurate and consistent control over the welding process than butt fusion machines, allowing the welder to deliver recordable welds with less bead on a more consistent basis. This type of fusion welding is not acceptable for use in systems that require in-place drainage and bioburden control. It may be used in single-use applications with approval of the end user.

Socket welding fusion is also a form of contact fusion welding in which the socket fitting and pipe both come in contact with a heating element. Using a bench type welder, and with the heating element preheated to the proper temperature, the pipe and fitting are forced axially on to and into the heating element, which lies between the two components. The ID surface of the socket fitting and OD surface of the mating pipe are then heated for a prescribed period of time. At the end of the heating cycle, the pipe and fitting are backed off of the heating element, the heating element removed, and the pipe inserted axially into the socket joint. Left in place, the two adjoining surfaces are then allowed to coalesce and cool. Upon completion of the process, there should be a fillet bead adhering to both the pipe wall and the end of the socket fitting along the full 360° circumference of the joint. This type of fusion weld is not acceptable for use in systems that require in-place drainage and bioburden control. It may be used in single-use applications with approval of the end user.

An acceptable beadless weld, as shown in Figure 4.4.1, is one in which the weld bead is flush with the internal and external surface of the components; has no cracks, crevices, pits, pores, voids, microbubbles, inclusions, discoloration in accordance with what is described in Figure 4.4.7, internal concavity within the tolerance described in Figure 4.4.8; and does not exceed misalignment as described in Figure 4.4.5.

Indicating improper pre- or postweld handling or resulting from the weld process itself, cracks and crevices, as shown in Figure 4.4.2, are not permitted in the weld zone.

Due in part to possible contaminants or improper preparation of the fusion area, pits and pores, as shown in Figure 4.4.3, are not permitted in the weld zone.

An occasional anomaly, resulting from shrinkage of the molten base material as it cools, is the forming of voids or microbubbles as represented in Figure 4.4.4. If they

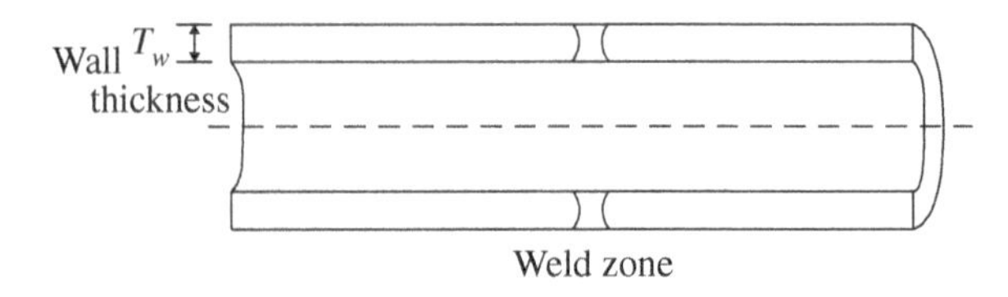

Figure 4.4.1 Acceptable weld profile for a beadless weld. *Source*: https://www.asme.org/products/codes-standards/bpe-2014-bioprocessing-equipment. © The American Society of Mechanical Engineers

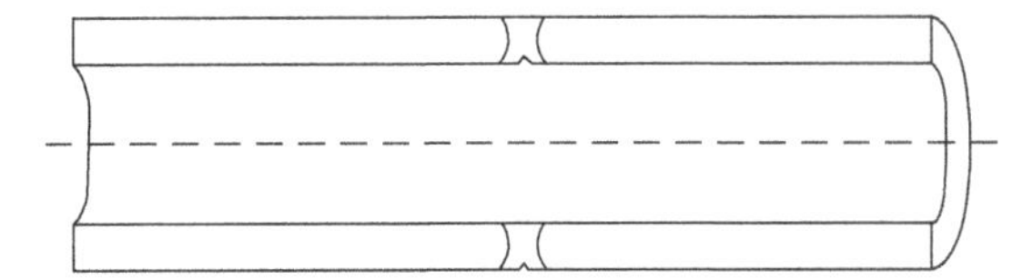

Figure 4.4.2 Unacceptable cracks or crevices in weld profile of a beadless weld. *Source*: https://www.asme.org/products/codes-standards/bpe-2014-bioprocessing-equipment. © The American Society of Mechanical Engineers

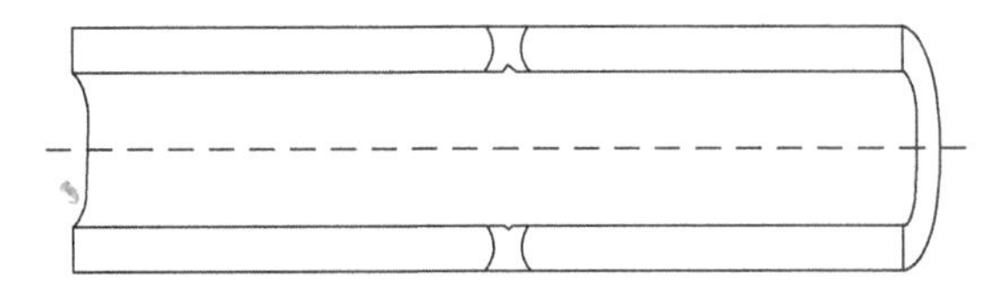

Figure 4.4.3 Unacceptable pits or pores in weld profile of a beadless weld. *Source*: https://www.asme.org/products/codes-standards/bpe-2014-bioprocessing-equipment. © The American Society of Mechanical Engineers

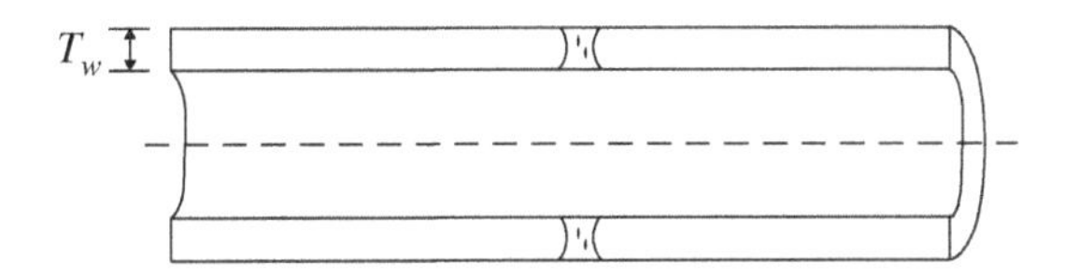

Figure 4.4.4 Unacceptable voids (microbubbles) in weld area of a beadless weld. *Source*: https://www.asme.org/products/codes-standards/bpe-2014-bioprocessing-equipment. © The American Society of Mechanical Engineers

do occur they will typically form inside the weld itself. Before these become a cause for rejection they should exceed either of the two measurable values. Voids or microbubbles can be grounds for rejection if either the size of a single void exceeds 10% of the nominal wall thickness or a combined total of the size of each void detected in a cross section of a cluster exceeds 10% of the nominal wall thickness.

Due in large measure to fitting tolerances, weld preparation and set up, misalignment between the two components being welded has to be held to acceptable tolerances. That misalignment, in accordance with Figure 4.4.5, cannot exceed 10% of the wall thickness (T_W) of the pipe or fitting. This takes into account pipe and fitting tolerances. Such set up measurements can only be made on the OD of the components. It is therefore recommended that if the weld joint is installed with the piping in a horizontal position, then the bottom OD of the piping should be aligned, allowing any offset difference in the components to accumulate at the top. If the weld is to be installed with the piping in a vertical position, the two joined pieces should be aligned concentrically about the centerline axis.

Any inclusion or dark speck, as represented in Figure 4.4.6, that is discovered in the fusion zone can be considered foreign matter and is unacceptable.

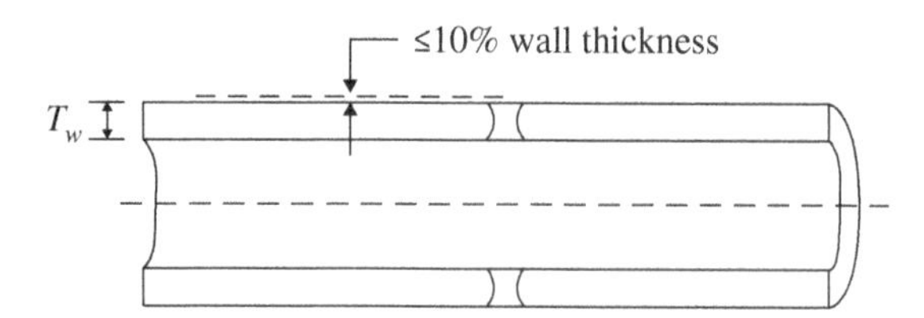

Figure 4.4.5 Misalignment tolerance in weld area of a beadless weld. *Source*: https://www.asme.org/products/codes-standards/bpe-2014-bioprocessing-equipment. © The American Society of Mechanical Engineers

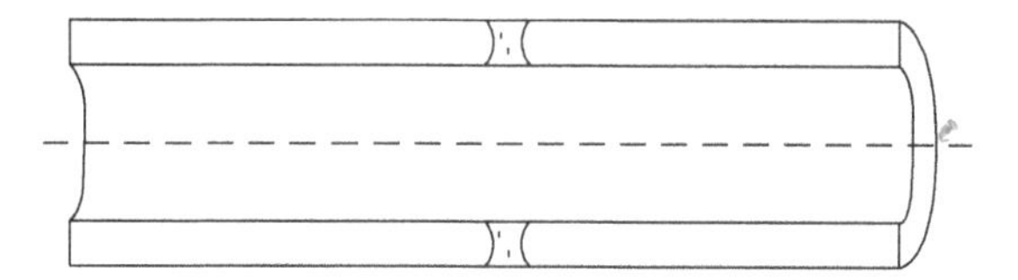

Figure 4.4.6 Unacceptable inclusions in weld area of a beadless weld. *Source*: https://www.asme.org/products/codes-standards/bpe-2014-bioprocessing-equipment. © The American Society of Mechanical Engineers

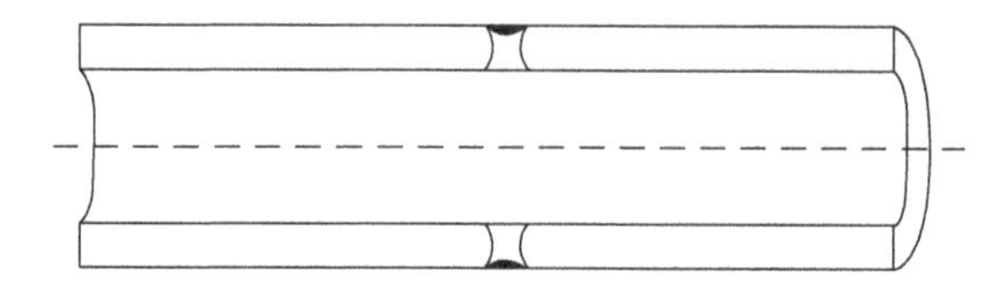

Figure 4.4.7 Discoloration in weld area of a beadless weld. *Source*: https://www.asme.org/products/codes-standards/bpe-2014-bioprocessing-equipment. © The American Society of Mechanical Engineers

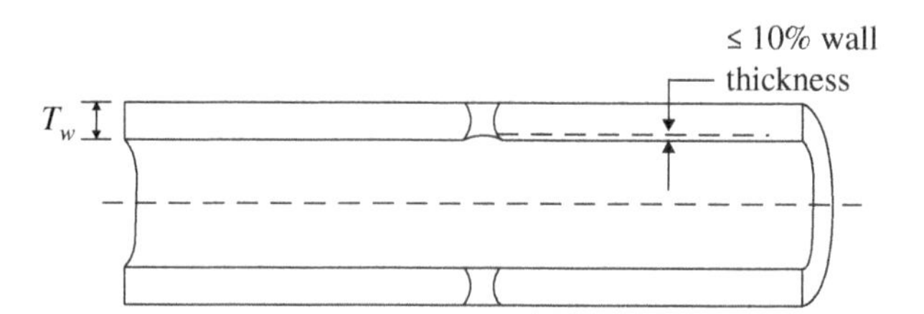

Figure 4.4.8 Tolerance for internal weld concavity of a beadless weld. *Source*: https://www.asme.org/products/codes-standards/bpe-2014-bioprocessing-equipment. © The American Society of Mechanical Engineers

Discoloration in the fusion zone, as represented in Figure 4.4.7, no darker than a straw color is acceptable. Discoloration considered to be brown in color to the darker hues of brown is not acceptable.

As seen in Figure 4.4.8, concavity on the ID of the weld cannot exceed 10% of the wall thickness (T_W) of the pipe or fitting.

4.5 Assembly and Installation

4.5.1 General

Installation of a piping system is accomplished by setting the main equipment in place. What can be considered "main" equipment are typically those items that are free-standing. Such items might include pressure vessels, tanks, large filters, pumps, etc., items that are self-supporting rather than being installed in and supported by the piping. Such items can be considered in-line components.

With the main equipment items in place, one of the other items that will need to be installed is the main pipe racks which can be referred to as secondary pipe supports. The secondary supports are the supports that carry multiple pipelines as the piping is routed out of the pipe racks for further distribution.

Then there are the individual pipe supports that are installed with the piping. In many cases the exact location of these individual supports will not be known until the piping is installed.

The final installation and assembly of high-purity piping, once the main equipment and supports are in place, are accomplished by either hygienic clamp joint assembly connections or by welding. Sections 4.2 and 4.3 discuss welding of the spool pieces in advance of installation and will be discussed further in the welding in place of those spool pieces. But first we will discuss installation and assembly with the BPE hygienic clamp joint assembly.

4.5.2 Characteristics of the Hygienic Clamp Joint

4.5.2.1 Joint Integrity

In assessing the integrity of the assembled hygienic clamp joint union, there are five basic elements to be considered:

1. Two mating ferrules
2. Circumferential clamp
3. Bolt or bolts
4. Gasket
5. Workmanship

The finer points of these five listed elements are as follows:

1 and 2 Ferrules and Clamp
In looking at the ferrule and its relative attributes to the assembly, there are two types found in the BPE Standard, Types A and B (Ref. Figures 4.5.1 and 4.5.2). Type A applies to ferrules in sizes 1″ and smaller. Type B ferrules apply to sizes 1″ and larger. The 1″ size overlap of the two types is due to a leak integrity issue having to do with the 1″ Type B ferrule; the specifics of which are not relative to this subject matter and will not be described in this report.

Figure 4.5.1 Type A ferrule

Figure 4.5.2 Type B ferrule

To mitigate the problem BPE extended the Type A ferrule design up through 1″ previously used only for ferrules 3/4″ and smaller. As shown in Figure 4.5.1, the 1″ and smaller Type A ferrules have a different gasket seat design than the 1″ and larger Type B ferrules as reflected in Figure 4.5.2. Since the 2009 edition of the BPE Standard, there has been a choice of the two types of sealing surfaces for the 1″ ferrules. What is represented in Figures 4.5.3 and 4.5.4 is the Type B ferrule.

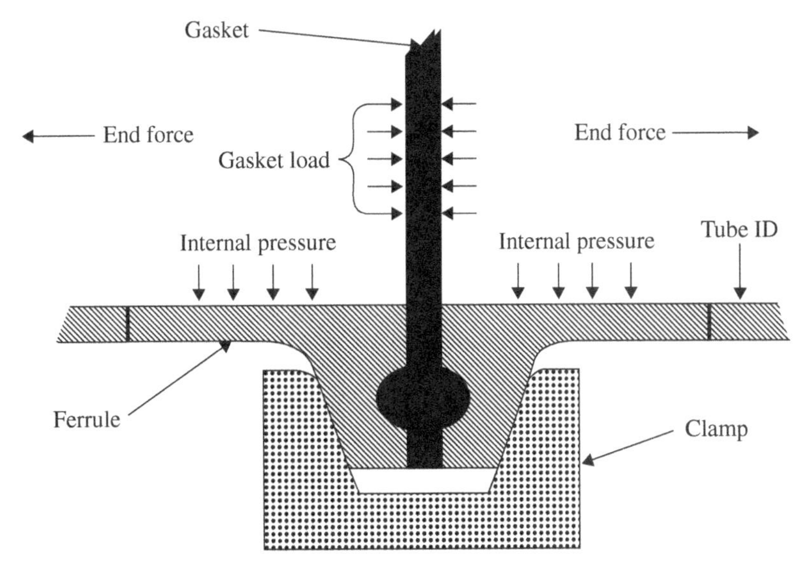

Figure 4.5.3 Forces acting for and against containment

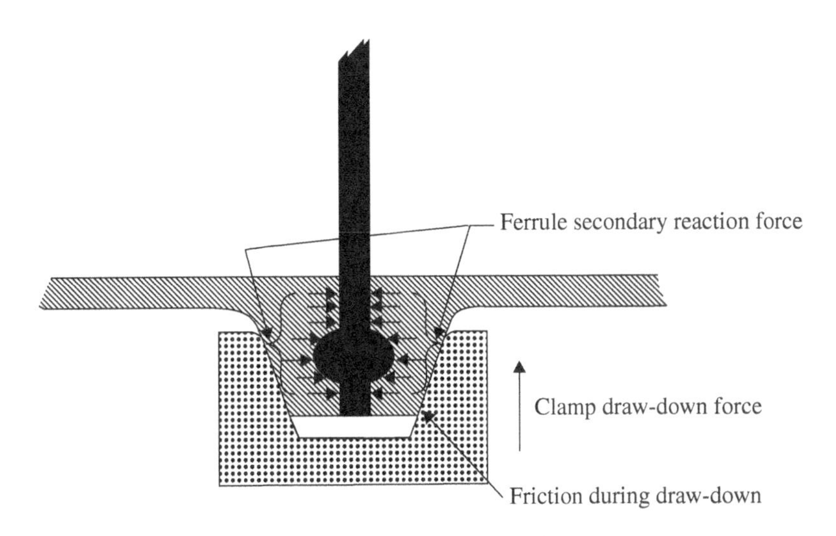

Figure 4.5.4 Forces acting to seal the joint

Figures 4.5.3 and 4.5.4 represent a segment of the clamp joint assembly consisting of the clamp, two ferrules (welded to tubing), and a gasket. In Figure 4.5.3 there are three forces working on this assembly. They include (i) hydrostatic end force, which is pressure of a confined fluid acting in parallel to the pipe axis to oppose the sealing force at the ferrule face; (ii) internal pressure, which is the force acting at right angles to the pipe axis attempting to overcome confinement; and (iii) gasket load, which is a function of the force created when a clamp draw down forces the ferrules to compress the gasket enabling the sealing surface to withstand internal fluid pressure.

In Figure 4.5.4 there are two forces at work. They include (i) clamp drawdown force, which is the force applied to the assembled ferrules through the act of drawing down the clamp over the ferrules with a bolt or bolts and (ii) ferrule secondary reaction force, which are the deflective opposing forces created by the drawdown of the clamp creating a preload on the gasket.

As seen in Figures 4.5.3 and 4.5.4, the two ferrules in the hygienic clamp joint assembly provide the assembly's sealing surfaces, which require a pliable material between the two surfaces in the form of a gasket. Gaskets and their material will be discussed later. In order to preload the gasket between the two ferrules, a clamp with a concaved wedge pattern is forced down over the two ferrules creating a drawdown force. This peripheral force, achieved in large part to resistance from the highly rigid form of the round tubular ferrule, is applied to the backs of the two ferrule flanges creating opposing deflective or secondary reactive forces on to both ferrule flanges and the gasket between them. The redirected force against the gasket becomes the preload force and has to be sufficient enough to contain the internal fluid pressure plus the hydraulic end force pressure attempting to separate the two ferrules.

For metallic ferrules and clamps, temperature does not become a factor in load calculations until it becomes elevated enough to begin depreciating the allowable stress value (S_a) of the metal. For 316 and 316L, that temperature is 399°F (203°C). At 400°F (204°C) and above, the allowable stress of this austenitic stainless steel then begins to depreciate making it, at those elevated temperatures, an essential factor in load calculations.

3. Bolts

There are two general types of hygienic clamps: those that are hinged and secured to the ferrule and gasket assembly with a single bolt and wing nut (Figure 4.5.5) and those that are made up of two separate pieces forming a peripheral clamp and secured to the ferrule and gasket assembly with two bolts and nuts (Figure 4.5.6). When assembling a hygienic clamp joint union, much of its theoretical sealing integrity is based on a predetermined torque value used in fastening the bolts. Sealing integrity refers to the ability of the gasket and ferrules to seal against the internal fluid pressure while not exceeding the intrusion/concavity tolerances specified in ASME BPE paragraph SG-4.2 and fig. SG-2.2.2-1.

The single wing nut clamp joint assembly is typically given a torque value of 30–50 in-lb (depending on gasket material), while the more robust two-bolt high-pressure clamp is given a value of 20 ft-lbs of torque. The assigned torque values are an indirect means of gaging proper compression (preload) on the gasket. And while using torque values to achieve proper gasket compression is certainly practical, it does have its issues.

In taking a page from the *Machinery's Handbook*, it states that "Laboratory tests have shown that whereas a satisfactory torque tension relationship can be established for a given set of conditions, a change of any of the variables, such as

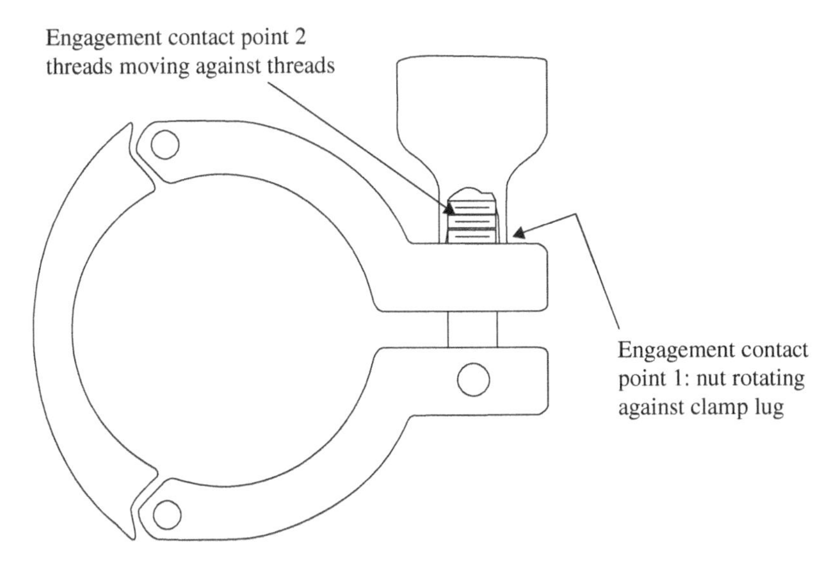

Figure 4.5.5 Three-piece double-pin clamp joint assembly

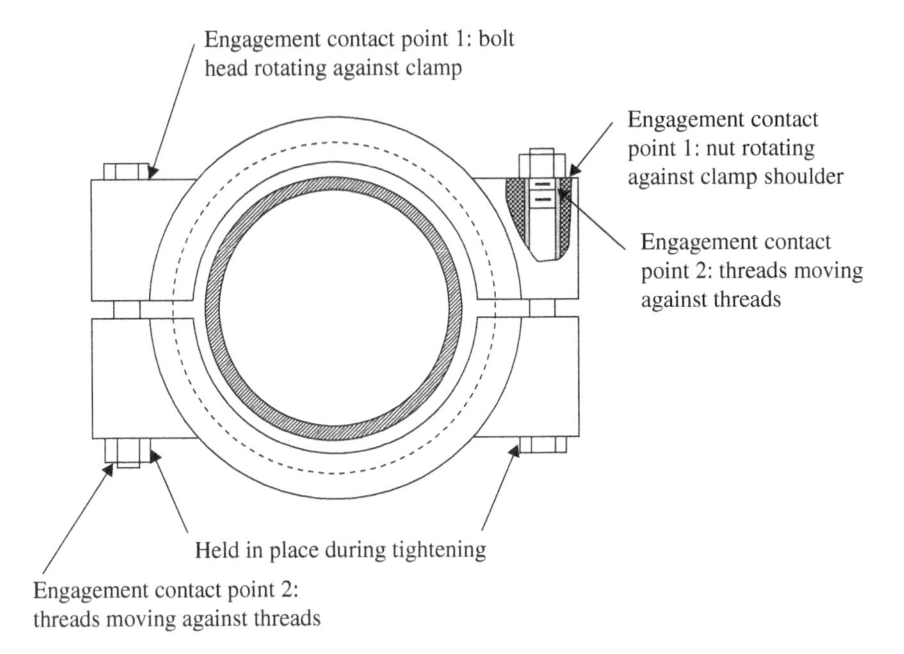

Figure 4.5.6 Two-bolt high-pressure clamp joint

fastener material, surface finish, and the presence or absence of lubrication, may severely alter the relationship. Because most of the applied torque is absorbed in intermediate friction, a change in the surface roughness of the bearing surfaces or a change in the lubrication will drastically affect the friction and thus the torque tension relationship." In a paraphrased statement found in Federal Standard FED-STD-H28A, it elaborates on the previous statement by further stating that "Thus, it must be recognized that a given torque will not always produce a definite stress in the bolt but will probably induce a stress that lies in a stress range that is satisfactory." These statements go to the heart of the fact that using a torque value to preload a hygienic clamp joint assembly is not a truly accurate mechanism for gaging load on a sealing surface but will instead achieve a load value approximate to a predetermined target value.

When tightening a bolt in order to seal an ASME B16.5 flange joint or an ASME BPE hygienic clamp joint, the installer is attempting to apply tension to the bolt or bolts used to create the sealing load. There is a direct correlation between bolt tension and the amount of load placed on the affected sealing surfaces whether it is a flange joint or a hygienic clamp joint. Bolt tension is directly related to the amount of compressive force applied to the gasket. That tension, while in the bolt's elastic range, creates a live load on the sealing surfaces which allows the joint to remain sealed throughout fluctuations of thermal cycles and transient dynamic surges unless, that is, bolt tension relaxes due to gasket creep or bolt stress exceeds the material's elastic range. But, even though gaging bolt stress may be a more accurate approach in determining gasket load, in field installations, it is not practical. To use such sophisticated and costly instrumentation as an ultrasonic stress analyzer to detect bolt tension when tightening bolts, thousands of bolts in many cases, cost and schedule make this method prohibitive. It is instead much more economical and practical to set a torque value and then use a torque wrench to meet that compressive load target value.

In using torque values as a target value, the gasket manufacturer has to first of all set a torque value that establishes the needed load or force required in tightening the nuts and bolts in order to achieve the gasket compression necessary to seal the joint. That predetermined torque value is then achieved during installation with the use of a torque wrench. But this can be a flawed method if not done properly.

The torque wrench measures and quantifies the energy required to tighten a nut to a predetermined torque value, which is given in in-lb or ft-lb values. Therefore the measurement of torque is neither an indication of the force applied at the nut, which does not necessarily translate into the force or compression load realized at the gasket, nor is it a direct correlation with the bolt tension values as mentioned previously.

In order for torque values to work properly for a given application, those values have to be predetermined through testing and qualifying by, in our case, the gasket manufacturer or a qualified third-party tester. However, the inherent problem with using torque values is that it is, as mentioned previously, an indirect and somewhat inaccurate method of measuring bolt tension, which is, as also mentioned, the less direct value associated with gasket compression. The problem is that when assembling possibly thousands of mechanical joints in the field it is very inefficient to gage bolt tension on thousands of bolts. By therefore being relegated to using a torque wrench against predetermined torque values, the process then becomes dependent, in large measure, upon replicating the method used initially to determine the proper preestablished torque values that will achieve the necessary gasket compression.

What occurs in the process of assembling a clamp joint is this: in tightening bolts there can be considerable variation and opportunity for loss in torque, or quantified force in the contact points between the bolt head, where the torque measurement is taken, and the gasket where the force is needed. This is due to what is referred to as a "nonlinear variation" caused by friction loss that occurs at these points between the bolt head (torque reading) and the gasket surface. The loss in force is due mainly to friction created at two points: the nut (engagement contact point 1 Figures 4.5.5 and 4.5.6) where as much as 50% or more of the torque value can be lost and also at the engagement threads (engagement contact point 2 Figures 4.5.1 and 4.5.2) where as much as 30% or more additional torque value can be lost leaving approximately 20% of the torque value or force, as measured on the readout dial of the torque wrench, actually being transferred via bolt tension to the gasket and ferrule sealing surface. There is also a resistant friction force created as the clamp moves down over the ferrule flanges (Ref. Figure 4.5.4). Since only the bolts are lubricated when assessing torque values for gaskets, lubricating this point of the clamp can be considered overlubricating and should not be done. When gasket manufacturers or fitting manufacturers insert a qualifying statement with their product such as "Torque values are based on nuts and bolts being lubricated and maintained in such a manner as to achieve a free-running condition," this same method of assembly has to be replicated as closely as possible during installation and joint makeup; otherwise the same degree of gasket compression cannot be expected to be achieved even though the torque readout value might be the same as the recommended value.

A lubricated 7/16″ dia. bolt at 20 ft-lbs of torque will create a bolt load of approximately 2800 lbs, thereby creating a bolt stress or tension of 30 000 psi. A nonlubricated 7/16″ dia. bolt tightened to a torque value of 20 ft-lbs will create a bolt load of approximately 1400 lbs, thereby creating a bolt stress of 15 000 psi. That is a 50% loss of force due to friction. If the specified force applied at the torque wrench is not being transferred efficiently through the bolt, then it is not transferring sufficient compression force to the gasket and sealing surfaces.

If the gasket and fitting manufacturers qualify the pressure containing integrity of their components by using a methodology in which torque values, using lubricated bolts and nuts, are used as determinant values instead of bolt tension values, do we need to concern ourselves with bolt tension values? No, not exactly. While bolt tension is certainly a key factor, what the installer needs to do is replicate the process by which the manufacturers have qualified the integrity of their product. This is why bolt torque requirements, rather than bolt tension values, are specified for installation requirements for the flange or clamp type mechanical joints in relation to the type of gasket material.

If the installer pulls a new clamp out of the box and installs it as is, the amount of torque load transferred to the gasket and ferrule seating surfaces as the wing nut or hex head bolt is tightened could be 50% or less of that indicated on the torque wrench dial. A lubricant applied to the bolt threads and other engagement contact points can allow more of the torque to transfer to the sealing surfaces where it is needed. But this too can be a double edged sword.

If lubrication is used too freely when assembling ASME B16.5 flange joints, you simply achieve greater compression at the gasket surface with no harm done, to a point. Doing the same with the hygienic clamp joint can cause the compression load to exceed the qualifying amount experienced by the gasket manufacturer due to the additional lubrication. The added lubrication, let's say at the clamp and ferrule contact points, makes the transfer of load from the torque wrench to the ferrule/gasket sealing surface much more efficient, thereby transferring a greater percentage of the torque value reflected on the torque wrench dial. When the gasket manufacturer warrants their gaskets to meet the tolerance requirements of ASME BPE Class I or Class II (Ref. Figures 4.5.7 and 4.5.8), they base this on an assembly procedure that included a lubricant on the clamp bolt(s) only. In order to achieve a comparable outcome, the field installation must be done in the same manner.

4. Gasket

A gasket's purpose is to provide a compressible surface by which flaws and deviations in a metallic (in this case) surface can be sealed against an internal fluid pressure or an external relief when under internal vacuum. As mentioned earlier, the gasket compression load, referring to Figure 4.5.3 shown previously, allowing for any surface imperfections on the sealing surface of the ferrules, must be sufficient to seal the joint against internal pressure in overcoming those hydrostatic forces attempting to separate the two ferrules.

In attempting to quantify the characteristics that surround the sealing mechanics of a flanged-type joint, we turn to Section VIII of the BPVC. From it we take gasket factors "y" and "m," which will be replaced or augmented by new gasket factors "Gb," "a," and "Gs." This change is the result of research and testing having been done over recent years by ASME's Pressure Vessel Research Council (PVRC). The "y" factor is defined as the minimum applied seating stress required to create a seal for a gasket between flange faces. The "m" factor, when multiplied by the internal factor, is the stress load necessary for the gasket to maintain a seal. The new constants, "Gb" and "a," in the PVRC methodology, represent the initial

SG-4.2 Static seal performance

Static seals shall meet the general performance requirements of SG-4.1.

Upon initial installation, a hygienic static seal shall provide a substantially flush interface with the hygienic clamp ferrules. Hygienic seals shall meet and be designated by one of the following intrusion categories when tested by the seal manufacturers:

(a) Intrusion category I. Seals having a maximum intrusion/recess of 0.025″ (0.6 mm).

(b) Intrusion category II. Seals having a maximum intrusion/recess of 0.008″ (0.2 mm).

Figure 4.5.7 ASME BPE-2014 paragraph SG-4.2

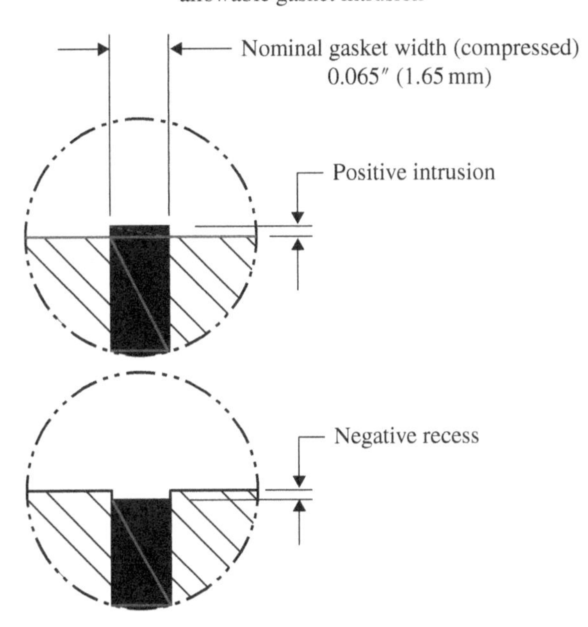

Figure 4.5.8 Allowable gasket misalignment Ref. BPE fig. SG-4.2-1. *Source*: https://www.asme.org/products/codes-standards/bpe-2014-bioprocessing-equipment. © The American Society of Mechanical Engineers

gasket compression characteristics as it relates to initial installation. "Gs" represents gasket unloading characteristics typically associated with operational behavior (fluctuation in gasket loading due to thermal cycling and dynamic effects). This report mentions the PVRC method for reference only and will stay with the "y" and "m" factors where needed.

While the aforementioned testing, research, and development pertain specifically to the ASME-type flange joint, this information can also be transposed for use with the hygienic clamp joint given that some of the same key principals apply to both types of mechanical joint designs—those key principals being that two flanged surfaces are forced together by two opposing perimetrical forces. The difference lies between the more direct bolt-up force that is applied to the ASME flange joint as opposed to the indirect force applied to the ferrule flanges affected by the use of a bolt load being applied to a circumferential clamp, which in turn applies a peripheral load to the two ferrule flanges.

5. Workmanship

The human element, no matter the amount of engineering and high degree of quality manufacturing all of it reinforced with written and/or oral instruction leading up to the installation, is where it can all come undone. It is imperative that any pressure containing mechanical joint be assembled in the proper manner with the proper tools. Turning down a wing nut bolt or a hex head bolt and nut with a pair of adjustable pliers, a screwdriver inserted into the hole of a wing nut bolt, or by any number of other unrecommended, but frequently used, assisted methods of tightening, is not good workmanship. It may be fast, expedient, and easy, but it is not a good workmanship and does not achieve proper results.

As mentioned earlier, in order to achieve the desired results, that which establishes the required torque value for an assembled joint on a repeatable basis, requires the same assembly method used to determine proper torque values in the first place. Such requirements as those established by the manufacturer have to be applied in the field in order to achieve the same result. The installer cannot expect to apply the same required load to the sealing surface as recommended by the gasket manufacturer if the bolt or bolts are not properly lubricated as recommended by the manufacturer. By not properly lubricating the bolt or bolts, much of the force registered on the torque wrench gage is transferred no further than the bolt threads and not to the sealing surfaces where it is needed.

To go further, workmanship of the joint assembly not only goes beyond simply bolt lubrication but it applies also to proper insertion of the gasket (ASME BPE fig. SG-2.2.2-1), alignment of the two mating ferrules, and placement and draw up of the clamp (ASME BPE fig. DT-2-1). Each of these elements, bolt tightening, gasket placement, ferrule alignment, and clamp installation, must be done properly. If any one of these elements is not executed in the proper manner, the integrity of the joint could be compromised.

4.6 The Piping Installation Process

4.6.1 Field Assembly and Installation (Stick Built)

With regard to piping, there are two basic methods to building a chemical process industry (CPI) facility such as a bioprocessing facility. One method is by designing and constructing a skid or modular unit off-site, transporting the unit to the job site, and then dropping it into place as a singular, stand-alone unit or as one part of a multiunit modular complex.

The other method is referred to colloquially as "stick-built" construction, in which piping is either fully fabricated in place or the piping, to the largest degree possible, is shop fabricated then delivered to the job site and installed as prefabricated sections or a combination of the two.

The prefabricated (prefab) spool pieces, whether delivered to the job site from a remote shop or moved to the install site from a site-located fabrication shop, will be marked or tagged with a spool piece number as described in Section 4.2. The weld ends and ferrule ends of any pipe openings on these spools should be capped or sealed in a manner that will prevent airborne particles or debris of any kind from entering the prefabricated piece of work. The end cap should also be compatible with the piping material and pliable as to not scratch, distort, or otherwise damage the ID or OD of the pipe or tubing.

It is recommended that spool pieces be staged for installation in an on-time fashion. This is to reduce their resident time in the staging area and therefore their risk of damage and/or debris getting inside the piping. The temporary on-site storage or laydown area should be located in a low traffic area away from construction activity as well as cables, chains, tools, lifting gear, metal tables, or pallets made of plain steel.

Thought and consideration should be given far in advance of these spool pieces reaching the job site in preparation for installation. The design and layout of piping is not guided by how that piping will be installed; that comes later in the process. The design of such piping systems instead takes into account a routing that gets each pipeline from point A to point B by the shortest practical route without creating dead legs, without obstructing access ways, by avoiding pockets, by integrating expansion/contraction flexibility, and by ensuring drainability and cleanability.

With the design and layout approved for construction, fabrication drawings can then be produced. At this stage the determination as to how the piping will be installed has to be made. This is done by segregating a pipeline into spool sections that can: (i) Lie reasonably flat on its transport bed. This, along with being properly secured for transportation, will allow each fabricated spool section to lay well during transportation—an effort at protecting the piping from becoming distorted in transit. (ii) Provide ease of installation. Multiple bends and long lengths in a spool piece creates difficulties when inserting it into its permanent location.

As seen in Figure 4.6.1, the pipeline that runs from pump P-1022 to reactor R-1023 is segregated into three spool piece sections. This is done to facilitate both transportation to the job site and placement of the piping into its permanent install

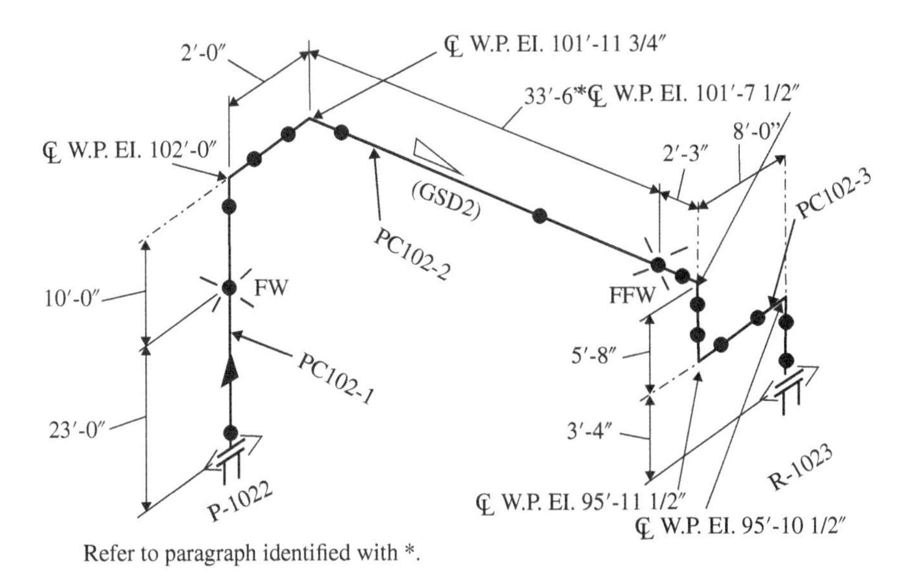

Figure 4.6.1 Simple isometric fabrication drawing

location. The spool sections are delineated by the indication of either an FW, an FFW, or an FCW, as further described in Section 4.2.1.

Due to the potential and likely buildup of tolerance deviations in the setting of equipment, nozzle locations on equipment, building walls and columns, cable tray installation, pipe supports, and so on, precise routings and locations of pipe in accordance with design drawings may not be tenable. Knowing this, the fabricator/installer will provide themselves some latitude in the fabrication of the piping to facilitate last minute in-place adjustments to the piping. This is accomplished by providing the added lengths of pipe as described in Section 4.2.1, when describing an FFW.

* The work point EL in Figure 4.6.1 has been rounded off from the mathematically calculated EL slope drop of 4 3/16″ to 4 1/4″ to show centerline work point ELs starting at 101′–11 3/4″ and sloping to 101′–7 1/2″. Working to within a 1/16″ (5 mm) creates an unnecessary complication when trying to set pipelines in the field. The BPE slope symbol of GSD2 indicates a minimum 1/8″/ft (10 mm/m) slope. This is the recommended minimum slope requirement for gravity drained process contact lines in accordance with BPE paragraph SD-2.4.3.3.

The GSD2 designation, as shown in table SD-2.4.3.1-1, is an indication of the "minimum" slope required. This implies that the designer has the flexibility or latitude to exceed that value if it becomes necessary or simply more practical to do that. Such rounding rules, as changing the 4 3/16″ drop to 4 1/4″, should be included in a design basis for a facility or project.

With the spool sections in place, the final welds can be made or the clamp joints can be assembled. In accordance with BPE paragraph SD-5.3.2.3.1, SD-4.3.1(b), and MJ-3.4, it is recommended that the FWs too be performed with an orbital welder to the extent possible. Where manual welds are necessary, they "...can be performed, but must be agreed to by the owner/user and contractor."

4.6.2 *As-Built and Other Drawings*

While this is not a requirement, it is highly recommended that any modifications made to the fabrication drawings during installation should be documented during the process of making and approving such changes to the piping. These documented changes are then applied to the fabrication drawings, which are then reviewed and approved and then issued "as built." These as-built drawings then become the final file drawings.

From the as-built drawings, duplicates can be made into "weld maps" and "slope maps." Documentation providing essential information is required for both welds and slope verification under BPE paragraph GR-5.2.1.1.1(c) weld map and (d) final slope check document. While it is not a requirement to document the information in this drawing format, it is arguably the more practical method.

An example of this type of documentation can be seen in Figures 4.6.2 and 4.6.3. In the weld map example of Figure 4.6.2, each weld is assigned a number and the initials of the welder for each weld is indicated. In this very simple

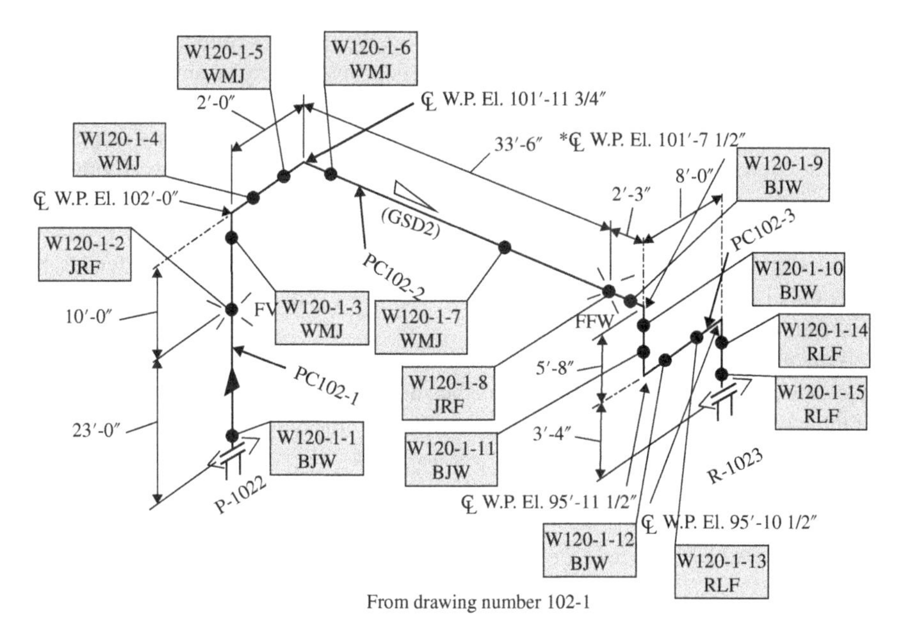

From drawing number 102-1

Figure 4.6.2 Weld map drawing

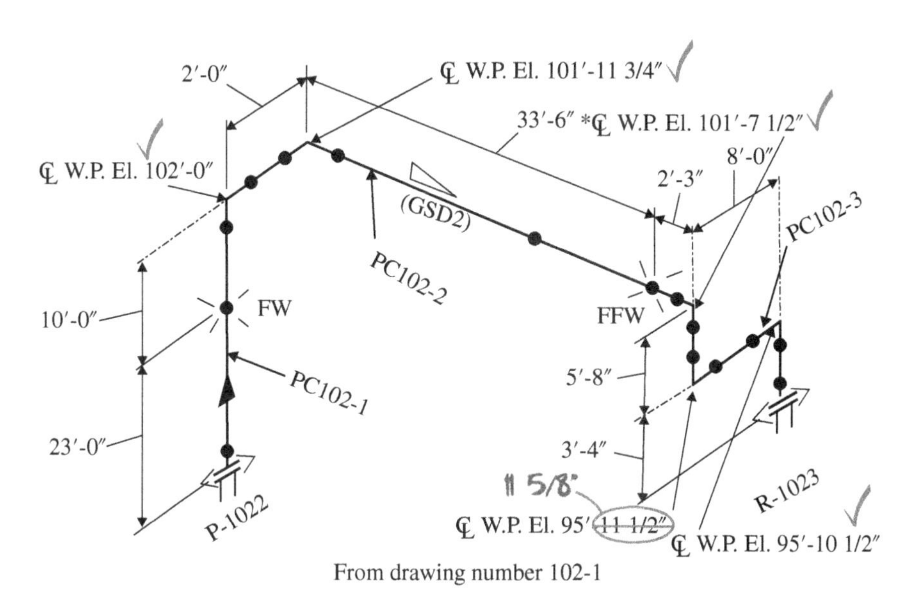

From drawing number 102-1

Figure 4.6.3 Slope map drawing

example, the weld number consists of the drawing number preceded by a "W" for weld with a sequence number following the drawing number. The sequence begins with the number "1" and increases in the direction of the fluid flow. The weld number itself is a line item in a weld log, an example of which can be seen in the

BPE Nonmandatory Appendix B, Form WEL-1 Weld and Examination/Inspection Log. The use of the form is not required but is provided only as an example.

In the slope map example of Figure 4.6.3, each EL point is verified, and the slope from point to point is verified in accordance with Nonmandatory Appendix C. The method described in Appendix C is a suggested method of verifying the slope indicated on the drawings. It is apparent in Figure 4.6.3 that all but one of the ELs agreed with those indicated on the drawing. The increase in EL added more slope to that short run of pipe and was therefore an acceptable deviation. The only modification for the as-built drawing was to change the EL to agree with the examiner's findings.

The weld map, weld log, and the slope map mentioned previously, along with the as-built drawings and all of the other documentation requirements listed in Table 4.3.1, will make up the turnover package that is handed to the owner's representative for their inspection and assessment.

4.6.3 Skid or Module Fabrication

Skid/module fabrication, when processing equipment size and logistics permit, is a much more efficient and controlled method of construction. Fabrication, assembly, and factory acceptance testing (FAT) of a piping system within the controlled environment of a shop facility has the potential to create the proper conditions for fabricating high-purity-type piping systems such as the filtration skid pictured in Figure 4.6.4.

That potential benefits in constructing a skid/module rest in:

- The ability of the designer to properly engineer a self-contained system package or a system made up of multiple skids or modules
- The craftsmanship of the welders and millwrights
- The ability to perform a well-orchestrated FAT
- The planning and logistics of delivering and setting the skid
- The ability to perform a well-orchestrated Site Acceptance Test (SAT)

Whether referred to as skids, super skids, or modules, hereafter referred to as skids, these units can be anything from simple modular pipe rack sections to complex packaged systems or system segments. Such units still require the same turnover documentation as listed in Table 4.3.1 and BPE paragraph GR-5.2.1.1.1, and they still require all of the examination, inspection, and testing prescribed to field installations.

BPE provides requirements and recommendations that pertain to skid design, which, in accordance with BPE paragraph SD-2.4.4.2(g), the hollow tube structural framework of skids shall be sealed to prevent water from getting inside the tubing. In addition, it is recommended per paragraph SD-5.2.1.2 that a common drain port for the skid be provided, as opposed to multiple drain points exiting the skid assembly having to be accommodated. It is recommended too, in accordance

Figure 4.6.4 Filtration skid. *Source*: http://cotterbrother.com. 2016 © Cotter Brothers Corporation

with Nonmandatory Appendix C, paragraph C-1(g), that the base of the skid framework be checked for level prior to verifying slope in the piping.

From the standpoint of fabrication, the same skills and craftsmanship required when fabricating and constructing a system in the field are required when fabricating and constructing a skid package. The main differences are that the crafts performing the work on a skid inside a shop do not have to contend with inclement or adverse weather conditions, depending on whether or not the facility structure is enclosed while their work is being done. All crafts working on a skid package work under the same management and supervision and therefore do not have to contend with other crafts from other contractors, working under other management, occupying the same workspace at the same time.

In looking at the intricate piping, instrumentation, electrical wiring, pumps, and heat exchanger in Figure 4.6.5, it is apparent that multiple crafts would be involved in building such a skid. Building a skid such as this in a shop all of the crafts such as pipe fitters, electricians, and millwrights would be coordinated by the same

Figure 4.6.5 Filtration skid close up. *Source*: http://cotterbrother.com. 2016 © Cotter Brothers Corporation

management team. If this work was done on-site, it would quite possibly be performed by multiple contractors that would have to coordinate with one another, which is not always as efficient as it should be.

Additionally, skid fabrication in a controlled environment is very conducive to high-purity fabrication. It allows for the work to be done in a much cleaner environment than what could be achieved on-site. This attributes greatly to cleanliness of the build.

5

Examination, Inspection, and Testing

5.1 Examination, Inspection, and Testing

In the designing, specifying, and constructing of a bioprocessing facility, there are an enormous number of factors that come into play. There are regulations that have to be followed, specific requirements for materials of construction that have to be adhered to, design nuances of piping systems and equipment that have to meet the all-important requirements for cleanability, and also the self-induced requirements that are put into place to make certain that all of this is accomplished in a well-orchestrated and documented fashion. The last line of defense in making certain that all of this comes together as specified is in the examination, inspection, and testing that takes place throughout the design, fabrication, and installation processes in the building of a bioprocessing facility.

There is frequently a degree of confusion with the semantics of the terms "examination" and "inspection." This is because the two terms inversely point to one another in their definitions—at least one form of their definitions. Those versions of their definitions can be found in the Oxford English dictionary and read as follows:

- Inspect: Examine something to ensure that it reaches an official standard—*Customs officers came aboard to inspect our documents.*
- Examine: Inspect something in detail to determine their nature or condition; investigate thoroughly—*We were forced to examine every item upon receipt.*

But, as mentioned in Section 1.5, in many cases codes and standards find it necessary to either better define a term for its particular use in the code or standard

Bioprocessing Piping and Equipment Design: A Companion Guide for the ASME BPE Standard, First Edition. William M. (Bill) Huitt.

or make clear its intended use by description. The BPE, boiler and pressure vessel (BPV) code, and B31.3 have elected to describe in detail what these terms mean rather than create definitions for the two terms.

In clarifying the intended use of the terms "examination" and "inspection," the BPE Standard defers to the BPV Code and the B31.3 Process Piping Code, wherein there is the following:

- For pressure vessels there is the "authorized inspector" (AI), defined, in part, in ASME BPV Code, Section VIII as:
  - "...an inspector regularly employed by an ASME accredited Authorized Inspection agency, i.e., the inspection organization of a state or municipality of the United States, a Canadian Province, or an insurance company authorized to write boiler and pressure vessel insurance..."
- For piping, tubing, and noncode vessels, an inspector is defined, according to B31.3, as the owner's inspector or the inspector's delegate who:
  - "...shall be the owner, an employee of the owner, an employee of an engineering or scientific organization, or of a recognized insurance or inspection company acting as the owner's agent. The owner's Inspector shall not represent nor be an employee of the piping manufacturer, fabricator, or erector unless the owner is also the manufacturer, fabricator, or erector."
- For piping, tubing, and noncode vessels, an examiner is defined, according to B31.3, as:
  - "...a person who performs quality control examinations for the manufacturer (for components only), fabricator, or erector."

When discussing testing we are referring to system leak testing. Leak testing is an integrity test, which verifies whether or not a piping system was installed properly and securely. Properly meaning that all components have been installed and all connections made. Securely, meaning that all mechanical type joints are connected with a preload force sufficient to seal them against test pressure leakage.

5.2 Examination

Examinations are quality control activities, like inspections, but are performed by manufacturers, fabricators, and installers of piping components and systems, whereas inspections, as we will discuss in the next segment, are performed by the owner's inspector. The Piping Code, ASME B31.3 in this case, had to differentiate between the two because both perform essentially the same function but do so from different perspectives; one is the maker and one the user. Because of these two different perspectives, they fall under different requirements according to the code.

Examiners then are those employed by the manufacturer to perform inspections on fittings such as elbows, tees, and crosses, as well as other components, at any point during production at a manufacturing facility. Examiners are also those employed by a piping fabricator to inspect the work of the fabrication and for the piping system installer for which they will inspect the installed systems for completion and accuracy. In most cases the fabricator and installer are one in the same.

BPE defers qualification requirements for examination personnel to the B31.3 Piping Code, in which paragraph 342.1 of the code states that:

> *Personnel performing nondestructive examination to the requirements of this Code shall be qualified and certified for the method to be utilized following a procedure as described in BPV Code, Section V, Article 1, T-120(e) or (f).*

The BPV Code, Section V, Article 1, T-120(e) and (f) state that:

> *(e) For those documents that directly reference this Article for the qualification of NDE personnel, the qualification shall be in accordance with their employer's written practice which must be in accordance with one of the following documents:*
>
> *(1) SNT-TC-1A, Personnel Qualification and Certification in Nondestructive Testing; or*
>
> *(2) ANSI/ASNT CP-189, ASNT Standard for Qualification and Certification of Nondestructive Testing Personnel*
>
> *(f) National or international central certification programs, such as the ASNT Central Certification Program (ACCP) or ISO 9712:2012-based programs, may be alternatively used to fulfill the training, experience, and examination requirements of the documents listed in (e) as specified in the employer's written practice.*

Currently the BPE examination requirements can be found in each section as those requirements relate to each of those sections. As an example, the minimum examination requirements stated in Part DT under paragraph DT-10 for fittings and process components include the following:

- 100% visual examination for the following:
  - Manufacturer's name, logo, or trademark
  - Alloy/material type
  - Description including size and configuration
  - Heat number/code
  - Process contact surface finish (SF) designation (only one SF designation allowed)
  - Reference to ASME BPE
    - ASME BPE Certificate of Authorization holders shall mark the reference to this standard by applying their ASME certification mark with BPE designator. Refer to Fig. CR-1-1.
    - Non-ASME BPE Certificate of Authorization holders shall only mark "BPE."
  - Pressure rating for valves
  - No damage or other noncompliances

From a physical examination standpoint, it is suggested that a percentage of each lot of material or components be examined by the manufacturer, installing contractor, inspection contractor, or owner/end user for the following:

- Wall thickness (for weld ends only)
- Outside diameter (OD) (for weld ends only)
- Surface finish (as specified)

The per lot examination percentage values are not specified by BPE but are instead left to be determined by the manufacturer, installing contractor, and inspection contractor in conjunction with the requirements of the owner/end user. Should the examination of any single lot of material reveal a defect, an additional 10% shall be selected from the same lot and examined for defects. Should the examination of this initial 10% extended examination group reveal an additional reject of the same defect, an additional 10% shall be selected from the same lot and examined. Should additional defects be found, then either 100% of the lot of material shall be examined from that point on or the entire lot of material simply rejected.

5.2.1 Weld Examination

Excluding manufactured longitudinal seam welds for tubing and fittings, the external surface of all fabricated welds shall be visually examined. And unless specified otherwise by the owner/user, tubing in a fluid service designated as high purity, in accordance with B31.3, does not require radiographic (RT), ultrasonic (UT), or in-process examination.

When agreed to between the fabrication contractor and the owner/end user, sample welds shall be made and submitted to the owner/end user and used to determine an agreed level of quality for welding. The prepared sample welds shall meet all of the acceptance criteria in Table MJ-8.4-1.

During weld production sample, welds shall be made on a regular basis to periodically ensure that the welding machine is operating as it should and that the purge gas is sufficient to prevent discoloration to the extent agreed to. It is recommended that sample welds be made at the following occasions, as a minimum:

- The beginning of each work shift
- Whenever the purge source bottle is changed
- When the automatic or machine welding equipment is changed in some manner

With regard to discoloration, the BPE Standard provides two sets of graphics, as seen in Figures 5.2.1 and 5.2.2, that depict photos of two short tubing sections, having an SF Ra of SF4 (15 μin max.), joined by the GTA welding method and then cut axially to reveal the inside surface of the weld of each section. Each sample represents a weld that has been welded while being subjected to a shielding gas containing varying levels of O_2.

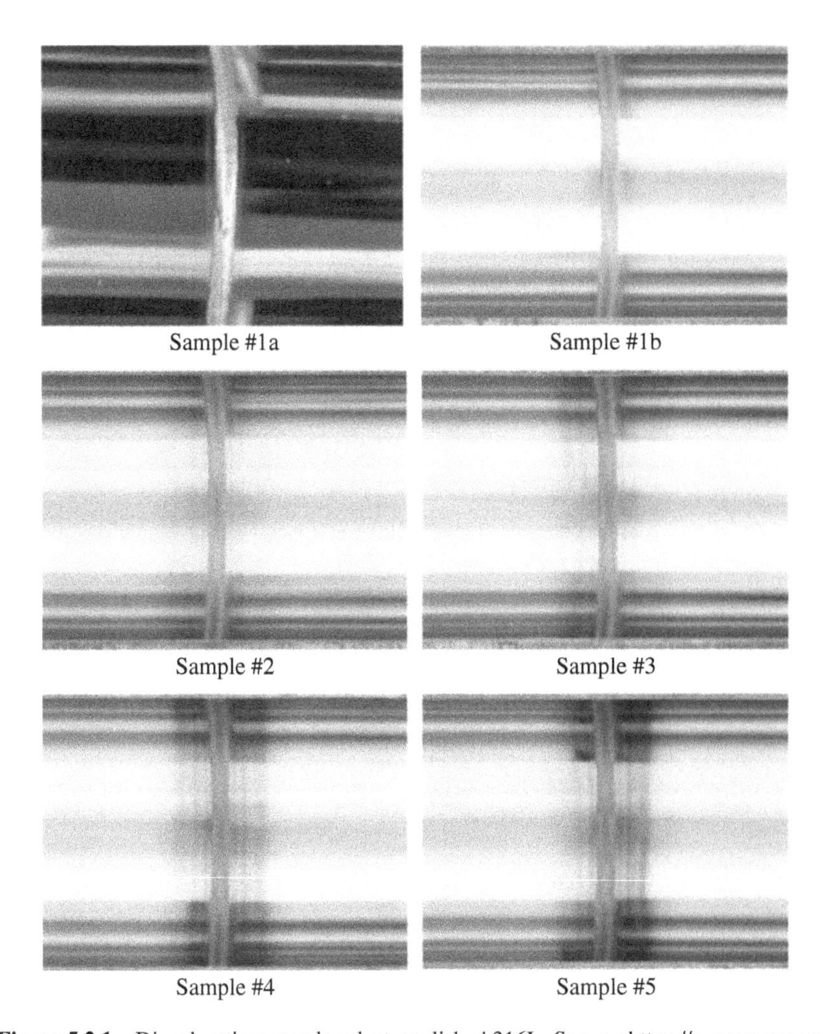

Figure 5.2.1 Discoloration samples electropolished 316L. *Source*: https://www.asme.org/products/codes-standards/bpe-2014-bioprocessing-equipment. © The American Society of Mechanical Engineers (*See plate section for color representation of this figure*)
Note: This graphic is not an acceptable substitute for the referenced insert found in the ASME BPE Standard

The purge gas was a mix of 95% argon and 5% hydrogen. During the welding of the samples in Figures 5.2.1 and 5.2.2, controlled amounts of O_2 were added to the purge gas mixture with the O_2 being measured on the downstream side of the weld. Electropolished weld samples #1 through #4 in Figure 5.2.1 are acceptable samples, #5 is not. Mechanically polished weld samples #1 through #3 in Figure 5.2.2 are acceptable samples, #4 and #5 are not .Following are the O_2 measurement readings in conjunction with the sample numbers represented in both figures:

- Sample #1a—10 ppm
- Sample #1b—10 ppm

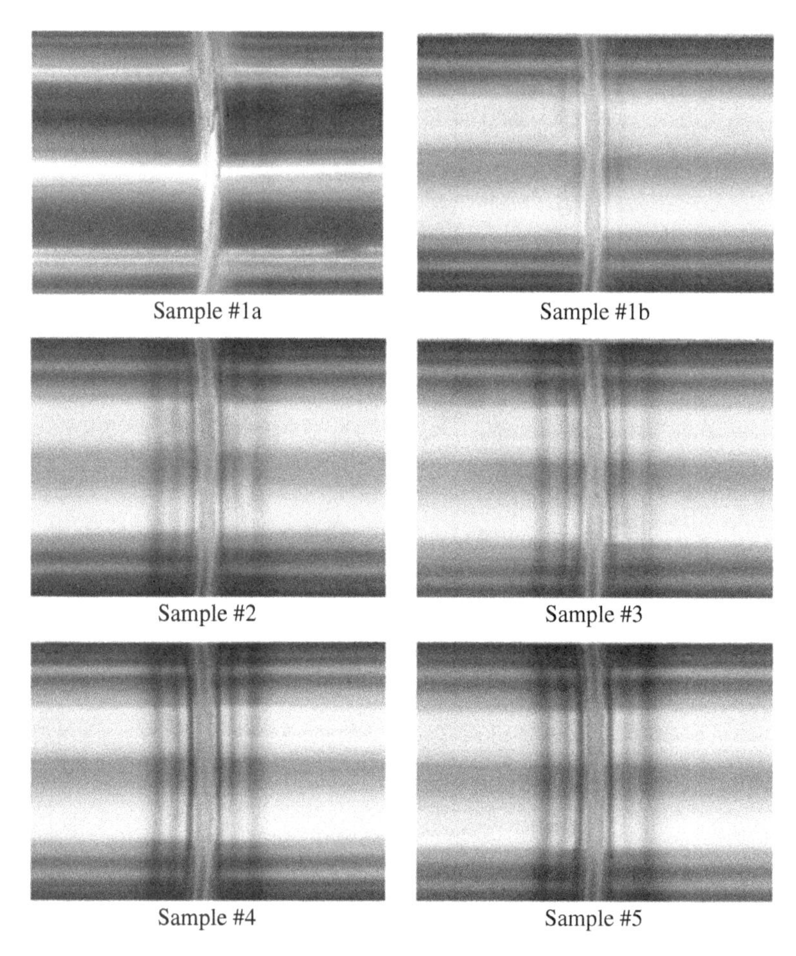

Figure 5.2.2 Discoloration samples of mechanically polished 316L SS. *Source*: https://www.asme.org/products/codes-standards/bpe-2014-bioprocessing-equipment. © The American Society of Mechanical Engineers (*See plate section for color representation of this figure*)
Note: This graphic is not an acceptable substitute for the referenced insert found in the ASME BPE Standard

- Sample #2—25 ppm
- Sample #3—35 ppm
- Sample #4—50 ppm
- Sample #5—80 ppm

This process of assessing acceptability of a weld by means of the degree of discoloration is founded on corrosion resistance testing that was done in the process of creating the samples pictured in Figures 5.2.1 and 5.2.2. The two tests used in this effort were ASTM G150, critical pitting temperature (CPT) test, and a modified ASTM G61, potentiodynamic polarization corrosion test.

The ASTM G150 test is used to determine the CPT of a material, which is the temperature above which pitting corrosion can initiate when in contact with a corrosive fluid. The higher the CPT value, the greater the materials resistance to corrosion.

The modified ASTM G61 test is used to determine a value referred to as E_{PIT}, which is a value that indicates a resistance to pitting corrosion at room temperatures. The higher the E_{PIT} value, the more resistant to pitting corrosion the material is. Both values reflect a corrosion resistance ranking enabling the engineer to make comparative assessments between different materials and between different material surfaces for their corrosion resistance capacity.

What occurs with the change in discoloration (heat tint), from no discoloration to the darker brown and blue tones, is that the corrosion resistance of the stainless steel becomes progressively diminished as the heat tint darkens. In using these tests to determine the corrosion resistance of the material in the heat-affected zone (HAZ) of the samples, it was learned that:

- Acceptable levels of discoloration in the HAZ of mechanically polished stainless steel were those that exhibited similar corrosion resistance to that of cold-rolled, mill-finished, 316L stainless steel.
- Acceptable levels of discoloration in the HAZ of electropolished stainless steel were those that exhibited similar corrosion resistance to that of unwelded, electropolished 316L stainless steel.
- The as-welded HAZ of mechanically polished tubing, given the same level of discoloration as electropolished tubing, will exhibit a lower resistance to pitting corrosion than that of the electropolished tubing.

These sample welds shall meet the acceptance criteria as listed in table MJ-8.4-1. Sample welds for polymeric tubing shall meet the acceptance criteria of table MJ-9.7.1-1. And likewise, for either metallic or nonmetallic tubing, sample coupons and welding machine printed or downloaded welding parameter records do not have to be retained after written acceptance by the owner, owner's representative, or inspector.

5.3 Inspection

Responsibility of an inspector, in accordance with B31.3 paragraph 340.2, is stated as follows:

> *It is the owner's responsibility, exercised through the owner's Inspector, to verify that all required examinations and testing have been completed and to inspect the piping to the extent necessary to be satisfied that it conforms to all applicable examination requirements of the Code and of the engineering design.*

It is therefore the responsibility of the inspector to oversee and ensure that all of the examination, testing, and documentation required by the code or specified standard in the fabrication, installation, and testing of piping systems are in accordance with the requirements of the specified code, standard, and owner's specifications.

BPE table GR-4.2-1 provides a long list of capabilities that an inspector must possess. The list is broken down into the different levels of qualification, making it apparent what capabilities are required at the different inspector levels. In table GR-4.2-1 the capability requirements for the lower tiered qualification levels extends into the upper tiered qualification levels.

BPE paragraph GR-4.2.1 identifies the four levels of qualification for an inspector. They include:

1. Trainee
2. Quality Inspector Delegate 1 (QID-1)
3. Quality Inspector Delegate 2 (QID-2)
4. Quality Inspector Delegate 3 (QID-3)

Each of the previously listed levels of qualification for an inspector is defined under BPE paragraph GR-4.2.1 as follows:

(a) **Trainee**. An individual who is not yet certified to any level shall be considered a trainee. Trainees shall work under the direction of a certified Quality Inspector Delegate and shall not independently conduct any tests or write a report of test results.
(b) **Quality Inspector Delegate 1 (QID-1)**. This individual shall be qualified to properly perform specific calibrations, specific inspections, and specific evaluations for acceptance or rejection according to written instructions. A QID-1 may perform tests and inspections according to the capabilities' requirements under the supervision of, at a minimum, a QID-2.
(c) **Quality Inspector Delegate 2 (QID-2)**. This individual shall be qualified to set up and calibrate equipment and to interpret and evaluate results with respect to applicable codes, standards, and specifications. The QID-2 shall be thoroughly familiar with the scope and limitations of the inspection they are performing and shall exercise assigned responsibility for on-the-job training and guidance of trainees and QID-1 personnel. A QID-2 may perform tests and inspections according to the capabilities' requirements.
(d) **Quality Inspector Delegate 3 (QID-3)**. This individual shall be capable of establishing techniques and procedures; interpreting codes, standards, specifications, and procedures; and designating the particular inspection methods, techniques, and procedures to be used. The QID-3 shall have sufficient practical background in applicable materials, fabrication, and product technology to establish techniques and to assist in establishing acceptance criteria when none are otherwise available. The QID-3 shall be capable of training personnel. A QID 3 may perform tests and inspections according to the capabilities' requirements.

As declared in B31.3 paragraph 340.3, "The owner's Inspector and the Inspector's delegates shall have access to any place where work concerned with the piping installation is being performed. This includes manufacture, fabrication, heat treatment, assembly, erection, examination, and testing of the piping. They shall have the right to audit any examination, to inspect the piping using any examination method specified by the engineering design, and to review all certifications and records necessary to satisfy the owner's responsibility stated in para. 340.2."

While it does not mention it, there are certain considerations that should be given in acquiring such stated access. Unless concerns arise, or other conditions prevail, demanding that agreements be made to the contrary, access to facilities in which the aforementioned work is performed should be during normal working hours with the courtesy of making an appointment for a prearranged (announced) visit to such facilities.

A scope of the inspection process is proprietary in nature, being based largely on an owner's history with the contractors; the type of work involved; what is demanded by regulations, codes, standards; and a sense of what might placate the owner's own sense of assurance of all requirements being met.

As mentioned earlier, it is not the inspector's place to inspect a standardized percentage of components or fabrication. It is their job to ensure that the examiner did that and that the examiner has documentation to state as much.

5.4 Leak Testing of Piping

One of the basic steps an owner has to perform is in the designation of all fluid services in a facility that falls under the requirements of the B31.3 *Process Piping* Code. Such a requirement to adhere to the B31.3 Piping Code can be by government regulation or by contractual stipulation. It requires that all fluid system piping, aside from those designated as plumbing, or facility services under ASME B31.9 *Building Services Piping* be assigned a fluid service designation as follows:

- (a) *Category D Fluid Service*: A fluid service in which all of the following apply:
    - (1) The fluid handled is nonflammable, nontoxic, and not damaging to human tissues as defined in paragraph 300.2
    - (2) The design gage pressure does not exceed 1035 kPa (150 psi)
    - (3) The design temperature is not greater than 186°C (366°F)
    - (4) The fluid temperature caused by anything other than atmospheric conditions is not less than −29°C (−20°F)
- (b) *Category M Fluid Service*: A fluid service in which both of the following apply:
    - (1) The fluid is so highly toxic that a single exposure to a very small quantity of the fluid, caused by leakage, can produce serious irreversible harm to persons on breathing or bodily contact, even when prompt restorative measures are taken.
    - (2) After consideration of piping design, experience, service conditions, and location, the owner determines that the requirements for normal fluid service do not sufficiently provide the leak tightness required to protect personnel from exposure.

- (c) *Elevated Temperature Fluid Service*: A fluid service in which the piping metal temperature is sustained equal to or greater than *Tcr* as defined in table 302.3.5, general note (b).
- (d) *High-pressure Fluid Service*: A fluid service for which the owner specifies the use of Chapter IX for piping design and construction (see also paragraph K300).
- (e) *High-purity Fluid Service*: A fluid service that requires alternative methods of fabrication, inspection, examination, and testing not covered elsewhere in the code, with the intent to produce a controlled level of cleanliness. The term thus applies to piping systems defined for other purposes as high purity, ultrahigh purity, hygienic, or aseptic.
- (f) *Normal Fluid Service*: A fluid service pertaining to most piping covered by this code, that is, not subject to the rules for Category D, Category M, Elevated Temperature, High-pressure, or High-purity Fluid Service.

Of the six fluid categories listed previously, only the High-purity Fluid Service can be considered an add-on to a basic category such as Category D, Category M, or Normal Fluid Service. Categorizing a fluid service establishes a level of system integrity and its associative level of examination/inspection. And while the High-purity Fluid Service designation does bring with it its own set of criteria, as identified in B31.3 Chapter X, there are additional considerations with regard to the fluid itself being a Category D, Category M, or Normal fluid in addition to that of a High-purity Fluid Service.

Simply designating a fluid service as high purity then is not sufficient in achieving proper leak testing, for example. If a High-purity Fluid Service also falls within the criteria for a Category D Fluid Service, then a hydrostatic test is not required. An initial service leak test would be sufficient. If the High-purity Fluid Service falls within the criteria for a Category M Fluid Service, then BPE paragraph UM345 would become a requirement. This leads to the point that when designating a fluid service as high purity that designation needs to be paired with a designation of Category D, Category M, or Normal.

In performing leak testing on high-purity systems, refer to Appendix A of this book and to the requirements of B31.3 Chapter X, paragraph U345 and UM345. In doing so the High-purity Fluid Services should also have the secondary fluid category, as mentioned previously, designated as well. Included in Appendix A, as attachments, are forms that can be used as is or replicated to document the testing.

6

Equipment and Component Quality

6.1 Assured Quality

Procuring quality equipment and components consistently from a reliable manufacturer or supplier is fundamental in the construction or maintenance of a bioprocessing facility. Finding out too late that an installed equipment item or a line of components that make up the already installed piping systems do not meet the specifications they were purchased against can impact a project or put a processing campaign at risk of failure.

From the standpoint of an owner/end user, it is a matter of understanding and being familiar with, not only the specifications and design requirements of the equipment, components, and material that is being procured, but also the method by which manufacturers and/or suppliers of these items are approved. The secondary component to this is in the process of receipt inspection when these items are delivered.

The vetting process, in assessing and approving manufacturers and suppliers, is a critical step in the overall procedure of specifying, procuring, and verification of purchased items. Along those lines, the BPE Standard has put in place its certification program—a program that audits applicants for BPE Certification to ensure that they have the proven ability to manufacture tubing and fitting, at this time, in a controlled, organized, and consistent manner.

6.2 BPE Certification

The BPE Certification program was initiated at the time of the 2009 edition of the Standard. However, it didn't go into effect until after the 2012 edition of the Standard was published. The first company to be audited and approved to be

Bioprocessing Piping and Equipment Design: A Companion Guide for the ASME BPE Standard, First Edition. William M. (Bill) Huitt.

certified as BPE compliant was United Industries, Inc., as indicated by their Certificate No. BPE-101. Their Certificate of Authorization was issued in January 2013. As you will notice in Figure 6.2.1c, there are currently five ASME BPE Certificate Holders with several more applicants in the pipeline to be reviewed for readiness.

Select	Company Name	Plant Address	City	State/Province	Country/Region	Certificate	Status
☐	Dockweiler AG	An der Autobahn 10/20 Mecklenburg-Vorpommern	Neustadt-Glewe		Germany	BPE- BPE-105	Active
☐	EGMO Ltd.	1 Hayotsrim St.	Nahariya		Israel	BPE- BPE-102	Active
☐	King Lai Hygienic Materials Co., Ltd	22 West Lufeng Road East Industrial Zone Kunshan Economic & Technical Development Zone	Jiangsu Province, P.R.		China	BPE- BPE-104	Active
☐	RathGibson	2505 Foster Avenue	Janesville	Wisconsin	United States	BPE- BPE-103	Active
☐	United Industries, Inc.	95 Lakeview Drive	Selmer	Tennessee	United States	BPE- BPE-101	Active

Figure 6.2.1 (a and b) Opening window of BPE Certificate Holder search and (c) list of current BPE Certificate Holders

To learn how to apply for certification, go to the following ASME web page:

https://www.asme.org/shop/certification-and-accreditation/bioprocessing-equipment-certification (*Note: If this address should change in the future, as they often do, simply do a search on the* www.asme.org *website.*)

An alphabetical listing of ASME BPE Certificate Holders can be found at the following website address:

http://caconnect.asme.org/CertificateHolderSearch/Index (*Note: If this address should change in the future, as they often do, simply do a search on the* www.asme.org *website.*)

In going to the Certificate Holder Search website address, you will be presented with the window as shown in Figure 6.2.1a.

Looking at the screen in Figure 6.2.1, click on the drop-down menu for "Certificate Type", and then locate and highlight "BPE." Click Search.

The screen shown in Figure 6.2.1c will appear with the list of BPE Certificate Holders.

Those BPE Certificate Holders listed are manufacturers who have been issued a Certificate of Authorization to mark their components with the ASME certification mark. Meaning that they have been audited and approved to mark their components certified as to being manufactured in accordance with all requirements of the ASME BPE Standard with respect to their line of products. They are thereby given the authority to mark their components as being certified to the ASME BPE Standard.

Having demonstrated their capability to provide items in conformance with the BPE Standard, they are thereby issued a Certificate of Authorization and are authorized to mark only those items covered under the certification with the certification mark and designator. Figure 6.2.2 is a graphic of those symbols.

And while many unauthorized manufacturers of tubing and fittings are currently producing such products and marking those products as BPE compliant, they are not certified under the rules of the Board on Conformity Assessment (BCA) for

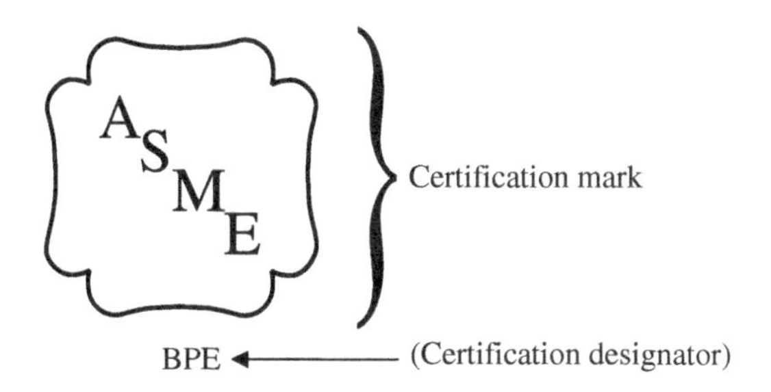

Figure 6.2.2 ASME certification mark and certification designator. *Source*: https://www.asme.org/products/codes-standards/bpe-2014-bioprocessing-equipment. © The American Society of Mechanical Engineers

BPE Certification. If the tubing or fitting is marked only with the letters "BPE" and not with the ASME certification mark, they are not BPE certified.

The ASME Certification program is governed by the BCA. To better understand how the certification program works, the BCA has published standard CA-1 conformity assessment requirements, Ref. Figure 6.2.3, that describes

ASME CA-1–2014
(Revision of ASME CA-1–2013)

Conformity
Assessment
Requirements

Figure 6.2.3 Cover of CA-1. *Source*: https://www.asme.org/products/codes-standards/bpe-2014-bioprocessing-equipment. © The American Society of Mechanical Engineers

the process and requirements of the certification program. CA-1 can be downloaded at:

https://www.asme.org/wwwasmeorg/media/ResourceFiles/Shop/Certification%20%26%20Accreditation/BPE-Certification/Conformity-Assessment-Requirements.pdf (*Note: If this address should change in the future, as they often do, simply do a search on the* www.asme.org *website.*)

6.3 A Quality Management System

A key element in obtaining a BPE Certificate of Authorization is in a company's development, organization, and content of their quality management system (QMS). The QMS establishes a written link between the operation and management of manufacturing a BPE compliant product with that of the requirements of the most current BPE Standard.

The Committee on BPE Certification (CBPEC) created a QMS checklist to assist the ASME BPE Survey Team in their survey of a manufacturer applying for BPE Certification. This checklist is general in nature in which not all line items will apply to all applicants. This checklist is a tool that can be used by a company in helping to prepare for application of a BPE Certification. The online checklist, which can currently be found at:

http://www.asme.org/wwwasmeorg/media/ResourceFiles/Shop/Certification%20%26%20Accreditation/BPE-Certification/BPE-Certification_Form_Quality-Management-System-Checklist.pdf, states the following in its introduction: (*Note: The previous address will change upon issue of the 2016 edition of the BPE. At that time simply do a search on the* www.asme.org *website.*)

> *In addition to assisting the ASME BPE Survey Team, this Checklist is provided to Applicants for their use in verifying and identifying the paragraph(s) where their QMS Manual is to be compared for compliance with the BPE Standard. The QMS Manual must contain the controls for implementing the Quality Management System, but it is not required to contain all the detailed programmatic requirements which will be found in working documents, such as, but not limited to, procedures and working instructions. The QMS Manual is not a reiteration of the requirements specified in the BPE Standard but rather an auditable Manual which identifies who, what, when, where & how the Quality Management System meets the current requirements of the BPE Standard.*

The process of filling out the online checklist (a copy can be printed for your records prior to submittal) will help a great deal in creating a company's QMS that will be acceptable when reviewed during the survey. It is not suggested or even recommended that a QMS resemble the checklist. In the process of applying for certification, the applicant is asked to submit their QMS for review and consideration. With the understanding that an ISO 9001 QMS is not the same as that of a BPE QMS, simply changing the title of an ISO 9001 QMS will typically not be approved.

To help with compiling a proper BPE QMS, the checklist used by the survey team will provide line by line key elements that will help tie in the requirements of the BPE with the process of manufacturing management and quality control. In doing this keep in mind that you only need to include those elements that pertain to your product and method of manufacture. The more organized and relevant the contents of a company's QMS, the better the chance for a survey to be performed in a more timely and productive manner. This will be the first task the survey team will tackle when the survey begins. They will go through the QMS manual and substantiate every reference made in connecting manufacturing and quality control elements back to the BPE Standard. That's in addition to everything else they will review in the QMS.

The survey itself, as stated in the checklist instructions, will take place in phases as follows:

1. Manual Review: Shall be performed by the Survey Team and observers authorized by ASME only, on the first day of the survey. This review will normally be held in the hotel remote from the Applicant's facility.
2. Entrance Meeting/Facility Tour: Will be held on the second day. The entrance meeting will provide the Applicant and the Survey Team an opportunity to: introduce themselves, review the Certificate and scope of activity applied for, and to establish the survey agenda. ASME fully encourages executive/senior management to attend the entrance meeting. During this meeting the Applicant may, if they wish, give a presentation of the company, products, personnel, etc.
3. Implementation: The Applicant is expected to demonstrate the implementation of the entire Quality Management System. If any deficiencies are discovered during the survey, they will be immediately identified to the Applicant, in order to provide them an opportunity to correct them prior to the conclusion of the survey.
4. Team Closed Meeting: This meeting will be held at the Applicant's facility prior to the exit meeting. This meeting will be attended only by the Survey Team members and observers authorized by ASME. During this meeting the Survey Team will review the results of the survey and vote on the recommendation that the team will present to ASME.
5. Exit Meeting: This meeting will be held with the Applicant's management and will review the results of the survey. ASME fully encourages executive/senior management to attend the exit meeting. If there were any deficiencies issued they will be reviewed and the Applicant advised of their status. The Survey Team's recommendation to ASME will be made known. The Applicant will be allowed to ask any questions relative or pertinent to the survey. At the conclusion of the exit meeting, the survey is officially ended and there will be no more discussion of the survey.

A broad-brush timeline, Ref. Figure 6.3.1, provides a graphic example of the steps taken in the process of becoming certified and the approximate length of

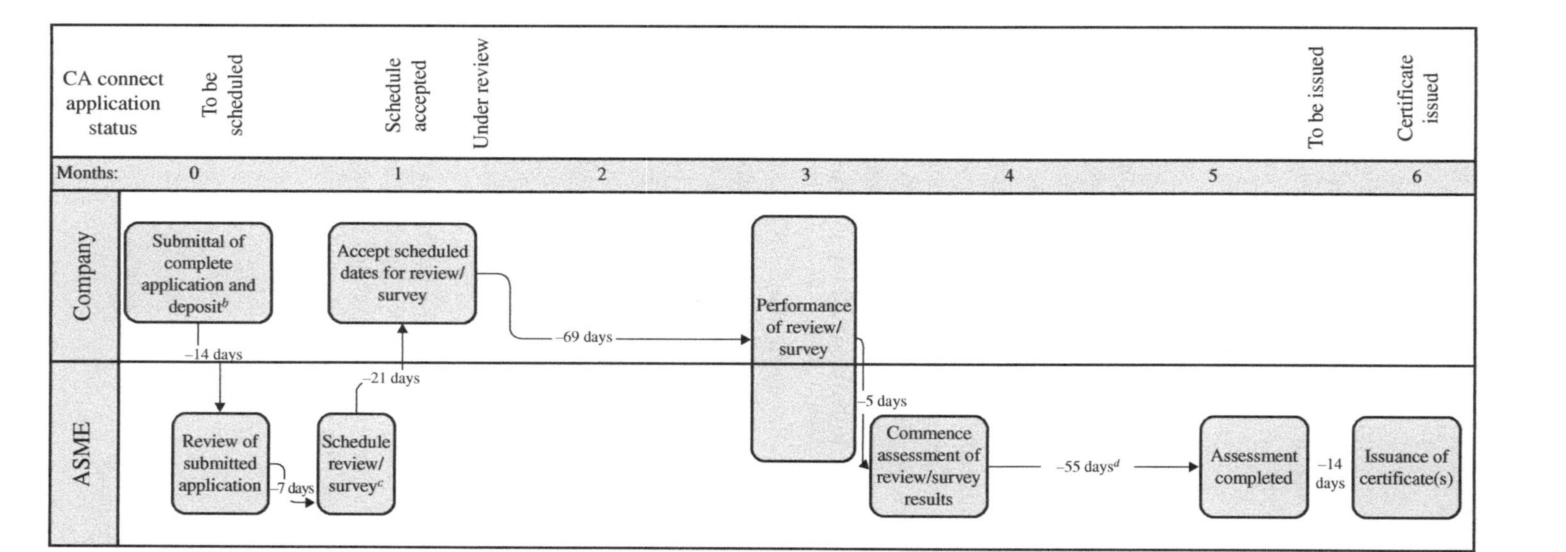

Notes:
[A] This timeline is based on submittal of application and deposit 6 months prior to expiration date of certificates and when ASME conducts the review/survey. Time periods are estimated and may vary for different applicants. If ASME does not conduct review, then the timeline will be 2–4 months. This timeline is subject to change.
[B] Completed application includes signed agreement form.
[C] Reviews/surveys are usually scheduled three months in advance. Scheduled dates may vary depending on applicants/team availability.
[D] Non-Boiler surveys require assessment by the responsible accreditation/certification committee which may take 3–4 weeks. Assessment of boiler reviews may take 2–3 weeks. It may take up to 30 days to receive the report for the review/survey.

Figure 6.3.1 General certification timeline. *Source*: https://www.asme.org/products/codes-standards/bpe-2014-bioprocessing-equipment. © The American Society of Mechanical Engineers

time the process might take. The graphic in Figure 6.3.1 can also be found on the ASME website at:

http://www.asme.org/wwwasmeorg/media/ResourceFiles/Shop/Certification%20%26%20Accreditation/Certificate%20Application/Timeline-of-the-Certification-Process.pdf (*Note: If this address should change in the future, as they often do, simply do a search on the* www.asme.org *website.*)

6.4 Purpose

Typically a fabricator or end user specifying material and components to be used in the fabrication of high-purity pharmaceutical or bioprocessing equipment and piping systems will develop a short list of approved manufacturers and suppliers. This short list of manufacturers and suppliers is typically developed at the expense of the fabricators and owners seeking reliable sources for the material and components needed to construct these high-purity processing systems.

The expense incurred by these companies is in the time and travel cost required in performing their own personal surveys of manufacturers and suppliers—a program in which a set of criteria has to be identified for the various types of material and components in order to have something to measure against when performing a survey. At that point it then becomes a process in which a manufacturer or supplier is surveyed against the stated criteria. While such individual and isolated programs were far from systematic or comprehensive in their survey assessments, they were a necessary evil. Meaning that, prior to the ASME BPE Certification, there was no alternative.

With this, the beginning stages of the BPE Certification program, becoming more and more established, bring with it a continuity and a sound basis of assessment in creating quality assurance throughout the engineering and constructing of bioprocessing facilities. The same survey team personnel used to perform surveys on BPVC Section VIII vessels have been trained to utilize their skills, knowledge, and experience in performing surveys on applicants for BPE Certification.

As shown in Figure 6.2.1c, a listing of BPE certified vendors can be easily selected from the online ASME Certificate Holder search engine to create a short list of vendors for a company's own use. The search list of certificate holders will continue to get more populated with not only tube and fitting manufactures but also the following in no particular order:

- Pressure vessel fabricators
- Valve manufacturers
- Seals and gasket manufacturers
- Polymeric tube and fittings manufacturers
- Packaged equipment fabricators

- Service providers (engineers, electropolishing, passivation, inspection, fabricators)
- Others

The quality assurance that comes from this BPE Certification program will have a positive impact on the industry worldwide. It's the final link in a chain that has created, and continues to create, continuity and assurances in the design, construction, and operation of pharmaceutical and bioprocessing facilities.

7

Design

7.1 BPE Scope of Design

In its introduction the ASME B31.3 Process Piping Code states that "The designer is cautioned that the Code is not a design handbook; it does not eliminate the need for the designer or for competent engineering judgment." That statement, while true enough for the Process Piping Code, does not apply to the BPE Standard. In fact, the BPE Standard essentially exists as a component standard, a design handbook, and an industry standard. If the BPE Standard had a focal point around which all other parts revolved, it would be Part SD—Systems Design. Figure 7.1.1 shows how all parts of the Standard are interconnected with Part GR—General Requirements, which encompasses all parts with all parts feeding into Part SD. All parts and all subcommittees are on equal footing, but each part interacts in different ways with their counterparts.

Part SD's scope states that it will "...provide requirements for the specification, design, fabrication, and verification of process equipment and systems that are fit for intended use and minimize risk to product quality." It goes on to say that it provides, "...design guidelines that should be applied at the discretion of the owner/user on the basis of assessed risk to the product. Figures in this Part are intended to illustrate accepted applications of general design principles and are not intended to limit alternate designs...." Additionally it "...encompasses requirements for equipment, process systems, and utilities that could potentially impact product quality."

Bioprocessing Piping and Equipment Design: A Companion Guide for the ASME BPE Standard, First Edition. William M. (Bill) Huitt.

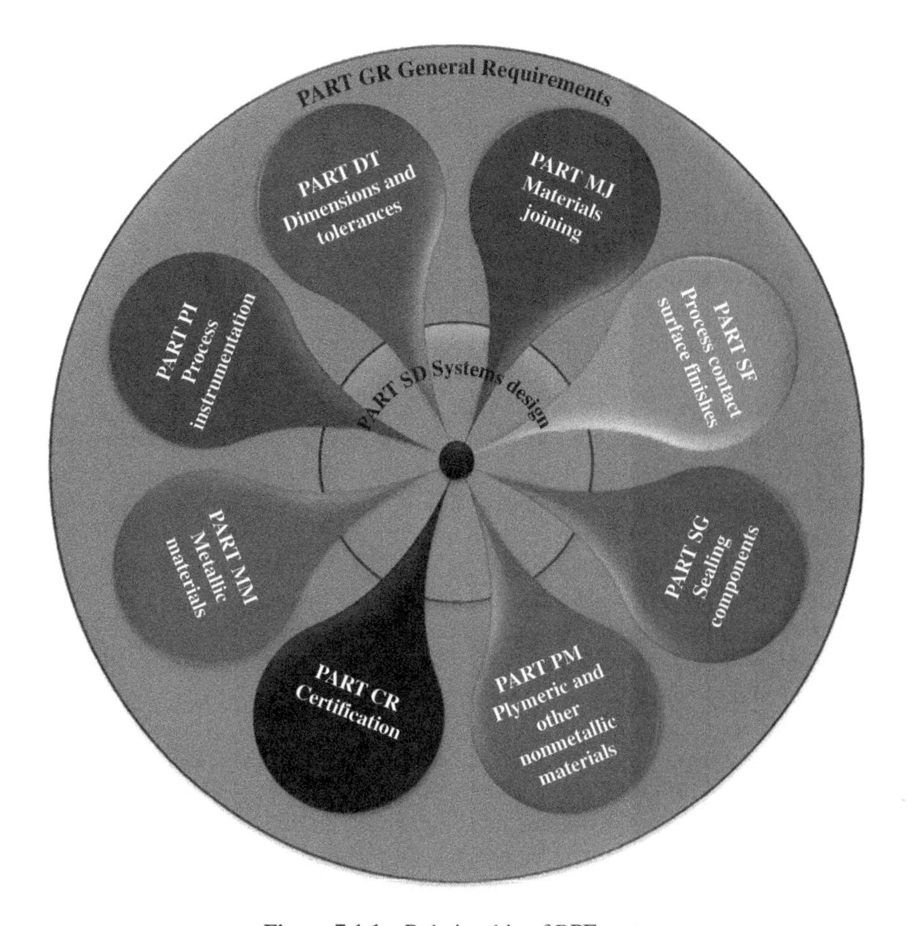

Figure 7.1.1 Relationship of BPE parts

7.2 Intent of Part SD

Part SD provides guidance and value-added design criteria that provides continuity of system design across the pharmaceutical and bioprocessing industry while leaving open the door to design innovation and proprietary needs. Part SD, like all parts in the BPE Standard, goes to great lengths in maintaining a pragmatic balance between what is required in achieving high-purity system design and what is provided as guidance and recommendations.

One of the major hurdles that existed in designing and in building a pharmaceutical facility prior to 1997, year of the first BPE publication, was in getting the component manufacturer, owner/user, designer, and constructor all on the same page in having a common understanding of requirements and ways in which to achieve them, meaning that the expectations of one were not necessarily the same as the others. A successful project, one that is completed on time, within budget, complying with all government regulations and meeting all expectations of the owner/user, was one in which the owner/user, on their own or through an intermediary, had to patch together a well-written and well-defined set of specifications and requirements to achieve their anticipated results. This quite frequently did not happen as planned or as hoped for.

The problem still exists today to some extent but not nearly to the degree that it did prior to 1997. The owner/user holds responsibility for compliance of a facility, because they are the owner. And unknowingly circumventing regulations or the intent of regulatory edicts is unacceptable in the eyes of the Food and Drug Administration (FDA). The owner is who the FDA will deal with in their inspection of a facility during commissioning, not the engineering firm and not the constructor.

That interaction, in addressing and complying with any Forms 483 from the FDA, will be dependent upon contractual obligations between the owner/user, engineering firm, and constructor. In other words, if the noncompliance issue was a fault of the installing contractor or a case of noncompliant design by the engineer, then the owner/user may get retribution by having them correct the issue as part of the contract. But the owner still carries the responsibility. They are still the responsible party that the FDA will look to for corrective action. And, unless there is definitive contractual requirements making the installing contractor, construction manager, or engineer responsible for compliance of their work, their boundaries of responsibility could become an issue.

In the planning and development of a pharmaceutical facility, it is essential that the owner/user ensure that all documents and procedures, from piping material specifications to layout drawings, convey, in a clear and concise manner, the expectation they have for the end result. This includes compliance requirements as well as craftsmanship in the construction. If this is not clear to the engineers and contractors designing and building a facility, the owner may not get what they expect.

The BPE Standard facilitates the dialog and understanding between the component manufacturer, owner/user, engineer, and constructor, as well as the FDA, by providing vetted consensus standard methodologies that bring all factions of a project team to the same table and same understanding, the intent being to coalesce industry practices, know-how, and understanding into a single accredited document. This breaks down the autonomy method of designing and constructing high-purity facilities and, as mentioned earlier, gets everyone on the same page. By simply providing the ability to reference a paragraph, figure, or table in the Standard, it mitigates confusion and misunderstandings.

7.3 It's a Bug's Life

7.3.1 Perspective on Bacteria

Designing high-purity piping systems for drug manufacturing entails having at least some basic understanding of what plant operations will have to contend with once the facility is built and in operation. This gets to the most elemental reason and purpose that guide not only FDA regulations but also design, fabrication, and component manufacturing. It is that of bacteria control, referred to as bioburden control, in other words, system cleanability.

The drugs that are to be manufactured could be an oral medication, a topical solution, or a parenteral drug. With drugs being applied by any of these three means, the opportunity for latent bacteria, if inadvertently contained in the drug, to attach itself to a human host is a very likely end result to a contaminated drug product and even more so with injectables.

Aside from the pressure containment aspect of designing such process systems, there is the overarching concern to design these systems to be highly cleanable with minimal dead legs. But if unavoidable keep dead legs within their maximum limits, as defined by the BPE Standard.

In order to gain some perspective as to what system design is working against and what system operators have to continually deal with, we are going to discuss "bugs," not the kind of bugs you can actually see, the kind with six or more legs, the kind with protruding mandibles, or who knows what else, but the kind of bugs you cannot see without highly sophisticated microscopes or even nanoscopes. I am referring to bacteria. In referencing Figures 7.3.1, 7.3.2, 7.3.3, 7.3.4, 7.3.5, 7.3.6, 7.3.7, 7.3.8, and 7.3.9, you will get some sense of what the FDA and the BPE are regulating, designing, and, in essence, fighting against.

There are 10 slides of artistic renderings of scanning electron microscope (SEM) micrographs depicting various viruses, bacteria, and cells at varying magnitude. The first of these, Figure 7.3.1, is showing a rhinovirus at approximately 30 nm across. As is implied by its name, this is a virus and not a bacterium. The main difference is that the virus has no cells while the bacteria are unicellular. And while the bacteria can exist and thrive in almost any environment, both hostile and passive, the virus requires a living host. The rhinovirus is the most common cause for the common cold.

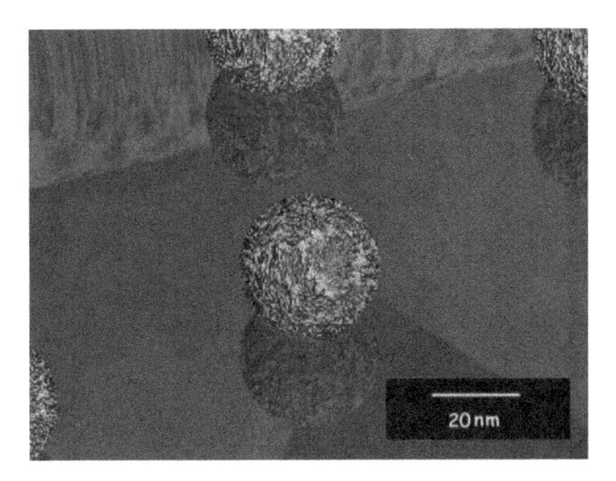

Figure 7.3.1 Rhinovirus at 1 000 000×

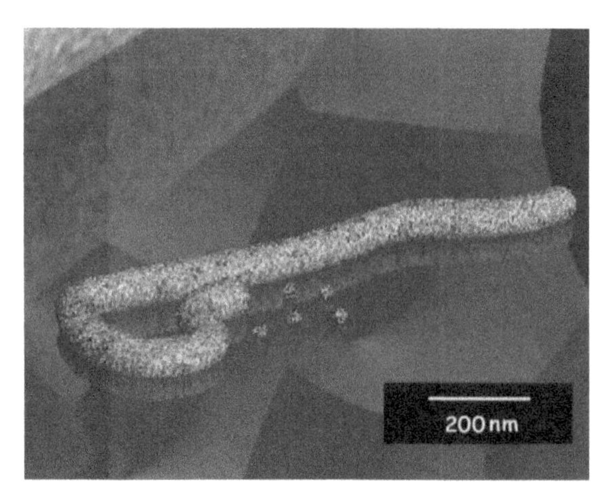

Figure 7.3.2 Ebola virus at 100 000×. *Source*: Cells Alive, http://www.cellsalive.com/. © 2015 James A. Sullivan

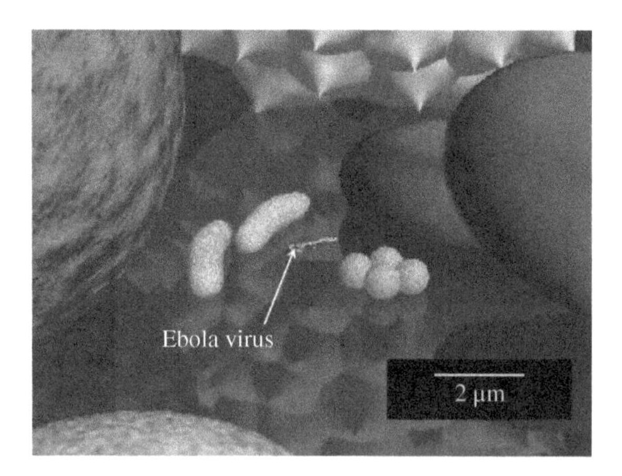

Figure 7.3.3 Various items at 10 000×. *Source*: Cells Alive, http://www.cellsalive.com/. © 2015 James A. Sullivan

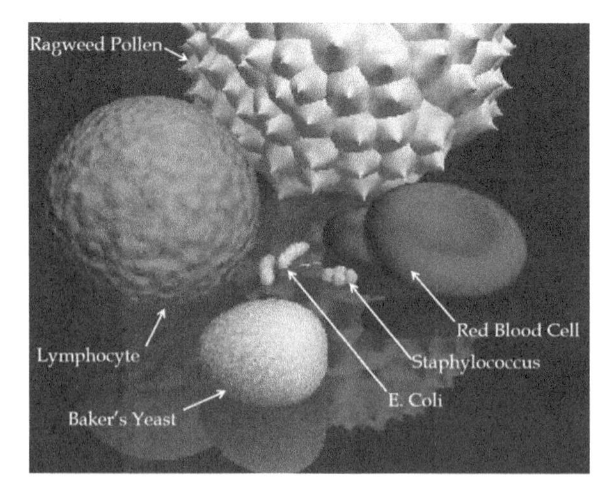

Figure 7.3.4 Various items at 7000×. *Source*: Cells Alive, http://www.cellsalive.com/. © 2015 James A. Sullivan

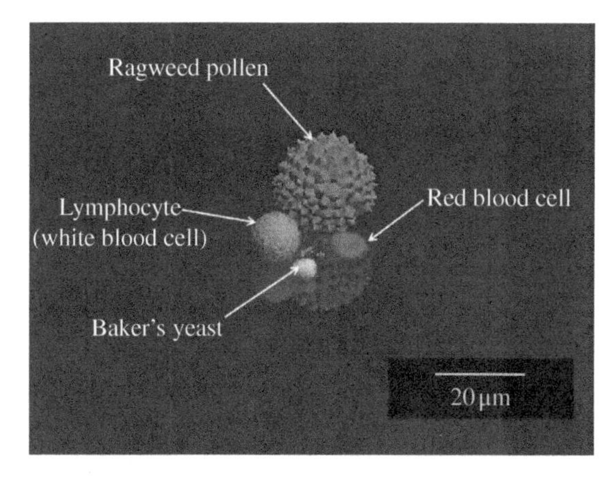

Figure 7.3.5 Various items at 1000×. *Source*: Cells Alive, http://www.cellsalive.com/. © 2015 James A. Sullivan

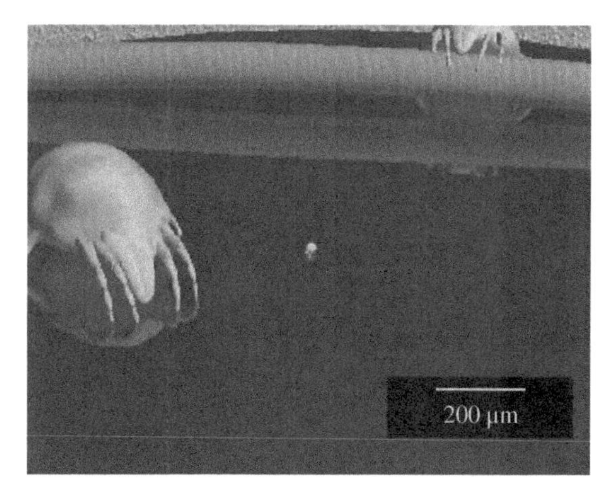

Figure 7.3.6 Various items at 100×. *Source*: Cells Alive, http://www.cellsalive.com/. © 2015 James A. Sullivan

Figure 7.3.7 Various items at 70×. *Source*: Cells Alive, http://www.cellsalive.com/. © 2015 James A. Sullivan

Figure 7.3.8 Various items at 10×. *Source*: Cells Alive, http://www.cellsalive.com/. © 2015 James A. Sullivan

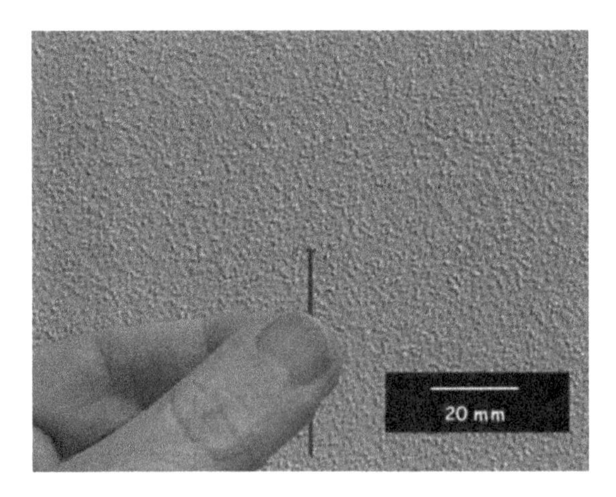

Figure 7.3.9 All on a 2 mm dia pin head. *Source*: Cells Alive, http://www.cellsalive.com/. © 2015 James A. Sullivan

Shown in Figure 7.3.2 is the *Ebola* virus at 100 000× with the rhinovirus alongside. Figure 7.3.3 represents a cluster of bacteria, viruses, cells, and pollen. As the magnification becomes less and less through Figures 7.3.4, 7.3.5, and 7.3.6, dust mites, at about 250–300 μm long, appear. In Figures 7.3.7, 7.3.8, and 7.3.9, it becomes apparent that these various bacteria, viruses, cells, pollens, and even dust mites are perched on the head of a pin.

Figure 7.3.10 shows a micrograph of a biofilm colony on the interior of a 316L stainless steel pipe. The striations are the result of mechanical polishing. The initial bacteria were able to set up house in that deep groove or gouge, which is visible in the micrograph. Once in place the colony of bacteria begins to multiply.

Figure 7.3.10 Biofilm at 2000×. *Source*: Reproduced with permission of Frank Riedewald (2004)

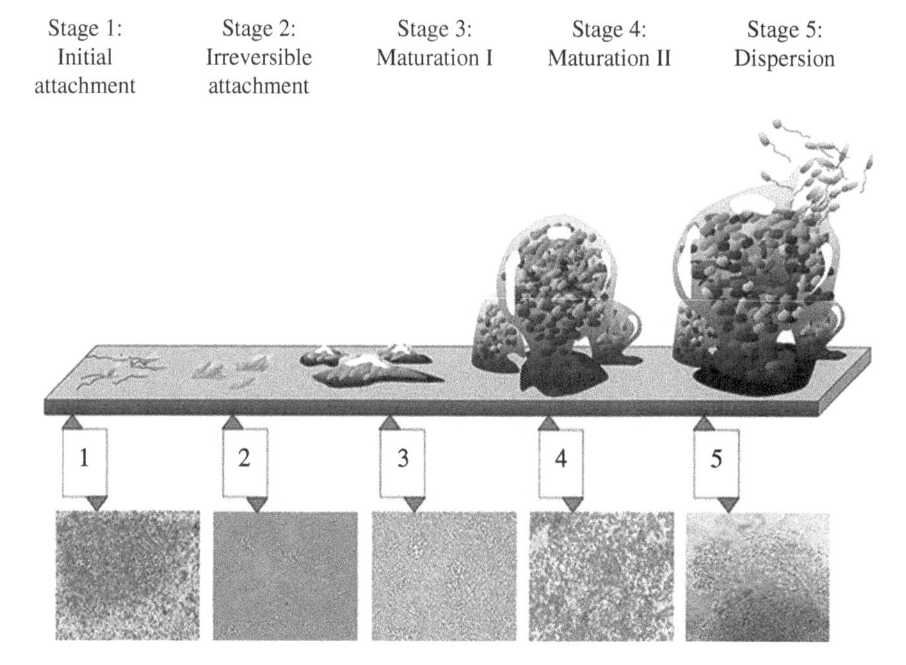

Figure 7.3.11 Biofilm life cycle. *Source*: Don Monroe, http://search.proquest.com/openview/f3f6bf29fc610d664df82f01f6576b7a/1?pq-origsite=gscholar. Used under CC-BY-SA 2.5

Figure 7.3.11, with courtesy of the CDC, represents the life cycle of a biofilm in five steps. In step 1 the resident plankton, or free-floating microorganism, finds a spot on the interior surface of a pipeline. This spot could be a microscopic pit in the pipe wall or a microscopic gouge from the mechanical polishing process, as shown in Figure 7.3.10. In step 2 it shows that the bacteria laid down a foundation that consists of a matrix of excreted polymeric compounds. This allows the bacteria to stabilize itself to the surface of the pipe wall and lay out the welcome mat to any other bacteria looking for a home. In step 3 the colony begins to grow and mature, becoming more difficult to remove. In step 4 the colony has continued to grow and expand the superstructure of the excreted polymeric compounds in the process in order to hold together its ever-increasing population. Step 5 sees the colony grow beyond what it can support in the way of population. This is the point at which the bacteria begin sloughing off to end up downstream in search of a new home.

In a 1990 Belgium study, by Erwin Vanhaecke, Ph.D., researchers discovered that bacteria can adhere to stainless steel, even electropolished stainless steel, within 30 sec. of being exposed to the material. And to compound matters, under perfect growth conditions, a bacterial cell divides into two daughter cells once every 20 min. This means that a single cell and its descendants will grow exponentially to more than two million cells in just 8 hr.

At the CDC in Atlanta, researchers took plastic pipes and filled them with water contaminated with two strains of bacteria. After allowing the bacteria to incubate for 8 weeks, the scientists emptied out the infested water and doused the pipes with germ-killing chemicals, including chlorine, for 7 days. They then refilled the pipes with sterile water and periodically sampled the "clean" water. The team reported that both strains of bacteria survived in the chemically treated pipes and reestablished colonies once again.

Shown in the Figure 7.3.12 micrograph is a Staphylococcus bacteria whose colony appears to be at or near step 5 of the CDC's lifecycle stages. This biofilm was located on the inner surface of a needleless connector (used to connect IV lines). Biofilms consist of many species of bacteria and archaea living within a matrix of excreted polymeric compounds. This matrix is a complex superstructure that

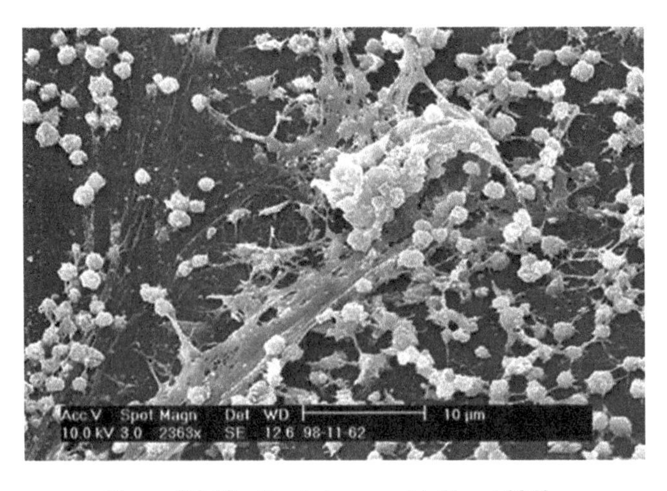

Figure 7.3.12 *Staphylococcus* biofilm at 2363×

protects the cells within it and facilitates communication among the cell population through chemical and physical signals. Some biofilm structures have been found to contain flow channels that help distribute nutrients and signaling molecules. This type of matrix is strong enough that in some cases, biofilms can exist long enough to become fossilized.

Since, from a practical standpoint, there are no foolproof designs by which bacteria can be totally prevented from getting into a closed and sealed fluid system, we are left with a few practical means by which to control this relentless onset. Some of the ways in which this can be done, again from a practical standpoint, is to comply with BPE paragraph SD-2.4.2 in which the following guidelines are stated:

- All surfaces shall be cleanable.
- All surfaces shall be accessible to the cleaning solutions and shall be accessible to establish and determine efficacy of the cleaning protocol.
- Fasteners or threads shall not be exposed to the process, steam, or cleaning fluids.
- No engraving or embossing of materials (for identification or traceability reasons) should be made on the process contact side.

Paragraph SD-2.4.2 goes further in subparagraph (b) to make clear additional key design characteristics needed for systems intended to incorporate a clean-in-place (CIP) protocol as follows:

- Internal horizontal surfaces should be minimized.
- The equipment shall be drainable and free of areas where liquids may be retained and where soil or contaminants could collect.
- Design of corners and radii should meet the requirements stated in paragraph SD-2.4.2(b)(3).

Additional needs for a hygienic system should take into consideration such things as:

- All pipes should be sloped to a drain point (paragraph SD-2.4.3.1).
- Microscopically smooth interior process contact surfaces.
- Filtration at all air breaks.
- Ensure against liquid holdup areas.

Having a new perspective of this microscopic nemesis, we will now see what these bacteria look like in relation to the inside of a piping system. These next figures, or micrographs, will provide us an up-close and personal look at a bacterium in residence.

Figure 7.3.13a is a micrograph of a 316L stainless steel coupon with a surface finish of Ra 22 μin. (0.57 μm). Looking at that same figure, there are three pits noted in that micrograph image. The pit in the center of the figure is enlarged in

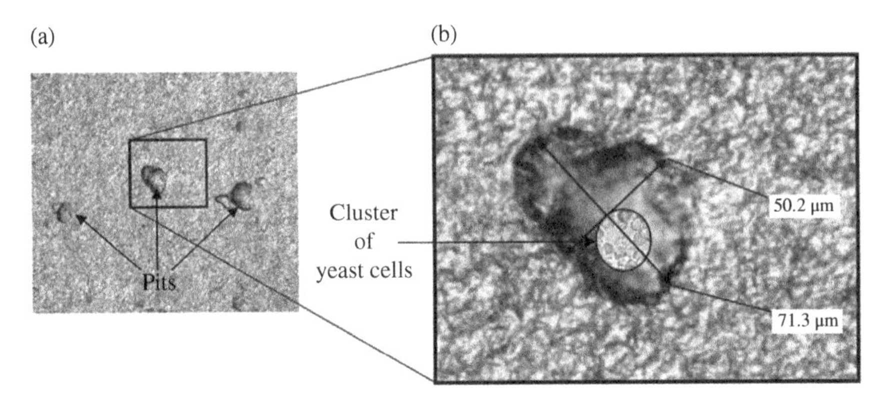

Figure 7.3.13 Cluster of yeast cells in pit. (a) Three pits in 316L stainless steel and (b) single pit with cluster of yeast cells. *Source*: http://www.dockweiler.com/en/. © Dockweiler AG. Reproduced with permission from Dr. Jan Rau

Figure 7.3.13b. Within the enlarged pit shown in Figure 7.3.13b is a cluster of yeast cells placed there to provide relative comparison with regard to the size of yeast cells, which range in size from 3 to 5 μin. and larger, in relation to that of an elongated pit that is, in this case, 2807 μin. (71 μm) long by 1976 μin. (50 μm) wide. A human hair, by contrast, would, with an average diameter of 3531 μin. (100 μm), cover the entire width of the detail box itself.

The encircled cluster of yeast cells contains approximately 30 cells. In doing a little guesstimation, based on the size and depth of the pit and the cells contained in the cluster, there could be well over 300 yeast cells contained in that one pit. With bacteria, also referred to as bioburden, which can be many times smaller than the yeast cell, there could be even more. When attempting to CIP between batches or at the end of a production campaign, the CIP process may not create sufficient turbulence in these pits to clean out the residue of cells, bacteria, or other contaminants that find a home in such surface recesses.

Introducing a new product batch into a piping system containing such holdover cells or malicious bacteria could jeopardize an entire product batch. In the pharmaceutical world, the possibility of such potential cross contamination is a major issue, in the biofuel and other such heavy biomanufacturing, not so much, that is, so long as that cross-contaminating residue does not jeopardize the next batch of cells entering the fermentation phase of a process system. If the next batch of cells coming into the system is contaminated by malicious bacteria that were inadvertently left behind in pits and other microscopic crevices, the entire process could be placed in jeopardy.

With that said, hopefully the stage is set for the following discussion, a discussion in which surface finish, microscopic crevices, minute gasket intrusion, pipe slope, weld tolerances, chemical properties of component material, and design criteria all play a major role in the creation of a piping system that is conducive to repeated bioprocessing manufacture of a product.

7.4 A Preamble to Design

Prior to getting into specifics, with regard to systems design, the following will touch on some general topics in an effort to provide somewhat of an overview of design.

7.4.1 Undeveloped Subject Matter

There are placeholders for topics within the BPE Standard providing what could be considered a prelude to what is coming in future issues of the Standard. Among those placeholder topics are the following:

- Bioburden control (Tentative BPE paragraph SD-2.2): This paragraph will contain guidelines on suggested methods and procedures on how to mitigate the onset of bioburden and how to remediate when it does occur.
- Bioburden reduction (Tentative BPE paragraph SD-2.3): This paragraph will contain guidelines, procedures, and possible requirements for methods and procedures that will assist the reader on how to reduce the possible onset of bioburden.
- Thermal sanitization (Tentative BPE paragraph SD-2.3.1): This paragraph will provide guidelines, procedures, and possible requirements for the application of temperature-related sanitization processes.
- Depyrogenation (Tentative BPE paragraph SD-2.3.1.2): Depyrogenation is the removal of pyrogens from a product or process solution. The process is typically identified with parenterals or injectable drug products. It will cover guidelines, procedures, and possible requirements for the application of temperature-related sanitization processes.
- Chemical sanitization (Tentative BPE paragraph SD-2.3.2): This paragraph will provide guidelines, procedures, and possible requirements for the application of chemicals used in the sanitization process.

The previous subject matter along with other interesting and much needed subject matter will continue to expand the content of the BPE Standard in future editions.

7.4.2 Containment

Chemicals, in any industry, are categorized as to their effect on the environment, their effect on human or animal life and tissue, their toxicity, and their volatility. Such categories will vary depending on the government organization (OSHA, EPA, DOT, etc.) or interest group they are evaluated under. In the bioprocessing industry, there is an additional categorizing element beyond that of toxicity and volatility. It is potency. In the manufacture of pharmaceuticals and to a larger degree, biologics, the potency of an active pharmaceutical ingredient (API) is of marked importance.

Both small-molecule drug compounds and typically more complex large molecule or biologics can have varying degrees of potency. Drug or API potency is quantified through occupational exposure limits (OELs), which are established through testing in an effort to create relative potency values that can be used in risk assessment analysis. Risk assessment with regard to drug manufacturing relates not only to creating a safe accident-free work environment in general but also to the processing and handling of high-potency API's (HPAPI's).

Depending on the OEL (potency) of a drug product, any human contact could adversely affect the health and well-being of the people operating such a facility. Knowing the degree of effect that coming in contact with an API or drug product would have upon skin contact, ingestion, or inhalation allows for engineering, design, operating, and handling protocols and containment to be applied respective of those OEL values.

While there are standardized methods for determining the OEL of an API and drug product, there are no standardized methods for categorizing those findings. In referring to Table 7.4.1, there are many drug manufacturers that have developed their own system of categorization. In fact many of those systems are very similar in both the identification of the categories as well as the values in which each category reflects an OEL. In referring to Table 7.4.1, the lower the OEL value, the higher the potency. Inversely, the lower the category number, the less potent a drug compound is.

In Table 7.4.1 is a listing of OELs and their relative categories identified as industry-accepted categories and National Institutes of Health (NIH) categories. The potency of a drug, or its OEL value, is inversely proportional to the size of

Table 7.4.1 Occupational exposure limits (OEL) and associated categories

OEL[a] (per m^3)	Industry-accepted categories	National Institutes of Health Categories		Category description
		Large scale[b]	Lab operations	
≥100 mg	↑	GLSP[c]	↑	Low toxicity
10 mg	1	BL1-LS	BL1	↓
1 mg	2	BL2-LS	BL2	↓
100 μm	3	BL3-LS	BL3	Intermediate toxicity
10 μm	4	Small molecule	BL4	↓
1 μm	4 or 5	and biologies	↓	Potent
100 ng	5	assessed as not	↓	Highly potent
10 ng	5 or 6	processed on a	↓	↓
≤1 ng	6	large-scale basis	↓	↓

[a]There are currently no standard dosage or OEL markers that delineate one category from another.
[b]LS in the following categories = large scale.
[c]Category GLSP = good large scale practice.

the listed dosage. In other words, the 10 mg dose has less potency than that of the 1 mg dose. The highest potency therefore, in Table 7.4.1, is the less than or equal to 1 ng dose.

Assimilating this type of information, the OEL of a drug product and its intermediates, in advance of designing an API facility, is necessary in that it helps in defining the degree of containment that will be necessary for processing and handling of the intended drug product and its intermediates. Knowing this information is essential early on in working through conceptual design and further on in detail design.

As indicated in Table 7.4.1, it is not practical to process Category 4 or BL4 drug products on a large-scale basis. Because of the inherent risk with potency levels in those lower OEL ranges (≤10 μm), control, from a personnel safety standpoint, is of paramount concern and is much too difficult with large-scale manufacturing. Processing an OEL Category 4 or higher drug product would most likely require the integration of isolators and glove boxes which would compromise and impugn the practicality of large-scale processing.

7.4.2.1 Quality by Design

Quality by Design (QbD) is a methodical and well-planned approach to designing anything from an automotive production facility to a bioprocess manufacturing facility. It is a methodology that introduces risk analysis into the mix of process analytical technology (PAT) and critical process parameters (CPPs). The FDA defines PAT essentially as a means by which the design, analysis, and control of pharmaceutical manufacturing can be accomplished through the measurement of CPPs, which affect the critical quality attributes (CQAs) of a process. It basically comes down to formulating a plan or methodology by which these basic measurable elements can be baked into the overall approach of drug development and facility design.

In March 2011, the FDA, together with the European Medicines Agency (EMA), began a parallel pilot assessment program that looked into the viability of QbD. Upon determining the potential viability of such methodology, it was agreed to extend their joint pilot program for two more years. The 2-year extension of the program began in April 2014.

The FDA looks at QbD as "…an approach to ensuring consistent drug quality through statistical, analytical and risk-management methodology in drug design, development and manufacturing." Such methodology in bioprocess manufacturing planning can help define not only the basic processing risk factors but also, by extension, the degree of containment that may be necessary.

Currently BPE paragraph SD-2.1 does nothing more than make mention of containment. It does not go into the specifics but merely states that "The containment level of the system or individual pieces of equipment should be specified and communicated by the owner/user." It is expected that this subject matter will be expanded upon in future publications. It is an area of design that might be difficult to regulate, but begs for standardization, if for no other reason than to get the industry on the same page with regard to potency categories and a better sense of containment requirements.

7.4.3 Working with BPE and B31.3

In using the BPE Standard, it is requisite that it be used in conjunction with the B31.3 Process Piping Code. But many users of these two documents have difficulty knowing when to use one or the other or understand fully how they coexist. In order to get a handle on this, it helps to have a general grasp of the fundamental basis for both the BPE and B31.3.

To begin with, content of the B31.3 Code is based on the basic and fundamental principle that ASME itself was founded on, that being the safety and integrity of pressure equipment and piping. ASME was founded in an effort to prevent pressure equipment and piping from essentially becoming the industrial time bombs they were at the turn of the twentieth century. This was the impetus for what created and gave direction to the ASME organization. From that came the Boiler and Pressure Vessel Code (BPVC) and the B31 series of piping codes, beginning with B31.1 Power Piping Code, the ancillary piping code to BPVC.

Growth of the chemical industry was recognized by ASME, and the B31.3 Process Piping Code was created. And while safety remains the bedrock of this organization, it has increasingly kept pace with industry and has continued to evolve into so much more.

Like so many of the other industry-related organizations that were born out of necessity during a fledgling and very hazardous industrial era during the late 1800s, ASME continued to reinvent itself throughout the twentieth century to become a global engineering force. It accomplished this by continuing to make itself relevant. Relevant, in that it was, and still is, able to adapt itself not only to evolving technologies but also to entire new industries, such as that of bioprocessing.

Referring to this book's introduction, paragraph B "Early History of the ASME BPE Standard," you can read in great detail, if you haven't already, about how a handful of ASME volunteers proposed the need for a new piping and equipment standard, one specific to the bioprocessing industry. Of course, out of that proposal came the BPE Standard.

The BPE was thus created out of a need for standardization in the bioprocessing industry that existed nowhere else at the time. This for an industry that was, during the 1980s and 1990s, beginning to grow at an increasingly high rate. From its point of inception, BPE was, and still remains, inexorably linked to B31.3.

While the BPE Standard covers system design, component dimensions and tolerances, and many other facets of high-purity design, B31.3 is, by reference, BPE's foundational handshake partner for piping safety and integrity criteria. This means that BPE, rather than duplicating requirements that are already expressed in B31.3, instead defers to the piping code for those same requirements or recommendations. Inversely, B31.3 defers to BPE for all high-purity design, fabrication, and component requirements. So where then do we draw the line between BPE and B31.3?

To begin with, there is no line to be drawn. The working relationship between BPE and B31.3 is not like mixing oil and water in which a line of demarcation is plain to see and made very apparent. It is instead an integral mixing of requirements, recommendations, and references in which many of the requirements are predetermined by the owner/user based on their determination of fluid categories found in B31.3. Figure 7.4.1 provides an overview as to how the two documents interact with one another. In using Figure 7.4.1 as a visual reference, the interapplication of the two documents can be explained in the following manner.

When designing, modifying, or expanding an API facility, it will be necessary to comply with the BPE Standard, not necessarily from a regulatory standpoint but rather from a contractual obligation or from the pragmatic standpoint of common sense. Simply put, there is no other industry standard available that will guide the engineer and constructor through the intricacies of high-purity design and the complexities of regulatory compliance issues like the BPE Standard.

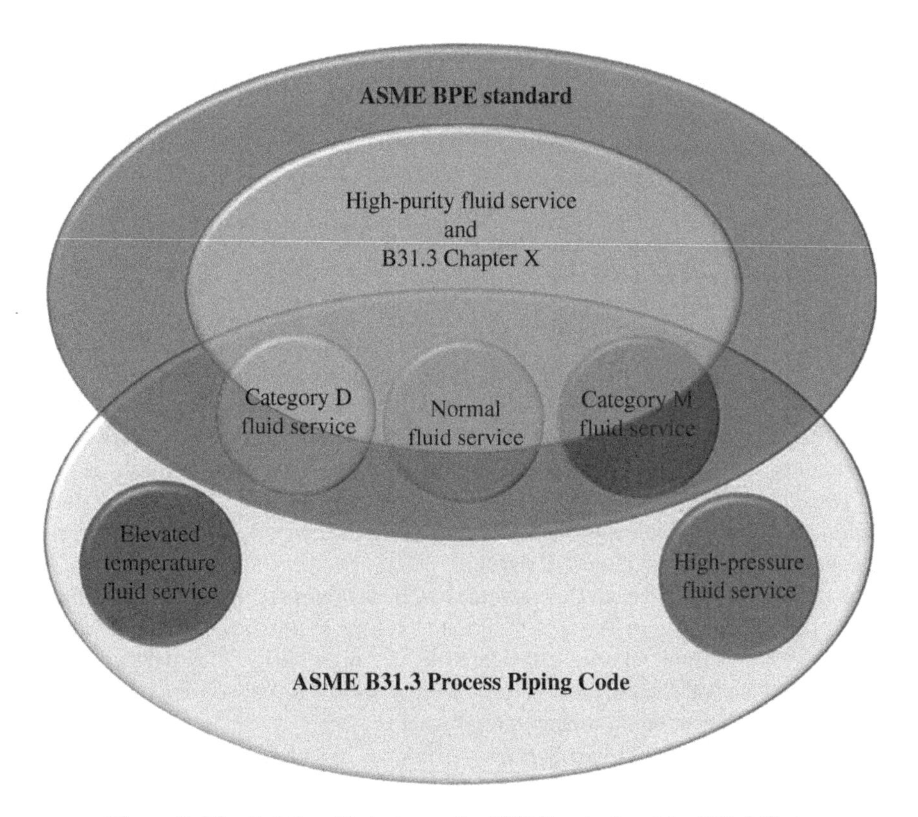

Figure 7.4.1 Relationship between the BPE Standard and the B31.3 Code

Within BPE, as mentioned earlier, are many references to both the ASME BPVC Section VIII for pressure vessels and to B31.3 for piping integrity requirements. Section VIII and B31.3 contain the base requirements and recommendations for pressure vessels and for pressure piping, respectively. Both of these codes hold all of the elements necessary to design, fabricate, install, examine, inspect, and test pressure vessels and pressure piping systems. What they do not have is the broad spectrum of requirements and recommendations for high-purity pressure vessel and piping design, fabrication, installation, examination, inspection, and testing. B31.3 Chapter X does give consideration to the safety and integrity aspects of high-purity piping but defers to BPE for the design, documentation, and regulatory requirements needed to achieve a high-purity piping system.

In an effort to clarify how BPE and B31.3 work seamlessly together, an explanation is offered in the following step-by-step process:

Step 1: In designing and building a Chemical Process Industry (CPI) facility, such as an API facility, it will be necessary to adopt the ASME B31.3 Process Piping Code through engineering fiat or to comply with government regulatory requirements. This is the only piping code that covers the wide range of fluid services, pressures, and temperatures found in such facilities.

Step 2: One of the requirements under B31.3 is the need to designate each fluid service in accordance with the fluid categories listed and defined within B31.3. Those categories include:

- Normal Fluid Service
- Category D Fluid Service
- Category M Fluid Service
- Elevated Temperature Fluid Service
- High-pressure Fluid Service
- High-purity Fluid Service

Step 3: Before designating fluid service categories, we need to, first of all, segregate the High-purity Fluid Service from the other five categories. This is because the High-purity Fluid Service category is what I will refer to as a secondary category, whereas the other five I will refer to as primary categories. Meaning that, while not too likely with the Elevated Temperature or High Pressure Fluid Service categories, as referenced in Figure 7.4.1, any of the primary fluid service categories may also carry the secondary category designation of High-purity Fluid Service.

Step 4: Having assigned each fluid service to a B31.3 fluid service category within an operating facility or for a project, the next step in this process would be to identify those fluid services considered to be "high purity." The High-purity Fluid Service is essentially an overlay category, which is assigned as a secondary designation. There could then potentially be:

- High-purity Normal Fluid Services
- High-purity Category D Fluid Services
- High-purity Category M Fluid Services

Table 7.4.2 Two sets of piping requirements in an API facility

General service piping			High-purity B31.3 Chapter X and BPE		
B31.3 fluid service categories			B31.3 fluid service categories		
Normal	Category D	Category M	Normal	Category D	Category M

Elevated temperature and high-pressure fluid service categories are typically not found in API-type facilities.

Step 5: Having identified those fluids considered "high purity," the designer is now left with two main groups of piping systems as referenced in Table 7.4.2: those designated as general service piping and those designated as high purity. In any case, the group of fluids with the added categorization of high purity will now be governed by the requirements of the BPE Standard in addition to that of B31.3 Chapter X. All other fluid services will be governed by B31.3.

Step 6: Stating that a group of piping systems categorized as high purity will be governed by the BPE Standard implies two things:

- The primary design, fabrication, and installation requirements will be determined by the BPE Standard.
- In designing such high-purity piping systems, the designer will defer first to the BPE Standard, which will in turn reference the B31.3 and Section VIII Codes as applicable.

Step 7: In summing up, for fluid services designated as high purity, Figure 7.4.2 depicts the various paragraphs that direct the user from the BPE Standard to the B31.3 and/or Section VIII Codes and so too from B31.3 to BPE. There is no reference flow as yet from the Section VIII Pressure Vessel Code to BPE. As an example, referring to Figure 7.4.2 again, you will find in BPE paragraph GR-2 a reference to B31.3. And likewise, in B31.3 you will find reference to BPE.

7.4.4 Fabrication

Fabrication of piping systems, including specifics on shop fabrication, field fabrication, and modular fabrication, will be discussed in detail in other chapters. These discussions will bring together aspects of welding, material selection, surface finish, installation, examination, inspection, and leak testing. As an overview on fabrication, what is required of the fabricator is a high degree of craftsmanship and an earnest desire to perform the work in a manner that complies with all applicable specifications and tolerances. In high-purity piping a weld is considered, not only for its strength and sealing ability, but also for the finish of the weld's contact surface. Any surface, including the surface of a weld that comes in contact with the process fluid, has to be cleanable down to a microscopic level. And this is where craftsmanship and regard for work product is essential.

You will find that throughout this entire process, from receipt of material and components to the final install and leak testing, a document trail will need to be

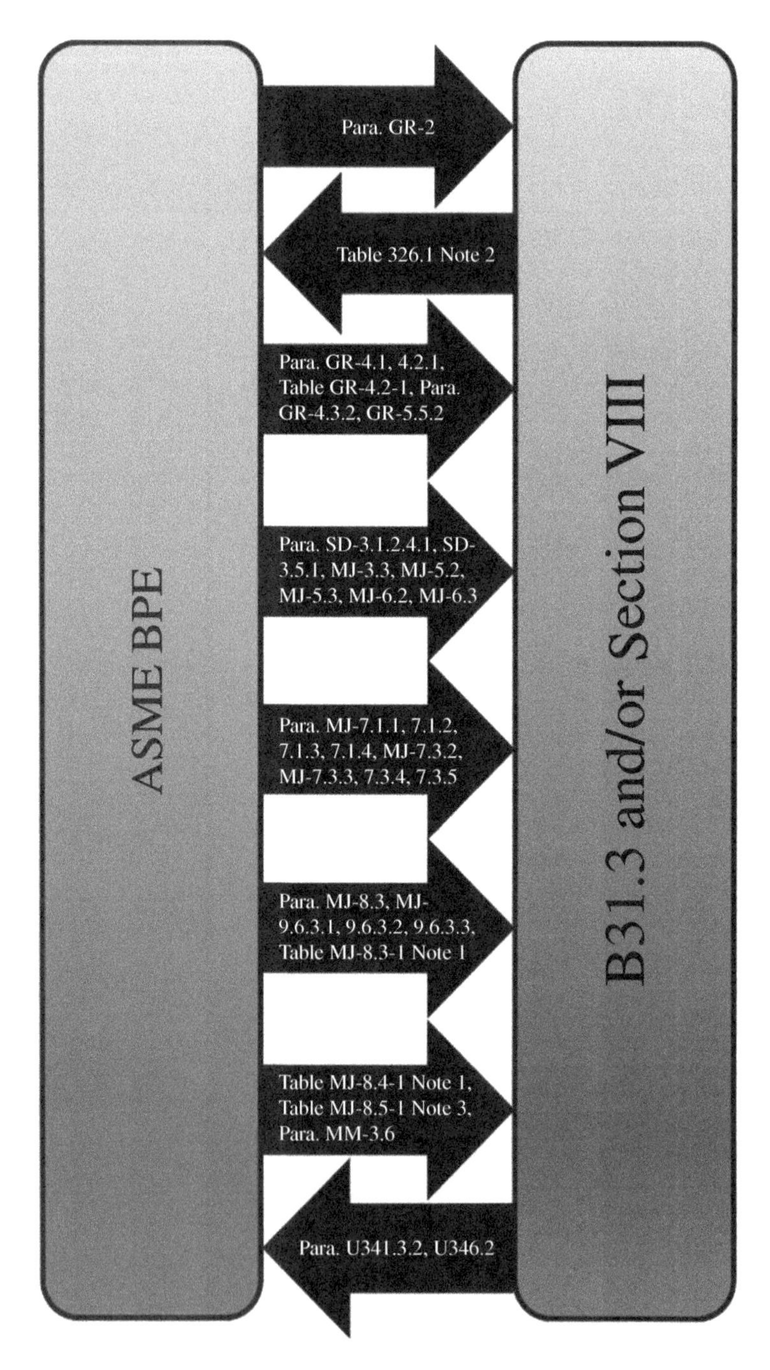

Figure 7.4.2 Interconnections of BPE and B31.3 Section VIII

created and maintained—documentation that not only provides traceability and certification of material and components but also verification and certification of everything from personnel involved in welding to signed leak test reports. These are documents the FDA inspectors will expect to find during an inspection. Not only will they expect to find such documentation, but they will expect them to be found easily and readily. This is not documentation left over at the end of a project that is placed in a cardboard file box and stuck in a warehouse to collect dust or to await the shredder.

In fact the owner/user should have, in place, a quality management system (QMS) that describes in great detail the quality control and assurance procedures that the process facility will operate under. Included in the QMS should be a listing of all documentation requirements for the new facility as well as documentation requirements expected during the operational life cycle of the process facility. The QMS manual will be discussed in much greater detail in Chapter 6.

7.4.5 Materials of Construction

For piping that handles process or product fluids in an API process facility, the range of acceptable material is, compared to typical industrial-type facilities, quite narrow. This is due primarily, since temperature and pressure are typically not at extremes in bioprocessing, to the need for material to be resolute in not contributing particulates or leaching material chemistry into the process or product fluid flow. The workhorse material in high-purity process and utility piping systems is austenitic ASTM A316L stainless steel. Beyond that primary material, there are a number of materials that are also listed as viable for use in the BPE Standard, such as:

- Wrought stainless steels:
  - Austenitics
  - Superaustenitics
  - Duplex
- Wrought nickel alloys
- Cast stainless steel:
  - Austenitics
  - Superaustenitics
  - Duplex
- Cast nickel alloys
- Wrought copper

It is the responsibility of the owner/end user or their representative to select a material suitable for use in specific process or product contact service. Suitability implies that the material selected will not leach into the fluid stream or adulterate the fluid in any way. It also means that the surface finish of the material will be such that it is cleanable and will not contain pit anomalies beyond what is specified in BPE table SF-2.2-1 as being acceptable.

The reason for such requirements is to prevent the use and installation, to a practical degree, of material that may contain surface anomalies that could provide microscopic recesses in which bacteria are able to establish colonies beyond the turbulence boundary of a flowing fluid stream. Such anomalies may contribute to cross contamination of chemicals and/or bacteria.

7.4.6 Cleanability and Drainability

One of the key aspects in achieving and maintaining high purity in an installed process system lies in a system's ability to be completely cleanable and/or sanitized. In order to achieve the appropriate degree of cleanability, a system's architecture has to include sufficient slope to allow residual fluids to flow by gravity, unencumbered to low points for evacuation. The ability for all piping and equipment to completely drain, with minimal to zero holdup volume, is an essential first step in an overall cleaning protocol. This too is a design element the FDA will be looking for in an inspection. That of a written cleaning/sanitization protocol with proof that such a protocol is actually adhered to and its adherence documented.

7.4.7 Bioprocessing System Boundaries

System boundary is an easy term to define but at times very difficult to delineate or describe. ISPE defines system boundary as, "A limit drawn around a system to logically define what is, and is not included in the system." A very simple and to the point definition. The difficulty, at times, comes when attempting to delineate, in relatively complex process systems, just where one system starts and another begins. Such demarcation or segregation of systems is, in many cases, at the root of establishing protocols and procedures, for developing material of construction requirements, and for system design.

At the heart of defining system boundaries lies the ability to better plan CIP/SIP circuits, identify strategic locations for filtration, identify demarcation points at which material of construction might need to change, better define containment requirements, and so on. This is a very essential part of design and postconstruction operation.

7.5 Design

Having provided something of an overview on a few topics, we will now get into specifics as they relate to design. As the diagram in Figure 7.1.1 indicates, design is where all of the elements of a bioprocessing system are brought together in concert with one another to create a complex and unified mechanical whole. It creates the mechanisms and metrics by which to design, construct, and operate a manufacturing facility intended to manufacture a product that consistently meets the demands of regulatory compliance on a repeatable basis. While the basic essentials and specifics of these various elements, such as welding and components, may lie in other sections of this book, the use and bringing together of these various elements resides within design. What this chapter will attempt to do is show how these elements are brought together in creating a processing system while leaving the specifics of these elements to be explained elsewhere in this book.

With design covering so many aspects of high purity, this chapter will be presented in such a way as to avoid becoming convoluted in bringing all of these design factions together. With that in mind, there will be a single focal point topic of which all aspects of system design will flow to and from. That single topic is cleanability.

Why cleanability? Cleanability seems to be such an unassuming topic to serve as a focal point, such an outlier of subject matter. Why should it garner so much attention and gain such interest as to be the focus of this dialog? Why indeed. Cleanability lies arguably at the heart of high-purity design.

Referring back to Section 7.3, it, in part, mentions the fact that bacteria, if present in a fluid, can adhere to stainless steel within 30 sec. of making contact. Based on such facts it has to then be assumed that no sooner is a system installed, cleaned, passivated, flushed, and purged, it is thereafter constantly under attack by bacteria and other foreign matter in general. Such an anticipated and ongoing battle to maintain purity requires a system that not only limits exposure to these foreign elements but is also designed in a manner that facilitates its cleaning—cleaning to a very high degree. Meaning then that while other aspects of design such as welding, material selection, surface finish, sealing capability, and so on are all an intrinsically major part of overall system design, the one factor these elements all have in common is the requisite to be cleanable.

7.5.1 The System

A piping system is the tubing, fittings, valves, instruments, and equipment all assembled in accordance with a designer's attempt to create a three-dimensional configuration that complies with a P&ID and its respective material and construction specifications. Such a design for a bioprocessing system will also have an overarching requirement to be cleanable from a CIP or SIP standpoint. This implies that a number of key attributes in the design of the components as well as their fabrication and assembly need to be considered, such as:

- Minimize the number of mechanical joints.
- Welded joints are preferred:
  - Orbital welds in particular
- Mechanical-type connections shall be of an agreed hygienic design
- Mechanical joints shall:
  - Be able to withstand CIP or SIP conditions
  - Have no crevices or hard-to-clean areas when assembled

- All components, when installed properly, shall be self-draining.
- The use of threaded joints shall be used only upon approval of the end user.
  - If used, such joints shall be used on a limited basis.
- The use of ASME-type flanged joints shall be used only upon approval of the owner/end user.
  - If used, it shall be done so on a limited basis.
  - It is acceptable, on a case-by-case basis, to use ASME-type flanged joints in clean steam distribution systems since such systems are continuously self-sanitizing while in operation.

That "overarching" concern for cleanability of process contact surfaces, alluded to earlier, translates into a determinant set of requirements for the following:

- Connections
- Fittings
- Tubing
- System configuration
- Flexible hoses
- Pumps
- Vessel design
- Agitators
- Mixers
- Transfer panels
- Filtration
- Spray devices
- Sampling
- Steam traps
- Check valves
- Orifice plates
- Relief devices
- Pressure regulators

Each of the previously listed line items will be discussed briefly in the paragraphs that follow, with greater detail provided throughout the book.

7.5.1.1 Connections, Fittings, and Tubing

It becomes apparent, as you read through the requirements described in the BPE Standard, that the cleanability aspect of component and assembly design is of paramount importance. It is a common thread throughout not only Part SD but also the entire Standard. One requisite in obtaining cleanability is in minimizing mechanical connections throughout a piping system. The implication is that welded (orbital) connections are the preferred method of connecting sections of tubing and components. Doing this helps mitigate or at the very least reduce the possibility of gasket intrusion, joint misalignment, improper clamp torque, and other anomalies that would otherwise compromise a system's integrity and/or cause holdup of a process liquid or create a breeding ground for microbial growth.

It therefore becomes necessary to utilize mechanical joint connections sparingly, default to orbital welded connections as much as practical, select fittings and tubing in material that is compatible with the process fluid chemistry, and is essentially pit-free, within the constraints given in BPE table SF-2.2-1 of the Standard.

7.5.1.2 System Configuration

In shaping the configuration of a three-dimensional high-purity piping system, drainability must always be at the forefront of design considerations. Anytime a system is shut down at the end of a campaign, for product changeover, for maintenance, or simply for periodic cleaning purposes, it will be necessary for the system to completely gravity drain prior to flushing the system, typically with DI water, and then to clean the system with a CIP solution or SIP the system.

In order to achieve drainability and thorough cleaning/sanitization of a piping system, the piping configuration has to be designed in such a manner as to have adequate slope in all piping and components and contain no dead legs that exceed those limits recommended by the BPE as stated in Part SD.

7.5.1.3 Flexible Hoses

Flexible hoses are typically specified for use in the following circumstances:

- Connecting to a vessel on load cells
- To isolate anticipated pipe vibration
- Ease of switching from one fluid source to another

These and other like reasons require the use of flexible hoses in an otherwise hard-piped system. In consideration of this, the use of flexible hoses spawns an added set of requirements. Such added requirements include:

- Drainable installation
- Ease of access
- Length of hose
- Hose construction

From a drainable standpoint, a flex hose should be installed so as to avoid either creating a pocket in the hose itself or allowing it to create hump in the hose causing holdup or backup in the pipeline.

Shown in Figures 7.5.1 (a modification of ASME BPE fig. SD-3.2.1-1(b)) and 7.5.2 (from ASME BPE fig. SD-3.2.1-1(b)) are two circumstances that should be avoided when designing the installation of a flex hose. Designing a horizontal

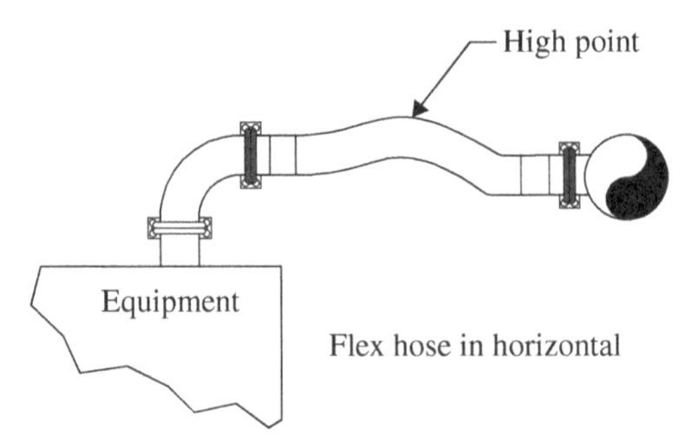

Figure 7.5.1 Kinked hose with rise. *Source*: https://www.asme.org/products/codes-standards/bpe-2014-bioprocessing-equipment. © The American Society of Mechanical Engineers

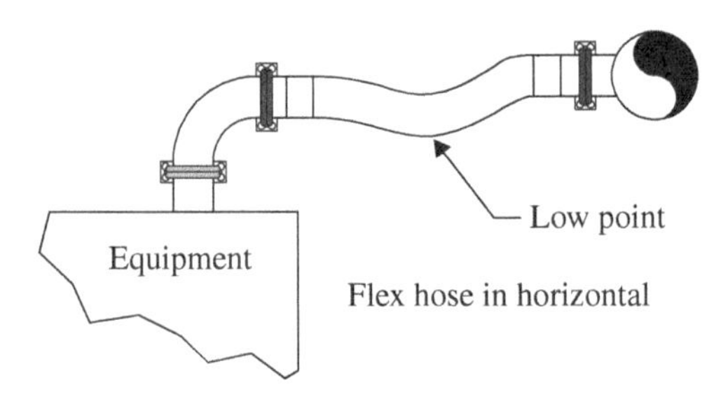

Figure 7.5.2 Hose pocket. *Source*: https://www.asme.org/products/codes-standards/bpe-2014-bioprocessing-equipment. © The American Society of Mechanical Engineers

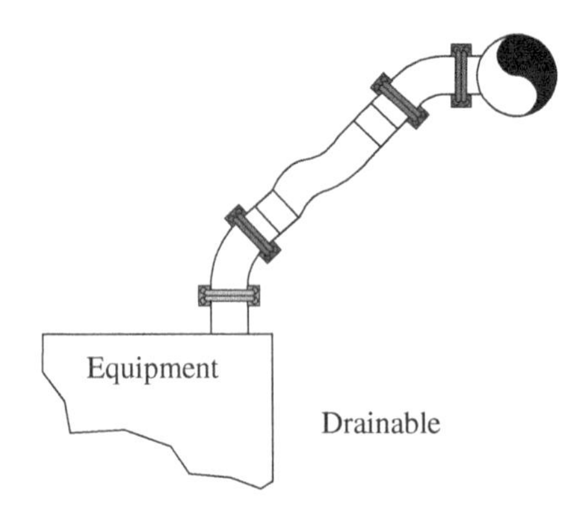

Figure 7.5.3 Proper hose installation. *Source*: https://www.asme.org/products/codes-standards/bpe-2014-bioprocessing-equipment. © The American Society of Mechanical Engineers

hose configuration can create two unwanted conditions. It can create the possibility of a kinked high point in the hose as shown Figure 7.5.1, or it can create the possibility of a low-point sag in the hose as shown Figure 7.5.2. Neither of these conditions is desirable but is certainly possible when designed in this manner.

A more appropriate design is represented in Figure 7.5.3 (from ASME BPE fig. SD-3.2.1-1(a)). This configuration allows for excellent drainage even if the hose has a convoluted liner. The possibility of kinking or sagging is all but eliminated.

Design must also consider the needed access to flex hoses to enable operations and maintenance the ability to work around the hose connections without being hindered by piping, valves, instruments, and equipment in close proximity, particularly if quick disconnects are used as connectors.

When specifying the length of hose, there are two basic criteria that should be considered. They are:

1. Make every attempt at specifying the same length for all hoses on a vessel. This allows maintenance the ability to reduce the number of spares they need to keep on hand. Rather than having to stock a spare for multiple lengths of hose, they can instead stock a couple of spares for each size and type knowing that they have the ability to swap out any one of a number of same size and type hoses.
2. Depending on the OD and construction of a flex hose, a short length of hose can be very ridged, with no real flexibility. If hoses are connected to a vessel on load cells, consideration should be given to the hoses' flexibility. Contact the hose vendor and get their recommendation on the minimum length the hose you are considering needs to be in order to minimize influence on the load cells.

From a high-purity design standpoint, one of the more crucial areas of concern in a hose assembly is the point at which the hose itself connects with the end connection component. This is a point at which, when not assembled correctly, can become a point of contention with regard to harboring unwanted bacteria or bioburden. When the surface finish of process contact material is held to only a few microns of roughness in order to mitigate, to the degree possible, the onset of bioburden and enhance cleanability, anything that would be counter to such process contact surface requirements, would be unacceptable.

It is therefore essential that the interface connection of the hose ID in mating up to the end connection barb is held tight and flush with the ID of the hose with a uniform compression force from the outer compression ring. As seen in Figure 7.5.4 (from ASME BPE fig. SD-3.2.1-1(c) as modified), the area of most concern is the encircled area of the Figure 7.5.4 graphic at which the end of the connection barb is compressed into the ID of the hose by the outer compression ring. If the tip of the connection barb is not fully flush with or slightly imbedded into the hose ID, it can create a point of entrapment for bacteria or particulate matter entrained in the process fluid.

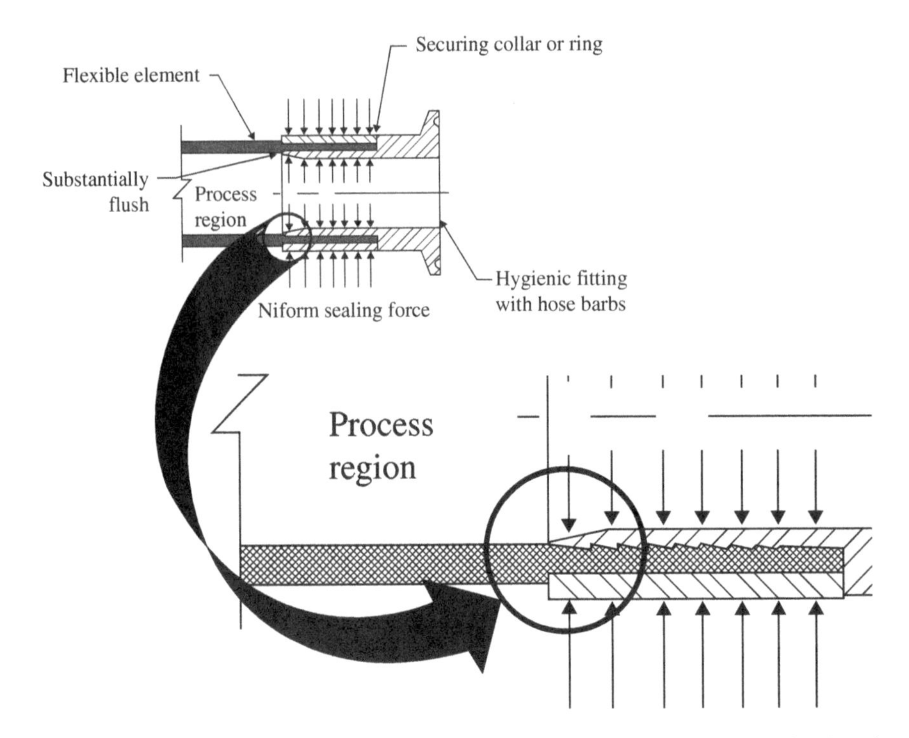

Figure 7.5.4 Proper integration of end connection. *Source*: https://www.asme.org/products/codes-standards/bpe-2014-bioprocessing-equipment. © The American Society of Mechanical Engineers

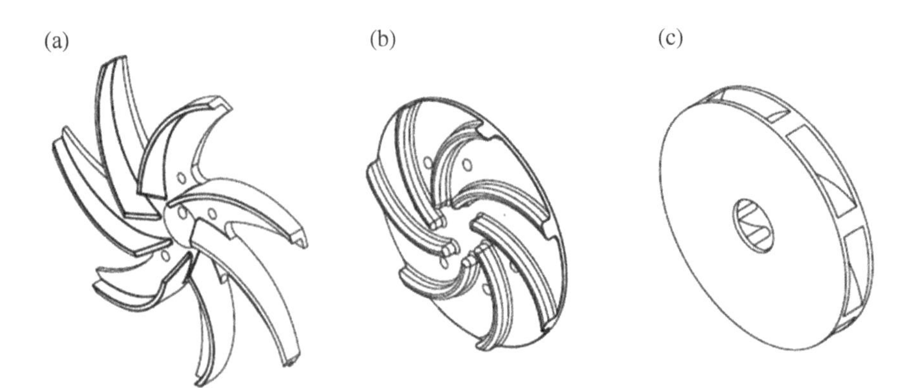

Figure 7.5.5 Pump impeller designs. (a) Open, (b) semiopen, and (c) shrouded/closed. *Source*: https://www.asme.org/products/codes-standards/bpe-2014-bioprocessing-equipment. © The American Society of Mechanical Engineers

7.5.1.4 Pumps

Depicted and discussed in the BPE Standard are three types of pumps:

1. Centrifugal
2. Positive displacement (PD)
3. Rotary lobe

In reality there are only two types of pumps listed: centrifugal displacement and PD. The rotary lobe pump is a type of PD pump. And while pumps applicable to hygienic service make up a very narrow band of pumps, there are other pumps out there that are also used in the pharmaceutical/bioprocessing industry. In order to see how all of the various pumps stack up and to see where high-purity-type pumps fit into the scheme of pump selection, Table 7.5.1 provides an alphabetical listing of pump types that are used across multiple industries.

The following list is comprised of pumps that function under two different mechanical methods of operation referred to as rotary and reciprocating. Within those two operating categories of pumps are the two basic forms of fluid motivation: centrifugal and PD. The centrifugal-type pump only operates as a rotary pump. The PD pump, on the other hand, can operate as either a rotary pump or a

Table 7.5.1 Pump types

ANSI pumps	Lobe pumps
Air Operated Double Diaphragm (AODD) pumps	Magnetic drive pumps
API process pumps	Metering pumps
Axial flow pumps	Multistage pumps
Bellow pumps	Peristaltic pumps
Booster pumps	Piston pumps
Canned motor pumps	Plunger pumps
Centrifugal pumps	Positive displacement pumps
Chopper pumps	Process pumps
Circulator pumps	Progressive cavity pumps
Concrete pumps	Reciprocating pumps
Cryogenic pumps	Regenerative turbine pumps
Diaphragm pumps	Rotary pumps
Drum pumps	Screw pumps
Emulsifier pumps	Self-priming pumps
End suction pumps	Slurry pumps
Fire pumps	Submersible pumps
Flexible impeller pumps	Trash pumps
Gear pumps	Vane pumps
Grinder pumps	Vertical sump pumps
Horizontal split case pumps	Vertical turbine pumps
Hygienic pumps	Well pumps
Jet pumps	

reciprocating pump. In clarifying this, the following is a list of rotary and reciprocating pumps:

- Rotary pumps, which includes:
  - Gear
  - Lobe
  - Peristaltic
  - Progressive cavity
  - Screw
  - Vane
- Reciprocating, which includes:
  - AODD
  - Bellows
  - Piston
  - Plunger
  - Hydraulic

While not a PD pump, the centrifugal pump is a rotary pump and is the most frequently used pump in the industry. It is available in multiple configurations including:

- Centrifugal
- ANSI (horizontal, end suction, single stage, dimensionally standardized)
- Emulsifier
- End suction
- Self-priming
- Fire
- Grinder
- Others

Both reciprocating and rotary pump designs lend themselves to high-purity applications as long as they meet the criteria described in paragraphs SD-3.3, SG-4.3.2, and 5.3. The term "process" pump is a broad term that includes such pump types as "ANSI," "API," and "hygienic" pumps.

- "ANSI" pumps are a single-stage, end suction, centrifugal pump designed and manufactured in accordance with ASME B73.1 "Specification for Horizontal End Suction Centrifugal Pumps for Chemical Process." I make the clarification of single stage in describing the ANSI pump type because there are other non-ANSI centrifugal pumps that are multistage centrifugal pump designs, which, because of the multistage design, do not comply with the B73.1 specification.
- The "American Petroleum Institute" (API) pumps are also centrifugal-type pumps but are considerably more varied than the ANSI-type pumps. API pumps are centrifugal pumps designed and manufactured in accordance with API

Standard 610 "Centrifugal Pumps for Petroleum, Petrochemical, and Natural Gas Industries." Unlike the ANSI-type pumps, API 610 does not specify standard dimensions for pumps. Because of this they do not have the interchangeability found in the ANSI pumps.

- Hygienic pumps are a group of pumps that can meet the demands of high-purity applications by means of material, cleanability, and ease of disassembly. This group of pumps includes, but is certainly not limited to, centrifugal, rotary-type PD pumps, reciprocating AODD, and bellows pumps. The sanitary centrifugal pumps that meet the requirements of B73.1 (ANSI pumps) should comply with P3-A Standard 003 "End Suction Centrifugal Pumps for Active Pharmaceutical Ingredients."

As mentioned earlier, the BPE Standard, in paragraphs SD-3.3, SG-4.3.2, and 5.3, provides criteria for high-purity pump design requirements. And even though Figure 7.5.5c (ASME BPE fig. SD-3.3.2.2-1) is represented in the BPE Standard, this closed impeller design is considered unacceptable, as stated in BPE paragraph SD-3.3.2.2(c). The unacceptability of the closed impeller design lies in the fact that its design does not allow it to be adequately cleaned. Neither does it allow sufficient access for examining the vanes between the two cover plates or shrouds for cleanliness.

In the following Figure 7.5.6 is a table from the 3-A Standard P3-A-003 "What to look for in P3-A-003 VS. ANSI/ASME B73.1" found in Appendix F of that standard. In it they list the differences between a typical centrifugal pump, as represented by ASME B73.1, and that of the high-purity requirements stated in P3-A-003.

7.5.1.5 Vessel Design

With regard to pressure vessel design for high-purity applications, there are essentially two layers of requirements that have to be met in their design, material, and fabrication. The first layer of requirements being that of pressure vessel design and fabrication requirements under the BPVC Section VIII or the EU Pressure Equipment Directive (PED).

As an example, if fabricating a pressure vessel to be used in a high-purity application in which the facility resides in a state or province that falls under the jurisdiction of the BPVC Section VIII or if a procurement contract proclaims that a pressure vessel is to be designed and fabricated in accordance with BPVC Section VIII, then all design and fabrication, first and foremost, are required to comply with the Section VIII requirements. However, this, in and of itself, does not meet the high-purity requirements of the BPE, FDA, or EHEDG.

The pressure vessel, without impacting or conflicting with the Section VIII requirements, should then meet the requirements as defined in Parts SD, MJ, and MM of the BPE Standard for design, fabrication, and material when designing and fabricating a high-purity pressure vessel. These are requirements additional to that of the BPVC Section VIII requirements enabling a vessel to meet the added demands and criteria needed for hygienic applications.

P3-A-003	ANSI/ASME B73.1
P3-A-003 clarifies B73.1 for use in API services with the interest to maintain product integrity.	B73.1 is a dimensional and feature focused broad-based specification designed to ensure dimensional interchangeability and operational reliability.
P3-A-003 Para. D2.2 • Casing drain connection, if specified, shall have a full penetration butt weld ground flush on the product contact surface. Para. D9-1 • There shall be no exposed threads on product contact surfaces.	B73.1 Para. 4.3.1. • Drain Connection Boss(es) Boss size shall accommodate ½ in. NPT min. Boss(es) shall be drilled and tapped when specified by customer.
P3-A-003 Para. D7– Gaskets • Gasket retaining grooves in product contacting surfaces. • Gaskets, when used within the pump, shall be self-positioning. Mating surfaces shall be within 1/32 inch.	B73.1 Para. 4.3.6. – Gaskets • The casing-to-cover gasket shall be confined on the atmospheric side to prevent blowout.
P3-A-003 Para. D11.2 • Enclosed impellers are not permitted.	B73.1 Para 4.4 Impeller • Para 4.4.1 Types: Impellers of open, semi-open, and closed designs are optional.
P3-A 003 Para D11.2 • The impellers shall be installed on the shaft in manner to isolate the product from the threads.	B73.1 Para 4.4.4 – Attachment • The impeller may be keyed or threaded to the shaft with rotation to tighten. Shaft threads and keyways shall be protected so they will not be wetted by the pumped liquid.
P3-A-003 Para C1.1 • Metals shall be specified by the user form among the metals listed in P3-A 002 (Sanitary/Hygienic Standards for Materials for Use in Process Equipment and Systems). This specification does not specify the ASTM number, only the Grade.	B73.1 Para 4.8 – Materials of Construction • Simply states that the pump should be available in D1 (A395); Carbon Steel (A216, WCB); 316SS (A744, CF8M); Alloy 20 (A744, CN7M); and other.
P3-A-003 Para C3.4 • Lubricants intended for use during assembly shall meet the requirements of P3-A-002 paragraph C8. • P3-A 003 Para D1 allows as cast but allows customer to specify improved finish.	• B73.1 does not address assembly lubricants and suppliers may use a wide variety of products, i.e., NeverSeize, Molykote, etc. • B73.1 does not address surface finish.
P3-A-003 Para C3 • Nonmetals requires that materials be suitable for cleaning cycles and meet GRAS requirements. • P3-A 003 Para C4 addresses these surfaces and prohibits painting of alloy parts.	• B73.1 does not address elastomers. • B73.1 does not address non-product contact surfaces and allows painting of alloy parts.
• P3-A-003 Para D4 and D5 require that product contact surfaces be accessible for cleaning (D4) and be drainable (D5). D6.7 requires drainability of the seal chamber. D8 addresses drainability of internal angels. D9 prohibits threads on product contact surfaces, etc.	• B73.1 does not address drainability or cleanability.
• P3-A-003 Appendix A (normative) is a P3-A specific data sheet that deletes the B73.1 options not used by P3-A-003 and adds features unique to P3-A, i.e., CIP/SIP and Product Contact Surface Finish.	• B73.1 Appendix A (Non-mandatory) is a very detailed 3-page data sheet.
• P3-A-003 Appendix C addresses engineering design and technical construction file required for authorization to the P3-A symbol.	• B73.1 Para 7 defines minimum required documentation. Includes basic drawings, curve, and instruction manual. Other data as requested by customer.

Figure 7.5.6 The high-purity differences between P3-A-003 and ASME B73.1. *Source*: 3-A Sanitary Standards, Inc., http://www.3-a.org/. © P3-A® End Suction Centrifugal Pumps for Active Pharmaceutical Ingredients, P3-A 003

The BPE Standard, in noting compliance to Section VIII, does so for pressure vessel joint design in paragraph MJ-3.2; procedure qualification in paragraph MJ-5.1; performance qualification in paragraph MJ-6.1; examination procedures in paragraph MJ-7.1.1; examination personnel requirements in MJ-7.2; examination, inspection, and testing requirements in paragraph MJ-7.3; and weld acceptance criteria in paragraph MJ-8.2.

In the design and fabrication of a pressure vessel, one intended for use in a high-purity application, it will be necessary to meet Section VIII integrity requirements. This includes material thickness, joint design, nozzle attachment design, and support design by applying the required strength calculations. At the same time, aspects of high-purity design requirements need to also be incorporated and made integral to the Section VIII requirements by complying with those of the BPE Standard. This includes such criteria that take into account the need for drainability and cleanability. By layering the BPE requirements on to those of Section VIII, the resulting pressure vessel can be code compliant for pressure containment and BPE compliant for high-purity application.

7.5.1.6 Transfer Panels

Transfer panels are defined in the BPE Standard as "a panel to which process and/or utilities are piped that mechanically precludes erroneous cross-connections." To be more explicit, these are typically flat stainless steel panels that contain an array of nozzles on the face side of the panel with piping, carrying process, and/or utility fluids to and away from the back of the panel, connected to those nozzles. The face of the panel contains labeling identifying any or all of the following: fluid service in or out and equipment item to or from.

An integral, but separate part of a transfer panel, are the "jumpers." These are U-bend piping spools (Ref. Figure 7.5.7a), often times elongated U-bends, that interconnect two nozzles on the panel in order to configure a fluid flow from one equipment item to another. Depending on the simplicity or complexity (Figure 7.5.8) of a transfer panel and the number of transfer configurations possible, there could be anywhere from one to several jumpers used to create multiple flow configurations. In many cases proximity switches (Ref. Figure 7.5.7a and b) may be included in a panel in order to signal the control room that a "jumper" is in place and has established the desired flow configuration.

As the BPE Standard explains under paragraph SD-3.7, the elevation of each nozzle, with respect to the transfer panel and the equipment it is associated with, shall be located so that each of the various jumper configurations are completely drainable, with the jumper(s) in place.

Labeling of the nozzles shall be done in accordance with BPE paragraph SD-2.4.4.2(i). In accordance with BPE paragraph SD-2.4.4.2(e), the nozzles and jumpers shall be designed in such a manner so as not to become an obstruction for the planned configuration of multiple jumper attachments at a single time. In other words, if multiple transfer panel configurations in use during a single campaign require some jumpers to crossover one another, this needs to be taken into consideration and jumpers designed accordingly.

In the following Figure 7.5.9, the proper concept of drainability for a looped manifold on a transfer panel is represented in two ways. Concept (a) represents a configuration in which liquids are drained away from the back of the transfer panel. In such configurations a low-point drain will have to exist at a point somewhere in the piping in the direction of the slope.

Concept (b) represents a configuration in which liquids are trained in the direction of the face of the transfer panel. In such configurations a low-point drain

(a)

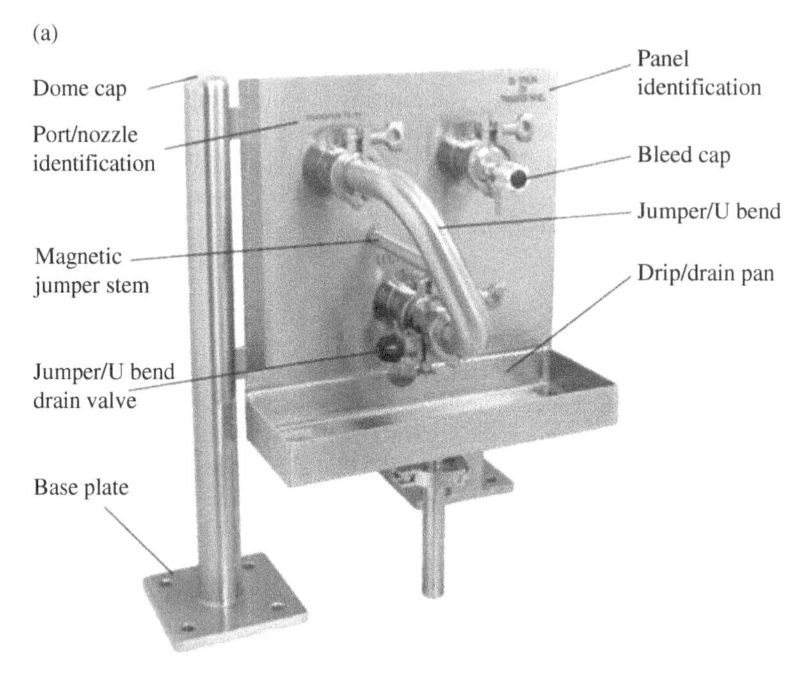

(b)

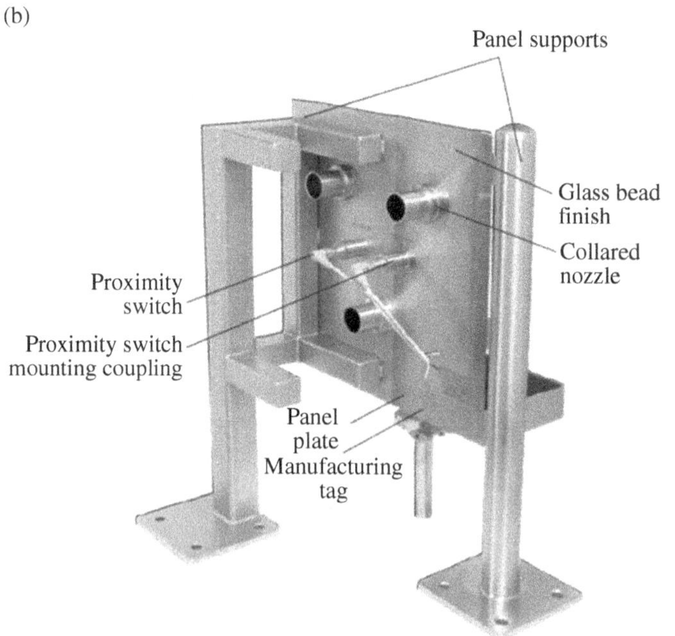

Figure 7.5.7 (a) Transfer panel front and its nomenclature and (b) transfer panel back and its nomenclature. *Source*: http://www.csidesigns.com. © 2004, Central States Industrial Equipment and Service, Inc.

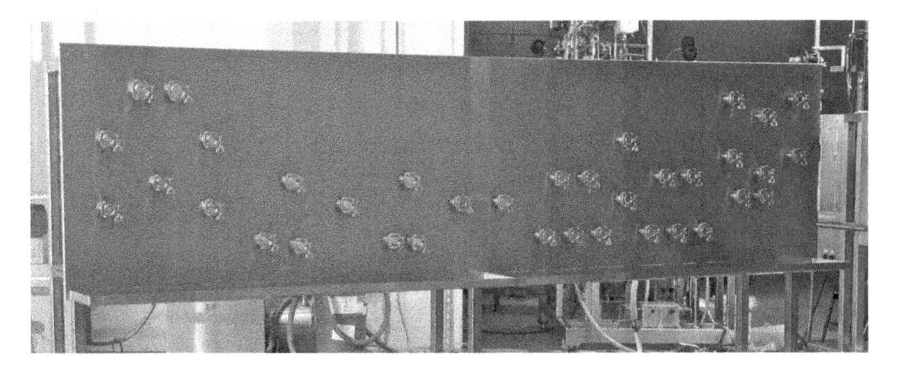

Figure 7.5.8 Large and complex transfer panel. *Source*: http://cotterbrother.com. © 2016 Cotter Brothers Corporation

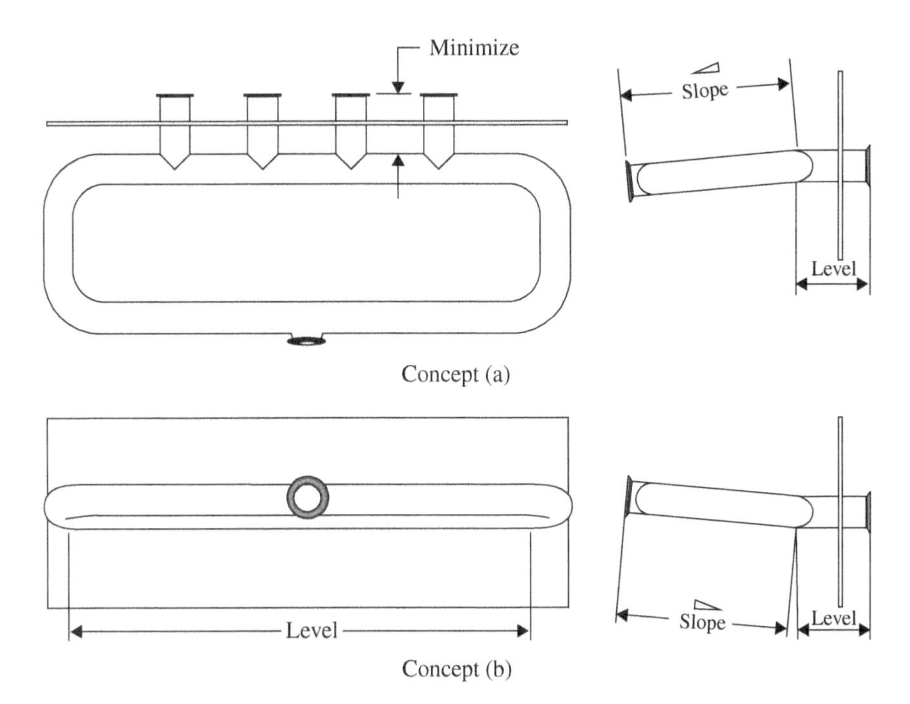

Figure 7.5.9 Acceptable design for a transfer panel looped manifold. *Source*: https://www.asme.org/products/codes-standards/bpe-2014-bioprocessing-equipment. © The American Society of Mechanical Engineers

will have to exist at the jumper as shown in Figure 7.5.10. But in no case should such manifold piping be installed level.

Transfer panel drain pans, when used, shall, in accordance with BPE paragraph SD-3.7.5, be an integral part of the fabrication. Such drain pans are intended to catch potential spills during the process of removing nozzle caps or jumpers.

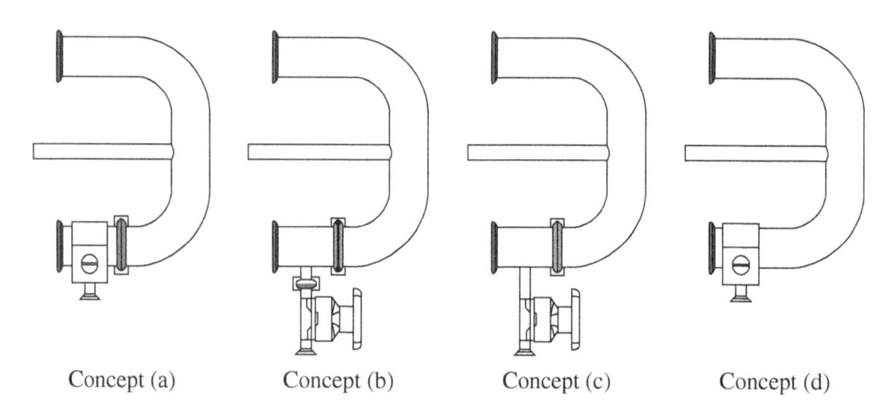

Figure 7.5.10 Acceptable design concepts for jumper drains. *Source*: https://www.asme.org/products/codes-standards/bpe-2014-bioprocessing-equipment. © The American Society of Mechanical Engineers

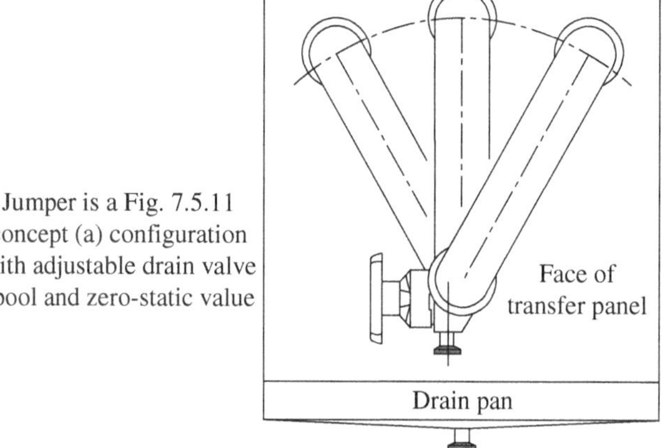

Figure 7.5.11 Multiple position jumpers with adjustable drain valve spool. *Source*: https://www.asme.org/products/codes-standards/bpe-2014-bioprocessing-equipment. © The American Society of Mechanical Engineers

The bottom of the pan shall have a minimum slope of 1/4″/ft (21 mm/m) to facilitate drainage to a process drain system.

Referring to Figure 7.5.10, low-point drains, when required to drain a system upon completion of a fluid transfer, is represented in four acceptable design configurations. As apparent in concepts (a), (b), and (c), the drains are attached to removable spool pieces. Concept (d) represents a jumper with a fixed position drain valve.

As the multitransfer locations of a jumper is rotated from position to position, as represented in Figure 7.5.11, the drain spool concepts of details (a), (b), and (c), shown in Figure 7.5.10, can be adjusted to always place the drain valve at the lowest point of the jumper. Detail (d) in Figure 7.5.10 represents a fixed and integral jumper drain valve acceptable for use on single position jumpers.

7.5.1.7 Proximity Switches

Proximity switches are used to provide an indication to a control room that a jumper is either installed or not installed. Design has the ability to interlock the proximity switch signal with that of system controls to prevent premature operation from occurring before the jumpers are in place. The preferred type of proximity switch, in accordance with BPE paragraph SD-3.7.6(b), is the magnetic type. This allows the unit to be mounted on the backside of the transfer panel with the ability to detect the magnetic jumper stem (Figure 7.5.7a and b) through the transfer panel. This alleviates the need to penetrate the panel itself.

8

BPE Appendices

8.1 Mandatory and Nonmandatory Appendices

The Bioprocess Equipment (BPE) Standard, like most other codes and standards, contains within its pages two sets of appendices: mandatory and nonmandatory. The purpose of such appendices is to provide space within the standard to elaborate on certain subject matter. As stated in Chapter 1 of this book, under Mandatory Appendix, "It is editorial policy to be concise, to the point, and without elaboration in the body of any standard or code when stating a requirement or recommendation."

The necessity in having to adhere to that mitigates that need to better explain why some things are required and other things merely a recommendation. This is where the appendices come in. The act of writing clarifications and explanations in an appendix and away from the body of the Standard allows for more freedom in elaborating on various topics beyond what can be provided for in the body of the Standard.

Some of this additional writing is simply educational or perhaps explanatory in stating what is truly meant or intended with regard to statements made in the body of the Standard. These are considered nonmandatory and therefore to be referred to and used at the reader's discretion. On the other hand, there are requirements that are expanded on in an appendix and intended as an extension of the body of the Standard. Such writing is therefore placed in the Mandatory Appendix.

Bioprocessing Piping and Equipment Design: A Companion Guide for the ASME BPE Standard, First Edition. William M. (Bill) Huitt.

8.2 Mandatory Appendices

8.2.1 Mandatory Appendix I: Submittal of Technical Inquiries to the BPE Committee

This appendix provides information on how to submit a technical inquiry to the committee for consideration and response. It explains the various types of inquiries that can be made such as requests for revisions or additions to the rules of the Standard, requests for Code Cases, and requests for interpretation. Each of these is described in this appendix.

Receiving a timely and effective response to an inquiry requires that the submittal of that inquiry be done in an appropriate manner. As outlined in Appendix I, such submittals should include the following:

- What is the purpose of the inquiry?
  - Are you requesting a revision be made to the current rules of a standard?
  - Are you requesting new rules be added to a standard?
  - Are you proposing a Code Case in which you are suggesting an alternative or addition to the current rules of a standard?
  - Is this a request for interpretation of the current rules in which you need clarification and better understanding of the meaning of the rule?
- Provide a brief narrative on the background and reason for the inquiry.
  - Briefly write about the reason you were prompted to submit your inquiry including references to the applicable editions, addenda, specific paragraphs, figures, and tables
- Face to face.
  - Depending on the complexity and potential impact an inquiry may have on the standard, you may be asked to make a formal presentation to better explain your inquiry and to answer questions before the respective committee. Travel is at the expense of the inquirer. Lack of attendance will not have any bearing on the final response or decision to accept or reject the inquiry by the respective committee.

To paraphrase Appendix I, an inquiry submittal should be in letter form. It can be written in legible handwritten form, but it is preferred that they be provided in typed or computer generated and printed form. The letter shall contain the name of the correspondent, postal address, telephone number, fax number (if available), and an e-mail address (if available).

The inquiry should be mailed to the following address:

Secretary
ASME BPE Committee
The American Society of Mechanical Engineers
Two Park Avenue
New York, NY 10016-5990

If the inquiry submittal is properly prepared, it will be submitted to the BPE Standards Committee or to the respective subcommittee for consideration and resolution. Should any questions arise during consideration of the inquiry, the committee may reach out to the inquirer for further clarification through any of the contact information provided. Upon reaching a resolution to the inquiry, a written response will be made to the inquirer as to that resolution.

8.2.2 Mandatory Appendix II: Standard Units

Appendix II provides a comparative listing of standard measurement units in US Customary and SI Units that are used in the BPE Standard.

8.3 Nonmandatory Appendices

8.3.1 Nonmandatory Appendix A—Commentary: Slag

The term "slag" is currently defined in BPE Part GR as "a concentration of nonmetallic impurities (often oxides or nitrides) that forms in the weld pool and solidifies on the underbead or weld top surface. Sometimes referred to as dross." These small anomalies tend to appear at the termination of a weld bead on the OD, ID, or both. They are generally a small, round, black spot that has proven to be rather innocuous.

8.3.2 Nonmandatory Appendix B: Material and Weld Examination/Inspection Documentation

Appendix B includes a set of three suggested forms as follows:

- Form MEL-1 Material Examination Log
- Form MER-1 Material Examination Record
- Form WEL-1 Weld and Examination/Inspection Log

While much, if not all of the entry titles on these forms is needed or required, the form template itself is merely a suggestion.

8.3.3 Nonmandatory Appendix C: Slope Measurement

Appendix C provides guidance on pipeline slope verification. This is an essential part of an overall installed system verification process.

8.3.4 Nonmandatory Appendix D: Rouge and Stainless Steel

Rouge, defined in BPE Part GR as, "a general term used to describe a variety of discolorations in high-purity stainless steel biopharmaceutical systems. It is composed of metallic (primarily iron) oxides and/or hydroxides. Three types of rouge have been categorized." Appendix D does an excellent job of not only describing this anomaly in its three classifications but of also explaining ways in which to detect and measure its presence in a fluid stream and methods by which to remediate its occurrence by classification.

In this appendix are tables that categorize the effect that certain actions, during fabrication or during operation, have on promoting or deterring the onset of rouge.

8.3.5 Nonmandatory Appendix E: Passivation Procedure Qualification

Passivation is an essential procedure that is performed on postfabricated piping systems and at times to remediate the process contact surface passivity of systems that have been in operation. It is defined in BPE Part GR as the "removal of exogenous iron or iron from the surface of stainless steels and higher alloys by means of a chemical dissolution, most typically by a treatment with an acid solution that will remove the surface contamination and enhance the formation of the passive layer."

Tables in this appendix provide information on passivation processes as well as evaluation tests for determining not only if a surface is passivated but also the thickness of the passive chromium layer.

8.3.6 Nonmandatory Appendix F: Corrosion Testing

Selecting a piping material that is not only suitable for the process fluid it is intended to handle but also to do so under the demanding requirements of a bioprocessing system is in itself very demanding. One of the several considerations in making such a selection is with respect to a material's corrosion resistance qualities. Appendix F discusses the methods by which materials are tested for their corrosion resistant qualities, the purpose of each test, the type of data obtained from each test, and the metal that is typically tested under each test procedure.

8.3.7 Nonmandatory Appendix G: Ferrite

The control of ferrite in the welds of 316L stainless steel in the as-welded condition is of significant importance due to the potential for a loss in corrosion resistance in the heat-affected zone of a weld. This appendix describes ferrite and discusses the effects, both positive and negative, it can have on 316L stainless steel material.

8.3.8 Nonmandatory Appendix H: Electropolishing Procedure Qualification

Electropolishing is defined in BPE Part GR as "a controlled electrochemical process utilizing acid electrolyte, DC current, anode, and cathode to smooth the surface by removal of metal." To create a hygienic environment in a piping system requires us to remove all of the hills and valleys found in a microscopic view of mill-finished steels. While mechanical polishing a metal's surface does help, it does not do it sufficiently enough. In order to accomplish this properly, it has to be done under a controlled procedure at a microscopic level. This can only be accomplished through electropolishing.

Appendix H explains the purpose for and the reasons why surfaces to be in contact with pharmaceutical processes are electropolished. It also explains the procedures used in performing electropolishing.

8.3.9 Nonmandatory Appendix I: Vendor Documentation Requirements for New Instruments

Appendix I describes the documentation requirements for instrumentation used in bioprocessing systems. Listed are "…the documents required to support commissioning/ qualification, installation, operation, and maintenance of instrumentation for the biopharmaceutical industry." Not all of the listed documentation will be required for each and every instrument. Only the documents pertaining to any particular instrument will be required for that instrument.

The table included in the appendix listed all of the various documentation and provides a definition for each to make clear what they are intended for.

8.3.10 Nonmandatory Appendix J: Standard Process Test Conditions (SPTC) for Seal Performance Evaluation

Appendix J currently addresses testing of seals or diaphragm valve seals in simulated SIP exposure only. This appendix is a very in-depth narrative on qualification testing of hygienic clamp gaskets and valve diaphragms. Within this narrative are graphics that help associate the terminology with the various components.

The testing protocols described with Appendix J include:

- SIP exposure testing on hygienic clamp unions
- SIP exposure testing on diaphragm valves
- Mechanical seal performance evaluation test
- Mechanical seal integrity test

The test protocols described are written in step-by-step fashion that are very clear and concise.

8.3.11 Nonmandatory Appendix K: Standard Test Methods for Polymers

Included in Appendix K is a listing of the various ASTM testing standards that apply to the testing of polymeric materials. It goes on to identify some of the immersion fluids used in testing. There are two tables contained in this appendix that include:

- Table K-3-1 Thermoset Polymer Test Properties
- Table K-4-1 Interpretation of Thermoset Material Property Changes

8.3.12 Nonmandatory Appendix L: Spray Device Coverage Testing

Spray device coverage testing is not a cleanability test. It is instead a test procedure intended to demonstrate the ability of the cleaning system to deliver a chemical cleaning solution to every square inch of process contact surface area of the targeted equipment. This is an aqueous riboflavin (vitamin B2) test that once applied and undergoes visual examination under long wavelength ultraviolet (UV) lighting. The riboflavin has a tendency to fluoresce under UV lighting, which highlights its spray pattern on vessel walls. Any areas not having been contacted by spray becomes very evident.

This appendix describes the riboflavin solution, preparation, application, rinse, inspection, acceptance criteria, and recommended documentation.

8.3.13 Nonmandatory Appendix M—Commentary: 316L Weld Heat-Affected Zone Discoloration Acceptance Criteria

This appendix explains the making and use of the photos seen in BPE figures MJ-8.4-2 and MJ-8.4-3. It explains the relationship between the surface finish of 316L stainless steel and the material's pitting resistance as well as its effect on discoloration.

8.3.14 Nonmandatory Appendix N: Guidance When Choosing Polymeric and Nonmetallic Materials

Appendix N contains a brief discussion on thermoplastic, thermoset, and single-use components.

8.3.15 Nonmandatory Appendix O: General Background/Useful Information for Extractables and Leachables

With regard to extractables and leachables, this appendix provides a listing of regulations, standards, guidelines, and organizations specific to that subject matter. It also touches on:

- Recommended conditions for an extractables study
- Recommended model process conditions for a bracketed study

8.3.16 Nonmandatory Appendix P: Temperature Sensors and Associated Components

Appendix P contains a somewhat brief but informative narrative on resistance temperature detectors (RTD's). It also gives mention to thermocouple sensors, bimetallic sensors, and liquid in a glass (LIG) sensors. It explains the RTD's:

- Accuracy
- Electronics
- Operational influences
- Environmental influences

This discussion also provides information on selection and maintenance of temperature sensors.

8.3.17 Nonmandatory Appendix Q: Instrument Receiving, Handling, and Storage

Appendix Q provides a quick step-by-step methodology for the receiving, handling, and storage of instrumentation.

8.3.18 Nonmandatory Appendix R: Application Data Sheet

Appendix R contains a form whose general note describes it in the following manner: "The purpose of the Application Data Sheet is to facilitate the communication of the service parameters, for a particular application, between the end-user and the manufacturer. It is also designed to act as a tool in the selection of proper materials for the process or utility service being defined. This Application Data Sheet will support paragraphs SD-3.4.2, SD-3.4.3, and SG-3.1."

8.3.19 Nonmandatory Appendix S—Polymer Applications: Chromatography Columns

Appendix S provides a brief general overview of chromatography columns.

8.3.20 Nonmandatory Appendix T: Guidance for the Use of US Customary and SI Units

Throughout the Standard both US customary and SI units are used for weights and measures. Appendix T explains the method used in converting these values and the method used in the rounding off of these values.

Figure 2.6.1 Class I and II rouge in pump casing. *Source*: Tom Hanks, CRB Consulting Engineers, Inc., presentation to ASME BPE

Bioprocessing Piping and Equipment Design: A Companion Guide for the ASME BPE Standard, First Edition. William M. (Bill) Huitt.

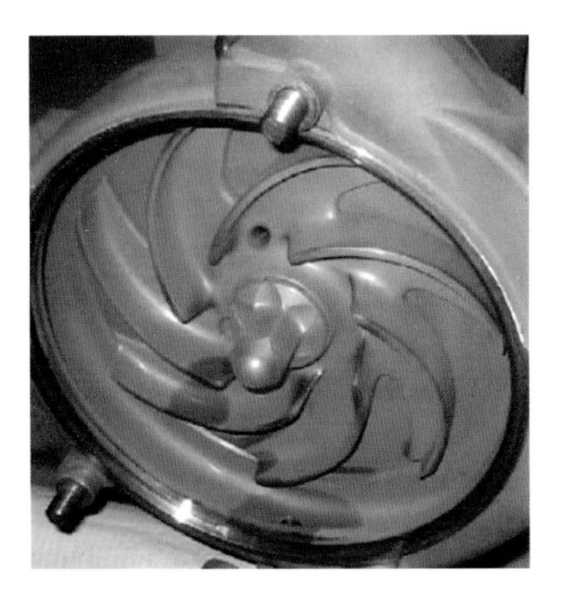

Figure 2.6.2 Class II rust-colored rouge in pump casing. *Source*: Tom Hanks, CRB Consulting Engineers, Inc., presentation to ASME BPE

Figure 2.6.3 Class II purple-colored rouge in pump casing. *Source*: Tom Hanks, CRB Consulting Engineers, Inc., presentation to ASME BPE

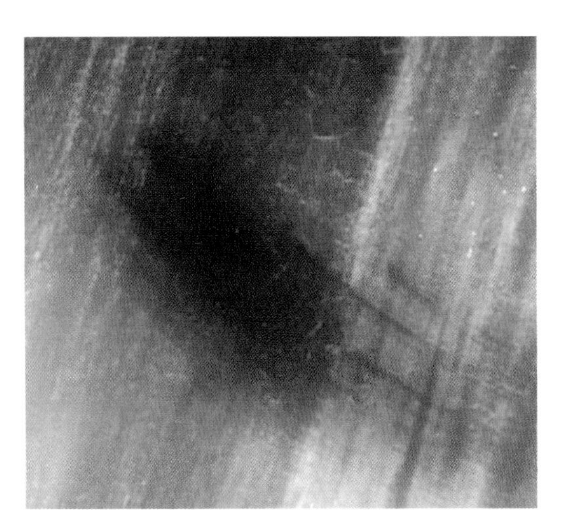

Figure 2.6.4 Class III rouge. *Source*: Tom Hanks, CRB Consulting Engineers, Inc., presentation to ASME BPE

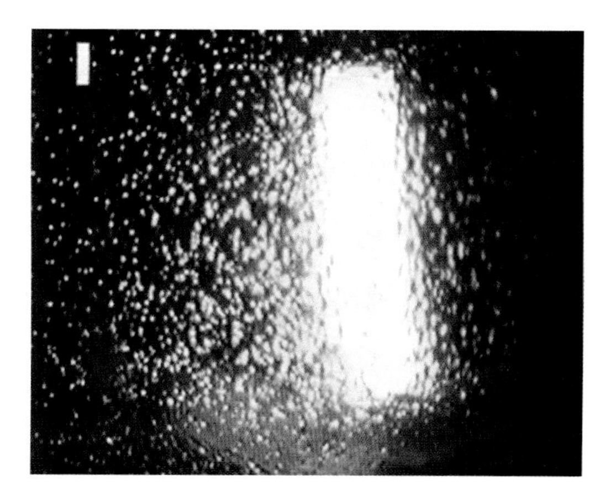

Figure 2.7.6 Example of frosting of electropolished 316L

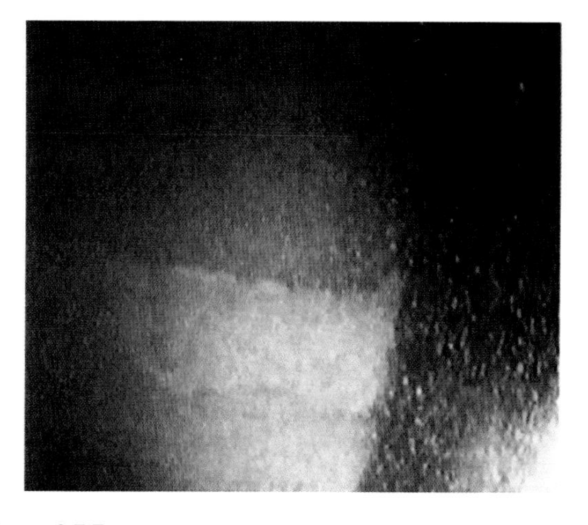

Figure 2.7.7 Example of cloudiness of electropolished 316L

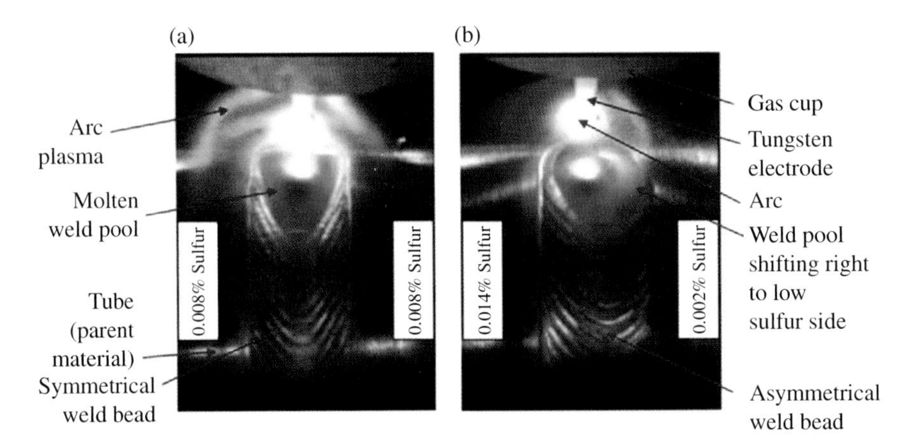

Figure 4.3.2 Effect of sulfur on welding 316L stainless tubing. (a) Symmetrical weld joint of two tubes having similar sulfur content. (b) Asymmetrical weld joint of two tubes having dissimilar sulfur content. *Source*: http://www.arcmachines.com/. © 2016 Arc Machines, Inc

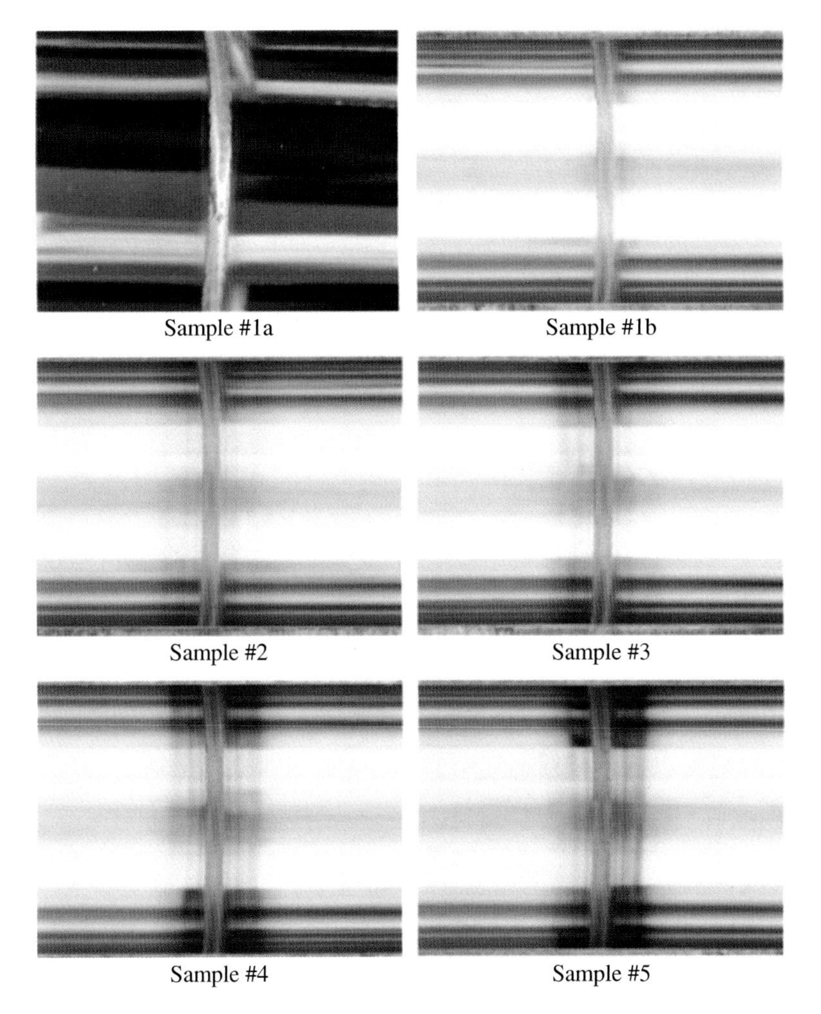

Figure 5.2.1 Discoloration samples electropolished 316L. *Source*: https://www.asme.org/products/codes-standards/bpe-2014-bioprocessing-equipment. © The American Society of Mechanical Engineers
Note: This graphic is not an acceptable substitute for the referenced insert found in the ASME BPE Standard

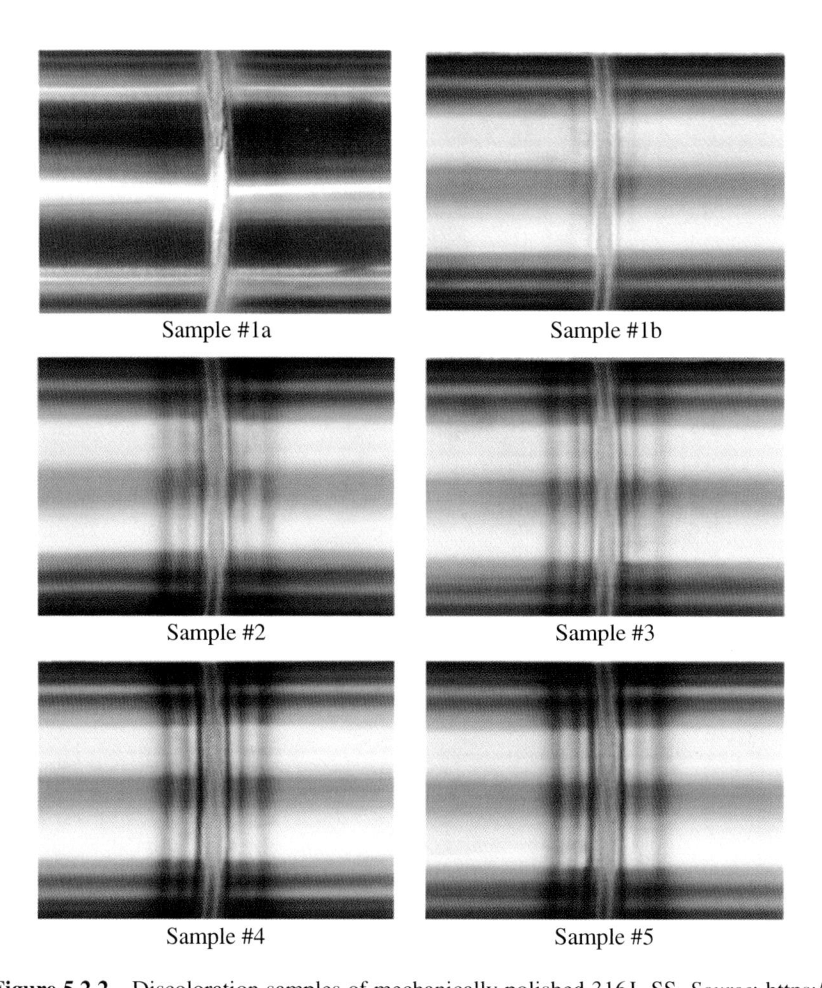

Figure 5.2.2 Discoloration samples of mechanically polished 316L SS. *Source*: https://www.asme.org/products/codes-standards/bpe-2014-bioprocessing-equipment. © The American Society of Mechanical Engineers

Note: This graphic is not an acceptable substitute for the referenced insert found in the ASME BPE Standard

Appendix A

Cleaning and Leak Testing Procedure

A.1 Introduction

Establishing an in-house procedure for cleaning and leak testing installed piping systems is an essential—not just for the engineering firm designing a facility or for the contractor fabricating and installing the piping systems. This type of procedure is an essential tool for the owner/user as well. Cleaning installed piping systems is a necessity, while leak testing is a requirement, if you are indeed adhering to a piping code.

Therefore the following is a suggested methodology and procedure for establishing a piping system cleaning and leak testing protocol. The methodology that follows has been developed and used by the author for decades. It adheres to the requirements of ASME B31.3 in describing leak test procedures required for all fluid service categories. However, the cleaning, passivation, and possible derouging requirements for high-purity fluid services are supplemental to normal cleaning and flushing requirements and are therefore not included in this Appendix, but can be found elsewhere in this book. There is ample information within the ASME Bioprocess Equipment (BPE) Standard to help create such proprietary cleaning requirements when paired with the advice of an expert in the field of surface finishes.

A.2 Scope

This procedure outlines specific requirements for the cleaning and leak testing of newly installed or modified piping systems. When applied to piping that is in service or has been in service, added precautionary measures and considerations

Bioprocessing Piping and Equipment Design: A Companion Guide for the ASME BPE Standard, First Edition. William M. (Bill) Huitt.

may need to be applied. This procedure also covers newly installed or modified sections of piping in an existing system. The requirements herein are based on code compliance and type of fluid service category intended for each system. Also represented and defined are the forms used to record all activity and test results during, and resulting from, this procedure. It is the intent of this procedure to provide concise, definitive guidelines for the cleaning and leak testing of all piping systems.

Piping systems that are covered by this procedure include:

- Installation of a new system using new or used components
- Alteration of an existing piping system:
  - Alteration is any change that affects the pressure-containing capability of the system. Some operational changes, such as an increase in the maximum allowable working pressure or design temperature, are considered alterations. A reduction in minimum temperature to the degree that, in accordance with the piping code, additional mechanical testing of system materials are required is also considered an alteration.
- Repair of an existing piping system:
  - Repair is any work necessary to restore a system to a safe operating condition, provided there is no deviation from the original design.

Part 1 General

1.1 Work Included

(A) Labor, Material, and Equipment. All labor, material, and equipment required to clean and leak test piping systems identified in a project piping line list shall be provided by the contractor.

1. Labor: Personnel employed by, or contracted by, contractor to perform the setup, cleaning, testing, and documentation under the scope of this section. This includes any preapproved third-party subcontractors used for any portion of the cleaning and testing procedure.
2. Material:
   a. Unless agreed to otherwise, the following material, as applicable, shall be provided by the contractor:
      - Rig assemblies for flushing and leak testing
      - Calibrated temporary gages
      - Temporary pressure relief valves
      - Approved cleaning chemicals
      - Hoses or temporary piping for supply and drain of flush and test fluids
      - Temporary blinds, plugs, pancakes, and spools for items removed for flushing and/or testing
      - Circulation pump, or circulation system, including cleaning solution holding tank
      - Waste containment (tank truck)

b. Unless agreed to otherwise, and given 24 h notice, the owner will make the following available:
   - Plant air and nitrogen
   - Potable water and/or deionized (DI) or better grade of water:
     - The use of DI water is required only for high-purity piping systems and is supplied by the owner only if available at the time of its need for flushing or testing.
     - Should DI water be unavailable at owner's site, the contractor shall arrange for temporary source of DI water to be brought on-site.
   - Handling of wastewater

(B) Temporary Removal of Equipment. The contractor shall be responsible for the temporary removal and control of in-line items listed in Attachment A, "Equipment Removed or Installed for Testing" and the installation of temporary closure pieces such as plugs, blind flanges, pancakes, and spool pieces in their place.

1. It shall be the responsibility of the contractor to control, tag, and store removed equipment, during their removal period, in a clean, dry area protected from the environment and possible damage from handling and construction activity.

(C) Reinstall Removed Equipment. After flushing and leak testing is complete, the contractor shall remove temporary closure pieces and reinstall permanent equipment and components.

(D) Documentation. Documentation, as described further in this procedure and its attachments, shall be completed and submitted as required.

1.2 Submittals

(A) Piping Test Circuits. A segment of a piping system, or an entire piping system that will be isolated and identified by number to undergo a single leak test, will have the extent of its leak test pressure boundaries marked on a P&ID and identified by number as a "leak test circuit."

(B) The contractor shall assemble at least two complete sets of project P&IDs to be used for the sole purpose of identifying the extent of each leak test circuit and identifying the number of each circuit. Each set of P&IDs used for this purpose shall be identified as "leak test circuits." One set shall be submitted to the owner, or owner's representative. Details for identifying and implementing the leak test circuit P&IDs, and their use, are as follows (*note: this can be done electronically, but is nonetheless described as using hard-copy P&IDs*):

1. Assemble at least two full sets of P&IDs and label them individually by hand or by stamp, "leak test circuits." For plumbing systems not having P&IDs, use plumbing layout drawings instead.
2. Indicate on each P&ID the limits of each test circuit by marking across the line at each limit with a very heavy mark:
   a. If the limit is a valve, then mark through the center of the valve.

3. At various points along the test circuit, put, in parenthesis, the test circuit number. The test circuit numbering system can be a simple sequencing, or it can be a number that signifies building, area, floor, or fluid system plus a set of sequence numbers. This can vary depending on the size and complexity of the project. The technique has to be decided well in advance of beginning this activity:
   a. Indicate the test circuit number at all points where a line comes on or goes off one drawing to or from another.
4. When preparing a copy of the test circuit for the leak test package, the following is required:
   a. Lay the entire P&ID on a copier and try to get as much of the test circuit in the 81/2″ × 11″ copy area as possible. In some cases it may take multiple copies to include the entire test circuit. (*Do not tape these together.*) An electronic snapshot works too.
   b. Highlight the test circuit on each copy with a yellow highlighter. Assemble multiple sheets in the order of flow to make it easy to follow.
   c. If the P&ID drawing number isn't visible, write in bold lettering the P&ID number in some convenient location on each copy.
5. The original copies will go with the leak test package to the owner.

(C) Leak Test Package. Receipt of the leak test package by the owner will serve as notice that the system indicated has been completely installed, cleaned, and tested by the contractor and has been turned over. The leak test package consists of the following documents:
1. Equipment Removed or Installed for Testing (Attachment A).
2. Leak Test Checklist (Attachment B).
3. Leak Test Data Form for Piping (Attachment C, D, E, or F).
4. 8 1/2″ × ″11″ copy of the area of the leak test circuit P&ID highlighting the applicable numbered test circuit. Use multiple copies when necessary.
5. If used, a copy of contractor's work package form associated with the test.

1.3 Codes and Standards

(A) Compliance. This project shall comply with, and adhere to, the requirements, guidelines, and stipulations of the following codes and standards in the cleaning and testing of all piping.
1. The following codes and standards are referred to herein. The latest editions of these codes and standards, at the time of project development, shall be used as applicable. Any exception shall be noted. Delete those codes and standards not used and insert codes or standards required but not listed:
   a. ASME B31.1: Power Piping
   b. ASME B31.3: Process Piping
   c. ASME B31.5: Refrigeration Piping
   d. ASME B31.9: Building Services Piping
   e. ASME BPE Standard

f. ASTM A380: Standard Practice for Cleaning, Descaling, and Passivation of Stainless Steel Parts, Equipment, and Systems
g. ASTM D6361: Selecting Cleaning Agents and Processes
h. ASTM G-121: Practices for Preparation of Contaminated Test Coupons for the Evaluation of Cleaning Agents for use in Oxygen-Enriched Systems and Components
i. ASTM G127: Standard Guide for the Selection of Cleaning Agents for Oxygen Systems
j. International Plumbing Code (IPC)
k. CGA G-4.1: Cleaning Equipment for Oxygen Service
l. CGA G-5.4: Standard for Hydrogen Piping Systems at Consumer Locations
m. NFPA 50A: Standard for Gaseous Hydrogen Systems at Consumer Sites
n. 29 CFR Part 1910.104: Oxygen
o. 40 CFR Part 82: Protection of Stratospheric Ozone; Refrigerant Recycling; Final Rule
p. All other governing codes and ordinances, including state, county, and municipality, as applicable

Part 2 Cleaning

2.1 General

(A) Preparation

1. The mechanical contractor shall furnish material and labor necessary to perform the cleaning of all piping systems within the scope of their work and this procedure.
2. At least one person shall be assigned to remain with the flush/test manifold at all times while system is being cleaned and flushed.
3. Untested joints shall not be properly shielded during the cleaning process.
4. Instrument lead lines or piping ends intended for attachment of instruments shall be capped, plugged, or blanked during the cleaning process. This may require that the instrument be disconnected and moved on its support, temporarily removed from the area, or scheduled for installation after the pressure test.
5. Lines containing check valves shall have the source of pressure on the upstream side of the valve. If this is not possible, swing checks shall be blocked open or their flapper removed, whereas ball and piston checks shall have their internal parts removed. All parts removed shall be put in a ziplock or other appropriate type bag and secured on or near the valve with identification as to which specific valve the parts are associated with. If none of this is practical, the check valve shall be blocked off or completely removed.

6. All pipe supports and guides shall be in place.
7. When owner provided, coordinate the use of flush water and air with the owner.
8. When owner provided, coordinate the need for flush water waste disposal with the owner.
9. With a minimum 24h notice, submit a request to the owner for the use of plant utilities and waste disposal along with the date and time the cleaning and flushing will be performed.
10. Temporarily remove or isolate all items listed in Attachment A, and close all openings with temporary and numbered plugs, blinds, caps, or spool pieces.
11. Temporarily remove all permanent strainer screens.
12. All system valves, unless indicated otherwise, shall be in their full open position, but not back-seated.
13. When using a liquid for cleaning check along the system for leaks during the cleaning process. Also, when necessary, prior to cleaning with a liquid, perform a low-pressure (~10psig) pneumatic test to verify sufficient tightness of the system.
14. While it is more beneficial to flush a system prior to testing in order to check and clear any potential blockage, during installation of large complex systems, it may be more practical to leak test smaller segments of the system prior to cleaning the entire system. When this occurs, it is acceptable to reverse the cleaning and testing process by leak testing the piping as smaller segments are completed. Once the entire system has been installed, cleaned, and flushed, it can be leak tested in its entirety. Take these alternatives into account when following the procedures in this section.

(B) Method and Hookup

1. There are two primary hookup configurations for cleaning and flushing with water and/or a cleaning solution:
   a. "Once through" the system to sewer or containment:
      (i) The "once-through" hookup consists of a valved connection at the supply source (water line, or a head tank with pump) and a temporary hose from the supply source to the flush/test manifold. The flush/test manifold, in the direction of flow, includes a block valve, valved connection for a hand pump (to achieve test pressure), a drain/vent valve, a block valve, drain/vent valve, pressure gage, temperature gage, and a flanged end that can be adapted to various end connections.
      (ii) The components that make up the flush/test manifold assembly shall be rated to accommodate the test pressures.
      (iii) The outlet of the test circuit is connected to a valve, a spool piece, and a temporary hose to carry the flush liquid to a sewer or to containment such as a tank truck:
         1. Destination of the waste will be determined by the plant rep.
      (iv) A conical-type strainer, installed in each outlet, is used during the final phase of the flush to determine if the system is clear of debris.

(v) The conical-type strainer, up through 4″ NPS, shall have 1/16″ dia. perforations at a density of 144 holes per square inch. For sizes 6″ NPS and above, they shall have 1/8″ diameter perforations at a density of 32 holes per square inch.

b. Circulated through the system with a temporary strainer:

(vi) The circulation hookup consists of a valved connection at the supply source (water line, or a head tank with a pump) and a temporary hose from the supply source to the flush/test manifold. The flush/test manifold, in the direction of flow, includes a block valve, drain/vent valve, block valve, pressure gage, tee to be installed in the pump discharge, and a sample valve installed in the discharge run of pipe. The flush/test manifold is connected to the discharge side of the circulation pump with a block valve between that connection and the pump. The suction loop is either assembled from hoses or temporary piping that connects the outlet of the system with the pump suction. In the suction loop is a connection for a head tank or other source containing a cleaning solution. The pump suction also contains a spool piece for the insertion of a conical-type strainer at the pump. Discharge of the pump connects to the upstream side of the piping system with temporary hoses or piping.

2. There is one hookup configuration for a pneumatic blowdown with air:
   a. The pneumatic hookup consists of a valved connection at the supply source and a temporary hose from the supply source to the blowdown manifold. The blowdown manifold, in the direction of flow, includes a block valve, valved connection (for a high-pressure cylinder with regulator to achieve test pressure), vent valve, a block valve, *relief valve, vent valve, pressure gage, temperature gage, and a flanged end that can be adapted to various end connections.
      (i) *Note: *The relief valve shall have a set pressure not higher than the test pressure plus the lesser of 50 psi or 10% of the test pressure.*
   b. The components that make up the blowdown manifold assembly shall be rated to accommodate the test pressures.
   c. The outlet of the test circuit is connected to a valve and a temporary hose to carry the blowdown exhaust to a safe area.
   d. A conical-type strainer, installed in each outlet, is used during the final phase of the blowdown to determine if the system is clear of debris.
   e. The conical-type strainer, up through 4″ NPS, shall have 1/16″ dia. perforations at a density of 144 holes per square inch. For sizes 6″ NPS and above, they shall have 1/8″ dia. perforations at a density of 32 holes per square inch.
3. Table A2.1 provides a profile of the different cleaning categories:
   a. Column title definitions:
      (i) *Category*: Cleaning categories, which are defined later.
      (ii) *Permanent*: The hookup to a flushing source is a temporary connection unless indicated with "Yes."

Table A2.1 Type of flush and hookup

Category	Permanent	Once through	Circulate	Pneumatic	Chlorinate	Description
C-1	Yes	Yes	—	—	—	Service water flush
C-2	—	Yes	—	—	—	Pot water flush
C-3	—	Yes	Yes	—	—	Pot water/chemical flush
C-4	Yes	—	—	Yes	—	Air purge
C-5	—	Yes	—	Yes	—	Air purge
C-6	Yes	Yes	—	—	—	DI water flush
C-7	Yes	Yes	—	—	Yes	Pot water flush/chlorinate
C-8	—	Yes	Yes	Yes (dry)	—	DI water/solvent flush
C-9	—	Yes	Yes	Yes (dry)	—	Solvent/DI water flush
C-10	—	—	Yes	Yes (BD)	—	Air purge/hyd fluid flush
C-11	—	Yes (drains)	—	—	—	Pot water flush
C-12	—	—	—	—	—	Visual only—atmos. vent

BD, blowdown.

(iii) *Once-through*: This describes the process of flushing through a system and directly to containment or sewer with no circulation.
(iv) *Circulate*: This is the process of circulating a cleaning solution through the system with the use of a circulation pump.
(v) *Pneumatic*: This cleaning process uses air or nitrogen to blow-down a system and/or dry it out.
(vi) *Chlorinate*: The process required to sterilize a system.

4. System volume can be calculated based on the following:
   a. Multiply the length of the pipe run by the volume indicated in Table A2.2, based on size and schedule of pipe.

(C) Clean and Flush Velocity. In order to dislodge, suspend, and remove debris in the piping system, such as dirt, metal filings, and weld spatter, it is necessary that water or air be forced through the piping system at a velocity sufficient to suspend that debris. The heaviest debris found in the system will be metal filings.

1. Depending on size and weight of the debris, metal filings will have a terminal midrange settling velocity, in water, of approximately 10 ft/s. Therefore, a flushing velocity of approximately 10 ft/s should be achieved during the flush. (This does not apply to acid cleaning.) Table A2.3 indicates the rate of flow required to achieve approximately 10 ft/s of velocity through various sizes and schedules of pipe.
2. Purging a piping system clear of debris with air requires a minimum velocity of approximately 25 ft/s. Table A2.4 indicates the rate of airflow required to achieve approximately 25 ft/s of velocity through various sizes and schedules of pipe.

(D) Categories. Because of the various cleaning requirements needed for the wide range of utility and process piping services, cleaning categories have been established in an effort to better define and identify the differing requirements needed. Each piping service, whether process or utility, should be associated with a particular cleaning category.

2.2 Cleaning Categories

(A) Category C-1. These systems shall be flushed with the fluid that the system is intended for. There shall be no hydrostatic or pneumatic leak test. An initial service leak test will be performed. Refer to test Category T-1.

1. Connect system to its permanent supply line. Include a permanent block valve at the supply line connection. All outlets shall have temporary hoses run to drain. Do not flush through coils, plates, or filter elements.
2. Using supply line pressure, flush system through all outlets until water is clear and free of any debris at all outlet points. Flush a quantity of fluid equal to at least three times that contained in the system. Use Table A2.2 to estimate volume in system.
3. These systems are required only to undergo an initial service leak test. During the flushing procedure, and as the system is placed into service, all joints shall be checked for leaks.

Table A2.2 Volume of water per lineal foot of pipe (qal.)

	Nominal pipe sizes (in.)														
Sch.	1/2	3/4	1	11/2	2	3	4	6	8	10	12	14	16	18	20	24
5s	0.021	0.035	0.058	0.129	0.207	0.455	0.771	1.68	—	—	—	—	—	—	—	—
20	—	—	—	—	—	—	—	—	2.71	4.31	6.16	7.34	9 70	12.4	15.2	22.2
40	0.016	0.028	0.045	0.106	0.176	0.386	0.664	1.51	2.61	4.11	5.84	9.22	9.22	14.5	14.5	—
80	0.012	0.023	0.037	0.093	0.154	0.3450	.60	1.36	—	—	—	—	—	—	—	—

Table A2.3 Rate of flushing liquid needed to maintain approximately 10 FPS velocity (GPM)

Pipe sch.	Pipe sizes (in.)														
	1/2	3/4	1	11/2	2	3	4	6	8	10	12	14	16	18	20
5s	12	20	34	77	123	272	460	1006	—	—	—	—	—	—	—
20	—	—	—	—	—	—	—	—	—	2575	3678	4385	5794	7400	9082
40	10	16	27	64	105	230	397	902	1560	2460	3493	4221	5514	6980	8676
80	7	13	22	55	92	—	—	—	—	—	—	—	—	—	—

Table A2.4 Rate of air flow to maintain approximately 25 FPS velocity (SCFS)

	Pipe sch.	Pipe sizes (in.)														
		1/2	3/4	1	11/2	2	3	4	6	8	10	12	14	16	18	20
Purge press. 15 psig	5s	0.14	0.23	0.39	0.86	1.39	3.06	5.17	—	—	—	—	—	—	—	—
	40	0.11	0.19	0.30	0.71	1.18	2.59	4.47	—	—	—	—	—	—	—	—
	80	0.08	0.15	0.25	0.62	1.04	2.32	4.03	—	—	—	—	—	—	—	—
Purge press. 50 psig	5s	0.30	0.51	0.84	1.88	3.02	6.67	11.3	—	—	—	—	—	—	—	—
	40	0.23	0.41	0.66	1.56	2.56	5.65	9.73	—	—	—	—	—	—	—	—
	80	0.18	0.33	0.55	1.35	2.26	5.05	8.79	—	—	—	—	—	—	—	—

4. Any leaks discovered during the flushing process, or during the process of placing the system into service, will require the system to be drained and repaired, after which the process will start over with step 2.

(B) Category C-2. These systems shall be flushed clean with potable water.

1. Hook up flush/test manifold at a designated inlet to the system, and temporary hose or pipe on the designated outlet(s) of the system.
2. Route temporary hose or pipe from potable water supply, approved by owner, and connect to flush/test manifold. Route test circuit outlet hose or pipe to sewer, or as directed by owner rep. Secure end of outlet.
3. Using the once-through procedure and the rate of flow in Table A2.3, perform an initial flush through the system with a quantity of potable water equal to at least three times that contained in the system. Use Table A2.2 to estimate volume in system.
4. Discharge to sewer, or as directed by owner rep.
5. After the initial flush, insert a conical strainer into the spool piece provided, located between the test circuit discharge and the outlet hose. Perform a second flush with a volume of potable water equal to at least that contained in the system.
6. After the second flush, pull the strainer and check for debris. If debris is found, repeat step 3. If no debris is found, the system is ready for leak testing.

(C) Category C-3. These systems shall be preflushed with potable water, cleaned with a preoperational cleaning solution and then a final rinse/neutralization, and leak tested with potable water. If it is determined that the system will be installed and tested in segments, the sequence of cleaning and testing can be altered to leak test prior to cleaning with the intent to clean the entire system once installed and tested.

1. Hook up flush/test manifold at a designated inlet to the system between the circulating pump discharge and the system inlet. Install a temporary hose or pipe on the designated outlet(s) of the system.
2. Route temporary hose or pipe from potable water supply, approved by owner, and connect to flush/test manifold. Route outlet hose or pipe to sewer, or as directed by owner rep.
3. Close valve between the circulating pump discharge and flush/test rig. Open valve between flush/test manifold and piping test circuit.
4. Using the once-through procedure and the rate of flow in Table A2.3, perform an initial flush through the system, bypassing the circulation pump, with a quantity of potable water equal to at least three times that contained in the system. Use Table A2.2 to estimate volume in system. Direct the discharge hose to a sewer.
   a. (*Note: During the water flush, check the system for leaks. Verify no leaks prior to introducing chemical cleaning solution to the piping system.*)
5. Discharge to sewer, or as directed by owner rep.

6. After completing the initial flush, drain remaining water in the system to sewer. Or, retain water if cleaning chemicals will be added to the circulating water.
7. Configure valves and hoses to circulate through pump. Connect head tank, or other source containing cleaning solution, to connection provided on circulation loop.
8. Fill the system with preoperational cleaning solution and circulate through the system for 48 h or as recommended. To minimize the possibility of corrosion to the piping, if anticipated, circulate cleaning agent at a low velocity rate prescribed by the cleaning agent manufacturer.
9. Drain cleaning solution to sewer or containment, as directed by owner.
10. Reconnect as before, for the once-through flush/neutralization, and flush to sewer with potable water using a quantity equal to three times that of the system volume. If the preoperational cleaning solution has a neutral pH, the rinse water will have to be visually examined for clarity. Rinse until clear. The rinse must be started in as short a time span as possible after flushing with the cleaning solution. If cleaning residue is allowed to dry on the interior pipe wall, it will be more difficult to remove with a simple flushing. The final rinse and neutralization must be accomplished before any possible residue has time to dry.
11. Upon completion of cleaning procedure, remove pump and temporary circulation loop. System is then ready for leak test.

(D) Category C-4. These systems shall be purged with air. There shall be no hydrostatic or pneumatic test. An initial service leak test will be performed.
1. For air systems, connect system to its permanent supply line. Include a permanent block valve at the supply line connection. All outlets shall have temporary hoses or piping run to a safe area and secured.
2. If the service is something other than air, then hook up blowdown/test manifold at a designated inlet to the system. Install temporary hose or pipe on all outlets of the system and secure the discharge.
3. Route temporary hose or pipe from an approved air supply and connect to blowdown/test manifold.
4. Using the once-through procedure and the rate of flow in Table A2.4, perform an initial blowdown through the system for approximately 12 s per every 100 ft of pipe, or a minimum of 12 s.
5. Discharge to a safe area as directed by owner rep.
6. After the initial blowdown, insert a conical strainer into the spool piece provided, located between the outlet hose and the piping system. Perform a second blowdown approximately 12 s for every 100 ft of pipe, or for a minimum of 12 s.
7. After the second blowdown, pull the strainer and check for debris. If debris is found, clean strainer and repeat from step 6. If no debris is found, the system is ready to be placed into service. Remove all temporary items and make all permanent connections.

8. These systems are required only to undergo an initial service leak test. As the systems are placed into service, all joints shall be checked for leaks. Refer to Part 3, Category T-2 or T-3 Testing.

(E) Category C-5. These systems shall be purged with air or nitrogen.

1. Hook up blowdown/test manifold at a designated inlet to the system. Install temporary hose(s) or pipe(s) on all outlets of the system to exhaust blowdown to a safe location. Secure the end of each hose or pipe.
2. Route temporary hose or pipe from an approved air supply and connect to blowdown/test manifold.
3. Using the once-through procedure and the rate of flow in Table A2.4, perform an initial blowdown through the system for approximately 12 s per every 100 ft of pipe, or for a minimum of 12 s.
4. Discharge to a safe area as directed by owner rep.
5. After the initial blowdown, insert a conical strainer into the spool piece provided, located between the outlet hose and the piping system. Perform a second blowdown approximately 12 s for every 100 ft of pipe, or for a minimum of 12 s.
6. After the second blowdown, pull the strainer and check for debris. If debris is found, clean strainer and repeat from step 5. If no debris is found, the system is ready to be leak tested.

(F) Category C-6. These systems shall be cleaned and flushed with DI or better water.

1. Temporarily connect DI water supply, or water of equal quality, to the supply side of the system. Connect a temporary hose to each outlet.
2. Using the once-through procedure and the rate of flow in Table A2.3, perform an initial flush through the system with a quantity of purified water equal to at least three times that contained in the system. Use Table A2.2 to estimate volume in system.
3. Discharge to sewer, or as directed by owner rep.
4. After the initial flush, insert a conical strainer into the end of each outlet between the outlet hose and the piping system. Perform a second flush with a volume of purified water equal to at least three times that contained in the system.
5. After the second flush, pull the strainer and check for debris. If debris is found, clean strainer and repeat step 3. If no debris is found, continue to step 6.
6. This paragraph is for Category D passivated systems only: Category D fluid systems are required only to undergo an initial service leak test. During the flushing process and as the systems are passivated, all joints shall be checked for leaks. Fluid systems categorized as normal are required to undergo hydrostatic leak testing.

(G) Category C-7. This category covers the flushing, sterilization, and service testing of potable water piping systems. Sanitization of potable water systems is typically regulated by local governments. Where such regulations deviate from the following, you are compelled to instead follow the rules of the local authority.

1. Connect system to its permanent potable water supply line. Include, in the direction of flow, a permanent block valve and a removable assembly that includes a check valve, block valve, valved connection for sterilizing solution, and a block valve.
2. Using the permanent potable water supply, flush system through all outlets until water is clear and free of any debris at all outlet points. Flush a quantity of potable water equal to at least three times that contained in the system. Use Table A2.2 to estimate volume in system.
3. Unless overruled by local plumbing code requirements, potable water systems are governed by the IPC and are therefore required only to undergo an in-service leak test. Based on those requirements, these systems will be placed into service after flushing and sanitizing procedures and upon approval from the Board of Health. During the flushing and sanitizing procedure, and as the systems are placed into service, all joints shall be checked for leaks.
4. Any leaks discovered during the flushing process, the sanitizing process, or the process of placing the system into service will require the system to be drained and repaired, after which the process will start over beginning with step 2.
5. After flushing, sterilize system with a solution containing 50 ppm (50 mg/L) of chlorine (liquid chlorine, Spec AWWA B301-59).
6. The sterilizing solution shall remain in the system for 24 h. For approximately 10 s every hour during this period, open and close valves and faucets.
7. Following the required standing time, flush the solution from the system with potable water until the residual chlorine content is not more than 0.2 ppm when tested at each outlet.
8. After previous requirements are satisfied, submit samples to the local Board of Health for approval. These pipelines are to be tagged "out of service" until approved for service by the Board of Health. Tags are to be provided by the contractor and shall be of a highly visible color and placed in locations that are readily visible at each use point.
9. Repeat sterilization if necessary until approval from the State Board of Health is obtained.
10. Provide 2 copies of Board of Health approval letter to the owner.
11. Once approved by the Board of Health, remove the temporary removable section at the supply line and insert a presanitized spool piece connecting the system permanent to its source. The system is now ready for service and does not require further leak testing.

(H) Category C-8. These systems shall be flushed with DI or better water, cleaned with a circulated solvent (e.g., terpene, aliphatic hydrocarbon, alcohol, ester, hydrochlorofluorocarbon, hydrofluorocarbon/silicon-based solvents) and given a final rinse with DI or better water.

1. Ensure that all components remaining in the system during the cleaning operation shall be compatible with the selected cleaning solvent.

2. Hook up flush/test manifold at a designated inlet to the system between the circulating pump discharge and the system inlet. Install temporary hose(s) or pipe on each outlet of the system.
3. Route temporary hose or pipe from DI water supply, approved by owner, and connect to flush/test manifold. Route temporary outlet hose or pipe to sewer, or as directed by owner rep.
4. Close valve between the circulating pump discharge and flush/test manifold. Open valve between flush/test manifold and piping system.
5. Using the once-through procedure and the rate of flow in Table A2.3, perform an initial flush through the system, bypassing the circulation pump, with a quantity of DI or better water equal to at least three times that contained in the system. Use Table A2.2 to estimate volume in system.
6. Discharge to sewer, or as directed by owner rep.
7. After the initial flush, drain remaining water in the system to sewer.
8. Configure valves and hoses to circulate through pump. Connect head tank, or other source containing cleaning agent, to connection provided on circulation loop. Install conical strainer.
9. After filling the system with the cleaning agent, circulate a quantity equal to three times that of the system volume.
10. Drain cleaning agent to containment, or as directed by owner. Remove and clean strainer and then reinstall.
11. Refill system with unused solvent and circulate a quantity equal to at least three times that of the system volume.
12. Take a sample of the solvent and visually check for color and residue. If color does not compare to unused solvent or if residue is detected, drain system, refill with unused solvent, and repeat step 11. If color is satisfactory and no residue detected drain system, remove strainer and proceed to step 13.
13. Prepare for final rinse by reconfiguring system for the once-through flush as before. Route outlet hose or pipe to sewer, or as directed by owner rep. Rinse immediately after cleaning. If cleaning residue is allowed to dry on the interior pipe wall, it will be more difficult to remove with a simple flushing. The final rinse and neutralization must be accomplished before any possible residue has time to dry.
14. Using the once-through procedure and the rate of flow in Table A2.3, flush all cleaning solution and residue from the system with a quantity of DI water equal to at least three times that contained in the system. Use Table A2.2 to estimate volume in system. Direct the discharge hose to a sewer, or as directed by owner rep.
15. After the final rinse, and immediately upon draining the system of all DI water, it must be thoroughly dried. Accomplish this by purging the system with instrument air. Connect an instrument air supply to the flush/test manifold and blowdown to a safe, open area. Do not discharge into an enclosed area. When drying out the system with instrument air, the humidity of the

purge outflow should be checked with a handheld hygrometer. Gauge the humidity level of the instrument air supply and use that as the benchmark for the discharge. Flow should continue until the outflow of instrument air is the same as the air supply.

16. When system has been thoroughly dried, and with instrument air still connected to the flush/test rig, remove outlet hose, the circulation pump, and all temporary hoses, and prepare for leak test.

(I) Category C-9. These systems shall be cleaned with a circulated solvent (e.g., terpene, aliphatic hydrocarbon, alcohol, ester, hydrochlorofluorocarbon, hydrofluorocarbon/silicon-based solvents) and given a final rinse with DI water.

1. Piping for oxygen service shall have been purchased precleaned for oxygen service by the manufacturer with special care taken during fabrication by the contractor to ensure a high degree of internal cleanliness. Therefore, a flush prior to the cleaning will not be required.
2. Ensure that all materials remaining in the system during the cleaning operation shall be compatible with the selected cleaning solvent.
3. Install circulation loop on system to be cleaned. Connect head tank, or other source containing cleaning agent, to connection provided on circulation loop. Install conical strainer.
4. After filling the system with the cleaning agent, circulate a quantity equal to at least three times that of the system volume.
5. Drain cleaning agent to containment, as directed by owner. Remove strainer, clean, and reinstall.
6. Refill system with unused solvent and circulate a quantity equal to at least three times that of the system volume.
7. Take a sample of the solvent and visually check for color and residue. If color does not compare to unused solvent or if residue is detected, drain system, refill with unused solvent, and repeat step 6. If color is satisfactory and no residue detected, drain system, remove and clean strainer, and proceed to step 8.
8. Disconnect pump and circulation loop. Flush/test manifold should remain installed.
9. Prepare for water flush by routing temporary hose or pipe from DI water supply, as approved by owner, and connect to flush/test manifold. Route outlet hose or pipe to sewer, or as directed by owner rep. Rinse immediately after cleaning. If cleaning residue is allowed to dry on the interior pipe wall, it will be more difficult to remove with a simple flushing. The final rinse and neutralization must be accomplished before any possible residue has time to dry.
10. Using the once-through procedure and the rate of flow in Table A2.3, flush all cleaning solution and residue from the system with a quantity of DI water equal to at least three times that contained in the system. Use Table A2.2 to estimate volume in system. Direct the discharge hose to a sewer, or as directed by owner rep.

11. After the final flush, and immediately upon draining the system of all DI water, it must be thoroughly dried. Accomplish this by purging the system with nitrogen. Connect a nitrogen supply to the flush/test rig and blowdown to a safe, open area. Do not discharge into an enclosed area.
12. When drying out the system with nitrogen, the humidity of the purge outflow should be checked with a handheld hygrometer. Gauge the humidity level of the nitrogen supply and use that as the benchmark for testing the outflow. Flow should continue until the outflow of nitrogen is the same as the nitrogen supply.
13. When system has been thoroughly dried, and with nitrogen still connected to the flush/test manifold, remove outlet hose and prepare for leak test.

(J) Category C-10. These systems shall be blown clear with air and flushed with hydraulic fluid.

1. Hook up blowdown/test manifold at a designated inlet to the system. Install temporary hose(s) or pipe on all outlets of the system to exhaust blowdown. Secure the end of each hose.
2. Route temporary hose or pipe from an approved air supply and connect to blowdown/test manifold.
3. Using the once-through procedure and the rate of flow in Table A2.4, blow down through the system for approximately 12 s per every 100 ft of pipe, or a minimum of 12 s.
4. Discharge to a safe area as directed by owner rep.
5. Install circulation loop on system to be cleaned. Connect head tank, or other source containing hydraulic fluid, to connection provided on circulation loop. Install 10 μm bag filter at the hydraulic flush return. A preheater will be required.
6. Preheat hydraulic fluid to 110°F, and then fill system.
7. After filling the system with the hydraulic fluid, circulate at a flow rate indicated in Table A2.3 for 30 min intervals.
8. Check filter bag for particles after each 30 min interval and clean if necessary. Continue this process until no particles are found in the filter bag.
9. Disconnect pump and circulation loop. Flush/test manifold should remain installed. Close all openings and prepare system for hydrostatic test.

(K) Category C-11. This category covers the flushing of sewers and drains.

1. Using potable water supplied by a handheld hose, flush each floor drain with a rate of flow the drain will accommodate without overflowing. The volume of water used shall be equal to three times that contained in the branch being flushed. Use Table A2.2 to estimate volume in system. Start with the lowest level of drain points.

(L) Category C-12. This category covers process and utility vents directly connected to no more than two tanks in close proximity to one another. It includes flange, threaded, and various types of coupled joints other than the initial equipment connection.

1. In lieu of a liquid flush or air blow, take added care during fabrication to have any threading lubricant, weld prep debris, or any other foreign matter cleaned from the interior of the pipe after each stage of fabrication.
2. During installation, the interior of each segment of pipe shall be visually inspected for sufficient cleanliness by the owner's representative prior to installation. If required, as determined by the owner's inspector, perform additional cleaning of the uninstalled pipe by pulling a cotton wipe (bundle if necessary) through the pipe until interior cleanliness is acceptable.
3. Upon completing the installation of a vent pipe, install a temporary cap, loose enough to slip on and off, over the discharge end of the pipe to prevent birds or debris from entering the vent. At the discretion of the owner, these temporary caps can be removed by the contractor at system turnover or plant turnover or left for removal by the owner at some later date.

2.3 Cleaning Solution and Corrosion Protection

(A) Cleaning Solution and Corrosion Inhibitors. When installing carbon steel pipe and copper tubing to be used in liquid services, it will be practical to circulate a corrosion inhibitor mixture immediately after cleaning. The corrosion inhibitor will effectively reduce the production of corrosion on new systems.

(B) Selecting an Inhibitor. The type of inhibitor used for carbon steel and copper piping systems should be a solution with a neutral pH able to clean the interior of rust, oil-based coatings, and other debris and passivate the interior surface of the pipe.

2.4 Post-Flushing Cleanup

(A) Cleanup. It is the owner's intent to maintain a safe working environment. Based on that regard, after cleaning and flushing is complete, at the end of the day, unless leak testing will be performed immediately after the flushing and cleaning process, the area involved shall be cleared of standing water, scattered tools, and all debris resulting from the cleaning and flushing activity.

Part 3 Testing

3.1 General

(A) Preparation
1. The mechanical contractor shall furnish material and labor necessary to perform all required leak testing requirements on piping systems within the scope of this procedure.
2. At least one person shall be assigned to remain with the test manifold at all times while system is being tested.
3. Joints not previously tested shall not be concealed during the testing process.

4. Underground or underslab piping, not previously tested, shall remain uncovered until leak testing has been completed.
5. Instrument lead lines or piping ends intended for attachment of instruments shall be capped, plugged, or blanked during the test. This may require that the instrument be disconnected and moved on its support, temporarily removed, or scheduled for installation after the leak test.
6. Lines containing check valves shall have the source of pressure on the upstream side of the valve. If this is not practical, swing checks shall be blocked open or their flapper removed, whereas ball and piston checks shall have their internal parts removed. All parts removed shall be put in a ziplock or other appropriate type bag and secured on or near the valve. If none of this is practical, the check valve shall be blocked off or completely removed.
7. Where a closed valve is intended to be a test boundary limit, the test pressure shall not exceed the pressure value used in its valve closure test, normally 110% of the ambient temperature rating (ASME a16.34, para 7.2). If the system test pressure is to be higher, the valve shall be excluded with a line blind or removed to allow the use of a blind flange on the end of the circuit.
8. All pipe supports and guides shall be in place.
9. If owner provided, coordinate the use of the testing fluid with the owner.
10. If owner provided, coordinate the need for liquid waste disposal with the owner.
11. With a minimum 24h notice, submit a request to the owner for the use of utilities and waste disposal along with the date and time the testing is scheduled to be performed.
12. Temporarily remove or isolate all items listed in Attachment A, and close all openings with plugs, blinds, caps, or spool pieces.
13. Temporarily remove all permanent strainer screens.
14. All system valves within the test circuit, unless indicated otherwise, shall be in their full open position, but not back-seated.
15. If heat treatment is required for any piping, it shall have been done prior to installation and testing.
16. During winter months, in unheated facilities, take special precautions to prevent freezing of the liquid test fluids.
17. During installation of large systems, it may be more practical to leak test smaller segments of the system prior to cleaning the entire system. When this occurs, joints previously tested can be covered with paint or insulation as required, even though they are subjected to repeated test pressures. Only the untested joints shall remain uncovered during testing. Temporary blind flanges, spools, or slip blinds used to seal off a test circuit for leak testing shall be painted a bright, visually noticeable color, and numbered for control (see Attachment A). These numbered items shall be listed and periodically inventoried throughout testing to insure that all are accounted for.

18. Refer also to the following paragraphs in ASME B31.3 as applicable:
    a. 345.2: General Requirements for Leak Tests
    b. 345.3: Preparation for Leak Test
    c. 346: Records

(B) Method and Hookup

1. There are five primary leak testing procedures and one alternative to leak testing:
   a. *Initial service leak test*: During the cleaning and flushing procedure, and after the system is placed into service, the test circuit is examined for leaks.
   b. *Hydrostatic leak test*: After cleaning and flushing, the system is leak tested in accordance with the following:
      (i) Not less than 1.5 time the system design pressure.
      (ii) If the design temperature is greater than the test temperature, the minimum test pressure shall be calculated in accordance with ASME B31.3 (eq (24):

$$P_{\mathrm{T}} = 1.5\frac{PS_{\mathrm{T}}}{S} \tag{24}$$

where

P = internal design gage pressure
P_{T} = minimum test gage pressure
S = allowable stress at component design temperature for the prevalent pipe material (see ASME B31.3 Table A-1)
S_{T} = allowable stress at test temperature for the prevalent pipe material (see Table A-1)

   a. *Pneumatic leak test*: After blowdown with air or nitrogen, the system is tested with air or nitrogen at a test pressure of 1.10 times design pressure, based on ASME B31.3 requirements:
      (i) When performing leak testing, using gases such as air or nitrogen on piping exposed to environmental conditions, there may be a wide fluctuation in internal pressure during the leak investigation due to radiant heat:
         - On cold days the volume of the test gas in the pipeline will contract, causing a drop in pressure.
         - On hot days the volume of the test gas in the pipeline will expand, causing an increase in pressure. The gas will, after a period of time, meet equilibrium with the environment and stabilize to a reasonable point.
      (ii) On cold days, pressurize to the test pressure and hold momentarily to determine if the cold ambient temperature will be sufficient to lower the pressure to the design pressure where it is held for examination. Either way, pressure shall be lowered to the design pressure value and maintained during examination.

   (iii) On hot days, pressurize to the test pressure and hold momentarily. Then drop to the design pressure value and hold for examination. As the pressure increases, release pressure as needed to maintain design pressure during examination.
   (iv) It is acceptable to exceed the test pressure and to fluctuate above the test pressure, but test pressure has to be achieved.
   (v) When holding for test pressure and design pressure, maintain pressure with a tolerance of minus 0, plus 10%.
  b. *Sensitive leak test*: This is a bubble test–direct pressure technique in accordance with the BPV Code Section V, Article 10, Appendix I. or another leak test method that has a demonstrated sensitivity of not less than 10^{-3} std ml/s under test conditions.
  c. *Static head test*: This applies to plumbing drains and vent stacks (floor drains, storm drains, sanitary waste) governed by the IPC, or the local plumbing code. The system is filled with water to a minimum of 10 ft at any point in the system except for the highest 10 ft section.
  d. *Alternative*: In lieu of performing a pneumatic or hydrostatic leak test, the following shall be done:
   (i) Weld examination of all welds not previously examined
   (ii) System flexibility analysis
   (iii) Sensitive leak test
2. There are two primary hookup configurations for leak testing:
  a. Hydrostatic testing
   (i) The hydrostatic testing rig consists of a valved connection at the water supply source and a temporary hose from the supply source to the flush/test rig. The flush/test rig, in the direction of flow, includes a block valve, valved connection for a hand pump (to achieve test pressure), a drain/vent valve, a block valve, drain/vent valve, pressure gage, temperature gage, and a flanged end that can be adapted to various end connections.
   (ii) The components that make up the flush/test rig assembly shall be rated to accommodate the test pressures.
  b. Pneumatic testing
   (i) The pneumatic test manifold consists of a valved connection at the supply source and a temporary hose from the supply source to the blowdown/test manifold. The blowdown/test manifold, in the direction of flow, includes a block valve, valved connection (for a high-pressure cylinder with regulator to achieve test pressure), vent valve, a block valve, *relief valve, vent valve, pressure gage, temperature gage, and a flanged end that can be adapted to various end connections.
    - *Note: *The relief valve shall have a set pressure not higher than the test pressure plus the lesser of 50 psi or 10% of the test pressure.*
    - The components that make up the blowdown/test manifold assembly shall be rated to accommodate the test pressures.

Table A3.1 Type of leak test

Category	Service test	Hydrostatic	Pneumatic	Sensitive	Static head	Alternative
T-1	Yes (liq)	—	—	—	—	—
T-2	Yes (gas)	—	—	—	—	—
T-3A, B, and C	—	Yes		—	—	—
T-4	—	—	Yes (air)	—	—	—
T-5	—	—	Yes (N_2)	—	—	—
T-6	—	—	—	Yes	—	—
T-7	—	Yes (hydraulic)	—	—	—	—
T-8	—	—	Yes	—	—	—
T-9	—	—	—	—	Yes	—
T-10	—	—	Yes	—	—	—
T-11	—	—	—	—	—	Yes
T-12	—	—	Yes*	—	—	—
T-13	—	—	Vacuum	—	—	—
T-14	—	Option	—	—	—	Option
T-15	—	—	At 5 psig	—	—	—

* Test for leaking refrigerant using a halide leak detector.

3. Table A3.1 provides a profile of the different testing categories.
   (A) Categories. Because of the various testing requirements needed for the wide range of utility and process piping services, and in an effort to better communicate those requirements, testing categories have been established and identified. Each piping service, whether process or utility, should be associated with a particular testing category.

3.2 Test Categories

1. In lieu of leak testing, visually examine all joints and support connections to verify proper and sufficient installation integrity. Ensure that all bolts are tight, all gaskets are installed, all threaded or coupled connections are secure, and all supports are properly installed. A leak test is not required.

(A) Category T-1. Systems under this category will utilize initial service leak testing.

1. After completion of the flushing and cleaning process, connect the system, if not already connected, to its permanent supply source and to all of its terminal points. Open the block valve at the supply line and gradually feed the liquid into the system, venting the system as necessary.
2. Start and stop the fill process to allow proper high point venting to be accomplished. Hold pressure to its minimum until the system has been completely filled and vented.

3. Once it is determined that the system has been filled and vented properly, gradually increase pressure until 50% of operating pressure is reached. Hold that pressure for 2 min to allow piping strains to equalize. Continue to supply the system gradually until full operating pressure is achieved.
4. During the process of filling the system, check all joints for leaks. Should leaks be found at any time during this process, stop the test, drain the system, repair leak(s), and begin again with step 1.
5. If, upon reaching full operating pressure there are no apparent leaks, the system is considered in-service and operating.
6. If the system is not placed into service or tested immediately after flushing and cleaning, and has set idle for a lengthy period of time (judgment call based on weather, material of construction, intended use, possibility of partial dismantling, etc.), it shall require a preliminary pneumatic test at the discretion of the owner. In doing so, air shall be supplied to the system to a pressure of 10 psig and held there for 15 min or longer to ensure that joints and components have not been tampered with and that the system is still intact. After this preliminary pressure check, proceed with step 1.
7. Record test procedure on Attachment F form.

(B) Category T-2. This category covers pneumatic piping systems categorized by ASME B31.3 as Category D fluid service and utilizing initial service leak testing.

1. After completion of the blowdown process, the system should be connected to its permanent supply source and to all of its terminal points. Open the block valve at the supply line and gradually feed the gas into the system.
2. Bring the pressure up to a point equal to the lesser of one-half the operating pressure or 25 psig. Make a preliminary check of all joints by sound or bubble test. If leaks are found, release pressure, repair leak(s), and begin again with step 1. If no leaks are identified, continue to step 3.
3. Continue to increase pressure in 25 psi increments, holding that pressure momentarily (2 min) after each increase to allow piping strains to equalize, until the operating pressure is reached.
4. Check for leaks by sound and/or bubble test. If leaks are found, release pressure, repair leak(s), and begin again with step 1. If no leaks are found, the system is ready for service.
5. Record test procedure on Attachment F form.

(C) Category T-3A. This category covers liquid piping systems categorized by ASME B31.3 as normal fluid service.

1. After completion of the flushing and cleaning process, with the flush/test manifold still in place and the temporary potable water supply still connected (reconnect if necessary), open the block valve at the supply line and complete filling the system with potable water.
2. Start and stop the fill process to allow proper high point venting to be accomplished. Hold pressure to its minimum until the system is completely filled and vented.

3. Once it is determined that the system has been filled and vented properly, gradually increase pressure until 50% of the test pressure is reached. Hold that pressure for 2 min to allow piping strains to equalize. Continue to supply the system gradually until test pressure is achieved.
4. During the process of filling the system, and increasing pressure to 50% of the test pressure, check all joints for leaks. Should any leaks be found, stop the test, drain system, repair leak(s), and begin again with step 1.
5. Once the test pressure has been achieved, hold it for a minimum of 30 min or until all joints have been checked for leaks. This includes valve and equipment seals and packing even though they should have already been factory tested.
6. If leaks are found, drain system as required, repair, and repeat from step 1. If no leaks are found, drain system and replace all items temporarily removed.
7. If the system is not placed into service or tested immediately after flushing and cleaning, and has set idle for a lengthy period of time (judgment call based on weather, material of construction, intended use, possibility of partial dismantling, etc.), it shall require a preliminary pneumatic test at the discretion of the owner. In doing so, air shall be supplied to the system to a pressure of 10 psig and held there for 15 min to ensure that joints and components have not been tampered with and that the system is still intact. After this preliminary pressure check, proceed with step 1.
8. Record all data and activities on Attachments A, B, and C.

(D) Category T-3B. Same as T-3A paragraphs 1 through 8 with the following added requirements:
1. After the final rinse, and immediately upon draining the system of all potable water, it must be thoroughly dried. Accomplish this by purging the system with dry, oil-free compressed air. Connect air supply to the flush/test manifold and blowdown to a safe, open area. Do not discharge into an enclosed area.
2. When drying out the system with air, the humidity of the purge outflow should be checked with a handheld hygrometer similar to TCI, Inc. model #8722. Gauge the humidity level of the air supply and use that as the benchmark. Flow should continue until the humidity level of the outflow air is the same as the air supply.

(E) Category T-3C. Same as T-3A and T-3B paragraphs 1 through 8 and 1 and 2, respectively, with the exception that DI or better water is used instead of potable water.

(F) Category T-4. This category covers pneumatic piping systems categorized by ASME B31.3 as normal fluid service. This procedure uses air as the test fluid.
1. Prior to initiating a pneumatic test of a system, clear all nonessential test personnel from the system area. All access points to the area shall then be mark with a no admittance caution tape until the leak test has been completed and pressure in the system has been reduced to atmospheric pressure.

2. If the system is not tested immediately after flushing and cleaning, and has set idle for a lengthy period of time (judgment call based on weather, material of construction, intended use, possibility of partial dismantling, etc.), it shall require a low-pressure preliminary pneumatic test at the discretion of the owner. In doing so, air shall be supplied to the system to a pressure of 10 psig and held there for 15 min to ensure that joints and components have not been tampered with and that the system is still intact. If the system shows no sign of leakage, based on a drop in pressure gage reading, proceed with step 3.
3. After completion of the blowdown process, with the blowdown/test manifold still in place and the temporary air supply still connected (reconnect if necessary), open the block valve at the supply line and begin increasing pressure in the system with air.
4. Increase pressure to the lesser of one-half the test pressure or 25 psig. At this pressure, perform a preliminary check of all joints by sound and/or by bubble test. If leaks are detected, release pressure, repair leak(s), and begin again with step 3. If no leaks are detected, continue to step 5.
5. Continue to increase pressure gradually in steps until the test pressure is reached. Hold the pressure at each step long enough to allow pipe strains to equalize.
6. Hold the test pressure for 5 min and then reduce the pressure to the design pressure. Hold the design pressure for a minimum of 30 min or until all joints are examined using the bubble test method.
7. Record all data and activities on Attachments A, B, and D.

(G) Category T-5. This category covers pneumatic piping systems categorized by ASME B31.3 as normal fluid service. This procedure uses nitrogen as the test fluid.

1. Prior to initiating a pneumatic test of a system, clear all personnel, not essential to the testing, from the system area. All access points to the area shall then be marked with a no admittance caution tape which shall remain in place until the pressure test has been completed and pressure in the system has been reduced to atmospheric pressure.
2. If the system is not tested immediately after flushing and cleaning, and has set idle for a lengthy period of time (judgment call based on weather, material of construction, intended use, possibility of partial dismantling, etc.), it shall require a low-pressure preliminary pneumatic test at the discretion of the owner. In doing so, nitrogen shall be supplied to the system to a pressure of 10 psig and held there for 15 min to ensure that joints and components have not been tampered with and that the system is still intact. If the system shows no sign of leakage, based on a drop in pressure gage reading, proceed with step 3.
3. After completion of the blowdown process, with the blowdown/test manifold still in place, connect a temporary nitrogen supply to the manifold, open the block valve at the supply line, and begin increasing pressure in the system with nitrogen.

4. Increase pressure to the lesser of one-half the test pressure or 25 psig. Hold that pressure and perform a preliminary check of all joints by sound and/ or by bubble test. If leaks are detected, release pressure, repair leak(s), and begin again with step 3. If no leaks are detected, continue to step 5.
5. Continue to increase pressure gradually in 25 psi increments until the test pressure is reached. Hold pressure at each step for approximately 1 min. This will allow pipe strains to equalize.
6. Hold the test pressure for 5 min and then reduce the pressure to the design pressure. Hold the design pressure for a minimum of 30 min, or until all joints are examined, using the bubble test method.
7. Record all data and activities on Attachments A, B, and D.

(H) Category T-6. This category covers normal fluid service systems with a high potential for leaking, or Category M fluid service systems as defined by ASME B31.3. This is a two-phase pneumatic leak test.

1. Prior to initiating a pneumatic test of a system, clear all personnel, not essential to the testing, from the system area. All access points to the area shall then be marked with a no admittance caution tape until the pressure test has been completed and pressure in the system has been reduced to atmospheric pressure.
2. If the system is not tested immediately after flushing and cleaning, and has set idle for a lengthy period of time (judgment call based on weather, material of construction, intended use, possibility of partial dismantling, etc.), it shall require a preliminary pneumatic test at the discretion of the owner. In doing so, nitrogen shall be supplied to the system to a pressure of 10 psig and held there for 15 min to ensure that joints and components have not been tampered with and that the system is still intact. If the system shows no sign of leakage, based on a drop in pressure gage reading, proceed with step 3. If leaks are discovered, depressurize, repair leaks, and continue to step 3.
3. After completion of the blowdown or flushing process, with the blowdown/test manifold still in place, connect a temporary nitrogen supply source, open the block valve at the supply line, and begin increasing pressure in the system with nitrogen.
4. Increase pressure to the lesser of one-half the test pressure or 25 psig. At this pressure perform a preliminary check of all joints by sound and/or by bubble test. If leaks are detected, release pressure, repair leak(s), and begin again with step 3. If no leaks are detected, continue to step 5.
5. Continue to increase pressure gradually in 25 psi increments until the test pressure is reached. Hold the pressure at each step for approximately 1 min. This will allow pipe strains to equalize.
6. Hold the test pressure for 5 min and then reduce the pressure to the design pressure. Hold the design pressure for a minimum of 30 min or until all joints are examined with the bubble test. If leaks are discovered, release pressure, repair leaks, and repeat from step 4. If no leaks are found, continue with step 7.

7. Release pressure and disconnect nitrogen source. Connect helium source to test manifold.
8. Open helium supply valve and bring helium pressure in the system to 15 psig. After reaching 15 psig, release pressure at one of the branch outlet points.
9. Do that for each branch of the system. Pressurize with helium and release pressure at each branch outlet in the system a minimum of three times.
10. After the last release, and with helium now throughout the system, pressurize to the lesser of 15 psig or 25% of design pressure. While holding at this pressure, perform a precursor leak check by sound if practical.
11. After the precursor leak check, continue to increase pressure gradually until the test pressure is reached. Once the test pressure has been reached, hold for 15 min and then reduce to design pressure before examining joints.
12. Using a helium probe (spectrometer), check each joint for leaks. Sensitivity of the leak test shall be not less than 10^{-3} atm·ml/s under test conditions.
13. If leaks are discovered, release pressure, repair, and repeat from step 8.
14. If no leaks are discovered, safely release the pressure and prepare for service.

(I) Category T-7. This category covers hydraulic piping systems categorized by ASME B31.3 as normal fluid service.

1. If the system is not tested immediately after flushing and cleaning, and has set idle for a lengthy period of time (judgment call based on weather, material of construction, intended use, possibility of partial dismantling, etc.), it shall require a preliminary pneumatic test at the discretion of the owner. Air shall be supplied to the system to a pressure of 10 psig and held there for 15 min to ensure that joints and components have not been tampered with and that the system is still intact. After this preliminary pressure check, if performed, proceed with step 2.
2. After completion of the clean and flush process, with the flush/test manifold still in place and the temporary hydraulic fluid source still connected (reconnect if necessary), open the block valve at the supply source and complete filling the system with filtered hydraulic fluid.
   a. *Note: Hydraulic fluid used for testing will remain in the system as the operating fluid. Therefore, all fluid used during this process shall be filtered through a 10 μm filter.*
3. Start and stop the fill process to allow proper high point venting to be accomplished. Hold pressure to its minimum until the system is completely filled and vented.
4. Once it is determined that the system has been filled and vented properly and the system is closed, gradually increase pressure until 50% of the test pressure is reached. Hold that pressure for approximately 2 min to allow piping strains to equalize. Continue to pressurize the system gradually until test pressure is reached.

5. During the step 3 process of filling the system and increasing pressure to 50% of the test pressure, check all joints for leaks. Should any leaks be found, drain system, repair leak(s), and begin again with step 1.
6. Once the test pressure has been reached, hold it for a minimum of 30 min or until all joints have been checked for leaks. This includes valve and equipment seals and packing.
7. If leaks are identified drain system, repair and repeat from step 1. If no leaks are found, drain system and replace all items temporarily removed. If there were no items removed, leave system full.
8. Record all data and activities on Attachments A, B, and C.

(J) Category T-8. This category covers plumbing drains and vent stacks governed by IPC and/or local plumbing code. It covers such systems as floor drains and storm drains.

1. Plug all vents and drains.
2. Force air into the system until a uniform pressure of 5 psig is achieved throughout the system.
3. Hold this pressure for a minimum of 15 min.
4. If pressure should drop, check each joint for leakage. After leaks are located, release pressure, repair leak, and repeat from step 2.
5. If no drop in pressure is indicated, release pressure and prepare for service.

(K) Category T-9. This category covers gravity process and sanitary waste lines and their vent stacks governed by the IPC and/or local plumbing code.

1. Cap, or plug the end of the system, at the point of connection with the existing system, or insert a line blind at the main trunk if the system is new connecting to an existing main. Make provisions for draining the system after the test.
2. Fill the system with water to the highest point creating a minimum static head pressure of 10 ft of water. The uppermost 10 ft will be subjected to a test pressure of less than 10 ft of static head.
3. After the system has been filled, maintain that head of water for a minimum of 15 min or until all joints are visually checked for leaks.
4. If leaks are found, drain system repair leak(s) and repeat from step 2. If no leaks are found, continue to step 5.
5. Drain water from the system leaving traps full (winter precautions are in effect).
6. As an optional precaution for large, complex systems, the owner may elect to perform a final smoke test of the system to validate trap seals.
7. Using smoke machines, introduce a pungent, thick smoke into a low point of the system until the smoke is visible at each vent stack. Plug each vent stack as the smoke begins to appear. This will force it to move through the system.
8. Once all vent stacks are plugged, achieve a pressure equal to a 1″ water column (248.8 Pa) and hold that pressure for a minimum of 15 min before beginning examination.
9. Check each P-trap location. If smoke appears at any of the drain points, add water to fill trap, or investigate reason for no seal.

(L) Category T-10. This category covers process and utility vacuum lines. Lab vacuum systems operate at negative pressures down to 730 mm Hg absolute.

1. Seal entire system and introduce air to a positive pneumatic pressure of 15 psig.
2. Close off air supply and hold pressure for a minimum of 15 min.
3. If pressure should drop, check each joint for leakage using a bubble test.
4. After locating leaks, release pressure, repair leaks, and repeat from step 2.
5. If no drop in pressure is indicated, release pressure and prepare for service.

(M) Category T-11. This category covers piping systems that have been altered or repaired and where it is not practical to undergo a hydrostatic or pneumatic leak test.

1. All new welds, including those used in the manufacture of welded pipe and fittings, which have not previously undergone hydrostatic or pneumatic leak tests, shall be examined and tested in the following manner:
   a. Circumferential, longitudinal, and spiral groove welds shall be 100% radiographically or ultrasonically examined.
   b. All welds, including structural attachment welds, not covered in (a) above, shall be examined using the liquid penetrant method or, for magnetic materials, the magnetic particle method.
   c. A flexibility analysis shall have been performed.
   d. A sensitive leak test shall be performed.

(N) Category T-12. This category covers refrigerant piping systems governed by ASME B31.5, International Mechanical Code, or American Society of Heating, Refrigerating, and Air Conditioning Engineers.

1. Evacuate the system (both sides of expansion valve) to 26″ Hg vacuum. Check to ensure that the entire system has been evacuated by using a vacuum/pressure gage.
2. After 26″ Hg vacuum is obtained, isolate system from vacuum source.
3. Hold system at 26″ Hg. vacuum for 20 min. A leak will result in loss of vacuum.
4. With sufficient, but not excessive force, tap brazed joints with a rubber mallet to disclose potential leaks that might start with expansion, contraction, or vibration.
5. If system holds 26″ Hg vacuum, disconnect vacuum source. Bleed in a small amount of refrigerant gas. (Do not allow liquid refrigerant to enter the system.) Turn suction shut off valve to midposition and open charge valve. Admit refrigerant gas to 10 psig.
6. Disconnect refrigerant supply and charge system with dry nitrogen gas to 250 psig. Soap test for leaks.
7. If leaks are discovered, evacuate entire system and repair. After repairs are made, repeat from step 1.
8. If system passes soap test check refrigerant system with a halide leak detector. (Caution: Refrigerant that might accumulate in the area can

interfere with the sensitive halide detector. Ventilate area to remove trace refrigerant.)

9. If leaks are discovered with the sniff test, release pressure in the system to make repairs. Repair brazed joints using new material.
10. After system is determined to be leak-free, evacuate the system to 29″ Hg vacuum. Maintain this pressure for 12 h, continuously, to remove moisture from the system.
11. After completion of previous step, back seat suction shutoff valve, close vacuum source valves, and disconnect vacuum source. System is ready to charge.

(O) Category T-13. This category covers vacuum drainage systems. Leak testing of these systems is governed by the IPC and/or local plumbing code.

1. Block all extremities of the system with caps, plugs, slip blinds, blind flanges, etc. in order to seal system.
2. Using a test manifold with a vacuum gage, pull a vacuum of 19″ Hg.
3. Hold vacuum for a minimum of 10 min.
4. If vacuum holds, system is ready for service. If a leak is indicated by a pressure rise on the vacuum gage, repair and repeat test.

(P) Category T-14. This category covers hot-tap connections as governed by ASME B31.3. Due to the varying conditions hot-tap installations can present, this category will provide three basic alternatives in providing documented assurances for verifying the integrity of the installation. The following alternatives are not prioritized. Each situation will dictate which alternative is more practical. These determinations need to be made well in advance of the actual work. All testing or examination shall be performed prior to making the hot tap. With flexibility analysis having been performed on the connecting system, including the hot-tap connection, any of the following alternative examinations may be performed:

1. Ferromagnetic materials:
   a. Perform a dry particle or wet fluorescent particle inspection on all untested welds or welds not previously examined.
2. Ferromagnetic materials and nonferromagnetic metals:
   a. Perform a dye penetrant examination on all untested welds or welds not previously examined.
3. Hydrostatic leak test:
   a. Do not use the hydrostatic leak test on live services with operating temperatures at or above 212°F, or at or below 32°F.
   b. After the hot-tap components are welded in place, and prior to mounting the hot-tap machine, install a blind flange or capped spool piece with a test manifold and separate vent valve attached.
   c. Connect water source, fill the cavity with water, venting air during the process, and bring to test pressure.
   d. Hold test pressure a minimum of 10 min.
   e. If no leaks are detected, release pressure, drain cavity, remove test assembly, and mount hot-tap machine.

(Q) Category T-15. This category covers process and utility vents that include flange, threaded, and various types of coupled joints other than the initial equipment connection.

3.3 Repair of Leaks

(A) Welded Joints
1. Conforming to ASME B31.3 paragraphs 328.6 and 341.3.3, repair defective welded joints by removing the defective weld metal to sound metal.
2. Repair of a weld shall not be attempted by adding weld metal to the defective area.
3. Repair welds shall be made by qualified welders or welding operators in accordance with the same weld procedure.
4. The same examination and qualification required for the original weld is required for the repair work.

(B) Threaded Joints
1. Leaking threaded joints shall be repaired by:
   a. Further tightening using discretion or
   b. Taking joint apart and replacing defective material
   c. Caulking that shall not be attempted for repairs to threaded joints

(C) Roll on Joints
1. Repair leaks by rerolling, realigning pipe, or new gaskets.
2. Do not weld.

(D) Plastic Joints
1. Remove leaking components and replace with new material.

3.4 Temporary Strainers

(A) Install
1. In pump suction lines not containing a permanently installed strainer, a temporary start-up strainer will be required.
2. After the flushing and testing of a test circuit containing a centrifugal or diaphragm pump, the contractor shall install a temporary strainer on the suction side of each pump after the system has been drained and is ready to be sealed up.
3. Each strainer shall be numbered, listed, and accounted for to indicate when they were installed and when they were removed by owner.
4. These strainers are to be removed by owner personnel at a time of their choosing.

3.5 Post-Leak Test Cleanup

(A) Cleanup. The owner requires that a safe working environment be maintained. Based on that requirement, after completion of cleaning and/or testing activities for each circuit, or at the end of the work day, the area involved shall be cleared of standing water, scattered tools, and all debris resulting from those cleaning and flushing activities.

Part 4 Special Procedures and Post-Leak Test Requirements

4.1 Special Procedures

(A) Double Containment Pipe. The integration of two pipelines, one inside the other, requires added consideration in their leak testing. The following procedure will establish a sequence of events in the testing of double containment piping systems. For details of the actual test procedure, refer to the proper testing category.

1. Definitions:
   a. *Primary pipe*: Also referred to as the carrier pipe. This is the pipe that is contained inside the containment pipe. It carries the service fluid.
   b. *Containment pipe*: Also referred to as the secondary containment pipe. This is the outer pipe that contains the carrier, or primary, pipe. This pipe captures and contains any leaking fluid that may escape the carrier pipe.
2. Sequence for testing double containment pipe systems:
   a. Perform a leak test, per the proper test category procedure, on the primary pipeline first.
   b. After testing of the primary pipeline is complete, reduce pressure to the design pressure in preparation for leak testing the containment pipe.
      (i) *Note: If the leak test performed on the primary pipeline was pneumatic, then the pressure will already be at the design pressure.*
   c. Perform a pneumatic leak test on the containment pipeline, per test Category T-4, using a test pressure not to exceed 15 psig.
   d. After testing is complete, release the test fluid from the primary pipeline followed by the test pressure from the containment pipe.

4.2 Post-Leak Test Procedures

(A) Post-Leak Test Procedures. After flushing and testing, some services require additional preparation prior to being ready for service. Services that require additional preparation include:
1. Hydrogen
2. Stainless steel high-purity services

(B) Hydrogen. After the leak test is completed, the hydrogen pipeline has to be made inert inside the expanse of the piping system. Oxygen content shall be made less than 1%. This can be achieved by an evacuation purge in the following manner:
1. Immediately after completing a pneumatic leak test with nitrogen, safely release the nitrogen pressure to atmosphere.
2. Connect a vacuum pump to the system and evacuate to 10 Torr (5.352″ of water). Record this and the following data on the Attachment G form.
3. Perform a pressure rise rate test under static conditions to ensure that the system is tight under vacuum by observing the rate of pressure rise within the system. A 1 Torr/min (0.5352″ of H_2O/min) rise for a 5 min period indicates good vacuum holding ability.

4. Backfill with nitrogen to atmospheric pressure, or a low positive pressure.
5. Evacuate the system again to 10 Torr (5.352″ of H_2O).
6. Three cycles of steps (4) and (5) are required to achieve an oxygen content of less than 1%. After the final evacuation, test for oxygen content. If necessary, continue with steps 4 and 5 until the oxygen level is brought to an acceptable level.
7. The system is now ready for hydrogen. If ready for service, the vacuum may be broken with hydrogen gas and brought to operating pressure. If the system will set idle for a period of time, that is, awaiting facility start-up, then backfill with helium to 10 psig and close up the system.

(C) Stainless Steel Hygienic Services

1. Perform post-leak testing requirements and passivation as necessary and in accordance with other prepared procedures.

Part 5 Documentation

5.1 Forms

The following attachments are forms that can be used in conjunction with the preceding leak testing protocols. There are many ways to comply with the requirements of ASME B31.3 Para. 345.2.7 Test Records when it states that "Records shall be made of each piping system during the testing, including (a) date of test, (b) identification of piping system tested, (c) test fluid, (d) test pressure, (e) certification of results by examiner." But these forms and the checklists extend much beyond those simple requirements in establishing added control and verification of performance.

(A) **Attachment A—Equipment Removed or Installed for Testing**. This form is used to list all in-line equipment that requires removal prior to cleaning and testing. It also includes a list of temporary blinds, plugs, caps, and spool pieces used to replace the removed item or block the ends of a test circuit. This creates a checklist and provides documented verification that these items were both removed and reinstalled. Following is a description of the input titles:

1. Project: Enter the identification number of the project.
2. Test circuit no: Enter the test circuit identification number.
3. Date: Enter the date the test was performed.

   Under **Equipment Temporarily Removed**

4. Item no: Enter the item number, such as PSV-1234.
5. Item: Enter description of item, such as "Pressure Relief Valve."
6. Removed: Enter the date item was removed.
7. Reinstalled: Enter the date item was installed back into the system.

   Under **Components Temporarily Installed**

8. Item no: All items require an ID number. Enter that number here.
9. Item: Describe the item, such as 3″ NPT, 150# flg. spool piece 8″ long.

10. Installed: Enter the date item was installed.
11. Replaced: Enter the date item was replaced by permanent item.
12. Contractor and date: Signature and date of contractor's representative indicating that they have firsthand knowledge that the activities on this form were completed as described.
13. Owner rep and date: Signature and date of owner's representative indicating that they reviewed the procedure and were present during the removal and installation of the listed items. And that the procedure was done properly.

Project Title: ____________
Project No: ____________
Project Location: ____________
Project Revision: ____________
Project Revision Date: ____________

Section No: ____________
Title: Cleaning and Testing of Pipe
Corporate Rev: ____________
Date of Corp. Rev: ____________
Responsible Engineer: ____________

ATTACHMENT A

Equipment Removed or Installed for Testing

Listed on this form is equipment that has to be temporarily removed to prevent being compromised or damaged during a leak test procedure. Also listed is equipment temporarily installed to replace removed items or to isolate or cap-off segments of a piping system during a test procedure. All temporarily removed items shall be stored in a safe and controlled area, protected from damage and debris during impoundment.

Project: ____________ Test Circuit No: ____________ Date: ____________

Equipment, Valves, and Instruments Temporarily Removed			
Item No.	Item	Removed*	Reinstalled*

**Enter date in space when confirmed.*

Blanks, Blinds, Spools, and Plugs Temporarily Installed			
Item No.**	Item	Installed*	Replaced*

**Enter date in space when confirmed.* ***Each item must have an Item No. for control.*

All permanent equipment removed for flushing and/or testing has been properly reinstalled. All temporary components installed for flushing and/or testing have been removed. As confirmed by:

____________ Contractor ____________ Date ____________ Owner Rep. ____________ Date

(B) **Attachment B—Leak Test Checklist**. This form is a step-by-step checklist for performing a leak test on a piping system. It is not used to record data but to verify that it was recorded. The contractor's representative shall write their initials at each step, where provided, indicating that that activity has been completed.

Project Title: ____________ Section No: ____________
Project No: ____________ Title: Cleaning and Testing of Pipe
Project Location: ____________ Corporate Rev: ____________
Project Revision: ____________ Date of Corp. Rev: ____________
Project Revision Date: ____________ Responsible Engineer: ____________

ATTACHMENT B

LEAK TEST CHECKLIST

Project:____________ Test Circuit:____________ Date of Test:____________

	Leak Test Checklist	**Initial**
1.	System Flush Completed	________
2.	Remove Equipment ID'd for Removal	________
3.	Isolate or Remove Safety Valves	________
4.	Align Valves for Test Procedure	________
5.	Notify Owner	________
6.	Test Piping Circuit	________

Test Results and Resolution Path

Test Ok
________ Record Data
________ Owner Verify
________ Drain System
________ Restore System

Failed Test ________
Problem Found ________
Drain System ________
Make Repair ________
Retest Circuit ________ (returns to Test Ok)

Reinstall Equipment	________
Remove Blanks, Pancakes, Caps	________
Reinstall strainer screens	________
Reinstall Safety Valves	________
Remove Temporary Spool Pieces	________
Reinstall Instruments	________
Final Check	________
System Test Completed	________
Drain System Completed	________

Note: After testing is complete, clean entire test area of any standing water or debris resulting from the testing procedure.

(C) **Attachments C, D, E, F, G, and H—Test Data Forms for Piping**. These forms provide all necessary test data for the type of test that pertains to a single test circuit. They set forth basic descriptive information on the test circuit and require that the data obtained from the testing process be entered in spaces provided.

Project Title: ____________
Project No: ____________
Project Location: ____________
Project Revision: ____________
Project Revision Date: ____________

Section No: ____________
Title: Cleaning and Testing of Pipe
Corporate Rev: ____________
Date of Corp. Rev: ____________
Responsible Engineer: ____________

ATTACHMENT C

HYDROSTATIC LEAK TEST DATA FORM FOR PIPING

Project:____________ Test Circuit No:____________ Date:____________

Location of Test: Bldg./Floor:____________ Nearest Column Coordinates:____________

Location Remarks:____________ Fluid Service Symbol:______

Piping Diagrams	Line Numbers

Hydrostatic Test Data

A.	Design pressure	______	PSIG (Obtain from Line List)
B.	Test pressure	______	PSIG (*1.5 x Design Pressure)(Also obtain from line list)
C.	Test Fluid	______	Yes/No for Water (If other than water, what ______)
D.	Start Pressure	______	PSIG (Recorded Test Pressure at start of test)
E.	Finish Pressure	______	PSIG (Recorded Test Pressure at end of test)
F.	Pressure Lost/Gained	______	PSIG (if more than 5 PSI, explain below)
G.	Acceptable Test Pressure	______	(Yes/No) Was test pressure within an acceptable range?
H.	Fluid temperature	______	(Test should be conducted with ambient temperature fluid)
I.	Start Time	______	24 hour clock
J.	Finish Time	______	24 hour clock
K.	Duration of test	______	Minutes (30 minutes minimum)
L.	Leaks Detected by Visual Inspection (Yes/No)	______	(If leaks detected repair and retest)

**For design temperatures in excess of test temperature see ASME B31.3 Para. 345.4.2 eq. (24)*

Additional Comments or Observations: ____________

This circuit has been tested, reassembled and secured.

____________	______	____________	______
Contractor	Date	Owner	Date

Project Title: ____________
Project No: ____________
Project Location: ____________
Project Revision: ____________
Project Revision Date: ____________

Section No: ____________
Title: Cleaning and Testing of Pipe
Corporate Rev: ____________
Date of Corp. Rev: ____________
Responsible Engineer: ____________

ATTACHMENT D

PNEUMATIC LEAK TEST DATA FORM FOR PIPING

Project: ____________ Test Circuit No: ____________ Date: ____________

Location of Test: Bldg./Floor: ____________ Nearest Column Coordinates: ____________

Location Remarks: ____________ Service Symbol: ____________

Piping Diagrams	Line Numbers

Pneumatic Test Data

A.	Design pressure	______	PSIG (Obtain from Line List)
B.	Test pressure	______	PSIG (110% of design pressure)(Also obtain from Line List)
	PSV* Set Point	______	PSIG (S.P.= Test PSI + [lesser of 10% of test or 50 PSI])
C.	Test gas	______	Should be dry compressed air. If other that air, enter type of gas and explain in comments section.
D.	Start Pressure	______	PSIG (Recorded Test Pressure at start of test)
E.	Finish Pressure	______	PSIG (Recorded Test Pressure at end of test)
F.	Pressure Lost/Gained	______	PSIG (if more than 5 PSIG, explain below)
G.	Acceptable Test Pressure	______	(Yes/No) Was test pressure within an acceptable range?
H.	Gas temperature	______	(Test should be conducted with ambient temperature gas)
I.	Start Time	______	24 hour clock
J.	Finish Time	______	24 hour clock
K.	Duration of test	______	Minutes (30 minutes minimum)
L.	Leaks Detected by Visual Inspection (Yes/No)	______	(If leaks detected repair and retest)

** Temporary PSV on test rig used for testing only in accordance with ASME B31.3. Para. 345.5.2.*

Additional Comments or Observations: ____________

____

____

____

This circuit has been tested, reassembled and secured.

____________ Contractor ____________ Date ____________ Owner ____________ Date

Project Title: ____________
Project No: ____________
Project Location: ____________
Project Revision: ____________
Project Revision Date: ____________

Section No: ____________
Title: Cleaning and Testing of Pipe
Corporate Rev: ____________
Date of Corp. Rev: ____________
Responsible Engineer: ____________

ATTACHMENT E

LEAK TEST DATA FORM FOR GRAVITY DRAIN PIPING

Project:____________ Test Circuit No:____________ Date:____________

Location of Test: Bldg./Floor:____________ Nearest Column Coordinates:____________

Location Remarks:____________ Service Symbol:____________

Piping Diagrams	Line Numbers

Gravity Drain and Waste Lines

General Instruction: Strike through the tests that are not being performed on this form (IE: if executing a Hydrostatic Test, strike through Air Test and Final Smoke Test)

Hydrostatic Test

1A	Low point head	______	Ft. of water (10 Ft. min.)
1B	Start Time	______	24 hour clock
1C	Finish Time	______	24 hour clock
1D	Duration of test	______	Minutes (30 minutes minimum)
1E	Visual Inspection for Leaks (NONE)	______	(If leaks detected, repair and retest)

Air Test

2A	Air pressure	______	PSIG (5 PSIG minimum)
2B	Start Time	______	24 hour clock
2C	Finish Time	______	24 hour clock
2D	Duration of test	______	Minutes (30 minutes minimum)
2E	Visual Inspection for leaks by soaping joints (NONE)	______	(If leaks detected, repair and retest)

Final Smoke Test*

3A	Air/Smoke Pressure	______	In. W.C. (1 in. minimum)
3B	Start Time	______	24 hour clock
3C	Finish Time	______	24 hour clock
3D	Duration of test	______	Minutes (30 minutes minimum)
3E	Visual inspection of smoke coming from traps	______	(If smoke present, fill trap and continue examination. If adding water to trap does not stop smoke, stop test and repair. Then retest)

**Perform test without purging trap seals.*

Record additional comments or observations on the back of this form:
This circuit has been tested, reassembled, secured.

____________ Contractor ______ Date ____________ Owner ______ Date

Project Title: ____________
Project No: ____________
Project Location: ____________
Project Revision: ____________
Project Revision Date: ____________

Section No: ____________
Title: Cleaning and Testing of Pipe
Corporate Rev: ____________
Date of Corp. Rev: ____________
Responsible Engineer: ____________

ATTACHMENT F

INITIAL SERVICE LEAK TEST DATA FORM FOR PIPING

Project:____________ Test Circuit No:____________ Date:____________

Location of Test: Bldg./Floor:____________ Nearest Column Coordinates:____________

Location Remarks:____________ Service Symbol:____________

Piping Diagrams	Line Numbers

Initial Service Test Data

A.	Operating pressure	______	PSIG (Obtain from Line List)
B.	Service Fluid	______	The test is performed using the service fluid.
C.	Start Time	______	24 hour clock
D.	Finish Time	______	24 hour clock
E.	Duration of test	______	Minutes (30 minute minimum)
F.	Leaks Detected by Visual Inspection (Yes/No)*	______	(If leaks detected repair and retest)

**For gaseous fluid services a bubble test should be used to detect leaks.*

Additional Comments or Observations: ____________

____

____

____

____

____

____

This circuit has been tested, reassembled, and secured.

____________	______	____________	______
Contractor	Date	Owner	Date

Project Title: ____________
Project No: ____________
Project Location: ____________
Project Revision: ____________
Project Revision Date: ____________

Section No: ____________
Title: Cleaning and Testing of Pipe
Corporate Rev: ____________
Date of Corp. Rev: ____________
Responsible Engineer: ____________

ATTACHMENT G

ALTERNATIVE LEAK TEST DATA FORM FOR PIPING

Note: This is a two-part test that is combined with the Sensitive Leak Test Attachment H

Project:____________ Test Circuit No:____________ Date:____________

Location of Test: Bldg./Floor:____________ Nearest Column Coordinates:____________

Location Remarks:____________ Service Symbol:____________

Piping Diagrams	Line Numbers

Groove Weld Examinations

A.	Circumferential (100%)	______	Enter RT or UT. Or enter N/A if appropriate.
B.	Longitudinal (100%)	______	Enter RT or UT. Or enter N/A if appropriate.
C.	Spiral (100%)	______	Enter RT or UT. Or enter N/A if appropriate.
D.	Fillet Welds*	______	Enter PT or MT. Or enter N/A if appropriate.
E.		______	
F.		______	

Flexibility Analysis

G.	System Analysis Performed (Yes/No)	______	Has flexibility analysis been performed on the respective piping system?

Leak Test

H.	Sensitive Leak Test	______	(Yes/No) Continue with Attachment H test.

**This includes structural attachment welds.*

Additional Comments or Observations: ____________

____

____

____

____

____

____

This circuit has been tested, reassembled, and secured.

____________ Contractor ____________ Date ____________ Owner ____________ Date

Project Title: ____________________
Project No: ____________________
Project Location: ____________________
Project Revision: ____________________
Project Revision Date: ____________________

Section No: ____________________
Title: Cleaning and Testing of Pipe
Corporate Rev: ____________________
Date of Corp. Rev: ____________________
Responsible Engineer: ____________________

ATTACHMENT H

SENSITIVE LEAK TEST DATA FORM FOR PIPING

Project:____________________ Test Circuit No:____________________ Date:____________________

Location of Test: Bldg./Floor:____________________ Nearest Column Coordinates:____________________

Location Remarks:____________________ Fluid Service Symbol:_______

Piping Diagrams	Line Numbers

Sensitive Leak Test Data

A.	Design pressure	________	PSIG (Obtain from Line List)
B.	Test pressure	________	PSIG (At minimum, the lesser of 105 kPa (15 psi) or 25% of design pressure)
C.	Test Gas	________	Enter 'air' or an inert gas used for testing
D.	Start Pressure	________	PSIG (Recorded Test Pressure at start of test)
E.	Finish Pressure	________	PSIG (Recorded Test Pressure at end of test)
F.	Pressure Lost/Gained	________	PSIG (if more than 5 PSI, explain below)
G.	Acceptable Test Pressure	________	(Yes/No) Was test pressure within an acceptable range?
H.	Fluid temperature	________	(Test should be conducted with ambient temperature fluid)
I.	Start Time	________	24 hour clock
J.	Finish Time	________	24 hour clock
K.	Duration of test	________	Minutes (30 minute minimum)
L.	Leaks Detected by Visual Inspection (Yes/No)	________	Perform Bubble Test—Direct Pressure Technique (If leaks detected repair and retest)

Additional Comments or Observations: ____

____

____

____

____

____

This circuit has been tested, reassembled and secured.

____________________	________	____________________	________
Contractor	Date	Owner	Date

Appendix B

Biotechnology Inspection Guide Reference Materials and Training Aids

The following is a US Food and Drug Administration (FDA) training guide used by the FDA in the training of their inspectors. Such training guides are used to prepare and provide FDA inspectors with a general knowledge and understanding of what it is they should be considering during inspection of an FDA-regulated manufacturing facility.

In many cases these are general considerations due in large part to the many and varied drugs and the proprietary processes required to manufacture them. These guides are intended to augment the education, experience, and accumulated knowledge each inspector already possesses. The inspection process and procedures by which they perform it leave room for on-site judgment consideration and assessment by the inspectors.

The purpose of including these FDA guides in this book is to provide the reader, designer, and engineer involved with the design and engineering of a bioprocessing facility some insight as to what the expectations of an FDA inspector might be. What expectations an FDA inspector might possess can most likely be attributed to their training and these guidelines. Providing these guidelines for the readers of this book, those of you designing, constructing, and operating FDA-regulated manufacturing facilities, will hopefully lessen the gap of understanding between you and the FDA inspector. Having some insight into the inspector's understanding and training helps one to be better prepared to meet those expectations when it's time for the inspection.

These inspection guidelines are reproduced here through the courtesy of the FDA. These and others also reside on the FDA web site at:

http://www.fda.gov/ICECI/Inspections/InspectionGuides/default.htm

Bioprocessing Piping and Equipment Design: A Companion Guide for the ASME BPE Standard, First Edition. William M. (Bill) Huitt.

Biotechnology Inspection Guide Reference Materials and Training Aids

November 1991
Division of Field Investigations (HFC-130)
Office of Regional Operations
Office of Regulatory Affairs
U.S. DEPARTMENT OF HEALTH AND HUMAN SERVICES
PUBLIC HEALTH SERVICE
FOOD AND DRUG ADMINISTRATION

Note: This document is reference material for investigators and other FDA personnel. The document does not bind FDA, and does no confer any rights, privileges, benefits, or immunities for or on any person(s).

Acknowledgements

This Guide was initiated by Robert C. Fish, Director, Division of Field Investigations (DFI). Mr. Fish asked Barbara-Helene Smith, Ph.D., DIB, CHI-DO, to chair a workgroup to develop inspectional guidelines for Investigators in the area of biotechnology. The workgroup, which also included Thaddeus T. Sze, Ph.D., Chemical Engineer, DFI, and Kim A. Rice, Supervisory Investigator, SEA-DO, prepared a draft document with information obtained from FDA Center and Field personnel who are actively involved in biotech inspections.

The document was reviewed and expanded upon during a 2 1/2 day workshop (May 29-31, 1991) attended by the following Center and Field personnel: Wendy Aaronson (CBER), Henry Avallone (NWK-DO), Yuan-Yuan Chiu, Ph.D. (CDER), Vitolis Vengris, D.V.M., Ph.D. (CVM), John Ingalls (BOS-DO), Rita Jhangiani (PHI-DO), George Kroehling (CDRH), Seth Pauker, Ph.D. (OB), Pearl Tanjuaquio (LOS-DO), Frank Twardochleb (CDRH) and Sylvia Yetts (DAL-DO).

We wish to express our appreciation to all who shared inspection reports and FDA-483s, contributed technical expertise, provided comments, and assisted in the preparation of this Guide. Special thanks to Ms. Kimberly Search (DFI) for her expert clerical assistance.

Contents

Appendix:

Biotechnology Inspection Guide for Investigators

Introduction

Biotechnology, defined as "the application of biological systems and organisms to technical and industrial processes", is not new. The use of yeast to ferment grain into alcohol has been ongoing for centuries. Likewise, farmers and breeders use a form of "genetic engineering" to produce improved crops and stock by selecting for desirable characteristics in plants and animals. Only recently have "new" biotechnology techniques enabled scientists to modify an organism's genetic material at the cellular or molecular level. These methods are more precise, but the results are similar to those produced with classical genetic techniques involving whole organisms. Biotechnology - derived products (BDP) used in this Guide refers to those products derived from the new biotechnology techniques.

The development of BDP and the inspection of the manufacture and control of these products offer many challenges. Because of the diversified manufacturing and control processes that are continuously being developed, considerable effort is required to achieve a level of technical competence to inspect these operations. Although the level of technology is increasing, it must be recognized that the same basic regulations and requirements are applicable to the manufacture and control of biotechnically derived substances and devices as for "conventionally" manufactured products.

The same criteria have been used for many years in the inspection of manufacturers of antibiotics, enzymes and other high molecular weight substances including insulin, heparin, and albumin. This Guide will address some of the basic problems identified during inspections of manufacturers of BDP. Production systems may include animals, cell clones (e.g. hybridomas), mammalian and insect cell cultures, yeast, and bacteria or combinations of these systems.

A. Objective

The major objective of an inspection is to determine whether the manufacturer is operating in a state of control and in compliance with the laws and regulations. The firm's commitment to quality is vital, regardless of the type of company or product that is being manufactured.

One important aspect of an inspection is to identify defective product, nonconforming product and system failures. The way in which companies investigate and correct objectionable conditions and deficient manufacturing and control systems is an important part of an inspection and typically illustrates the level of quality within a firm.

B. Inspection Team

As with pre-approval inspections of human and veterinary drugs and devices, it is recommended that inspections of biotech firms be conducted by teams, with the lead Investigator being responsible for the overall conduct of the inspection. Analysts (Chemists and/or Microbiologists), Computer Specialists, and Engineers can participate in all or parts of the inspection. Prior to the inspection, the "team" should discuss the duties of the particular team members.

C. Inspection Approach

The biotech inspection is also a product-specific inspection. As with any inspection, coverage is generally an audit and is not all inclusive. Thus, validation data for all systems, processes, controls and test procedures cannot be reviewed. However, specific detailed coverage should be given to a few systems or controls. A flow chart from the application document or from the firm should be obtained prior to or early in the inspection and the specific manufacturing steps should be reviewed with the manufacturer's responsible personnel.

Cell Culture and Fermentation

A. Master Cell Bank and Working Cell Bank

The starting material for manufacturing BDP includes the bacterial, yeast, insect or mammalian cell culture which expresses the protein product or monoclonal antibody of interest. The cell seed lot system is used by manufacturers to assure identity and purity of the starting raw material. A cell seed lot consists of aliquots of a single culture. The master cell bank (MCB) is derived from a single colony (bacteria, yeast) or a single eukaryotic cell, stored cryogenically to assure genetic stability and is composed of sufficient ampules of culture to provide source material for the working cell bank (WCB). The WCB is defined as a quantity of cells derived from one or more ampules of the MCB, stored cryogenically and used to initiate the production batch.

1. Origin and History

   Because genetic stability of the cell bank during storage and propagation is a major concern, it is important to know the origin and history (number of passages) of both the MCB and WCB. A MCB ampule is kept frozen or lyophilized and only used once. Occasionally, a new MCB may be generated from a WCB. The new MCB should be tested and properly characterized. For biological products, a product license application or amendment must be submitted and approved before a new MCB can be generated from a WCB.
2. Characterization and Qualifying Tests

   Information about the construction of the expression vector, the fragment containing the genetic material that encodes the desired product, and the relevant genotype and phenotype of the host cell(s) are submitted as part of a product application. The major concerns of biological systems are genetic stability of cell banks during production and storage, contaminating microorganisms, and the presence of endogenous viruses in some mammalian cell lines. As part of the application document, manufacturers will submit a description of all tests performed to characterize and qualify a cell bank.

It must be emphasized that the tests required to characterize a cell bank will depend on the intended use of the final product, the host/expression system and the method of production including the techniques employed for purification of the product. In addition, the types of tests may change as technology advances.

The MCB is rigorously tested. The following tests are generally performed, but are not limited to:

a. Genotypic characterization by DNA fingerprinting
b. Phenotypic characterization by nutrient requirements, isoenzyme analysis, growth and morphological characteristics
c. Reproducible production of desired product
d. Molecular characterization of vector/cloned fragment by restriction enzyme mapping, sequence analysis
e. Assays to detect viral contamination
f. Reverse transcriptase assay to detect retroviruses
g. Sterility test and mycoplasma test to detect other microbial contaminants

It is not necessary to test the WCB as extensively as the MCB; however, limited characterization of a WCB is necessary. The following tests are generally performed on the WCB, but this list is not inclusive:

a. Phenotypic characterization
b. Restriction enzyme mapping
c. Sterility and mycoplasma testing
d. Testing the reproducible production of desired product

3. Storage Conditions and Maintenance

   The MCB and WCB must be stored in conditions that assure genetic stability. Generally, cells stored in liquid nitrogen or its vapor phase are stable longer than cells stored at –70 C. In addition, it is recommended that the MCB and WCB be stored in more than one location in the event that a freezer malfunctions.

4. Inspection Approach

   a. Verify that the written procedures reflect accurately what is submitted in the application document. b. Determine that batch records follow written procedures. c. Determine the identity and traceability of the MCB/WCB. d. Check the conditions of storage at each location. e. Check the accessibility to MCB and WCB. Determine if there are security measures and accountability logs. f. Document any samples of the MCB/WCB that failed to meet all specifications, especially if they have been released for use.

B. Media

(1) Raw Materials

Raw materials used to prepare the media must be carefully selected to provide the proper rate of growth and the essential nutrients for the organisms producing the desired product. Raw materials should not contain any undesirable and toxic components that may be carried through the cell culture, fermentation and the purification process to the finished product. Water is an important component of the media and the quality of the water will depend on the recombinant system used, the phase of manufacture and intended use of the product. Raw materials considered to be similar when supplied by a different vendor should meet acceptance criteria before use. In addition, a small scale pilot run followed by a full-scale production run is recommended when raw materials from a different vendor are used, to assure that growth parameters, yield, and final product purification remain the same.

(2) Bovine Serum

Most mammalian cell cultures require serum for growth. Frequently, serum is a source of contamination by adventitious organisms, especially mycoplasma, and firms must take precautions to assure sterility of the serum. Some Brazilian bovine serum (BBS) have been contaminated with hoof and mouth disease. Also make sure that the serum is indeed bovine serum and not derived from human sources.

There is an additional concern that bovine serum may be contaminated with bovine spongiform encephalopathy (BSE) agent. BSE is a slow disease which has been detected in herds from the United Kingdom. Because there is no sensitive in vitro assay to detect the presence of this agent, it is essential that the manufacturers know the source of the serum and request certification that the serum does not come from areas where BSE is endemic. Other potential sources of BSE may be proteases and other enzymes derived from bovine sources. Biological product manufacturers have been requested to determine the origin of these materials used in manufacturing.

(3) Sterilization

The media used must be sterilized. A sterilized in place (SIP) or a continuous sterilizing system (CSS) process is usually used. Any nutrients or chemicals added beyond this point must be sterile. Air lines must include sterile filters.

(4) Inspection Approach

a. Determine the source of serum.
b. Confirm that the sterilization cycle has been properly validated to ensure that the media will be sterile.
c. Verify that all raw materials have been tested by quality control. Determine the origin of all bovine material.
d. Document instances where the media failed to meet all specifications.
e. Verify that expired raw materials have not been used in manufacture.
f. Check that media and other additives have been properly stored.

C. Culture Growth

(1) Inoculation and Aseptic Transfer

Bioreactor inoculation, transfer, and harvesting operations must be done using validated aseptic techniques. Additions or withdrawals from industrial bioreactors are generally done through steam sterilized lines and steam-lock assemblies. Steam may be left on in situations for which the heating of the line or bioreactor vessel wall would not be harmful to the culture.

(2) Monitoring of Growth Parameters and Control

It is important for a bioreactor system to be closely monitored and tightly controlled to achieve the proper and efficient expression of the desired product. The parameters for the fermentation process must be specified and monitored. These may include: growth rate, pH, waste byproduct level, viscosity, addition of chemicals, density, mixing, aeration, foaming, etc. Other factors which may affect the finished product include shear forces, process-generated heat, and effectiveness of seals and gaskets.

Many growth parameters can influence protein production. Some of these factors may affect deamidation, isopeptide formation, or host cell proteolytic processing. Although nutrient-deficient media are used as a selection mechanism in certain cases, media deficient in certain amino acids may cause substitutions. For example, when E. coli is starved of methionine and/or leucine while growing, the organism will synthesize norleucine and incorporate it in a position normally occupied by methionine, yielding an analogue of the wild-type protein. The presence of these closely related products will be difficult to separate chromatographically; this may have implications both for the application of release specifications and for the effectiveness of the product purification process.

Computer programs used to control the course of fermentation, data logging, and data reduction and analysis should be validated.

(3) Containment

Bioreactor systems designed for recombinant microorganisms require not only that a pure culture is maintained, but also that the culture be contained within the systems. The containment can be achieved by the proper choice of a host-vector system that is less capable of surviving outside a laboratory environment and by physical means, when this is considered necessary.

Revision of Appendix K of the NIH Guidelines (1991) reflects a formalization of suitable containment practices and facilities for the conduct of large-scale experiments involving recombinant DNA-derived industrial microorganisms. Appendix K replaces portions of Appendix G when quantities in excess of 10 liters of culture are involved in research or production. For large-scale research or production, four physical containment levels are established: GLSP, BL1-LS, BL2-LS and BL3-LS.

GLSP

(Good Large-Scale Practice) level of physical containment is recommended for large-scale research of production involving viable, nonpathogenic and nontoxigenic recombinant strains derived from host organisms that have an extended history or safe large scale use. The GLSP level of physical containment is recommended for organisms such as those that have built-in environmental limitations that permit optimum growth in the large scale setting but limited survival without adverse consequences in the environment.

BL1-LS
(Biosafety Level 1 -Large Scale) level of physical containment is recommended for large-scale research or production of viable organisms containing recombinant DNA molecules that require BL1 containment at the laboratory scale.

BL2-LS
Level of physical containment is required for large-scale research or production of viable organisms containing recombinant DNA molecules that require BL2 containment at the laboratory scale.

BL3-LS
Level of physical containment is required for large-scale research or production of viable organisms containing recombinant DNA molecules that require BL3 containment at the laboratory scale.

No provisions are made at this time for large-scale research or production of viable organisms containing recombinant DNA molecules that require BL4 containment at the laboratory scale.

(4) Contamination Control
There should be no adventitious organisms in the system during cell growth. Contaminating organisms in the bioreactor may adversely affect both the product yield and the ability of the downstream process to correctly separate and purify the desired protein. The presence or effects of contaminating organisms in the bioreactor can be detected in a number of ways -growth rate, culture purity, bacteriophage assay, and fatty acid profile.
a. Verify that there are written procedures to assure absence of adventitious agents and criteria established to reject contaminated runs.
b. Review cell growth records and verify that the production run parameters are consistent with the established pattern. c. Review written procedures to determine what investigations and corrective actions will be performed in the event that growth parameters exceed established limits.
c. Review written procedures to determine what investigations and corrective actions will be performed in the event that growth parameters exceed established limits.
d. Assure proper aseptic techniques during cell in5. Inspection Approach:
e. Determine that appropriate in-process controls are utilized prior to further processing.

Ascites Production

Monoclonal antibodies can be produced in cell culture or in the abdomen of a mouse. There are unique critical points in ascites production that should be examined.

A. Mouse Colony

(1) Characterization and Control of the Mouse Colony
Characterization and control of the mouse colony used to produce ascites are critical. The type of mouse, source, vendor, and certification that the colony is free from viral disease should be recorded. Animals used in production should be quarantined and inspected daily for a period of a week to assure that mice remain in good health and meet all acceptance criteria. Mice should be observed daily during production. There should be strict SOPs to remove any mouse that does not remain in overt good health during quarantine and production.

(2) Animal Quarters/Environmental Controls
Strict attention to the animal quarters is necessary to assure that mice remain free from disease, especially viruses that commonly infect colonies. To prevent contamination of colonies housed in different rooms, it is a good idea for people to wear disposable gloves, lab coats, head coverings and booties so that these items can be changed before entering another room. Animal quarters and cages must be kept in sanitary condition.

B. Manufacturing Processes

(1) Animal Identification
Individual mice must be identified so that a record of the number of times a mouse has been tapped and the amount of fluid obtained from each tap can be accurately maintained.

(2) Tapping Procedure
Injection of mice and removal of ascites fluid should be done in a clean environment such as under a unidirectional hood or at a station that will protect the mice from infectious agents. There should be written procedures that describes the tapping process. A different needle for each mouse is recommended to prevent the possibility of transmitting infections from other mice. There should also be written procedures for handling needles with strict adherence to biohazardous containment to prevent cross contamination.

(3) Storage and Pooling of Ascites
In addition, there should be written procedures that describes the storage temperatures and conditions before processing. This will include establishing a time limitation on collection and processing. Pooling of ascites is acceptable, but there should be written procedures that describes how a pool is made (how many and which mice make up a pool) and records must accurately reflect what makes up the pool. Thus, if it is discovered that an animal is infected, the records will reflect which pool contains the infected animal's ascites fluid.

(4) Purification
Pristane is sometimes used to prime mice and enhance ascites production. For parenteral products, the firm must demonstrate that the purification process will remove pristane. This should not be a concern for products used in in vitro diagnostic devices.

(5) Inspection Approach
   a. Review SOPs to assure adequate controls for quarantining and accepting mice, housing and caring for mice, mice identification, maintaining a clean environment to prevent viral infection of colony, disposing unhealthy mice, and processing of ascites fluid.
   b. Review records to assure that animals are in good health and are observed daily during the quarantine period and production.
   c. Verify the presence of a qualified animal care staff.

Extraction, Isolation and Purification

A. Introduction

Once the fermentation process is completed, the desired product is separated, and if necessary, refolded to restore configurational integrity, and purified. For recovery of intracellular proteins, cells must be disrupted after fermentation. This is done by chemical, enzymatic or physical methods. Following disruption, cellular debris can be removed by centrifugation or filtration. For recovery of extracellular protein, the primary separation of product from producing organisms is accomplished by centrifugation or membrane filtration. Initial separation methods, such as ammonium sulfate precipitation and aqueous two-phase separation, can be employed following centrifugation to concentrate the products. Further purification steps primarily involve chromatographic methods to remove impurities and bring the product closer to final specifications.

B. Process Types

(1) Extraction and Isolation
   a. Filtration -Ultrafiltration is commonly used to remove the desired product from the cell debris. The porosity of the membrane filter is calibrated to a specific molecular weight, allowing molecules below that weight to pass through while retaining molecules above that weight.
   b. Centrifugation - Centrifugation can be open or closed. The adequacy of the environment must be evaluated for open centrifugation.

(2) Purification - The purification process is primarily achieved by one or more column chromatography techniques.
   a. Affinity Chromatography
   b. Ion-Exchange Chromatography (IEC)
   c. Gel filtration
   d. Hydrophobic Interaction Chromatography (HIC)
   e. Reverse-Phase HPLC

C. Description/Written Procedures

All separation and purification steps should be described in detail and presented with flow charts. Adequate descriptions and specifications should be provided for all equipment, columns, reagents, buffers and expected yields. When applicable, written procedures should be compared with the application documents submitted to the Agency. In-process storage conditions and quality control assays should be reviewed.

D. Process Validation

FDA defined process validation in the May 1987 "Guideline on General Principles of Process Validation" as follows:

Validation-establishing documented evidence which provides a high degree of assurance that a specific process will consistently produce a product meeting its pre-determined specifications and quality attributes.

(1) Documentation

We expect to see documentation that justifies the process and demonstrates that the process works consistently. For biological products, all validation data are submitted and reviewed and the specifications are established and approved as part of the product licensing application (PLA).

Manufacturers should have validation reports for the various key process steps. For example, if an ion-exchange column is used to remove endotoxins, there should be data documenting that this process is consistently effective. By determining endotoxin levels before and after processing, a manufacturer should be able to demonstrate the validity of this process. It is important to monitor the process before, during, and after to determine the efficiency of each key purification step. "Spiking" the preparation with a known amount of a contaminant to demonstrate its removal may be a useful method to validate the procedure.

(2) Validation

Typically, manufacturers develop purification processes on a small scale and determine the effectiveness of the particular processing step. When scale-up is performed, allowances must be made for several differences when compared with the laboratory-scale operation. Longer processing times can affect product quality adversely, since the product is exposed to conditions of buffer and temperature for longer periods. Product stability, under purification conditions, must be carefully defined. Manufacturers should define the limitations and effectiveness of the particular step. Process validation on the production size batch will then compare the effect of scale-up. Manufacturers may sometimes use development data on the small scale for validation. However, it is important that validation be performed on the production size batches.

Process validation and/or reports for the validation of some of the purification processes should be reviewed. Additionally, the controls and tests used to assure the consistency of the process should also be reviewed.

Often columns are regenerated to allow repeated use. Proper validation procedures should be performed and the process should be periodically monitored for chemical and microbial contamination.

(3) Follow Up Investigations

Manufacturers occasionally reject the product following the purification process. As with other regulated products, it is expected that reports of investigations be complete and relate to other batches. For example, during one inspection it was noted that approximately six batches of a BDP were rejected because of low potency and high levels of impurities. The problem was attributed to a column and all of the batches processed on the column were rejected. It should be pointed out that any batch failing specifications should be investigated.

It is, therefore, important to identify defective product so that the specific manufacturing and control systems can be given more detailed inspectional coverage.

E. Process Water/Buffers/WFI

The quality of water should depend on the intended use of the finished product. For example, CBER requires Water for Injection (WFI) quality for process water. On the other hand, for in-vitro diagnostics purified water may suffice. For drugs, the quality of water required depends on the process. Also, because processing usually occurs cold or at room temperature, the self-sanitization of a hot WFI system at 75 to 80 C is lost.

For economic reasons, many of the biotech companies manufacture WFI by reverse osmosis rather than by distillation. Most of these systems have been found to be contaminated. Typically, they employ plastic pipe (PVC) and non-sealed storage tanks, which are difficult to sanitize. Any threads or drops in a cold system provide an area where microorganisms can lodge and multiply. Some of the systems employ a terminal sterilizing filter. However, the primary concern is endotoxins, and the terminal filter may merely serve to mask the true quality of the WFI used. The limitations of relying on a 0.1 ml sample of WFI for endotoxins from a system should also be recognized. The system should be designed to deliver high purity water, with the sample merely serving to assure that it is operating adequately. As with other WFI systems, if cold WFI water is needed, point-of-use heat exchangers can be used.

Buffers can be manufactured as sterile, non-pyrogenic solutions and stored in sterile containers. Some of the smaller facilities have purchased commercial sterile, non-pyrogenic buffer solutions.

The production and/or storage of non-sterile water that may be of reagent grade or used as a buffer should be evaluated from both a stability and microbiological aspect.

WFI systems for BDP are the same as WFI systems for other regulated products. As with other heat sensitive products, cold WFI is used for formulation. Cold systems are prone to contamination. The cold WFI should be monitored both for endotoxins and microorganisms. Validation data and reports of monitoring should be reviewed.

F. Plant Environment

Microbiological quality of the environment during various processing steps is a concern. As the process continues downstream, increased consideration should be given to environmental controls and monitoring. The environment and areas used for the isolation of the BDP should also be controlled to minimize microbiological and other foreign contaminants. The typical isolation of BDP should be of the same control as the environment used for the formulation of the solution prior to sterilization and filling.

Cleaning Procedure

Validation of the cleaning procedures for the processing of equipment, including columns, should be carried out. This is especially critical for a multi-product facility. The manufacturer should have determined the degree of effectiveness of the cleaning procedure for each BDP or intermediate used in that particular piece of equipment.

Validation data should verify that the cleaning process will reduce the specific residues to an acceptable level. However, it may not be possible to remove absolutely every trace of material, even with a reasonable number of cleaning cycles. The permissible residue level, generally expressed in parts per million (ppm), should be justified by the manufacturer. Cleaning should remove endotoxins, bacteria, toxic elements, and contaminating proteins, while not adversely affecting the performance of the column. Specific inspectional coverage for cleaning should include:

A. Detailed Cleaning Procedure

There should be a written equipment cleaning procedure that provides details of what should be done and the materials to be utilized. Some manufacturers list the specific solvent for each BDP and intermediate.

For stationary vessels, often clean-in-place (CIP) apparatus may be encountered. For evaluation of these systems, diagrams will be necessary, along with identification of specific valves.

B. Sampling Plan

After cleaning, there should be some routine testing to assure that the surface has been cleaned to the validated level. One common method is the analysis of the final rinse water or solvent for the presence of the cleaning agents last used in that piece of equipment. There should always be direct determination of the residual substance.

C. Analytical Method/Cleaning Limits

Part of the answer to the question, "how clean is clean?" is, "how good is your analytical system?" The sensitivity of modern analytical apparatus has lowered some detection thresholds below parts per million (ppm), down to parts per billion (ppb).

The residue limits established for each piece of apparatus should be practical, achievable, and verifiable. When reviewing these limits, ascertain the rationale for establishment at that level. The manufacturer should be able to document, by means of data, that the residual level permitted has a scientifically sound basis.

Another factor to consider is the possible non-uniform distribution of the residue on a piece of equipment. The actual average residue concentration may be more than the level detected.

Processing and Filling

A. Processing

Most BDP cannot be terminally sterilized and must be manufactured by aseptic processing. The presence of process related contaminants in a product or device is chiefly a safety issue. The sources of contaminants are primarily the cell substrate (DNA, host cell proteins, and other cellular constituents, viruses), the media (proteins, sera, and additives) and the purification process (process related chemicals, and product related impurities).

Because of stability considerations, most BDP are either refrigerated or lyophilized. Low temperatures and low moisture content are also deterrents to microbiological proliferation. For the validation of aseptic processing of the non-preserved single dose biopharmaceutical (that is aseptically filled) stored at room temperature as a solution, the limitations of 0.1% media fill contamination rate should be recognized.

Media fill data and validation of the aseptic manufacturing process should be reviewed during an inspection. Some BDP may not be very stable and may require gentle mixing and processing. Whereas double filtrations are relatively common for aseptically filled parenterals, single filtration at low pressures are usually performed for BDP. It is for this reason that manufacturing directions be specific, with maximum filtration pressures given.

The inspection should include a review of manufacturing directions in batch records to assure that they are complete and specific.

The environment and accessibility for the batching of the non-sterile BDP should be controlled. Because many of these products lack preservatives, inherent bacteriostatic, or fungistatic activity, bioburden before sterilization should be low and-the bioburden should be determined prior to sterilization of these bulk solutions and before filling. Obviously, the batching or compounding of these bulk solutions should be controlled in order to prevent any potential increase in microbiological levels that may occur up to the time that the bulk solutions are filtered (sterilized). One concern with any microbiological level is the possible increase in endotoxins that may develop. Good practice for the compounding of these products would also include batching in a controlled environment and in sealed tanks, particularly if the solution is to be stored prior to sterilization. Good practice would also include limitations on the length of manufacturing time between formulation and sterilization.

B. In-Process Quality Control

In-process testing is an essential part of quality control and ensures that the actual, real-time performance of an operation is acceptable. Examples of in-process controls are: stream parameters, chromatography profiles, protein species and protein concentrations, bioactivity, bioburden, and endotoxin levels. This set of in-process controls and the selection of acceptance criteria require coordination with the results from the validation program.

C. Filling

The filling of BDP into ampules or vials presents many of the same problems as with the processing of conventional products. In established companies these issues are relatively routine. However, for the new BDP facility, attempting to develop and prove clinical effectiveness and safety along with validation of sterile operations, equipment and systems, can be a lengthy process, particularly if requirements are not clearly understood.

The batch size of a BDP, at least when initially produced, likely will be small. Because of the small batch size, filling lines may not be as automated as for other products typically filled in larger quantities. Thus, there is more involvement of people filling these products, particularly at some of the smaller, newer companies.

Problems that have been identified during filling include inadequate attire; deficient environmental monitoring programs; hand-stoppering of vials, particularly those that are to be lyophilized; and failure to validate some of the basic sterilization processes. Because of the active involvement of people in filling and aseptic manipulations, the number of persons involved in these operations should be minimized, and an environmental program should include an evaluation of microbiological samples taken from people working in aseptic processing areas. This program along with data should be reviewed during the inspection.

Another concern about product stability is the use of inert gas to displace oxygen during both the processing and filling of the solution. As with other products that may be sensitive to oxidation, limits for dissolved oxygen levels for the solution should be established. Likewise, validation of the filling operation should include parameters such as line speed and location of filling syringes with respect to closure, to assure minimal exposure to air (oxygen) for oxygen-sensitive products. In the absence of inert gas displacement, the manufacturer should be able to demonstrate that the product is not affected by oxygen. These data may be reviewed during an inspection (These data are evaluated as part of a Product Licensing Application (PLA) review).

Typically, vials to be lyophilized are partially stoppered by machine. However, some filling lines have been observed that utilize an operator to place each stopper on top of the vial by hand. The concern is the immediate avenue of contamination offered by the operator. The observation of operators and active review of filling operations should be performed.

Another major concern with the filling operation of a lyophilized product is assurance of fill volumes. Obviously, a low fill would represent a sub-potency in the vial. Unlike a powder or liquid fill, a low fill would not be readily apparent after lyophilization, particularly for a product where the active ingredient may be only a milligram. Because of the clinical significance, sub-potency in a vial potentially can be a very serious situation, clinically.

Again, the inspection should include the observation and the review of filling operations, not only regarding aseptic practices, but also for fill uniformity.

D. Lyophilization

Many products are lyophilized for stability concerns. Unfortunately, GMP aspects of the design of lyophilizers have lagged behind the sterilization and control technology employed for other processing equipment. It is not surprising that many problems with the lyophilization process have been identified.

These problems are not limited to BDP but generally pertain to lyophilization of all products including BDP. A detailed discussion of lyophilization and controls can be found in Inspection Technical Guide No. 43, issued 4/18/86.

Laboratory Controls

During the inspection of the firm's laboratory facility, the following areas should be reviewed and any deficiencies should be documented:

A. Training

Laboratory personnel should be adequately trained for the jobs they are performing.

B. Equipment Maintenance/Calibration/Monitoring

Firms should have documentation and schedules for maintenance, calibration, and monitoring of laboratory equipment involved in the measurement, testing and storage of raw materials, product, samples, and reference reagents.

All laboratory methods should be validated with the equipment and reagents specified in the test methods. Changes in vendor and/or specifications of major equipment/reagents would require revalidation.

C. Method Validation

Firms should have raw data to support validation parameters in submitted applications.

D. Standard/Reference Material

Reference standards should be well characterized and documented, properly stored, secured, and utilized during testing.

E. Storage of Labile Components

Laboratory cultures and reagents, such as enzymes, antibodies, test reagents, etc., may degrade if not held under proper storage conditions.

F. Laboratory SOPs

Procedures should be written, applicable and followed. Quality control samples should be properly segregated and stored.

Testing

The following tests may be applicable to component, in process, bulk and/or final product testing. The tests that are needed will depend on the process and the intended use of the product.

A. Quality

(1) Color/Appearance/Clarity
(2) Particulate Analysis
(3) pH Determination
(4) Moisture Content
(5) Host Cell DNA

B. Identity

A single test for identity may not be sufficient. Confirmation is needed that the methods employed are validated. Availability of reference material should be checked. A comparison of the product to the reference preparation in a suitable bioassay will provide additional evidence relating to the identity and potency of the product.
Tests that may be encountered:

1. Peptide Mapping (reduced/non-reduced)
2. Gel Electrophoresis
   - SDS PAGE
   - Isoelectric Focusing (IEF)
   - Immunoelectrophoresis
3. 2-Dimensional Electrophoresis
4. Capillary Electrophoresis
5. HPLC (Chromographic Retention)
   - Immunosassay
   - ELISA
   - Western Blot
   - Radioimmunoassay
6. Amino Acid Analysis
7. Amino Acid Sequencing
8. Mass Spectroscopy
9. Molecular Weight (SDS PAGE)
10. Carbohydrate Composition Analysis (glycosylation)

C. Protein Concentration/Content

Tests that may be encountered:

1. Protein Quantitations
2. Lowry
3. Biuret Method

4. UV Spectrophotometry
5. HPLC
6. Amino Acid Analysis
7. *Partial Sequence Analysis

D. Purity

"Purity" means relative freedom from extraneous matter in the finished product, whether or not harmful to the recipient or deleterious to the product. Purity includes, but is not limited to, relative freedom from residual moisture or other volatile substances and pyrogenic substances. Protein impurities are the most common contaminants. These may arise from the fermentation process, media or the host organism. Endogenous retroviruses may be present in hybridomas used for monoclonal antibody production. Specific testing for these constituents is imperative in in vivo products. Removal of extraneous antigenic proteins is essential to assure the safety and the effectiveness of the product.
Tests that may be encountered:

1. Tests for Protein Impurities:
   a. Electrophoresis
      - SDS PAGE
      - IEF
      - 2-Dimensional Electrophoresis
   b. Peptide Mapping
   c. Multiantigen ELISA
   d. HPLC Size Exclusion HPLC Reverse Phase HPLC
2. Tests for Non-Protein Impurities:
   a. DNA Hybridization
   b. HPLC
   c. Pyrogen/Endotoxin Testing
      - U.S.P. Rabbit Pyrogen Test
      - Limulus Amebocyte Lysate (LAL) E
      - endogenous Pyrogen Assay

Pyrogen Contamination - Pyrogenicity testing should be conducted by injection of rabbits with the final product or by the limulus amebocyte lysate (LAL) assay. The same criteria used for acceptance of the natural product should be used for the biotech product.

The presence of endotoxins in some in vitro diagnostic products may interfere with the performance of the device. Also, it is essential that in vivo products be tested for pyrogens. Certain biological pharmaceuticals are pyrogenic in humans despite having passed the LAL test and the rabbit pyrogen test. This phenomenon may be due to materials that appear to be pyrogenic only in humans. To attempt to predict whether human subjects will experience a pyrogenic response, an endogenous pyrogen assay is used. Human blood mononuclear cells are cultured in vitro with the final product, and the cell culture fluid

is injected into rabbits. A fever in the rabbits indicates the product contains a substance that may be pyrogenic in humans.
Tests that may be encountered:

1. U.S.P. Rabbit Pyrogen Test
2. Limulus Amebocyte Lysate (LAL)
3. Assay Endogenous Pyrogen Assay

Viral Contamination - Tests for viral contamination should be appropriate to the cell substrate and culture conditions employed. Absence of detectable adventitious viruses contaminating the final product should be demonstrated.
Tests that may be encountered:

1. Cytopathic effect in several cell types
2. Hemabsorption Embryonated Egg Testing
3. Polymerase Chain Reaction (PCR)
4. Viral Antigen and Antibody Immunoassay
5. Mouse Antibody Production (MAP)

Nucleic Acid Contamination - Concern about nucleic acid impurities arises from the possibility of cellular transformation events in a recipient. Removal of nucleic acid at each step in the purification process may be demonstrated in pilot experiments by examining the extent of elimination of added host cell DNA. Such an analysis would provide the theoretical extent of the removal of nucleic acid during purification.

Direct analyses of nucleic acid in several production lots of the final product should be performed by hybridization analysis of immobilized contaminating nucleic acid utilizing appropriate probes, such as nick-translated host cell and vector DNA. Theoretical concerns regarding transforming DNA derived from the cell substrate will be minimized by the general reduction of contaminating nucleic acid.

1. DNA Hybridization (Dot Blot)
2. Polymerase Chain Reaction (PCR)

Protein Contamination
Tests that may be encountered for product-related proteins:

1. SDS PAGE
2. PLC
3. IEF

Tests that may be encountered for foreign proteins:

1. Immunoassays
2. Radioimmunoassays

3. ELISA
4. Western Blot
5. SDS Page
6. 2-Dimensional Electrophoresis

Microbial Contamination - Appropriate tests should be conducted for microbial contamination that demonstrate the absence of detectable bacteria (aerobes and anaerobes), fungi, yeast, and mycoplasma, when applicable.
Tests that may be encountered:

1. U.S.P. Sterility Test
2. Heterotrophic Plate Count and Total Yeasts and Molds
3. Total Plate Count
4. Mycoplasma Test
5. LAL/Pyrogen

Chemical Contaminants - Other sources of contamination must be considered, e.g., allergens, petroleum oils, residual solvents, cleaning materials, column leachable materials, etc.

E. Potency (Activity)
"Potency" is interpreted to mean the specific ability or capacity of the product, as indicated by appropriate laboratory tests or by adequately controlled clinical data obtained through the administration of the product in the manner intended, to produce a given result. Tests for potency should consist of either in vitro or in vivo tests, or both, which have been specifically designed for each product so as to indicate its potency. A reference preparation for biological activity should be established and used to determine the bioactivity of the final product. Note: Where applicable, in-house biological potency standards should be cross-referenced against international (World Health Organization (WHO), National Institute of Biological Standards and Control (NIBSC)) or national (National Institutes of Health (NIH), National Cancer Institute (NCI), Food and Drug Administration (FDA)) reference standard preparations, or USP standards.
Tests that may be encountered:

1. Validated method of potency determination
   - Whole Animal Bioassays
   - Cell Culture Bioassays
   - Biochemical/Biophysical Assays
   - Receptor Based Immunoassays
2. Potency Limits
3. Identification of agents that may adversely affect potency
4. Evaluation of functional activity and antigen/antibody specificity
   - Various immunodiffusion methods (single/double)
   - Immunoblotting/Radio-or Enzyme-linked Immunoassays
5. HPLC-validated to correlate certain peaks to biological activity

F. Stability

"Stability" is the capacity of a product to remain within specifications established to ensure its identity, strength, quality, purity, safety, and effectiveness as a function of time. Studies to support the proposed dating period should be performed on the final product. Real-time stability data would be essential to support the proposed dating period. Testing might include stability of potency, pH, clarity, color, particulates, physiochemical stability, moisture and preservatives. Accelerated stability testing data may be used as supportive data. Accelerated testing or stress tests are studies designed to increase the ratio of chemical or physical degradation of a substance or product by using exaggerated storage conditions. The purpose is to determine kinetic parameters to predict the tentative expiration dating period. Stress testing of the product is frequently used to identify potential problems that may be encountered during storage and transportation and to provide an estimate of the expiration dating period. This should include a study of the effects of temperature fluctuations as appropriate for shipping and storage conditions. These tests should establish a valid dating period under realistic field conditions with the containers and closures intended for the marketed product.

Some relatively fragile biotechnically-derived proteins may require gentle mixing and processing and only a single filtration at low pressure. The manufacturing directions must be specific with maximum filtration pressures given in order to maintain stability in the final product. Products containing preservatives to control microbial contamination should have the preservative content monitored. This can be accomplished by performing microbial challenge tests (i.e. U.S.P. Antimicrobial Preservative Effectiveness Test) or by performing chemical assays for the preservative. Areas that should be addressed are:

1. Effective monitoring of the stability test environment (i.e. light, temperature, humidity, residual moisture);
2. Container/closure system used for bulk storage (i.e. extractables, chemical modification of protein, change in stopper formulations that may change extractable profile);
3. Identify materials that would cause product instability and test for presence of aggregation, denaturation, fragmentation, deamination, photolysis, and oxidation;
4. Tests to determine aggregates or degradation products.

Tests that may be encountered:

1. SDS PAGE
2. IEF
3. HPLC
4. Ion Exchange Chromatography
5. Gel Filtration
6. Peptide Mapping
7. Spectrophotometric Methods
8. Potency Assays

9. Performance Testing
10. 12-Dimensional Electrophoresis

G. Batch To Batch Consistency

The basic criterion for determining that a manufacturer is producing a standardized and reliable product is the demonstration of lot-to-lot consistency with respect to certain predetermined release specifications.

1. Uniformity: identity, purity, functional activity
2. Stability: acceptable performance during shelf life, precision, sensitivity, specificity

Environmental Coverage

Environmental/biocontainment coverage for biotechnology facilities should be conducted as part of regular GMP inspections, particularly pre-approval or prelicensing inspections. FDA is responsible under the National Environmental Policy Act (NEPA) for ascertaining the environmental impact that may occur due to the manufacture, use, and disposal of FDA regulated products. No other federal or state regulatory agency can be informed by FDA of the existence of an unapproved product application. Consequently, FDA must also make sure that the product sponsor is conducting investigations safely.

A. Environmental Assessments

Typically, a product sponsor describes environmental control measures in environmental assessments (EAs) that are part of the product application. When the product is approved, the EA is released to the public. FDA must be able to verify the accuracy and the appropriateness of the information contained in the EA. The Investigator should have a copy of the firm's environmental assessment, addressing the manufacture of the product that is the subject of the GMP inspection. The EA should be requested from the originating office if it has not been provided.

B. Inspection Approach

(1) Review the NIH Guidelines for Recombinant DNA Research (1987, 1988, 1991). Pay particular attention to Appendix K (1991), regarding the establishment of guidelines for the level of containment appropriate to Good Industrial Large Scale Practices (see references).
(2) Determine that the equipment and controls described in the EA as part of the biocontainment and waste processing systems are validated to operate to the standards; the equipment is in place, is operating, and is properly maintained. Such equipment may include, for example, HEPA filters, spill collection tanks with heat or hypochlorite treatment, and diking around bioreactors and associated drains. SOPs should be in use for the cleanup of spills, for actions

to be taken in the case of accidental exposure of personnel, for opening and closing of vessels, for sampling and sample handling, and for other procedures which involve breaching containment or where exposure to living cells may occur.

(3) Determine if there is a workplace and/or environmental monitoring program designed to verify that organisms are subject to appropriate biocontainment practices. Review SOPs for the sampling, isolation, counting, and reporting of results. Obtain copies of relevant SOPs and monitoring data for inclusion in reports to headquarters.

(4) Ask for and obtain copies of all federal, state and local permits governing emissions and occupational safety for the facility being inspected. Determine whether any of the permits have expired and whether there is any action pending relating to violations of the permits.

(5) For facilities in foreign countries, the same procedures should be followed. Compliance with the requirement of the foreign country should be demonstrated.

APPENDIX: For this guideline

A. Flow Chart

Note: Flow Chart no longer exists

B. Test Methods

Affinity Chromatography – A chromatography separation method based on a chemical interaction specific to the target species. Types of affinity methods are: biosorption site recognition (e.g., monoclonal antibody, protein A); hydrophobic interaction -contacts between non-polar regions in aqueous solutions; dye-ligand specific binding of macromolecules to triazine and triphenylmethane dyes; metal chelate - matrix bound chelate complexes with target molecule by exchanging low molecular weight metal bound ligands; and covalent - disulfide bonding reversible under mild conditions.

Amino Acid Composition Analysis – Used to determine the amino acid composition and/or the protein quantity. A two-step process involving a complete hydrolysis (chemical or enzymatic) of the protein into its component amino acids followed by chromatographic separation and quantitation via HPLC. The complete amino acid composition of the peptide or protein should include accurate values for methionine, cysteine, and tryptophan. The amino acid composition presented should be the average of at least three (3) separate hydrolysates of each lot number. Integral values for those amino acid residues generally found in low quantities, such as tryptophan and/or methionine, could be obtained and used to support arguments of purity.

Amino Acid Sequencing – A partial sequencing (8– 15 residues) of amino acids within a protein or polypeptide by either amino-terminal or carboxy-terminal sequencing. This method is done to obtain information about the primary structure of the protein, its homogeneity, and the presence or absence of polypeptide cleavages. The sequence data determined by HPLC analysis is presented in tabular form and should include the total yield for every amino acid at each sequential cleavage cycle. Full sequence is often done by sequencing the peptide fragments isolated from HPLC fractionation.

Capillary Electrophoresis – Used as a complement to HPLC, particularly for peptide mapping. This technique is faster and will often separate peptides that co-elute using HPLC. Separation is accomplished by relative mobility of the peptides in a buffer in response to an electrical current.

Carbohydrate Analysis – Used to determine the consistency of the composition of the covalently bound monosaccharides in glycoproteins. Unlike the polypeptide chain of the glycoprotein where production is controlled by the genetic code, the oligosaccharides are synthesized by posttranslational enzymes. Microheterogeneity of the carbohydrate chains is common. Determination can be accomplished on underivatized sugars after hydrolysis by HPLC separation with pulsed amperometric detection or by gas chromatography after derivatization.

Circular Dichroism – With optical rotary dispersion, one of the optical spectrophotometric methods used to determine secondary structure and to quantitate the specific structure forms (a-helix, B-pleated sheet, and random coil) within a protein. The resultant spectra are compared to that of the natural protein form or to the reference standard for the recombinant.

DNA Hybridization (Dot Blot) Analysis – Detection of DNA to the nanogram level using hybridization of cellular DNA with specific DNA probes. Manifestation can be by 32P-labeling, chemiluminescence, chromogenic or avidin-biotin assays.

Edman Degradation – A type of protein sequencing from the amino-terminus.

Electrophoresis – Methods in which molecules or molecular complexes are separated on the basis of their relative ability to migrate when placed in an electric field. An analyte is placed on an electrophoretic support, then separated by charge (isoelectric focusing) or by molecular weight (SDS-PAGE). Visualization is accomplished by staining of the protein with nonselective (Coomassie Blue) or selective (silver) staining techniques.The dye-binding method using Coomassie blue is a quantifiable technique when a laser densitometer is used to read the gels. The silver stain method is much more sensitive and therefore used for detection of low levels of protein impurities, but due to variability of staining from protein to protein, it cannot be used for quantitation.

Two-dimensional Gel Electrophoresis – A type of electrophoresis in which proteins are separated first in one direction by charge followed by a size separation in the perpendicular direction.

Enzyme-linked Immunosorbent Assay (ELISA) – A multiantigen test for unknown residual (host) cellular protein and confirmation of desired protein. It may be used to determine the potency of a product. It is extremely specific and sensitive, basically simple, and inexpensive. It requires a reference standard preparation of host cell protein impurities to serve as an immunogen for preparation of polyclonal antibodies used for the assay.

Endogenous Pyrogen Assay – An in vitro assay based on the release of endogenous pyrogen produced by endotoxin from human monocytes. This assay appears to be more sensitive than the USP Rabbit Pyrogen Test, but is much less sensitive than the LAL assay. It does have the advantage that it can detect all substances that cause a pyrogenic response from human monocytes.

Gel Permeation or Filtration (Size Exclusion) Chromatography (GPC, GFC or SEC) – A separation method based on the molecular size or the hydrodynamic volume of the components being separated. This can be accomplished with the proteins in their natural state or denatured with detergents.

High Performance Liquid Chromatography (HPLC) – An instrumental separation technique used to characterize or to determine the purity of a BDP by passing the product (or its component peptides or amino acids) in liquid form over a chromatographic column containing a solid support matrix. The mode of separation, i.e. reversed phase, ion exchange, gel filtration, or hydrophobic interaction, is determined by the column matrix and the mobile phase. Detection is usually by UV absorbance or by electrochemical means.

Hydrophobic Interaction Chromatography (HIC) – HIC is accomplished in high salt medium by binding the hydrophobic portions of a protein to a slightly hydrophobic surface containing such entities as phenyl, or short-chain hydrocarbons. The protein can be eluted in a decreasing salt gradient, with the most hydrophobic proteins eluting from the column last.

Immunoassay – A qualitative or quantitative assay technique based on the measure of interaction of high affinity antibody with antigen used to identify and quantify proteins.

Immunoblotting – A technique for transferring antibody/antigen from a gel to a nitrocellulose filter on which they can be complexed with their complementary antigen/antibody.

Immunodiffusion (single) – An identity diffusion technique whereby the product (antigen) is placed in a well cut into a medium such as agar containing its complementary antibody. The product diffuses into the medium forming a ring shaped precipitate whose density is a function of antigen concentration.

Immunodiffusion (double, Ouchterlony technique) – A technique in which an antigen and antibody are placed in two adjacent wells cut into a medium such as agar. As they diffuse through the medium, they form visible precipitation lines of antigen/antibody complexes at the point where the respective concentrations are at the optimum ratio for latice formation.

Ion Exchange Chromatography (IEC) – A gradient driven separation based on the charge of the protein and its relative affinity for the chemical backbone of the column. Anion/cation exchange is commonly used for proteins.

Isoelectric Focusing (IEF) – An electrophoretic method which separates proteins by their pI. They move through a pH gradient medium in an electric field until they are located at their isoelectric point where they carry no net charge. Prior to reaching their pI, protein mobility also depends upon size, conformation, steepness of pH gradient, and the voltage gradient. This method is used to detect incorrect or altered forms of a protein as well as protein impurities.

Limulus Amoebocyte Lysate Test (LAL) – A sensitive test for the presence of endotoxins using the ability of the endotoxin to cause a coagulation reaction in the blood of a horseshoe crab. The LAL test is easier, quicker, less costly and much more sensitive that the rabbit test, but it can detect only endotoxins and not all types of pyrogens and must therefore be thoroughly validated before being used to replace the USP Rabbit Pyrogen test. Various forms of the LAL test include a gel clot test, a colormetric test, a chromogenic test, and a turbidimetric test.

Mass Spectrometry – A technique useful in primary structure analysis by determining the molecular mass of peptides and small proteins. Often used with peptide mapping to identify variants in the peptide composition. Useful to locate disulfide bonds and to identify post-translational modifications.

Northern Blot – Technique for transferring RNA fragments from an agarose gel to a nitrocellulose filter on which they can be hybridized to a complementary DNA.

Peptide Mapping – A powerful technique which involves the breakdown of proteins into peptides using highly specific enzymes. The enzymes cleave the proteins at predictable and reproducible amino acid sites and the resultant peptides are separated via HPLC or electrophoresis. A sample peptide map is compared to a map done on a reference sample as a confirmational step in the identity profiling of a product. It is also used for confirmation of disulfide bonds, location of carbohydrate attachment, sequence analysis, and for identification of impurities and protein degradation.

Polymerase Chain Reaction (PCR) – In vitro technique for amplifying nucleic acid. The technique involves a series of repeated cycles of high temperature denaturation, low temperature oligonucleotide primer annealing and intermediate temperature chain extension. Nucleic acid can be amplified a million-fold after 25– 30 cycles.

Protein Quantification – Quantitation of the total amount of protein can be done by a number of assays. There is no one method that is better than the rest; each has its own disadvantages ranging from the amount of protein required to do the test to a problem with variability between proteins. Some of the types include Lowry, Bicinchonic Acid (BCA), Bradford, Biuret, Kjeldahl, Ultraviolet spectroscopy.

Protein Sequencing – (See Amino Acid Sequencing).

Rabbit Pyrogen Test. U.S.P. – An assay for the presence of pyrogens (not restricted to endotoxins as is the LAL test) involving the injection of the test material into rabbits that are well controlled and of known history. The rabbits are then monitored for a rise in temperature over a period of three hours.

Radioimmunoassay (RIA) – A generic term for immunoassays having a radioactive label (tag) on either the antigen or antibody. Common labels include I125 and H3 which are used for assay detection and quantitation. Classical RIA's are competitive binding assays where the antigen and tagged antigen compete for a limited fixed number of binding sites on the antibody. The antibody bound tagged complex is inversely proportional to the concentration of the antigen.

Reverse Phase Chromatography – A chromatographical separation method based on a column stationary phase coated to give non-polar hydrophobic surface. Analyte retention is proportional to hydrophobic reactions between solute and surface. Retention is roughly proportional to the length of the bonded carbon chain.

SDS PAGE (Sodium Dodecyl Sulfate Polyacrylamide Gel Electrophoresis) – An electrophoretic separation of proteins based on their molecular weights. A uniform net negative charge is imposed on the molecules by the addition of SDS. Under these conditions, migration toward the anode through a gel matrix allows separation via size, not charge, with the smaller molecules migrating the longest distance. This technique is not reliable for sizes below a MW of ca. 8000. Proteins are observed via Coomassie blue or silver staining or can be further transferred to membranes for antigen/antibody specificity testing.

Southern Blot – Technique for transferring DNA fragments from an agarose gel to a nitrocellulose filter on which they can be hybridized to a complementary DNA.

UV Spectroscopy – A quantitation technique for proteins using their distinctive absorption spectra due to the presence of side-chain chromophores (phenylalanine, tryptophan, and tyrosine). Since this absorbance is linear, highly purified proteins can be quantitated by calculations using their molar extinction coefficient.

Western Blot – This test is used to detect contaminating cell substrates and to evaluate recombinant polypeptides. After electrophoretic separation, the negatively charged proteins (the antigens) are electrophoretically transferred from the polyacrylamide gel onto a nitrocellulose membrane positioned on the anode side of the gel. Following incubation of the membrane with a specific antibody, they are labeled with another anti-antibody for detection.

C. Glossary

Adventitious Organism – Bacteria, yeast, mold, mycoplasma or viruses that can potentially contaminate prokaryote or eukaryote cells used in production. Potential sources of adventitious organisms include the serum used in cell culture media, persistently or latently infected cells, or the environment.

Affinity – The thermodynamic quantity defining the energy interaction or binding of two molecules, usually that of antibody with its corresponding antigenic determinant.

Antibody (Immunoglobulin) – A protein molecule having a characteristic structure consisting of two types of peptide chains: heavy (H) and light (L). Antibodies contain areas (binding sites) that specifically fit to and can bind to its corresponding determinant site on an antigen, which has induced the production of that antibody by the B-lymphocytes and plasma cells in a living species.

Antigen – Substance, usually a foreign protein or carbohydrate, which when introduced into a organism, activates specific receptors on the surface immuno-competent T and B lymphocytes. After interaction between antigen and receptors, there usually will be induction of an immune response, i.e. production of antibodies capable of reacting specifically with determinant sites on the antigen.

Antigenic Determinant – The specific part of a structure of an antigen which will induce an immune response, i.e. will fit to the receptors on T and B lymphocytes and will also be able to react with the antibodies produced.

Antiserum – Blood serum which contains antibodies against a particular antigen (or immunogen). This frequently means serum from an animal that has been inoculated with the antigen.

Ascites – Liquid accumulations in the peritoneal cavity. Monoclonal antibodies can be purified from the ascites of mice that carry a transplanted hybridoma.

Association Constant – A reaction between antibody and its determinant which comprises a measure of affinity. The constant is quantitated by mass action law rate constants for association and for dissociation.

Autoradiography – Detection of radioactively labelled molecules on X-ray film.

Avidity – The total binding strength between all available binding sites of an antibody molecule and the corresponding determinants present on antigen.

Bacteriophage – A virus that attacks bacteria. The lambda bacteriophage is frequently used as a vector in recombinant gene experiments.

Binding Site – The part of the antibody molecule that will specifically bind antigen.

Bioactivity – The level of specific activity or potency as determined by animal model, cell culture, or in vitro biochemical assay.

Biological Containment – Characteristics of an organism that limit its survival and/or multiplication in the environment.

Biological Response Modifier – Generic term for hormones, neuroactive compounds, and immunoreactive compounds that act at the cellular level; many are possible candidates for biotechnological production.

Bioreactor – A vessel in which the central reactions of a biotechnological process takes place. Typically the vessel contains microbes grown under controlled conditions of temperature, aeration, mixing, acidity and sterility.

Biosensors – The powerful recognition systems of biological chemicals (enzymes, antibodies, DNA) are coupled to microelectronics to enable rapid, accurate low-level detection of such substances as sugars and proteins (such as hormones) in body fluids, pollutants in water and gases in air.

Calibrator – A term in clinical chemistry commonly referring to the standard used to "calibrate" an instrument or used in construction of a standard (calibrator) curve.

Cell Culture – The in-vitro growth of cells isolated from multicellular organisms. These cells are usually of one type.

Cell Differentiation – The process whereby descendants of a common parental cell achieve and maintain specialization of structure and function.

Cell Fusion – The formation of a hybrid cell with nuclei and cytoplasm from different cells, produced by fusing two cells of the same or different species.

Cell Line – Cells that acquire the ability to multiply indefinitely in-vitro.

Chemotaxis – Net oriented movement in a concentration gradient of certain compounds. Various sugars and amino acids can serve as attractants while some substances such as acid or alkali serve as repellants in microbial chemotaxis. White blood cells and macrophages demonstrate chemotactic movement in the presence of bacterial products, complement proteins and antigen activated T cells to contribute to the local inflammatory reaction and resistance to pathogens.

Cistron – The smallest unit of genetic material which is responsible for the synthesis of a specific polypeptide.

Clone – A cell line stemming from a single ancestral cell and normally expressing all the same genes. If this is a B lymphocyte clone, they will normally produce identical antibodies, i.e. monoclonal antibodies.

Codon – Group of three nucleotide bases in DNA or RNA that determines the composition of one amino acid in "building" a protein and also can code for chain termination.

Cohesive Termini – DNA molecule with single-stranded ends with exposed (cohesive) complementary bases.

COMPLEMENTARY DNA (Cdna) – DNA that is complementary to messenger RNA; used for cloning or as a probe in DNA hybridization studies.

Cosmid – A vector that is similar to a plasmid but it also contains the cohesive sites (cos site) of bacteriophage lambda to permit insertion of large fragments of DNA and in vitro packaging into a phage.

Cross Reaction – Antibodies against an antigen A can react with other antigens if the latter has one or more determinants in common with the determinants present on the antigen A or carry one or more determinants that are structurally very similar to the determinants present on antigen A.

Cytokine – Small, non-immunoglobulin proteins produced by monocytes and lymphocytes that serve as intercellular communicators after binding to specific receptors on the responding cells. Cytokines regulate a variety of biological activities.

Cytopathic Effect – Morphological alterations of cell lines produced when cells are infected with a virus. Examples of cytopathic effects include cell rounding and clumping, fusion of cell membranes, enlargement or elongation of cells, or lysis of cells.

Cytotoxic – Damaging to cells.

Denaturation – Unfolding of a protein molecule into a generally bio-inactive form. Also the disruption of DNA duplex into two separate strands.

Dna (Deoxyribonucleic Acid) – The basic biochemical component of the chromosomes and the support of heredity. DNA contains the sugar deoxyribose and is the nucleic acid in which genetic information is stored (apart from some viruses).

Dna Cloning – Production of many identical copies of a defined DNA fragment.

Dna Library – Set of cloned DNA fragments which together represent the entire genome or the transcription of a particular tissue.

Dna Polymerase – An enzyme which catalyzes the synthesis of double-stranded DNA from single-stranded DNA.

Dna Synthesis – The formation of DNA by the sequential addition of nucleotide bases.

Dnase – An enzyme which produces single-stranded nicks in DNA. DNase is used in nick translation.

Elution – The removal of adsorbed material from an adsorbent such as the removal of a product from an enzyme bound on a column.

Endonuleases – Enzymes which cleave bonds within nucleic acid molecules.

Endotoxin – A heat-stable lipopolysaccharide associated with the outer membrane of certain gram-negative bacteria. It is not secreted and is released only when the cells are disrupted. When injected into humans, endotoxins produce a febrile response, leading to severe clinical problems, including death. An endotoxin unit (EU) is defined in comparison to the current USP Reference Standard Lot EC-5. One vial of lot EC-5 contains 10,000 EU. The official test for endotoxin is found in the USP.

Enzymes – Proteins that act as a catalyst in biochemical reactions.

Exonucleases – Enzymes that catalyze the removal of nucleotides from the ends of a DNA molecule.

Fermentation – An anaerobic bioprocess. Fermentation is used in various industrial processes for the manufacture of products such as alcohols, acids, and cheese by the action of yeasts, molds, and bacteria. The fermentation process is used also in the production of monoclonal antibodies.

Fusion Of Protoplasts – Fusion of two cells whose walls have been eliminated, making it possible to redistribute the genetic heritage of micro-organisms.

Gene – The basic unit of heredity, which plays a part in the expression of a specific characteristic. The expression of a gene is the mechanism by which the genetic information that it contains is transcribed and translated to obtain a protein. A gene is a part of the DNA molecule that directs the synthesis of a specific polypeptide chain. It is composed of many codons. When the gene is considered as a unit of function in this way, the term cistron is often used.

Gene Transfer – The use of genetic or physical manipulation to introduce foreign genes into a host cells to achieve desired characteristics in progeny.

Genetic Engineering – A technique used to modify the genetic information in a living cell, reprogramming it for a desired purpose (such as the production of a substance it would not naturally produce).

Genome – All the genes carried by a cell.

Glycoprotein – Protein to which groups of sugars become attached. Human blood group proteins, cell wall proteins and some hormones are examples of glycoproteins.

Glycosylation – The covalent attachment of sugars to an amino acid in the protein portion of a glycoprotein.

Hapten – A low molecular weight substance that alone can react with its corresponding antibody. In order to be immunogenic, haptens are bonded to molecules having molecular weights greater than 5000. An example would be the hapten digoxin covalently bonded to bovine serum albumin, forming the digoxin-BSA immunogen.

High Affinity Antibody – Antibodies with a high affinity for antigen. These antibodies are predominantly IgG, and produced during a secondary response to antigen. Cells producing a high affinity antibody can be triggered by low concentration of antigen.

High Performance Liquid Chromatography (Hplc Or Lc) – (See Test Methods)

HostA cell whose metabolism is used for the growth and reproduction of a virus, plasmid, or other form of foreign DNA.

Hybridoma Technology – Fusion between an antibody forming cell (lymphocyte) and a malignant myeloma cell ("immortal"), which will result in a continuously growing cell clone (hybridoma), that can produce antibodies of a single specificity.

Immunoassay Specificity – A performance characteristic determined by conducting cross-reactivity studies with structurally similar substances that may be present in the analyte matrix. Specificity studies are determined with each new lot of polyclonal antibodies used in the immunoassay. For monoclonal antibody, each subsequent new lot is usually characterized by biochemical and biophysical techniques in lieu of comprehensive specificity studies.

Immunoelectrophoresis (Iep) – (See Test Methods - Immunodiffusion (double, Ouchterlony techiques))

Immunotoxin – Monoclonal antibodies coupled with toxins that are capable of delivering the toxin moiety to a target cell.

In Situ Hybridization – Hybridization with an appropriate probe carried out directly on a chromosome preparation or histological section.

In Vitro – Biological reactions taking place outside the body in an artificial system.

In Vivo – Biological reaction taking place inside a living cell or organism.

Inducer – A chemical or conditional change that activates the expression leading to the production of a desired product. A small molecule which interacts with a regulator protein and triggers gene transcription.

Ligase – Enzyme used to join DNA molecules.

Locus – The site of a gene on a chromosome.

Lymphokines – Substances released predominantly from T-lymphocytes after reaction with the specific antigen. Lymphokines are biologically highly active and will cause chemotaxis and activation of macrophages and other cell mediated immune reactions. Gamma-interferon is a lymphokine.

Lysis – The process whereby a cell wall breakdown occurs releasing cellular content into the surrounding environment. Destruction of bacteria by infective phage.

Master Cell Bank (Mcb) – A cell seed lot consisting of aliquots of a single culture (in most cases, expanded from a single cell) and stored cryogenically to assure genetic stability. There should be sufficient ampules of the MCB to provide the source material for a working seed bank.

MESSENGER RNA (Mrna) – RNA that serves as the template for protein synthesis; it carries the transcribed genetic code from the DNA to the protein synthesizing complex to direct protein synthesis.

Microheterogeneity – Slight differences in large, complex macromolecules that result in a population of closely related but not identical structures. Protein microheterogeneity can arise from many sources: genetic variants, proteolytic activity in cells, during translation into protein, during attachment of sugars and during commercial production.

Monoclonal Antibodies – Antibodies that are produced by a cellular clone and are all identical.

Mutagenesis – The induction of genetic mutation by physical or chemical means to obtain a characteristic desired by researchers.

Mutation – A change in the genetic material, either of a single base pair (point mutation) or in the number or structure of the chromosomes.

Myeloma – Tumor cell line derived from a lymphocyte.

Nick Translation – In vitro method used to introduce radioactively labelled nucleotides into DNA.

Nick – A break in the sugar-phosphate backbone of a DNA or RNA strand.

Oligonucleotides – Short segments of DNA or RNA, i.e.; a chain of a few nucleotides.

Operator Gene – A gene which switches on adjacent structural gene(s).

Operon – Complete unit of bacterial gene expression consisting of a regulator gene(s), control elements (promoter and operator) and adjacent structural gene(s).

Pathogen – A disease-producing agent, usually restricted to a living agent, such as a bacterium or virus.

Peptide Bond – Chemical bond between the carboxyl (-COOH) group of one amino acid and the amino (-NH2) group of another.

Plaque – Clear area in a plated bacterial culture due to lysis by a phage.

Plasmid – An extrachromosomal, self-replicating, circular segment of DNA; plasmids (and some viruses) are used as "vectors" for cloning DNA in bacterial "host" cells.

Polyclonal – Derived from different types of cells.

Prokaryote – An organism (e.g. bacterium, virus, blue-green algae) whose DNA is not enclosed within a nuclear membrane.

Protein – A polypeptide consisting of amino acids. In their biologically active states, proteins function as catalysts in metabolism and, to some extent, as structural elements of cells and tissues.

Pyrogenicity – The tendency for some bacterial cells or parts of cells to cause inflammatory reactions in the body, which may detract from their usefulness as pharmaceutical products.

Recombinant Dna Dna – that contains genes from different sources that have been combined by methods of genetic engineering as opposed to traditional breeding experiments.

Restriction Map – Linear arrangement of various restriction enzyme sites.

Restriction Site – Base sequence recognized by an enzyme.

Retrovirus Rna – virus which replicates via conversion into a DNA duplex.

Reverse Transcriptase – An enzyme that catalyzes the synthesis of DNA from RNA.

Rna (Ribonucleic Acid) – Basic biochemical component of the chromosome that is found mainly in the nucleolus and ribosomes. Messenger RNA transfers genetic information from the nucleus to the ribosomes in the cytoplasm and also acts as a template for the synthesis of polypeptides. Transfer RNA transfers activated amino acids from the cytoplasm to messenger RNA.

Rna Polymerase – An enzyme that catalyzes the synthesis of RNA in transcription.

Sodium Dodecyl Sulfate Polyacrylamide Gel Electroporesis (Sds-Page) – (See Test Methods)

Strain – A group of organisms of the same species having distinctive characteristics, but not usually considered a separate breed or variety.

T-Helper Cells – T-lymphocytes with the specific capacity to help other cells, such as B-lymphocytes, to make antibodies. T helper cells are also required for the induction of other T-lymphocyte activities. Synonym is T inducer cell, T4 cell, or CD 4 lymphocyte.

T-Suppressor Cells – T-lymphocytes with specific capacity to inhibit T-helper cell function.

Transcription – The first stage in the expression of a gene by means of genetic information being transmitted from the DNA in the chromosomes to messenger RNA.

Translation – The second stage in the expression of a gene by means of genetic information being transmitted from the mRNA to the synthesis of protein.

Vector – A plasmid, phage or cosmid into which foreign DNA may be inserted for cloning.

Western Blot – (See Test Methods).

Working Cell Bank (Wcb) – A quantity of cells derived from one or more ampules of the Master Cell Bank and used to initiate the production batch.

D. References

Books and Literature

1. Antebi, E. and Fishlock, D., *Biotechnology - Strategies for Life.* MIT Press, Cambridge, MA (1986).
2. Avallone, H.L., Beatrice, M.G., and Sze, T.T. Food and Drug Administration Inspection and Licensing of Manufacturing Facilities. Drug Biotechnology Regulation, Y.-y.H. Chiu and J.L. Gueriguian, Ed., 315–340 (1991).
3. Biotechnology Quality Control Training Course Manual by PMA (1991).
4. Campbell, A.M., *Laboratory Techniques in Biochemistry and Molecular Biology - Monoclonal Antibody Technology, Volume 13*, Elsevier, New York (1986).
5. Chiu, Y.- y.H. Validation of the Fermentation Process for the Production of Recombinant DNA Drugs. *Pharm. Technol* **12**:132. (1988).
6. Chiu, Y.- y.H. Review and Discussion of Special Chemical and Pharmaceutical Requirements in the U.S. for Biotechnology Products. *Drug Information Journal* **23**:47 (1989).
7. Davis, B.D., Dulbecco, R., Eisen, H. and Ginsberg, H. in *Microbiology*, 3rd edition. Maryland, Harper & Row (1980).
8. *Dorland's Pocket Medical Dictionary* (23rd Edition), W. B. Saunders Company (1982).
9. Emery, A.E.H., *An Introduction to Recombinant DNA*, John Wiley and Sons, New York (1984).
10. "*Genetic Engineering, A Natural Science*," Monsanto Company, St. Louis, Mo. (1985).
11. Hanson, L.A. and Wigzell, H., *Immunology*, Butterworths, Boston (1985).
12. Smith, B.- H. FDA Enforcement in Bioprocessing Facilities. ASTMSTP 1051 W.C. Hyer, Jr., Ed., 152–157 (1990).
13. Tetzlaff, R.F. FDA Regulatory Inspections of Aseptic Processing Facilities. Aseptic Pharmaceutical Manufacturing, W.P. Olson and M.J. Groves, Ed., 367–401 (1987).
14. "What Is Biotechnology?" Industrial Biotechnology Association (1984).

FDA and Other Government Publications

1. Biotech Inspection Outline (1988).
2. Cytokine and Growth Factor Pre-approval Trial Information Package (1990).
3. Federal Register Coordinated Framework for Regulation of Biotechnology. 51:23303-23393 (1986).
4. Guide to Inspection of Computerized Systems in Drug Processing (1983).
5. Guideline for Submitting Documentation for the Stability of Human Drugs and Biologics (1987).
6. Guideline for the Manufacture of In Vitro Diagnostic Products (1990).
7. Guideline on the General Principles of Process Validation (1987).
8. Guidelines on Validation of the Limulus Amebocyte Lysate Test as an End-Product Endotoxin Test for Human and Animal Parenteral Drugs, Biologics and Medical Devices (1987).
9. Inspection Technical Guide No. 43 Lyophilization of Parenterals (1986).
10. Interferon Test Procedures: Points to be Considered in the Production and Testing of Interferon intended for Investigational Use in Humans (1983).
11. National Institutes of Health. Recombinant DNA Research; Action under Guidelines. Fed. Reg. 52:31848-31850 (1987) 53:43410-43411 (1988); 56:33174-33183 (1991).
12. Points to Consider in the Characterization of Cell Lines to Produce Biological Products (1987).
13. Points to Consider in the Collection, Processing, and Testing of Ex-Vivo Activated Mononuclear Leucocytes for Human Use (1989).
14. Points to Consider in the Manufacture and Clinical Evaluation of In Vitro Tests to Detect Antibodies to the Human Immunodeficiency Virus (1989).
15. Points to Consider in the Manufacturing of In Vitro Monoclonal Products Subject of Licensure (1983).
16. Consider in the Manufacturing of Monoclonal Antibody Products for Human Use (1987).
17. Points to Consider in the Production and Testing of New Drugs and Biologics Produced by Recombinant DNA Technology (1985).
18. Recommended Test Procedures for Mycoplasmas (1988).

Appendix C

Guide to Inspections of High Purity Water Systems

The following is a US Food and Drug Administration (FDA) training guide used by the FDA in the training of their inspectors. Such training guides are used to prepare and provide FDA inspectors with a general knowledge and understanding of what it is they should be considering during inspection of an FDA-regulated manufacturing facility.

In many cases these are general considerations due in large part to the many and varied drugs and the proprietary processes required to manufacture them. These guides are intended to augment the education, experience, and accumulated knowledge each inspector already possesses. The inspection process and procedures by which they perform it leave room for on-site judgment consideration and assessment by the inspectors.

The purpose of including these FDA guides in this book is to provide the reader, designer, and engineer involved with the design and engineering of a bioprocessing facility some insight as to what the expectations of an FDA inspector might be. What expectations an FDA inspector might possess can most likely be attributed to their training and these guidelines. Providing these guidelines for the readers of this book, those of you designing, constructing, and operating FDA-regulated manufacturing facilities, will hopefully lessen the gap of understanding between you and the FDA inspector. Having some insight into the inspector's understanding and training helps one to be better prepared to meet those expectations when it's time for the inspection.

These inspection guidelines are reproduced here through the courtesy of the FDA. These and others also reside on the FDA web site at:

http://www.fda.gov/ICECI/Inspections/InspectionGuides/default.htm

Bioprocessing Piping and Equipment Design: A Companion Guide for the ASME BPE Standard, First Edition. William M. (Bill) Huitt.

Guide to Inspections of High Purity Water Systems

Note: This document is reference material for investigators and other FDA personnel. The document does not bind FDA, and does no confer any rights, privileges, benefits, or immunities for or on any person(s).

This guide discusses, primarily from a microbiological aspect, the review and evaluation of high purity water systems that are used for the manufacture of drug products and drug substances. It also includes a review of the design of the various types of systems and some of the problems that have been associated with these systems. As with other guides, it is not all-inclusive, but provides background and guidance for the review and evaluation of high purity water systems. The Guide to Inspections of Microbiological Pharmaceutical Quality Control Laboratories (May, 1993) provides additional guidance.

I. System Design

One of the basic considerations in the design of a system is the type of product that is to be manufactured. For parenteral products where there is a concern for pyrogens, it is expected that Water for Injection will be used. This applies to the formulation of products, as well as to the final washing of components and equipment used in their manufacture. Distillation and Reverse Osmosis (RO) filtration are the only acceptable methods listed in the USP for producing Water for Injection. However, in the bulk Pharmaceutical and Biotechnology industries and some foreign companies, Ultra Filtration (UF) is employed to minimize endotoxins in those drug substances that are administered parenterally.

For some ophthalmic products, such as the ophthalmic irrigating solution, and some inhalation products, such as Sterile Water for Inhalation, where there are pyrogen specifications, it is expected that Water for Injection be used in their formulation. However, for most inhalation and ophthalmic products, purified water is used in their formulation. This also applies to topicals, cosmetics and oral products.

Another design consideration is the temperature of the system. It is recognized that hot (65–80°C) systems are self-sanitizing. While the cost of other systems may be less expensive for a company, the cost of maintenance, testing and potential problems may be greater than the cost of energy saved. Whether a system is circulating or one-way is also an important design consideration. Obviously, water in constant motion is less liable to have high levels of contaminant. A one-way water system is basically a "dead-leg".

Finally, and possibly the most important consideration, is the risk assessment or level of quality that is desired. It should be recognized that different products require different quality waters. Parenterals require very pure water with no endotoxins. Topical and oral products require less pure water and do not have a requirement for endotoxins. Even with topical and oral products there are factors that dictate different qualities for water. For example, preservatives in antacids are marginally effective, so more stringent microbial limits have to be set. The quality control department should assess each product manufactured with the water from their system and determine the microbial action limits based on the most microbial sensitive product. In lieu of stringent water action limits in the system the manufacturer can add a microbial reduction step in the manufacturing process for the sensitive drug product(s).

II. System Validation

A basic reference used for the validation of high purity water systems is the Parenteral Drug Association Technical Report No. 4 titled, "Design Concepts for the Validation of a Water for Injection System." The introduction provides guidance and states that, "Validation often involves the use of an appropriate challenge. In this situation, it would be undesirable to introduce microorganisms into an on-line system; therefore, reliance is placed on periodic testing for microbiological quality and on the installation of monitoring equipment at specific checkpoints to ensure that the total system is operating properly and continuously fulfilling its intended function."

In the review of a validation report, or in the validation of a high purity water system, there are several aspects that should be considered. Documentation should include a description of the system along with a print. The drawing needs to show all equipment in the system from the water feed to points of use. It should also show all sampling points and their designations. If a system has no print, it is usually considered an objectionable condition. The thinking is if there is no print, then how can the system be validated? How can a quality control manager or microbiologist know where to sample? In those facilities observed without updated prints, serious problems were identified in these systems. The print should be compared to the actual system annually to insure its accuracy, to detect unreported changes and confirm reported changes to the system.

After all the equipment and piping has been verified as installed correctly and working as specified, the initial phase of the water system validation can begin. During this phase the operational parameters and the cleaning/sanitization procedures and frequencies will be developed. Sampling should be daily after each step in the purification process and at each point of use for two to four weeks. The sampling procedure for point of use sampling should reflect how the water is to be drawn e.g. if a hose is usually attached the sample should be taken at the end of the hose. If the SOP calls for the line to be flushed before use of the water from that point, then the sample is taken after the flush. At the end of the two to four week time period the firm should have developed it's SOPs for operation of the water system.

The second phase of the system validation is to demonstrate that the system will consistently produce the desired water quality when operated in conformance with the SOPs. The sampling is performed as in the initial phase and for the same time period. At the end of this phase the data should demonstrate that the system will consistently produce the desired quality of water.

The third phase of validation is designed to demonstrate that when the water system is operated in accordance with the SOPs over a long period of time it will consistently produce water of the desired quality. Any variations in the quality of the feedwater that could affect the operation and ultimately the water quality will be picked up during this phase of the validation. Sampling is performed according to routine procedures and frequencies. For Water for Injection systems the samples should be taken daily from a minimum of one point of use, with all points of use tested weekly. The validation of the water system is completed when the firm has a full years-worth of data.

While the above validation scheme is not the only way a system can be validated, it contains the necessary elements for validation of a water system. First, there must be data to support the SOPs. Second, there must be data demonstrating that the SOPs are valid and that the system is capable of consistently producing water that meets the desired specifications. Finally, there must be data to demonstrate that seasonal variations in the feedwater do not adversely affect the operation of the system or the water quality.

The last part of the validation is the compilation of the data, with any conclusions into the final report. The final validation report must be signed by the appropriate people responsible for operation and quality assurance of the water system.

A typical problem that occurs is the failure of operating procedures to preclude contamination of the system with non-sterile air remaining in a pipe after drainage. In a system illustrated as in Figure 1, (below) a typical problem occurs when a washer or hose connection is flushed and then drained at the end of the operation. After draining, this valve (the second off of the system) is closed. If on the next day or start-up of the operation the primary valve off of the circulating system is opened, then the non-sterile air remaining in the pipe after drainage would contaminate the system. The solution is to pro-vide for operational procedures that provide for opening the secondary valve before the primary valve to flush the pipe prior to use.

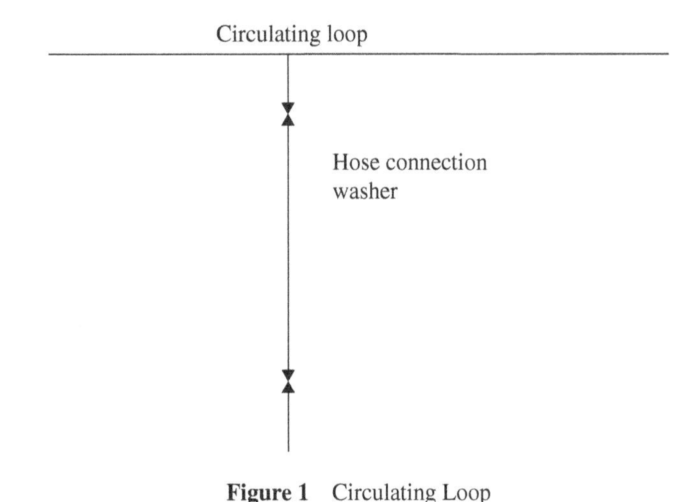

Figure 1 Circulating Loop

Another major consideration in the validation of high purity water systems is the acceptance criteria. Consistent results throughout the system over a period of time constitute the primary element.

III. Microbial Limits

Water For Injection (WFI) Systems

Regarding microbiological results, for Water For Injection, it is expected that they be essentially sterile. Since sampling frequently is performed in non-sterile areas and is not truly aseptic, occasional low level counts due to sampling errors may occur. Agency policy, is that less than 10 CFU/100ml is an acceptable action limit. None of the limits for water are pass/fail limits. All limits are action limits. When action limits are exceeded the firm must investigate the cause of the problem, take action to correct the problem and assess the impact of the microbial contamination on products manufactured with the water and document the results of their investigation.

With regard to sample size, 100–300 mL is preferred when sampling Water for Injection systems. Sample volumes less than 100 mL are unacceptable.

The real concern in WFI is endotoxins. Because WFI can pass the LAL endotoxin test and still fail the above microbial action limit, it is important to monitor WFI systems for both endotoxins and microorganisms.

Purified Water Systems

For purified water systems, microbiological specifications are not as clear. USP XXII specifications, that it complies with federal Environmental Protection Agency regulations for drinking water, are recognized as being minimal specifications. There have been attempts by some to establish meaningful microbiological specifications for purified water. The CFTA proposed a specification of not more than 500 organisms per ml. The USP XXII has an action guideline of not greater than 100 organisms per ml. Although microbiological specifications have been discussed, none (other than EPA standards) have been established. Agency policy is that any action limit over 100 CFU/mL for a purified water system is unacceptable.

The purpose of establishing any action limit or level is to assure that the water system is under control. Any action limit established will depend upon the overall purified water system and further processing of the finished product and its use. For example, purified water used to manufacture drug products by cold processing should be free of objectionable organisms. We have defined "objectionable organisms" as any organisms that can cause infections when the drug product is used as directed or any organism capable of growth in the drug product. As pointed out in the Guide to Inspections of Microbiological Pharmaceutical Quality Control Laboratories, the specific contaminant, rather than the number is generally more significant.

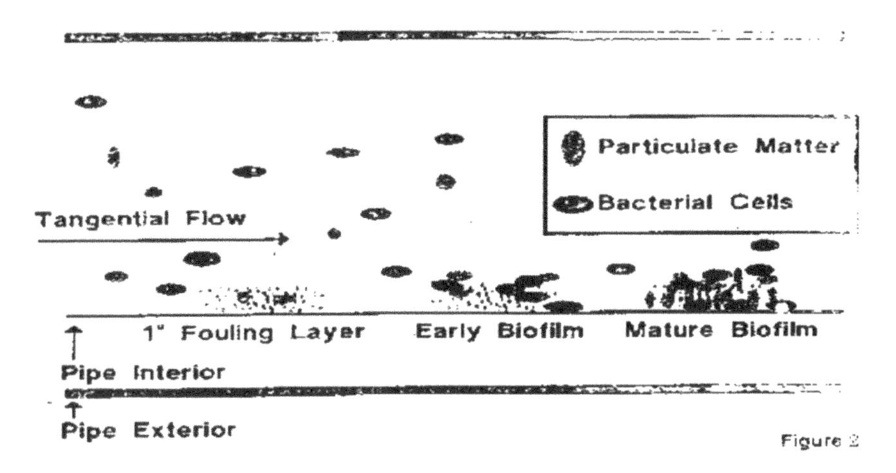

Figure 2 Organisms in a Water System

Organisms exist in a water system either as free floating in the water or attached to the walls of the pipes and tanks. When they are attached to the walls they are known as biofilm, which continuously slough off organisms. Thus, contamination is not uniformly distributed in a system and the sample may not be representative of the type and level of contamination. A count of 10 CFU/mL in one sample and 100 or even 1000 CFU/mL in a subsequent sample would not be unrealistic.

Thus, in establishing the level of contamination allowed in a high purity water system used in the manufacture of a non-sterile product requires an understanding of the use of the product, the formulation (preservative system) and manufacturing process. For example, antacids, which do not have an effective preservative system, require an action limit below the 100 CFU/mL maximum.

The USP gives some guidance in their monograph on Microbiological Attributes of Non-Sterile Products. It points out that, "The significance of microorganisms in non-sterile pharmaceutical products should be evaluated in terms of the use of the product, the nature of the product, and the potential harm to the user." Thus, not just the indicator organisms listed in some of the specific monographs present problems. It is up to each manufacturer to evaluate their product, the way it is manufactured, and establish an acceptable action level of contamination, not to exceed the maximum, for the water system, based on the highest risk product manufactured with the water.

IV. Water for Injection Systems

In the review and evaluation of Water For Injection systems, there are several concerns.

Pretreatment of feedwater is recommended by most manufacturers of distillation equipment and is definitely required for RO units. The incoming feedwater quality may fluctuate during the life of the system depending upon seasonal variations and other external factors beyond the control of the pharmaceutical facility. For example, in the spring (at least in the N.E.), increases in gram negative organisms have been known. Also, new construction or fires can cause a depletion of water stores in old mains which can cause an influx of heavily contaminated water of a different flora.

A water system should be designed to operate within these anticipated extremes. Obviously, the only way to know the extremes is to periodically monitor feedwater. If the feedwater is from a municipal water system, reports from the municipality testing can be used in lieu of in-house testing.

V. Still

Figures 3–5 represent a typical basic diagram of a WFI system. Most of the new systems now use multi-effect stills. In some of the facilities, there has been evidence of endotoxin contamination. In one system this occurred, due to malfunction of the feedwater valve and level control in the still which resulted in droplets of feedwater being carried over in the distillate.

In another system with endotoxin problems, it was noted that there was approximately 50 liters of WFI in the condenser at the start-up. Since this water could lie in the condenser for up to several days (i.e., over the weekend), it was believed that this was the reason for unacceptable levels of endotoxins.

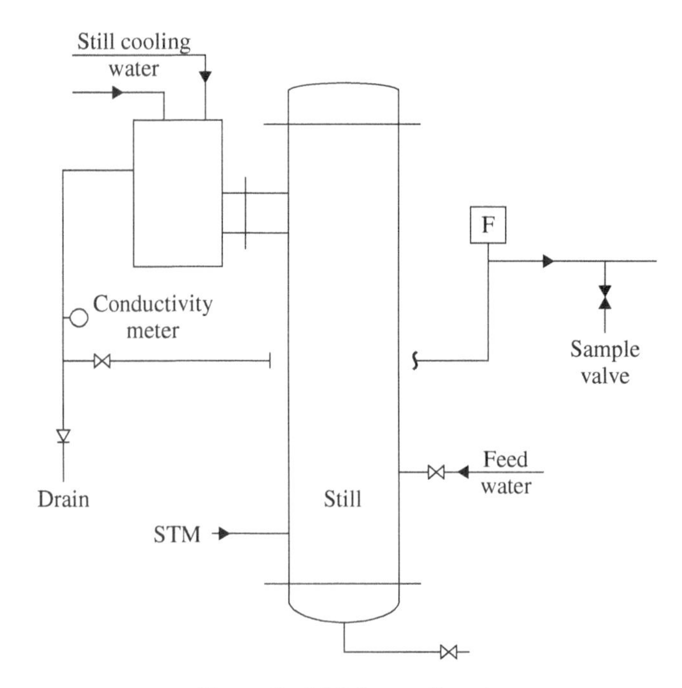

Figure 3 WFI System Type A

More common, however, is the failure to adequately treat feedwater to reduce levels of endotoxins. Many of the still fabricators will only guarantee a 2.5 log to 3 log reduction in the endotoxin content. Therefore, it is not surprising that in systems where the feedwater occasionally spikes to 250 EU/ml, unacceptable levels of endotoxins may occasionally appear in the distillate (WFI). For example, recently three new stills, including two multi-effect, were found to be periodically yielding WFI with levels greater than .25 EU/ml. Pretreatment systems for the stills included only deionization systems with no UF, RO or distillation. Unless a firm has a satisfactory pretreatment system, it would be extremely difficult for them to demonstrate that the system is validated.

The above examples of problems with distillation units used to produce WFI, point to problems with maintenance of the equipment or improper operation of the system indicating that the system has not been properly validated or that the initial validation is no longer valid. If you see these types of problems you should look very closely at the system design, any changes that have been made to the system, the validation report and the routine test data to determine if the system is operating in a state of control.

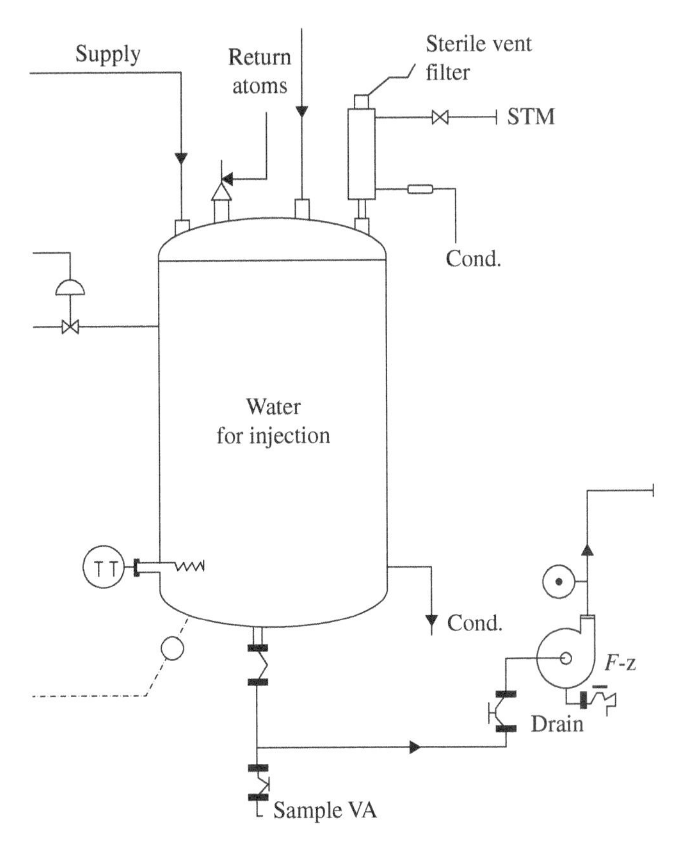

Figure 4 WFI System Type B

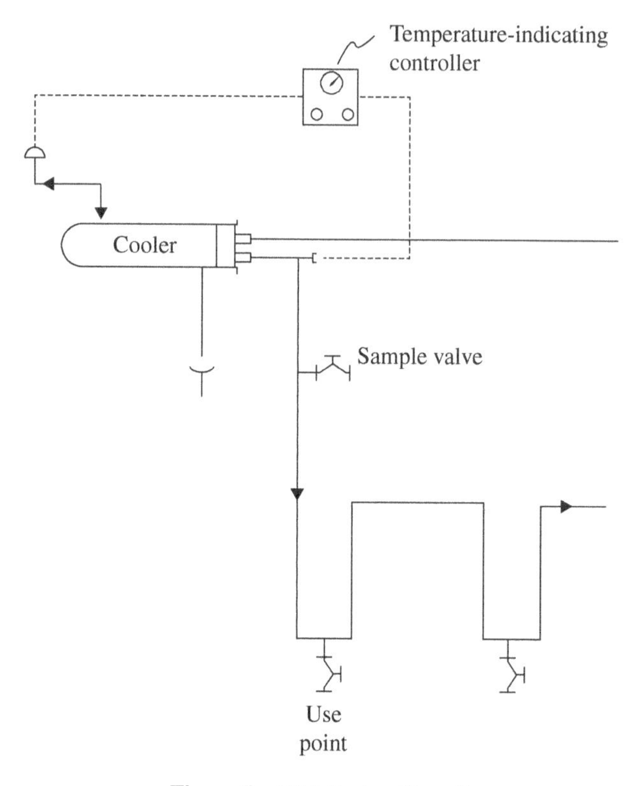

Figure 5 WFI System Type C

Typically, conductivity meters are used on water systems to monitor chemical quality and have no meaning regarding microbiological quality.

Figures 3–5 also show petcocks or small sampling ports between each piece of equipment, such as after the still and before the holding tank. These are in the system to isolate major pieces of equipment. This is necessary for the qualification of the equipment and for the investigation of any problems which might occur.

VI. Heat Exchangers

One principal component of the still is the heat exchanger. Because of the similar ionic quality of distilled and deionized water, conductivity meters cannot be used to monitor microbiological quality. Positive pressure such as in vapor compression or double-tubesheet design should be employed to prevent possible feedwater to distillate contamination in a leaky heat exchanger.

An FDA Inspectors Technical Guide with the subject of "Heat Exchangers to Avoid Contamination" discusses the design and potential problems associated with heat exchangers. The guide points out that there are two methods for preventing contamination by leakage. One is to provide gauges to constantly monitor pressure differentials to ensure that the higher pressure is always on the clean fluid side. The other is to utilize the double-tubesheet type of heat exchanger.

In some systems, heat exchangers are utilized to cool water at use points. For the most part, cooling water is not circulated through them when not in use. In a few situations, pinholes formed in the tubing after they were drained (on the cooling water side) and not in use. It was determined that a small amount of moisture remaining in the tubes when combined with air caused a corrosion of the stainless steel tubes on the cooling water side. Thus, it is recommended that when not in use, heat exchangers not be drained of the cooling water.

VII. Holding Tank

In hot systems, temperature is usually maintained by applying heat to a jacketed holding tank or by placing a heat exchanger in the line prior to an insulated holding tank.

The one component of the holding tank that generates the most discussion is the vent filter. It is expected that there be some program for integrity testing this filter to assure that it is intact. Typically, filters are now jacketed to prevent condensate or water from blocking the hydrophobic vent filter. If this occurs (the vent filter becomes blocked), possibly either the filter will rupture or the tank will collapse. There are methods for integrity testing of vent filters in place.

It is expected, therefore, that the vent filter be located in a position on the holding tank where it is readily accessible.

Just because a WFI system is relatively new and distillation is employed, it is not problem-free. In an inspection of a manufacturer of parenterals, a system fabricated in 1984 was observed. Refer to Figure 6. While the system may appear somewhat complex on the initial review, it was found to be relatively simple. Figure 7 is a schematic of the system. The observations at the conclusion of the inspection of this manufacturer included, "Operational procedures for the Water For Injection system failed to provide for periodic complete flushing or draining. The system was also open to the atmosphere and room environment. Compounding equipment consisted of non-sealed, open tanks with lids. The Water for Injection holding tank was also not sealed and was never sampled for endotoxins." Because of these and other comments, the firm recalled several products and discontinued operations.

VIII. Pumps

Pumps burn out and parts wear. Also, if pumps are static and not continuously in operation, their reservoir can be a static area where water will lie. For example, in an inspection, it was noted that a firm had to install a drain from the low point in a pump housing. Pseudomonas sp. contamination was periodically found in their water system which was attributed in part to a pump which only periodically is operational.

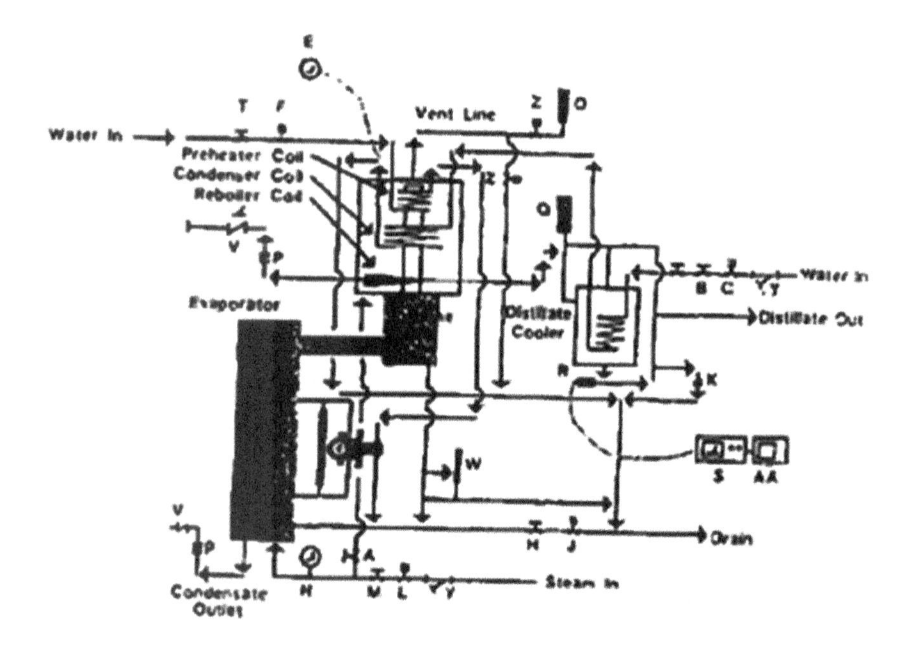

Figure 6 Older WFI System

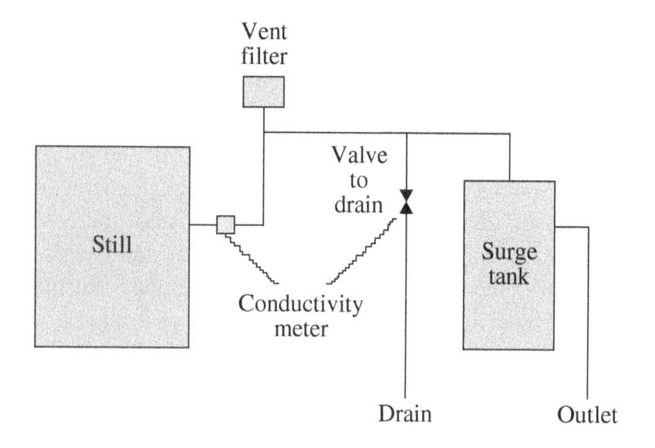

Figure 7 Schematic of Older WFI System

IX. Piping

Piping in WFI systems usually consist of a high polished stainless steel. In a few cases, manufacturers have begun to utilize PVDF (polyvinylidene fluoride) piping. It is purported that this piping can tolerate heat with no extractables being leached. A major problem with PVDF tubing is that it requires considerable support. When this tubing is heated, it tends to sag and may stress the weld (fusion) connection and result in leakage. Additionally, initially at least, fluoride levels are high.

This piping is of benefit in product delivery systems where low level metal contamination may accelerate the degradation of drug product, such as in the Biotech industry.

One common problem with piping is that of "dead-legs". The proposed LVP Regulations defined dead-legs as not having an unused portion greater in length than six diameters of the unused pipe measured from the axis of the pipe in use. It should be pointed out that this was developed for hot 75–80°C circulating systems. With colder systems (65–75°C), any drops or unused portion of any length of piping has the potential for the formation of a biofilm and should be eliminated if possible or have special sanitizing procedures. There should be no threaded fittings in a pharmaceutical water system. All pipe joints must utilize sanitary fittings or be butt welded. Sanitary fittings will usually be used where the piping meets valves, tanks and other equipment that must be removed for maintenance or replacement. Therefore, the firm's procedures for sanitization, as well as the actual piping, should be reviewed and evaluated during the inspection.

X. Reverse Osmosis

Another acceptable method for manufacturing Water for Injection is Reverse Osmosis (RO). However, because these systems are cold, and because RO filters are not absolute, microbiological contamination is not unusual. Figure 8 shows a system that was in use several years ago. There are five RO units in this system which are in parallel. Since RO filters are not absolute, the filter manufacturers

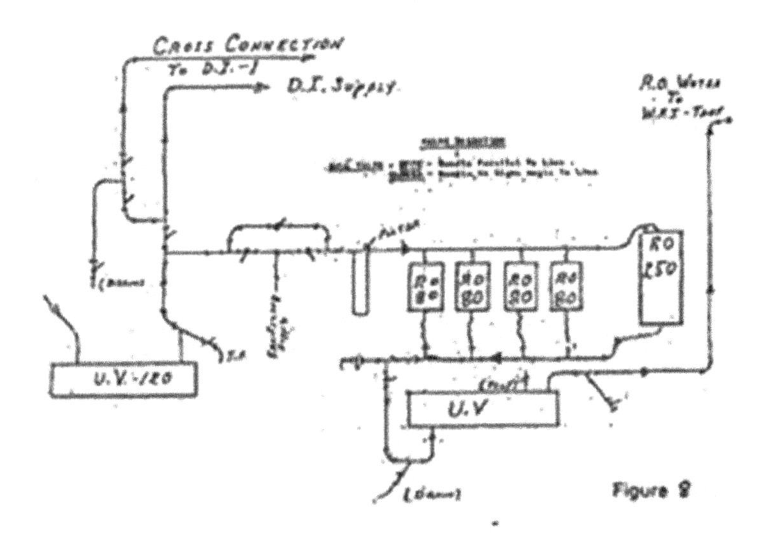

Figure 8 Older RO Unit

recommend that at least two be in series. The drawing also illustrates an Ultraviolet (UV) light in the system downstream from the RO units. The light was needed to control microbiological contamination.

Also in this system were ball valves. These valves are not considered sanitary valves since the center of the valve can have water in it when the valve is closed. This is a stagnant pool of water that can harbor microorganisms and provide a starting point for a biofilm.

As an additional comment on RO systems, with the recognition of microbiological problems, some manufacturers have installed heat exchangers immediately after the RO filters to heat the water to 75–80°C to minimize microbiological contamination.

With the development of biotechnology products, many small companies are utilizing RO and UF systems to produce high purity water. For example, Figure 9 illustrates a wall mounted system that is fed by a single pass RO unit.

As illustrated, most of these systems employ PVC or some type of plastic tubing. Because the systems are typically cold, the many joints in the system are subject to contamination. Another potential problem with PVC tubing is extractables. Looking at the WFI from a system to assure that it meets USP requirements without some assurance that there are no extractables would not be acceptable.

The systems also contain 0.2 micron point of use filters which can mask the level of microbiological contamination in the system. While it is recognized that

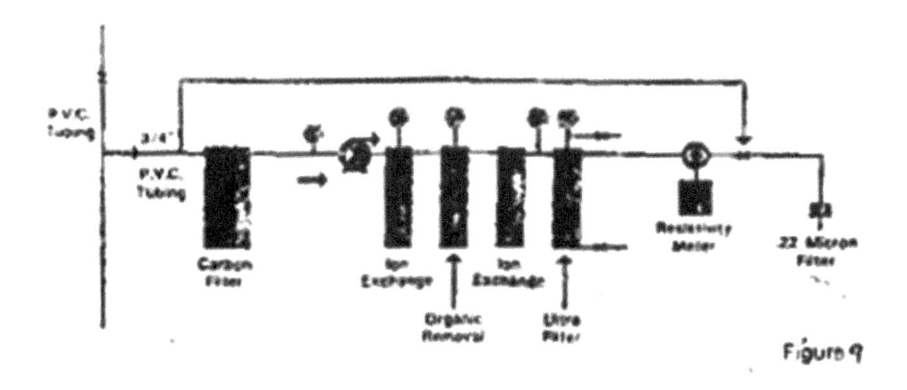

REVERSE OSMOSIS SYSTEM - BIOTECH FACILITY

DATE	BIOBURDEN (CFU/100 ML)	LAL (EU/ML)
3/10 (9 AM)	1	<.1
3/11 (9 AM)	29	<.1
3/11 (10 AM)	>300	1.0
3/11 (11 AM)	120	<.1
3/11 (12 AM)	115	<.1

Figure 9 RO Unit

endotoxins are the primary concern in such a system, a filter will reduce microbiological contamination, but not necessarily endotoxin contamination. If filters are used in a water system there should be a stated purpose for the filter, i.e., particulate removal or microbial reduction, and an SOP stating the frequency with which the filter is to be changed which is based on data generated during the validation of the system.

As previously discussed, because of the volume of water actually tested (.1ml for endotoxins vs. 100ml for WFI), the microbiological test offers a good index of the level of contamination in a system. Therefore, unless the water is sampled prior to the final 0.2 micron filter, microbiological testing will have little meaning.

At a re-inspection of this facility, it was noted that they corrected the deficient water system with a circulating stainless steel piping system that was fed by four RO units in series. Because this manufacturer did not have a need for a large amount of water (the total system capacity was about 30 gallons), they attempted to let the system sit for approximately one day. Figure 9 shows that at zero time (at 9 AM on 3/10), there were no detectable levels of microorganisms and of endotoxins. After one day, this static non-circulating system was found to be contaminated. The four consecutive one hour samples also illustrate the variability among samples taken from a system. After the last sample at 12 PM was collected, the system was re-sanitized with 0.5% peroxide solution, flushed, recirculated and resampled. No levels of microbiological contamination were found on daily samples after the system was put back in operation. This is the reason the agency has recommended that non-recirculating water systems be drained daily and water not be allowed to sit in the system.

XI. Purified Water Systems

Many of the comments regarding equipment for WFI systems are applicable to Purified Water Systems. One type system that has been used to control microbiological contamination utilizes ozone. Figure 10 illustrates a typical system. Although the system has purported to be relatively inexpensive, there are some problems associated with it. For optimum effectiveness, it is required that dissolved ozone residual remain in the system. This presents both employee safety problems and use problems when drugs are formulated.

Published data for Vicks Greensboro, NC facility showed that their system was re-contaminated in two to three days after the ozone generator was turned off. In an inspection of another manufacturer, it was noted that a firm was experiencing a contamination problem with Pseudomonas sp. Because of potential problems with employee safety, ozone was removed from the water prior to placing it in their recirculating system. It has been reported that dissolved ozone at a level of 0.45 mg/liter will remain in a system for a maximum of five to six hours.

Another manufacturer, as part of their daily sanitization, removes all drops off of their ozonated water system and disinfects them in filter sterilized 70% isopropyl alcohol. This manufacturer has reported excellent microbiological

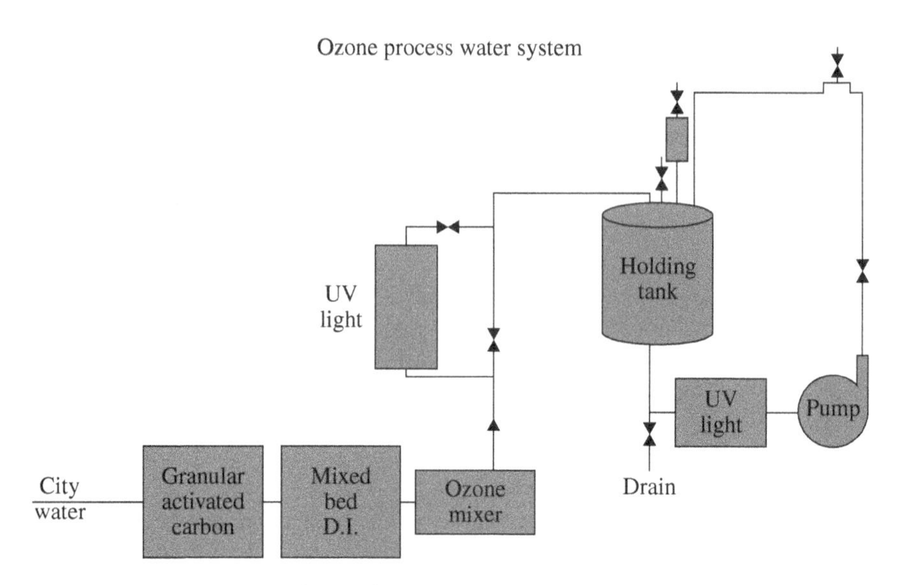

Figure 10 Purified Water System

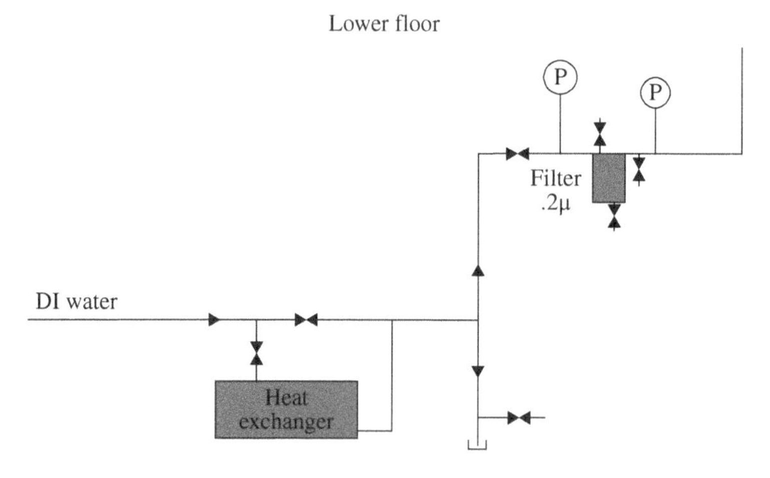

Figure 11 Problematic Purified Water System_1

results. However, sampling is only performed immediately after sanitization and not at the end of operations. Thus, the results are not that meaningful.

Figure 11 and Figure 12 illustrate another purified water system which had some problems. Unlike most of the other systems discussed, this is a one-way and not recirculating system. A heat exchanger is used to heat the water on a weekly basis and sanitize the system. Actually, the entire system is a "dead-leg."

Figure 11 also shows a 0.2 micron in line filter used to sanitize the purified water on a daily basis. In addition to the filter housing providing a good environment for microbiological contamination, a typical problem is water hammer that

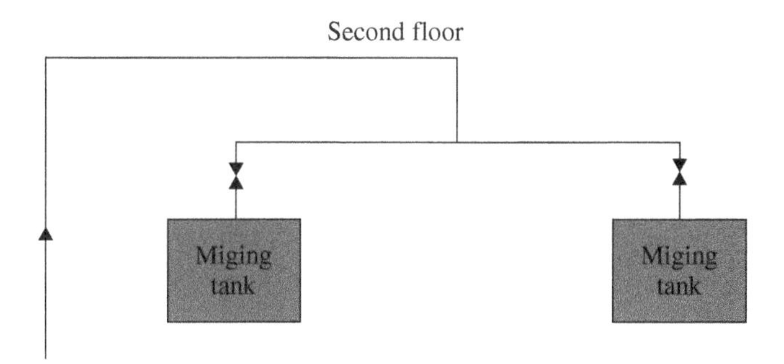

Figure 12 Problematic Purified Water System_2

can cause "ballooning" of the filter. If a valve downstream from the filter is shut too fast, the water pressure will reverse and can cause "ballooning". Pipe vibration is a typical visible sign of high back pressure while passage of upstream contaminants on the filter face is a real problem. This system also contains several vertical drops at use points. During sanitization, it is important to "crack" the terminal valves so that all of the elbows and bends in the piping are full of water and thus, get complete exposure to the sanitizing agent.

It should be pointed out that simply because this is a one-way system, it is not inadequate. With good Standard Operational Procedures, based on validation data, and routine hot flushing of this system, it could be acceptable. A very long system (over 200 yards) with over 50 outlets was found acceptable. This system employed a daily flushing of all outlets with 80°C water.

The last system to be discussed is a system that was found to be objectionable. Pseudomonas sp. found as a contaminant in the system (after FDA testing) was also found in a topical steroid product (after FDA testing). Product recall and issuance of a Warning Letter resulted. This system shown in Figure 13 is also one-way that employs a UV light to control microbiological contamination. The light is turned on only when water is needed. Thus, there are times when water is allowed to remain in the system. This system also contains a flexible hose which is very difficult to sanitize. UV lights must be properly maintained to work. The glass sleeves around the bulb(s) must be kept clean or their effectiveness will decrease. In multi-bulb units there must be a system to determine that each bulb is functioning. It must be remembered that at best UV light will only kill 90% of the organisms entering the unit.

XII. Process Water

Currently, the USP, pg. 4, in the General Notices Section, allows drug substances to be manufactured from Potable Water. It comments that any dosage form must be manufactured from Purified Water, Water For Injection, or one of the forms of Sterile Water. There is some inconsistency in these two statements, since Purified

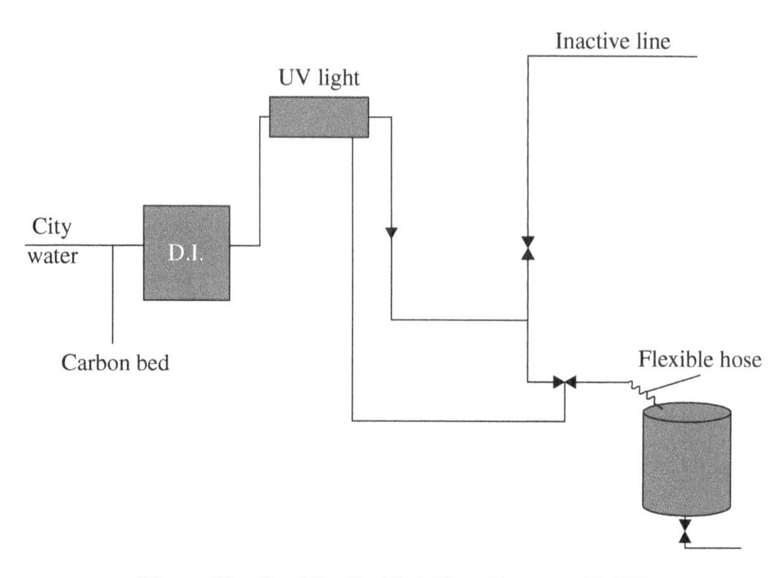

Figure 13 One Way Purified Water System with UV

Water has to be used for the granulation of tablets, yet Potable Water can be used for the final purification of the drug substance.

The FDA Guide to Inspection of Bulk Pharmaceutical Chemicals comments on the concern for the quality of the water used for the manufacture of drug substances, particularly those drug substances used in parenteral manufacture. Excessive levels of microbiological and/or endotoxin contamination have been found in drug substances, with the source of contamination being the water used in purification. At this time, Water For Injection does not have to be used in the finishing steps of synthesis/purification of drug substances for parenteral use. However, such water systems used in the final stages of processing of drug substances for parenteral use should be validated to assure minimal endotoxin/microbiological contamination.

In the bulk drug substance industry, particularly for parenteral grade substances, it is common to see Ultrafiltration (UF) and Reverse Osmosis (RO) systems in use in water systems. While ultrafiltration may not be as efficient at reducing pyrogens, they will reduce the high molecular weight endotoxins that are a contaminant in water systems. As with RO, UF is not absolute, but it will reduce numbers. Additionally, as previously discussed with other cold systems, there is considerable maintenance required to maintain the system.

For the manufacture of drug substances that are not for parenteral use, there is still a microbiological concern, although not to the degree as for parenteral grade drug substances. In some areas of the world, Potable (chlorinated) water may not present a microbiological problem. However, there may be other issues. For example, chlorinated water will generally increase chloride levels. In some areas, process water may be obtained directly from neutral sources.

In one inspection, a manufacturer was obtaining process water from a river located in a farming region. At one point, they had a problem with high levels of pesticides which was a run-off from farms in the areas. The manufacturing process and analytical methodology was not designed to remove and identify trace pesticide contaminants. Therefore, it would seem that this process water when used in the purification of drug substances would be unacceptable.

XIII. Inspection Strategy

Manufacturers typically will have periodic printouts or tabulations of results for their purified water systems. These printouts or data summaries should be reviewed. Additionally, investigation reports, when values exceed limits, should be reviewed.

Since microbiological test results from a water system are not usually obtained until after the drug product is manufactured, results exceeding limits should be reviewed with regard to the drug product formulated from such water. Consideration with regard to the further processing or release of such a product will be dependent upon the specific contaminant, the process and the end use of the product. Such situations are usually evaluated on a case-by-case basis. It is a good practice for such situations to include an investigation report with the logic for release/rejection discussed in the firm's report. End product microbiological testing, while providing some information should not be relied upon as the sole justification for the release of the drug product. The limitations of microbiological sampling and testing should be recognized.

Manufacturers should also have maintenance records or logs for equipment, such as the still. These logs should also be reviewed so that problems with the system and equipment can be evaluated.

In addition to reviewing test results, summary data, investigation reports and other data, the print of the system should be reviewed when conducting the actual physical inspection. As pointed out, an accurate description and print of the system is needed in order to demonstrate that the system is validated.

Appendix D

Guide to Inspections of Lyophilization of Parenterals

The following is a US Food and Drug Administration (FDA) training guide used by the FDA in the training of their inspectors. Such training guides are used to prepare and provide FDA inspectors with a general knowledge and understanding of what it is they should be considering during inspection of an FDA-regulated manufacturing facility.

In many cases these are general considerations due in large part to the many and varied drugs and the proprietary processes required to manufacture them. These guides are intended to augment the education, experience, and accumulated knowledge each inspector already possesses. The inspection process and procedures by which they perform it leave room for on-site judgment consideration and assessment by the inspectors.

The purpose of including these FDA guides in this book is to provide the reader, designer, and engineer involved with the design and engineering of a bioprocessing facility some insight as to what the expectations of an FDA inspector might be. What expectations an FDA inspector might possess can most likely be attributed to their training and these guidelines. Providing these guidelines for the readers of this book, those of you designing, constructing, and operating FDA-regulated manufacturing facilities, will hopefully lessen the gap of understanding between you and the FDA inspector. Having some insight into the inspector's understanding and training helps one to be better prepared to meet those expectations when it's time for the inspection.

These inspection guidelines are reproduced here through the courtesy of the FDA. These and others also reside on the FDA web site at:

http://www.fda.gov/ICECI/Inspections/InspectionGuides/default.htm

Bioprocessing Piping and Equipment Design: A Companion Guide for the ASME BPE Standard, First Edition. William M. (Bill) Huitt.

Guide to Inspections of Lyophilization of Parenterals

Note: This document is reference material for investigators and other FDA personnel. The document does not bind FDA, and does no confer any rights, privileges, benefits, or immunities for or on any person(s).

Introduction

Lyophilization or freeze drying is a process in which water is removed from a product after it is frozen and placed under a vacuum, allowing the ice to change directly from solid to vapor without passing through a liquid phase. The process consists of three separate, unique, and interdependent processes; freezing, primary drying (sublimation), and secondary drying (desorption).

The advantages of lyophilization include:

- Ease of processing a liquid, which simplifies aseptic handling
- Enhanced stability of a dry powder
- Removal of water without excessive heating of the product
- Enhanced product stability in a dry state
- Rapid and easy dissolution of reconstituted product
- Disadvantages of lyophilization include:
- Increased handling and processing time
- Need for sterile diluent upon reconstitution
- Cost and complexity of equipment

The lyophilization process generally includes the following steps:

- Dissolving the drug and excipients in a suitable solvent, generally water for injection (WFI).
- Sterilizing the bulk solution by passing it through a 0.22 micron bacteria-retentive filter.
- Filling into individual sterile containers and partially stoppering the containers under aseptic conditions.
- Transporting the partially stoppered containers to the lyophilizer and loading into the chamber under aseptic conditions.
- Freezing the solution by placing the partially stoppered containers on cooled shelves in a freeze-drying chamber or pre-freezing in another chamber.
- Applying a vacuum to the chamber and heating the shelves in order to evaporate the water from the frozen state.
- Complete stoppering of the vials usually by hydraulic or screw rod stoppering mechanisms installed in the lyophilizers.

There are many new parenteral products, including anti-infectives, biotechnology derived products, and in-vitro diagnostics which are manufactured as lyophilized products. Additionally, inspections have disclosed potency, sterility and stability problems associated with the manufacture and control of lyophilized products. In order to provide guidance and information to investigators, some industry procedures and deficiencies associated with lyophilized products are identified in this Inspection Guide.

It is recognized that there is complex technology associated with the manufacture and control of a lyophilized pharmaceutical dosage form. Some of the important aspects of these operations include: the formulation of solutions; filling of vials and validation of the filling operation; sterilization and engineering aspects of the lyophilizer; scale-up and validation of the lyophilization cycle; and testing of the end product. This discussion will address some of the problems associated with the manufacture and control of a lyophilized dosage form.

Product Type/Formulation

Products are manufactured in the lyophilized form due to their instability when in solution. Many of the antibiotics, such as some of the semi-synthetic penicillins, cephalosporins, and also some of the salts of erythromycin, doxycycline and chloramphenicol are made by the lyophilization process. Because they are antibiotics, low bioburden of these formulations would be expected at the time of batching. However, some of the other dosage forms that are lyophilized, such as hydrocortisone sodium succinate, methylprednisolone sodium succinate and many of the biotechnology derived products, have no antibacterial effect when in solution.

For these types of products, bioburden should be minimal and the bioburden should be determined prior to sterilization of these bulk solutions prior to filling. Obviously, the batching or compounding of these bulk solutions should be controlled in order to prevent any potential increase in microbiological levels that may occur up to the time that the bulk solutions are filtered (sterilized). The concern with any microbiological level is the possible increase in endotoxins that may develop. Good practice for the compounding of lyophilized products would also include batching in a controlled environment and in sealed tanks, particularly if the solution is to be held for any length of time prior to sterilization.

In some cases, manufacturers have performed bioburden testing on bulk solutions after pre-filtration and prior to final filtration. While the testing of such solutions may be meaningful in determining the bioburden for sterilization, it does not provide any information regarding the potential formation or presence of endotoxins. While the testing of 0.1 ml samples by LAL methods of bulk solution for endotoxins is of value, testing of at least 100 ml size samples prior to pre-filtration, particularly for the presence of gram negative organisms, would be of greater value in evaluating the process. For example, the presence of Pseudomonas sp. in the bioburden of a bulk solution has been identified as an objectionable condition.

Filling

The filling of vials that are to be lyophilized has some problems that are somewhat unique. The stopper is placed on top of the vial and is ultimately seated in the lyophilizer. As a result the contents of the vial are subject to contamination until they are actually sealed.

Validation of filling operations should include media fills and the sampling of critical surfaces and air during active filling (dynamic conditions).

Because of the active involvement of people in filling and aseptic manipulations, an environmental program should also include an evaluation of microbiological levels on people working in aseptic processing areas. One method of evaluation of the training of operators working in aseptic processing facilities includes the surface monitoring of gloves and/or gowns on a daily basis. Manufacturers are actively sampling the surfaces of personnel working in aseptic processing areas. A reference which provides for this type of monitoring is the USP XXII discussion of the Interpretation of Sterility Test Results. It states under the heading of "Interpretation of Quality Control Tests" that review consideration should be paid to environmental control data, including...microbial monitoring, records of operators, gowns, gloves, and garbing practices. In those situations in which manufacturers have failed to perform some type of personnel monitoring, or monitoring has shown unacceptable levels of contamination, regulatory situations have resulted.

Typically, vials to be lyophilized are partially stoppered by machine. However, some filling lines have been noted which utilize an operator to place each stopper on top of the vial by hand. At this time, it would seem that it would be difficult for a manufacturer to justify a hand-stoppering operation, even if sterile forceps are employed, in any type of operation other than filling a clinical batch or very small number of units. Significant regulatory situations have resulted when some manufacturers have hand-stoppered vials. Again, the concern is the immediate avenue of contamination offered by the operator. It is well recognized that people are the major source of contamination in an aseptic processing filling operation. The longer a person works in an aseptic operation, the more microorganisms will be shed and the greater the probability of contamination.

Once filled and partially stoppered, vials are transported and loaded into the lyophilizer. The transfer and handling, such as loading of the lyophilizer, should take place under primary barriers, such as the laminar flow hoods under which the vials were filled. Validation of this handling should also include the use media fills.

Regarding the filling of sterile media, there are some manufacturers who carry out a partial lyophilization cycle and freeze the media. While this could seem to greater mimic the process, the freezing of media could reduce microbial levels of some contaminants. Since the purpose of the media fill is to evaluate and justify the aseptic capabilities of the process, the people and the system, the possible reduction of microbiological levels after aseptic manipulation by freezing would not be warranted. The purpose of a media fill is not to determine the lethality of freezing and its effect on any microbial contaminants that might be present.

In an effort to identify the particular sections of filling and aseptic manipulation that might introduce contamination, several manufacturers have resorted to expanded media fills. That is, they have filled approximately 9000 vials during a media fill and segmented the fill into three stages. One stage has included filling of 3000 vials and stoppering on line; another stage included filling 3000 vials, transportation to the lyophilizer and then stoppering; a third stage included the filling of 3000 vials, loading in the lyophilizer, and exposure to a portion of the nitrogen flush and then stoppering. Since lyophilizer sterilization and sterilization of the nitrogen system used to backfill require separate validation, media fills should primarily validate the filling, transportation and loading aseptic operations.

The question of the number of units needed for media fills when the capacity of the process is less than 3000 units is frequently asked, particularly for clinical products. Again, the purpose of the media fill is to assure that product can be aseptically processed without contamination under operating conditions. It would seem, therefore, that the maximum number of units of media filled be equivalent to the maximum batch size if it is less than 3000 units.

After filling, dosage units are transported to the lyophilizer by metal trays. Usually, the bottom of the trays are removed after the dosage units are loaded into the lyophilizer. Thus, the dosage units lie directly on the lyophilizer shelf. There have been some situations in which manufacturers have loaded the dosage units on metal trays which were not removed. Unfortunately, at one manufacturer, the trays warped which caused a moisture problem in some dosage units in a batch.

In the transport of vials to the lyophilizer, since they are not sealed, there is concern for the potential for contamination. During inspections and in the review of new facilities, the failure to provide laminar flow coverage or a primary barrier for the transport and loading areas of a lyophilizer has been regarded as an objectionable condition. One manufacturer as a means of correction developed a laminar flow cart to transport the vials from the filling line to the lyophilizer. Other manufacturers building new facilities have located the filling line close to the lyophilizer and have provided a primary barrier extending from the filling line to the lyophilizer.

In order to correct this type of problem, another manufacturer installed a vertical laminar flow hood between the filling line and lyophilizer. Initially, high velocities with inadequate return caused a contamination problem in a media fill. It was speculated that new air currents resulted in rebound contamination off the floor. Fortunately, media fills and smoke studies provided enough meaningful information that the problem could be corrected prior to the manufacture of product. Typically, the lyophilization process includes the stoppering of vials in the chamber.

Another major concern with the filling operation is assurance of fill volumes. Obviously, a low fill would represent a subpotency in the vial. Unlike a powder or liquid fill, a low fill would not be readily apparent after lyophilization particularly for a biopharmaceutical drug product where the active ingredient may be only a milligram. Because of the clinical significance, sub-potency in a vial potentially can be a very serious situation.

For example, in the inspection of a lyophilization filling operation, it was noted that the firm was having a filling problem. The gate on the filling line was not coordinated with the filling syringes, and splashing and partial filling was occurring. It was also observed that some of the partially filled vials were loaded into the lyophilizer. This resulted in rejection of the batch.

On occasion, it has been seen that production operators monitoring fill volumes record these fill volumes only after adjustments are made. Therefore, good practice and a good quality assurance program would include the frequent monitoring of the volume of fill, such as every 15 minutes. Good practice would also include provisions for the isolation of particular sections of filling operations when low or high fills are encountered.

There are some atypical filling operations which have not been discussed. For example, there have also been some situations in which lyophilization is performed on trays of solution rather than in vials. Based on the current technology available, it would seem that for a sterile product, it would be difficult to justify this procedure.

The dual chamber vial also presents additional requirements for aseptic manipulations. Media fills should include the filling of media in both chambers. Also, the diluent in these vials should contain a preservative. (Without a preservative, the filling of diluent would be analogous to the filling of media. In such cases, a 0% level of contamination would be expected.)

Lyophilization Cycle and Controls

After sterilization of the lyophilizer and aseptic loading, the initial step is freezing the solution. In some cycles, the shelves are at the temperature needed for freezing, while for other cycles, the product is loaded and then the shelves are taken to the freezing temperature necessary for product freeze. In those cycles in which the shelves are precooled prior to loading, there is concern for any ice formation on shelves prior to loading. Ice on shelves prior to loading can cause partial or complete stoppering of vials prior to lyophilization of the product. A recent field complaint of a product in solution and not lyophilized was attributed to preliminary stoppering of a few vials prior to exposure to the lyophilization cycle. Unfortunately, the firm's 100% vial inspection failed to identify the defective vial. Typically, the product is frozen at a temperature well below the eutectic point.

The scale-up and change of lyophilization cycles, including the freezing procedures, have presented some problems. Studies have shown the rate and manner of freezing may affect the quality of the lyophilized product. For example, slow freezing leads to the formation of larger ice crystals. This results in relatively large voids, which aid in the escape of water vapor during sublimation. On the other hand, slow freezing can increase concentration shifts of components. Also, the rate and manner of freezing has been shown to have an effect on the physical form (polymorph) of the drug substance.

It is desirable after freezing and during primary drying to hold the drying temperature (in the product) at least 4-5° below the eutectic point. Obviously, the manufacturer should know the eutectic point and have the necessary instrumentation to assure the uniformity of product temperatures. The lyophilizer should also have the necessary instrumentation to control and record the key process parameters. These include: shelf temperature, product temperature, condenser temperature, chamber pressure and condenser pressure. The manufacturing directions should provide for time, temperature and pressure limits necessary for a lyophilization cycle for a product. The monitoring of product temperature is particularly important for those cycles for which there are atypical operating procedures, such as power failures or equipment breakdown.

Electromechanical control of a lyophilization cycle has utilized cam-type recorder-controllers. However, newer units provide for microcomputer control of the freeze drying process. A very basic requirement for a computer controlled process is a flow chart or logic. Typically, operator involvement in a computer controlled lyophilization cycle primarily occurs at the beginning. It consists of loading the chamber, inserting temperature probes in product vials, and entering cycle parameters such as shelf temperature for freezing, product freeze temperature, freezing soak time, primary drying shelf temperature and cabinet pressure, product temperature for establishment of fill vacuum, secondary drying shelf temperature, and secondary drying time.

In some cases, manufacturers have had to continuously make adjustments in cycles as they were being run. In these situations, the lyophilization process was found to be non-validated.

Validation of the software program of a lyophilizer follows the same criteria as that for other processes. Basic concerns include software development, modifications and security. The Guide to Inspection of Computerized Systems in Drug Processing contains a discussion on potential problem areas relating to computer systems. A Guide to the Inspection of Software Development Activities is a reference that provides a more detailed review of software requirements.

Leakage into a lyophilizer may originate from various sources. As in any vacuum chamber, leakage can occur from the atmosphere into the vessel itself. Other sources are media employed within the system to perform the lyophilizing task. These would be the thermal fluid circulated through the shelves for product heating and cooling, the refrigerant employed inside the vapor condenser cooling surface and oil vapors that may migrate back from the vacuum pumping system.

Any one, or a combination of all, can contribute to the leakage of gases and vapors into the system. It is necessary to monitor the leak rate periodically to maintain the integrity of the system. It is also necessary, should the leak rate exceed specified limits, to determine the actual leak site for purposes of repair.

Thus, it would be beneficial to perform a leak test at some time after sterilization, possibly at the beginning of the cycle or prior to stoppering. The time and frequency for performing the leak test will vary and will depend on the data developed during the cycle validation. The pressure rise found acceptable at validation should be used to determine the acceptable pressure rise during production. A limit and what action is to be taken if excessive leakage is found should be addressed in some type of operating document.

In order to minimize oil vapor migration, some lyophilizers are designed with a tortuous path between the vacuum pump and chamber. For example, one fabricator installed an oil trap in the line between the vacuum pump and chamber in a lyophilizer with an internal condenser. Leakage can also be identified by sampling surfaces in the chamber after lyophilization for contaminants. One could conclude that if contamination is found on a chamber surface after lyophilization, then dosage units in the chamber could also be contaminated. It is a good practice as part of the validation of cleaning of the lyophilization chamber to sample the surfaces both before and after cleaning.

Because of the lengthy cycle runs and strain on machinery, it is not unusual to see equipment malfunction or fail during a lyophilization cycle. There should be provisions in place for the corrective action to be taken when these atypical situations occur. In addition to documentation of the malfunction, there should be an evaluation of the possible effects on the product (e.g., partial or complete meltback. Refer to subsequent discussion). Merely testing samples after the lyophilization cycle is concluded may be insufficient to justify the release of the remaining units. For example, the leakage of chamber shelf fluid into the chamber or a break in sterility would be cause for rejection of the batch.

The review of Preventive Maintenance Logs, as well as Quality Assurance Alert Notices, Discrepancy Reports, and Investigation Reports will help to identify problem batches when there are equipment malfunctions or power failures. It is recommended that these records be reviewed early in the inspection.

Cycle Validation

Many manufacturers file (in applications) their normal lyophilization cycles and validate the lyophilization process based on these cycles. Unfortunately, such data would be of little value to substantiate shorter or abnormal cycles. In some cases, manufacturers are unaware of the eutectic point. It would be difficult for a manufacturer to evaluate partial or abnormal cycles without knowing the eutectic point and the cycle parameters needed to facilitate primary drying.

Scale-up for the lyophilized product requires a knowledge of the many variables that may have an effect on the product. Some of the variables would include freezing rate and temperature ramping rate. As with the scale-up of other drug products, there should be a development report that discusses the process and logic for the cycle. Probably more so than any other product, scale-up of the lyophilization cycle is very difficult.

There are some manufacturers that market multiple strengths, vial sizes and have different batch sizes. It is conceivable and probable that each will have its own cycle parameters. A manufacturer that has one cycle for multiple strengths of the same product probably has done a poor job of developing the cycle and probably has not adequately validated their process. Investigators should review the reports and data that support the filed lyophilization cycle.

Lyophilizer Sterilization/Design

The sterilization of the lyophilizer is one of the more frequently encountered problems noted during inspections. Some of the older lyophilizers cannot tolerate steam under pressure, and sterilization is marginal at best. These lyophilizers can only have their inside surfaces wiped with a chemical agent that may be a sterilant but usually has been found to be a sanitizing agent. Unfortunately, piping such as that for the administration of inert gas (usually nitrogen) and sterile air for backfill or vacuum break is often inaccessible to such surface "sterilization" or treatment. It would seem very difficult for a manufacturer to be able to demonstrate satisfactory validation of sterilization of a lyophilizer by chemical "treatment".

Another method of sterilization that has been practiced is the use of gaseous ethylene oxide. As with any ethylene oxide treatment, humidification is necessary. Providing a method for introducing the sterile moisture with uniformity has been found to be difficult.

A manufacturer has been observed employing Water For Injection as a final wash or rinse of the lyophilizer. While the chamber was wet, it was then ethylene oxide gas sterilized. As discussed above, this may be satisfactory for the chamber but inadequate for associated plumbing.

Another problem associated with ethylene oxide is the residue. One manufacturer had a common ethylene oxide/nitrogen supply line to a number of lyophilizers connected in parallel to the system. Thus, there could be some ethylene oxide in the nitrogen supply line during the backfilling step. Obviously, this type of system is objectionable.

A generally recognized acceptable method of sterilizing the lyophilizer is through the use of moist steam under pressure. Sterilization procedures should parallel that of an autoclave, and a typical system should include two independent temperature sensing systems. One would be used to control and record temperatures of the cycle as with sterilizers, and the other would be in the cold spot of the chamber. As with autoclaves, lyophilizers should have drains with atmospheric breaks to prevent back siphonage.

As discussed, there should also be provisions for sterilizing the inert gas or air and the supply lines. Some manufacturers have chosen to locate the sterilizing filters in a port of the chamber. The port is steam sterilized when the chamber is sterilized, and then the sterilizing filter, previously sterilized, is aseptically connected to the chamber. Some manufacturers have chosen to sterilize the filter and downstream piping to the chamber in place. Typical sterilization-in-place of filters may require steaming of both to obtain sufficient temperatures. In this type of system, there should be provisions for removing and/or draining condensate. The failure to sterilize nitrogen and air filters and the piping downstream leading into the chamber has been identified as a problem on a number of inspections.

Since these filters are used to sterilize inert gas and/or air, there should be some assurance of their integrity. Some inspections have disclosed a lack of integrity testing of the inert gas and/or air filter. The question is frequently asked how often should the vent filter be tested for integrity? As with many decisions made by manufacturers, there is a level of risk associated with the operation, process or system, which only the manufacturer can decide. If the sterilizing filter is found to pass the integrity test after several uses or batches, then one could claim its integrity for the previous batches. However, if it is only tested after several batches have been processed and if found to fail the integrity test, then one could question the sterility of all of the previous batches processed. In an effort to minimize this risk, some manufacturers have resorted to redundant filtration.

For most cycles, stoppering occurs within the lyophilizer. Typically, the lyophilizer has some type of rod or rods (ram) which enter the immediate chamber at the time of stoppering. Once the rod enters the chamber, there is the potential for contamination of the chamber. However, since the vials are stoppered, there is no avenue for contamination of the vials in the chamber which are now stoppered. Generally, lyophilizers should be sterilized after each cycle because of the potential for contamination of the shelf support rods. Additionally, the physical act of removing vials and cleaning the chamber can increase levels of contamination.

In some of the larger units, the shelves are collapsed after sterilization to facilitate loading. Obviously, the portions of the ram entering the chamber to collapse the shelves enters from a non-sterile area. Attempts to minimize contamination have included wiping the ram with a sanitizing agent prior to loading. Control aspects have included testing the ram for microbiological contamination, testing it for residues of hydraulic fluid, and testing the fluid for its bacteriostatic effectiveness. One lyophilizer fabricator has proposed developing a flexible "skirt" to cover the ram.

In addition to microbiological concerns with hydraulic fluid, there is also the concern with product contamination.

During steam sterilization of the chamber, there should be space between shelves that permit passage of free flowing steam. Some manufacturers have placed "spacers" between shelves to prevent their total collapse. Others have resorted to a two phase sterilization of the chamber. The initial phase provides for sterilization of the shelves when they are separated. The second phase provides for sterilization of the chamber and piston with the shelves collapsed.

Typically, biological indicators are used in lyophilizers to validate the steam sterilization cycle. One manufacturer of a Biopharmaceutical product was found to have a positive biological indicator after sterilization at 121°C for 45 minutes. During the chamber sterilization, trays used to transport vials from the filling line to the chamber were also sterilized. The trays were sterilized in an inverted position on shelves in the chamber. It is believed that the positive biological indicator is the result of poor steam penetration under these trays.

The sterilization of condensers is also a major issue that warrants discussion. Most of the newer units provide for the capability of sterilization of the condenser along with the chamber, even if the condenser is external to the chamber. This provides a greater assurance of sterility, particularly in those situations in which there is some equipment malfunction and the vacuum in the chamber is deeper than in the condenser.

Malfunctions that can occur, which would indicate that sterilization of the condenser is warranted, include vacuum pump breakdown, refrigeration system failures and the potential for contamination by the large valve between the condenser and chamber. This is particularly true for those units that have separate vacuum pumps for both the condenser and chamber. When there are problems with the systems in the lyophilizer, contamination could migrate from the condenser back to the chamber. It is recognized that the condenser is not able to be sterilized in many of the older units, and this represents a major problem, particularly in those cycles in which there is some equipment and/or operator failure.

As referenced above, leakage during a lyophilization cycle can occur, and the door seal or gasket presents an avenue of entry for contaminants. For example, in an inspection, it was noted that during steam sterilization of a lyophilizer, steam was leaking from the unit. If steam could leak from a unit during sterilization, air could possibly enter the chamber during lyophilization.

Some of the newer lyophilizers have double doors - one for loading and the other for unloading. The typical single door lyophilizer opens in the clean area only, and contamination between loads would be minimal. This clean area, previously discussed, represents a critical processing area for a product made by aseptic processing. In most units, only the piston raising/lowering shelves is the source of contamination. For a double door system unloading the lyophilizer in a non-sterile environment, other problems may occur. The non-sterile environment presents a direct avenue of contamination of the chamber when unloading, and door controls similar to double door sterilizers should be in place.

Obviously, the lyophilizer chamber is to be sterilized between batches because of the direct means of contamination. A problem which may be significant is that of leakage through the door seal. For the single door unit, leakage prior to stoppering around the door seal is not a major problem from a sterility concern, because single door units only open into sterile areas. However, leakage from a door gasket or seal from a non-sterile area would present a significant microbiological problem. In order to minimize the potential for contamination, it is recommended that the lyophilizers be unloaded in a clean room area to minimize contamination. For example, in an inspection of a new manufacturing facility, it was noted that the unloading area for double door units was a clean room, with the condenser located below the chamber on a lower level.

After steam sterilization, there is often some condensate remaining on the floor of the chamber. Some manufacturers remove this condensate through the drain line while the chamber is still pressurized after sterilization. Unfortunately, some manufacturers have allowed the chamber to come to and remain at atmospheric pressure with the drain line open. Thus, non-sterile air could contaminate the chamber through the drain line. Some manufacturers have attempted to dry the chamber by blowing sterile nitrogen gas through the chamber at a pressure above atmospheric pressure.

In an inspection of a biopharmaceutical drug product, a Pseudomonas problem probably attributed to condensate after sterilization was noted. On a routine surface sample taken from a chamber shelf after sterilization and processing, a high count of Pseudomonas sp. was obtained. After sterilization and cooling when the chamber door was opened, condensate routinely spilled onto the floor from the door. A surface sample taken from the floor below the door also revealed Pseudomonas sp. contamination. Since the company believed the condensate remained in the chamber after sterilization, they repiped the chamber drain and added a line to a water seal vacuum pump.

Finished Product Testing For

Lyophilized Products

There are several aspects of finished product testing which are of concern to the lyophilized dosage form. These include dose uniformity testing, moisture and stability testing, and sterility testing.

(a) Dose Uniformity

The USP includes two types of dose uniformity testing: content uniformity and weight variation. It states that weight variation may be applied to solids, with or without added substances that have been prepared from true solutions and freeze-dried in final containers. However, when other excipients or other additives are present, weight variation may be applied, provided there is correlation with the sample weight and potency results. For example, in the determination of potency, it is sometimes common to reconstitute and assay the entire contents of a vial without knowing the weight of the sample. Performing the assay in this manner will provide information on the label claim of a product, but without knowing the sample weight will provide no information about dose uniformity. One should correlate the potency result obtained form the assay with the weight of the sample tested.

(b) Stability Testing

An obvious concern with the lyophilized product is the amount of moisture present in vials. The manufacturer's data for the establishment of moisture specifications for both product release and stability should be reviewed. As with other dosage forms, the expiration date and moisture limit should be established based on worst case data. That is, a manufacturer should have data that demonstrates adequate stability at the moisture specification.

As with immediate release potency testing, stability testing should be performed on vials with a known weight of sample. For example, testing a vial (sample) which had a higher fill weight (volume) than the average fill volume of the batch would provide a higher potency results and not represent the potency of the batch. Also, the expiration date and stability should be based on those batches with the higher moisture content. Such data should also be considered in the establishment of a moisture specification.

For products showing a loss of potency due to aging, there are generally two potency specifications. There is a higher limit for the dosage form at the time of release. This limit is generally higher than the official USP or filed specification which is official throughout the entire expiration date period of the dosage form. The USP points out that compendial standards apply at any time in the life of the article.

Stability testing should also include provisions for the assay of aged samples and subsequent reconstitution of these aged samples for the maximum amount of time specified in the labeling. On some occasions, manufacturers have established expiration dates without performing label claim reconstitution potency assays at the various test intervals and particularly the expiration date test interval. Additionally, this stability testing of reconstituted solutions should include the most concentrated and the least concentrated reconstituted solutions. The most concentrated reconstituted solution will usually exhibit degradation at a faster rate than less concentrated solutions.

(c) Sterility Testing

With respect to sterility testing of lyophilized products, there is concern with the solution used to reconstitute the lyophilized product. Although products may be labeled for reconstitution with Bacteriostatic Water For Injection, Sterile Water For Injection (WFI) should be used to reconstitute products. Because of the potential toxicities associated with Bacteriostatic Water For Injection, many hospitals only utilize WFI. Bacteriostatic Water For Injection may kill some of the vegetative cells if present as contaminants, and thus mask the true level of contamination in the dosage form.

As with other sterile products, sterility test results which show contamination on the initial test should be identified and reviewed.

Finished Product Inspection - Meltback

The USP points out that it is good pharmaceutical practice to perform 100% inspection of parenteral products. This includes sterile lyophilized powders. Critical aspects would include the presence of correct volume of cake and the cake appearance. With regard to cake appearance, one of the major concerns is meltback.

Meltback is a form of cake collapse and is caused by the change from the solid to liquid state. That is, there is incomplete sublimation (change from the solid to vapor state) in the vial. Associated with this problem is a change in the physical form of the drug substance and/or a pocket of moisture. These may result in greater instability and increased product degradation.

Another problem may be poor solubility. Increased time for reconstitution at the user stage may result in partial loss of potency if the drug is not completely dissolved, since it is common to use in-line filters during administration to the patient.

Manufacturers should be aware of the stability of lyophilized products which exhibit partial or complete meltback. Literature shows that for some products, such as the cephalosporins, that the crystalline form is more stable than the amorphous form of lyophilized product. The amorphous form may exist in the "meltback" portion of the cake where there is incomplete sublimation

Glossary

Atmosphere, The Earth – The envelope of gases surrounding the earth, exerting under gravity a pressure at the earth's surface, which includes by volume 78% nitrogen, 21% oxygen, small quantities of hydrogen, carbon dioxide, noble gases, water vapor, pollutants and dust.

Atmospheric Pressure – The pressure exerted at the earth's surface by the atmosphere. For reference purposes a standard atmosphere is defined as 760 torr or millimeters of mercury, or 760,000 microns.

Backstreaming – A process that occurs at low chamber pressures where hydrocarbon vapors from the vacuum system can enter the product chamber.

Blank-Off Pressure – This is the ultimate pressure the pump or system can attain.

Blower (see Mechanical Booster Pump) – This pump is positioned between the mechanical pump and the chamber. It operates by means of two lobes turning at a high rate of speed. It is used to reduce the chamber pressure to less than 20 microns.

Breaking Vacuum – Admitting air or a selected gas to an evacuated chamber, while isolated from a vacuum pump, to raise the pressure towards, or up to, atmospheric.

Circulation Pump – A pump for conveying the heat transfer fluid.

CONDENSER (Cold Trap) – In terms of the lyophilization process, this is the vessel that collects the moisture on plates and holds it in the frozen state. Protects the vacuum pump from water vapor contaminating the vacuum pump oil.

Condenser/Receiver – In terms of refrigeration, this unit condenses (changes) the hot refrigerant gas into a liquid and stores it under pressure to be reused by the system.

Cooling – The lowering of the temperature in any part of the temperature scale.

Conax Connection – A device to pass thermocouple wires through and maintain a vacuum tight vessel.

Contamination – In the vacuum system, the introduction of water vapor into the oil in the vacuum pump, which then causes the pump to lose its ability to attain its ultimate pressure.

Defrosting – The removal of ice from a condenser by melting or mechanical means.

Degree Of Crystallization – The ratio of the energy released during the freezing of a solution to that of an equal volume of water.

Degree Of Supercooling – The number of degrees below the equilibrium freezing temperature where ice first starts to form.

Desiccant – A drying agent.

Dry – Free from liquid, and/or moisture.

Drying – The removal of moisture and other liquids by evaporation.

Equilibrium Freezing Temperatures – The temperature where ice will form in the absence of super-cooling.

Eutectic Temperature – A point of a phase diagram where all phases are present and the temperature and composition of the liquid phase cannot be altered without one of the phases disappearing.

Expansion Tank – This tank is located in the circulation system and is used as a holding and expansion tank for the transfer liquid.

Filter Or Filter/Drier – There are two systems that have their systems filtered or filter/dried. They are the circulation and refrigeration systems. In the newer dryers this filter or filter/dryer is the same, and can be replaced with a new core.

Free Water – The free water in a product is that water that is absorbed on the surfaces of the product and must be removed to limit further biological and chemical reactions.

Freezing – This is the absence of heat. A controlled change of the product temperature as a function of time, during the freezing process, so as to ensure a completely frozen form.

Gas Ballast – Used in the vacuum system on the vacuum pump to decontaminate small amounts of moisture in the vacuum pump oil.

GAS BLEED (Vacuum Control) – To control the pressure in the chamber during the cycle to help the drying process. In freeze-drying the purpose is to improve heat-transfer to the product.

Heat Exchanger – This exchanger is located in the circulation and refrigeration systems and transfers the heat from the circulation system to the refrigeration system.

Heat Transfer Fluid – A liquid of suitable vapor pressure and viscosity range for transferring heat to or from a component, for example, a shelf or condenser in a freeze-dryer. The choice of such a fluid may depend on safety considerations. Diathermic fluid.

Hot Gas Bypass – This is a refrigeration system. To control the suction pressure of the BIG FOUR (20–30 Hp) compressors during the refrigeration operation.

Hot Gas Defrost – This is a refrigeration system. To defrost the condenser plates after the lyophilization cycle is complete.

Ice – The solid, crystalline form of water.

Inert Gas – Any gas of a group including helium, radon and nitrogen, formerly considered chemically inactive.

Interstage – In a two stage compressor system, this is the cross over piping on top of the compressor that connects the low side to the high side. One could also think of it as low side, intermediate, and high side.

Interstage Pressure Regulating Valve – This valve controls the inter-stage pressure from exceeding 80–90 PSI. This valve opens to suction as the inter-stage pressure rises above 80–90 PSI.

Lexsol – A heat transfer fluid (high grade kerosene).

LIQUID SUB-COOLER HEAT EXCHANGER (See Sub-Cooled Liquid) – The liquid refrigerant leaving the condenser/receiver at cooling water temperature is sub-cooled to a temperature of +15°F (-10°C) to -15°F (-25°C).

Lyophilization – A process in which the product is first frozen and then, while still in the frozen state, the major portion of the water and solvent system is reduced by sublimation and desorption so as to limit biological and chemical reactions at the designated storage temperature

MAIN VACUUM VALVE (See Vapor Valve) – This valve is between the chamber and external condenser to isolate the two vessels after the process is finished. This is the valve that protects the finished product.

Matrix – A matrix, in terms of the lyophilization process, is a system of ice crystals and solids that is distributed throughout the product.

MECHANICAL BOOSTER PUMP (See Blower) – A roots pump with a high displacement for its size but a low compression ratio. When backed by an oil-seal rotary pump the combination is an economical alternative to a two-stage oil-sealed rotary pump, with the advantage of obtaining a high vacuum.

Mechanical Vacuum Pump – The mechanical pumping system that lowers the pressure in the chamber to below atmospheric pressure so that sublimation can occur.

MELTING TEMPERATURE (Melt-Back) – That temperature where mobile water first becomes evident in a frozen system.

MICRON (See Torr) – A unit of pressure used in the lyophilization process. One micron = one Mtorr or 25,400 microns = 1" Hg., or 760,000 microns = one atmosphere.

Noncondensables – A mixture of gases such as nitrogen, hydrogen, chlorine, and hydrocarbons. They may be drawn into the system through leaks when part of the system is under a vacuum. Their presence reduces the operating efficiency of the system by increasing the condensing pressure.

Nucleation – The formation of ice crystals on foreign surfaces or as a result of the growth of water clusters.

Oil-Mist Filter – In vacuum terminology a filter attached to the discharge (exhaust) of an oil-sealed rotary pump to eliminate most of the "smoke" of suspended fine droplets of oil which would be discharged into the environment.

Oil Sealed Rotary Pump –A standard type of mechanical vacuum pump used in freeze-drying with a high compression ratio but having a relatively low displacement (speed) for its size. A two-stage pump is effectively two such pumps in series and can obtain an ultimate vacuum.

Oil Separator – Separates the oil from the compressor discharge gas and returns the oil through the oil float trap and piping to the compressor crankcase.

Real Leak – A real leak is a source of atmospheric gases resulting from a penetration through the chamber.

Reconstitute – The dissolving of the dried product into a solvent or diluent.

Relief Valve – Used for safety purposes to prevent damage in case excessive pressure is encountered.

Rotary Vane Pump – A mechanical pumping system with sliding vanes as the mechanical seal. Can be single or two stages.

Shelf Compressor (Controlling Compressor) – Used for controlling the shelf temperature, either cooling or from overheating.

Self Liquid Heat Exchanger – The transfer of heat from the shelf fluid to the refrigeration system through tubes in the exchanger causing compressor suction gas to warm.

Shelves – In terms of the lyophilization process, they are a form of heat exchanger, within the chamber, that have a serpentine liquid flow through them, entering one side and flowing to the other side. They are located in the circulation system.

Single Stage Compressor – This is a normal type compressor used in refrigeration. In the lyophilization process it is used to control the shelf temperature, both for cooling and keeping the shelf temperature from overheating using a temperature controller.

Silicone Oil – A heat transfer fluid.

Sterilization – The use of steam and pressure to kill any bacteria that may be able to contaminate that environment or vessel.

Sublimation – The conversion of a material from a solid phase directly to a vapor phase, without passing through the liquid phase. This is referred to as the primary drying stage.

SUB-COOLED LIQUID (See Liquid Sub-Cooler Heat Exchanger) – The liquid refrigerant is cooled through an exchanger so that it increases the refrigerating effect as well as reduces the volume of gas flashed from the liquid refrigerant in passage through the expansion valve.

Suction Line Accumulator – To provide adequate refrigerant liquid slug protection (droplets of liquid refrigerant) from returning to the compressor, and causing damage to the compressor.

Tce – Trichloroethylene - A heat transfer fluid.

Temperature – The degree of hotness or coldness of a body.

Thermocouple – A metal-to-metal contact between two dissimilar metals that produces a small voltage across the free ends of the wire.

Thermostatic Expansion Valve – An automatic variable device controlling the flow of liquid refrigerant.

Torr (See Micron) – A unit of measure equivalent to the amount of pressure in 1000 microns.

TWO STAGE COMPRESSOR (See Interstage) – This is a specially built compressor. Its function is to be able to attain low temperatures by being able to operate at low pressures. It is two compressors built into one. A low stage connected internally and a high stage connected externally with piping, called interstage.

Unloading Valve – This valve connects the interstage with suction to equalize both pressures during pump-down.

Vacuum – Strictly speaking, a space in which the total pressure is less than atmospheric.

Vacuum Control (Gas Bleed) – To assist in the rate of sublimation, by controlling the pressure in the lyophilizer.

Vacuum Pump – A mechanical way of reducing the pressure in a vessel below atmospheric pressure to where sublimation can occur. There are three types of pumps, rotary vane, rotary piston and mechanical booster.

Vapor Baffle – A target shaped object placed in the condenser to direct vapor flow and to promote an even distribution of condensate.

Vacuum Valves – The vacuum valves used are of a ball or disk type that can seal without leaking. The ball types are used for services to the chamber and condenser. They are also used for drains and isolation applications. The disk types are used in the vacuum line system and are connected to the vacuum pump, chamber and condenser.

Vapor Valve (See Main Vacuum Valve) – The vacuum valve between the chamber and external condenser. When this valve is closed the chamber is isolated from the external condenser. Also known as the main vapor valve.

Vial – A small glass bottle with a flat bottom, short neck and flat flange designed for stoppering.

Virtual Leak – In the vacuum system a virtual leak is the passage of gas into the chamber from a source that is located internally in the chamber.

Appendix E

Guide to Inspections and Validation of Cleaning Processes

The following is a US Food and Drug Administration (FDA) training guide used by the FDA in the training of their inspectors. Such training guides are used to prepare and provide FDA inspectors with a general knowledge and understanding of what it is they should be considering during inspection of an FDA-regulated manufacturing facility.

In many cases these are general considerations due in large part to the many and varied drugs and the proprietary processes required to manufacture them. These guides are intended to augment the education, experience, and accumulated knowledge each inspector already possesses. The inspection process and procedures by which they perform it leave room for on-site judgment consideration and assessment by the inspectors.

The purpose of including these FDA guides in this book is to provide the reader, designer, and engineer involved with the design and engineering of a bioprocessing facility some insight as to what the expectations of an FDA inspector might be. What expectations an FDA inspector might possess can most likely be attributed to their training and these guidelines. Providing these guidelines for the readers of this book, those of you designing, constructing, and operating FDA-regulated manufacturing facilities, will hopefully lessen the gap of understanding between you and the FDA inspector. Having some insight into the inspector's understanding and training helps one to be better prepared to meet those expectations when it's time for the inspection.

These inspection guidelines are reproduced here through the courtesy of the FDA. These and others also reside on the FDA web site at:

http://www.fda.gov/ICECI/Inspections/InspectionGuides/default.htm

---

Bioprocessing Piping and Equipment Design: A Companion Guide for the ASME BPE Standard, First Edition. William M. (Bill) Huitt.

Guide to Inspections and Validation of Cleaning Processes

Note: This document is reference material for investigators and other FDA personnel. The document does not bind FDA, and does no confer any rights, privileges, benefits, or immunities for or on any person(s).

I. Introduction

Validation of cleaning procedures has generated considerable discussion since agency documents, including the Inspection Guide for Bulk Pharmaceutical Chemicals and the Biotechnology Inspection Guide, have briefly addressed this issue. These Agency documents clearly establish the expectation that cleaning procedures (processes) be validated.

This guide is designed to establish inspection consistency and uniformity by discussing practices that have been found acceptable (or unacceptable). Simultaneously, one must recognize that for cleaning validation, as with validation of other processes, there may be more than one way to validate a process. In the end, the test of any validation process is whether scientific data shows that the system consistently does as expected and produces a result that consistently meets predetermined specifications.

This guide is intended to cover equipment cleaning for chemical residues only.

II. Background

For FDA to require that equipment be clean prior to use is nothing new, the 1963 GMP Regulations (Part 133.4) stated as follows "Equipment *** shall be maintained in a clean and orderly manner ***." A very similar section on equipment cleaning (211.67) was included in the 1978 CGMP regulations. Of course, the main rationale for requiring clean equipment is to prevent contamination or adulteration of drug products. Historically, FDA investigators have looked for gross insanitation due to inadequate cleaning and maintenance of equipment and/or poor dust control systems. Also, historically speaking, FDA was more concerned about the contamination of nonpenicillin drug products with penicillins or the cross-contamination of drug products with potent steroids or hormones. A number of products have been recalled over the past decade due to actual or potential penicillin cross-contamination.

One event which increased FDA awareness of the potential for cross contamination due to inadequate procedures was the 1988 recall of a finished drug product, Cholestyramine Resin USP. The bulk pharmaceutical chemical used to produce the product had become contaminated with low levels of intermediates and degradants from the production of agricultural pesticides. The cross-contamination in that case is believed to have been due to the reuse of recovered solvents. The recovered solvents had been contaminated because of a lack of control over the reuse of solvent drums. Drums that had been used to store recovered solvents from a pesticide production process were later used to store recovered solvents used for the resin manufacturing process. The firm did not have adequate controls over these solvent drums, did not do adequate testing of drummed solvents, and did not have validated cleaning procedures for the drums.

Some shipments of this pesticide contaminated bulk pharmaceutical were supplied to a second facility at a different location for finishing. This resulted in the contamination of the bags used in that facility's fluid bed dryers with pesticide contamination. This in turn led to cross contamination of lots produced at that site, a site where no pesticides were normally produced.

FDA instituted an import alert in 1992 on a foreign bulk pharmaceutical manufacturer which manufactured potent steroid products as well as non-steroidal products using common equipment. This firm was a multi-use bulk pharmaceutical facility. FDA considered the potential for cross-contamination to be significant and to pose a serious health risk to the public. The firm had only recently started a cleaning validation program at the time of the inspection and it was considered inadequate by FDA. One of the reasons it was considered inadequate was that the firm was only looking for evidence of the absence of the previous compound. The firm had evidence, from TLC tests on the rinse water, of the presence of residues of reaction byproducts and degradants from the previous process.

III. General Requirements

FDA expects firms to have written procedures (SOP's) detailing the cleaning processes used for various pieces of equipment. If firms have one cleaning process for cleaning between different batches of the same product and use a different process for cleaning between product changes, we expect the written procedures to address these different scenario. Similarly, if firms have one process for removing water soluble residues and another process for non-water soluble residues, the written procedure should address both scenarios and make it clear when a given procedure is to be followed. Bulk pharmaceutical firms may decide to dedicate certain equipment for certain chemical manufacturing process steps that produce tarry or gummy residues that are difficult to remove from the equipment. Fluid bed dryer bags are another example of equipment that is difficult to clean and is often dedicated to a specific product. Any residues from the cleaning process itself (detergents, solvents, etc.) also have to be removed from the equipment.

FDA expects firms to have written general procedures on how cleaning processes will be validated.

FDA expects the general validation procedures to address who is responsible for performing and approving the validation study, the acceptance criteria, and when revalidation will be required.

FDA expects firms to prepare specific written validation protocols in advance for the studies to be performed on each manufacturing system or piece of equipment which should address such issues as sampling procedures, and analytical methods to be used including the sensitivity of those methods.

FDA expects firms to conduct the validation studies in accordance with the protocols and to document the results of studies.

FDA expects a final validation report which is approved by management and which states whether or not the cleaning process is valid. The data should support a conclusion that residues have been reduced to an "acceptable level."

IV. Evaluation of Cleaning Validation

The first step is to focus on the objective of the validation process, and we have seen that some companies have failed to develop such objectives. It is not unusual to see manufacturers use extensive sampling and testing programs following the cleaning process without ever really evaluating the effectiveness of the steps used to clean the equipment. Several questions need to be addressed when evaluating the cleaning process. For example, at what point does a piece of equipment or system become clean? Does it have to be scrubbed by hand? What is accomplished by hand scrubbing rather than just a solvent wash? How variable are manual cleaning processes from batch to batch and product to product? The answers to these questions are obviously important to the inspection and evaluation of the cleaning process since one must determine the overall effectiveness of the process. Answers to these questions may also identify steps that can be eliminated for more effective measures and result in resource savings for the company.

Determine the number of cleaning processes for each piece of equipment. Ideally, a piece of equipment or system will have one process for cleaning, however this will depend on the products being produced and whether the cleanup occurs between batches of the same product (as in a large campaign) or between batches of different products. When the cleaning process is used only between batches of the same product (or different lots of the same intermediate in a bulk process) the firm need only meet a criteria of, "visibly clean" for the equipment. Such between batch cleaning processes do not require validation.

1. Equipment Design

Examine the design of equipment, particularly in those large systems that may employ semi-automatic or fully automatic clean-in-place (CIP) systems since they represent significant concern. For example, sanitary type piping without ball valves should be used. When such non-sanitary ball valves are used, as is common in the bulk drug industry, the cleaning process is more difficult.

When such systems are identified, it is important that operators performing cleaning operations be aware of problems and have special training in cleaning these systems and valves. Determine whether the cleaning operators have knowledge of these systems and the level of training and experience in cleaning these systems. Also check the written and validated cleaning process to determine if these systems have been properly identified and validated.

In larger systems, such as those employing long transfer lines or piping, check the flow charts and piping diagrams for the identification of valves and written cleaning procedures. Piping and valves should be tagged and easily identifiable by the operator performing the cleaning function. Sometimes, inadequately identified valves, both on prints and physically, have led to incorrect cleaning practices.

Always check for the presence of an often critical element in the documentation of the cleaning processes; identifying and controlling the length of time between the end of processing and each cleaning step. This is especially important for topicals, suspensions, and bulk drug operations. In such operations, the drying of residues will directly affect the efficiency of a cleaning process.

Whether or not CIP systems are used for cleaning of processing equipment, microbiological aspects of equipment cleaning should be considered. This consists largely of preventive measures rather than removal of contamination once it has occurred. There should be some evidence that routine cleaning and storage of equipment does not allow microbial proliferation. For example, equipment should be dried before storage, and under no circumstances should stagnant water be allowed to remain in equipment subsequent to cleaning operations.

Subsequent to the cleaning process, equipment may be subjected to sterilization or sanitization procedures where such equipment is used for sterile processing, or for nonsterile processing where the products may support microbial growth. While such sterilization or sanitization procedures are beyond the scope of this guide, it is important to note that control of the bioburden through adequate cleaning and storage of equipment is important to ensure that subsequent sterilization or sanitization procedures achieve the necessary assurance of sterility. This is also particularly important from the standpoint of the control of pyrogens in sterile processing since equipment sterilization processes may not be adequate to achieve significant inactivation or removal of pyrogens.

2. *Cleaning Process Written*

Procedure and Documentation

Examine the detail and specificity of the procedure for the (cleaning) process being validated, and the amount of documentation required. We have seen general SOPs, while others use a batch record or log sheet system that requires some type of specific documentation for performing each step. Depending upon the complexity of the system and cleaning process and the ability and training of operators, the amount of documentation necessary for executing various cleaning steps or procedures will vary.

When more complex cleaning procedures are required, it is important to document the critical cleaning steps (for example certain bulk drug synthesis processes). In this regard, specific documentation on the equipment itself which includes information about who cleaned it and when is valuable. However, for relatively simple cleaning operations, the mere documentation that the overall cleaning process was performed might be sufficient.

Other factors such as history of cleaning, residue levels found after cleaning, and variability of test results may also dictate the amount of documentation required. For example, when variable residue levels are detected following cleaning, particularly for a process that is believed to be acceptable, one must establish the effectiveness of the process and operator performance. Appropriate evaluations must be made and when operator performance is deemed a problem, more extensive documentation (guidance) and training may be required.

3. Analytical Methods

Determine the specificity and sensitivity of the analytical method used to detect residuals or contaminants. With advances in analytical technology, residues from the manufacturing and cleaning processes can be detected at very low levels. If levels of contamination or residual are not detected, it does not mean that there is no residual contaminant present after cleaning. It only means that levels of contaminant greater than the sensitivity or detection limit of the analytical method are not present in the sample. The firm should challenge the analytical method in combination with the sampling method(s) used to show that contaminants can be recovered from the equipment surface and at what level, i.e. 50% recovery, 90%, etc. This is necessary before any conclusions can be made based on the sample results. A negative test may also be the result of poor sampling technique (see below).

4. Sampling

There are two general types of sampling that have been found acceptable. The most desirable is the direct method of sampling the surface of the equipment. Another method is the use of rinse solutions.

(a) Direct Surface Sampling
Determine the type of sampling material used and its impact on the test data since the sampling material may interfere with the test. For example, the adhesive used in swabs has been found to interfere with the analysis of samples. Therefore, early in the validation program, it is important to assure that the sampling medium and solvent (used for extraction from the medium) are satisfactory and can be readily used.
Advantages of direct sampling are that areas hardest to clean and which are reasonably accessible can be evaluated, leading to establishing a level of contamination or residue per given surface area. Additionally, residues that are "dried out" or are insoluble can be sampled by physical removal.

(b) Rinse Samples
Two advantages of using rinse samples are that a larger surface area may be sampled, and inaccessible systems or ones that cannot be routinely disassembled can be sampled and evaluated.

A disadvantage of rinse samples is that the residue or contaminant may not be soluble or may be physically occluded in the equipment. An analogy that can be used is the "dirty pot." In the evaluation of cleaning of a dirty pot, particularly with dried out residue, one does not look at the rinse water to see that it is clean; one looks at the pot.

Check to see that a direct measurement of the residue or contaminant has been made for the rinse water when it is used to validate the cleaning process. For example, it is not acceptable to simply test rinse water for water quality (does it meet the compendia tests) rather than test it for potential contaminates.

(c) Routine Production In-Process Control

Monitoring - Indirect testing, such as conductivity testing, may be of some value for routine monitoring once a cleaning process has been validated. This would be particularly true for the bulk drug substance manufacturer where reactors and centrifuges and piping between such large equipment can be sampled only using rinse solution samples. Any indirect test method must have been shown to correlate with the condition of the equipment. During validation, the firm should document that testing the uncleaned equipment gives a not acceptable result for the indirect test.

V. Establishment of Limits

FDA does not intend to set acceptance specifications or methods for determining whether a cleaning process is validated. It is impractical for FDA to do so due to the wide variation in equipment and products used throughout the bulk and finished dosage form industries. The firm's rationale for the residue limits established should be logical based on the manufacturer's knowledge of the materials involved and be practical, achievable, and verifiable. It is important to define the sensitivity of the analytical methods in order to set reasonable limits. Some limits that have been mentioned by industry representatives in the literature or in presentations include analytical detection levels such as 10 PPM, biological activity levels such as 1/1000 of the normal therapeutic dose, and organoleptic levels such as no visible residue.

Check the manner in which limits are established. Unlike finished pharmaceuticals where the chemical identity of residuals are known (i.e., from actives, inactives, detergents) bulk processes may have partial reactants and unwanted by-products which may never have been chemically identified. In establishing residual limits, it may not be adequate to focus only on the principal reactant since other chemical variations may be more difficult to remove. There are circumstances where TLC screening, in addition to chemical analyses, may be needed. In a bulk process, particularly for very potent chemicals such as some steroids, the issue of by-products needs to be considered if equipment is not dedicated. The objective of the inspection is to ensure that the basis for any limits is scientifically justifiable.

VI. Other Issues

a. Placebo Product

In order to evaluate and validate cleaning processes some manufacturers have processed a placebo batch in the equipment under essentially the same operating parameters used for processing product. A sample of the placebo batch is then tested for residual contamination. However, we have documented several significant issues that need to be addressed when using placebo product to validate cleaning processes.

One cannot assure that the contaminant will be uniformly distributed throughout the system. For example, if the discharge valve or chute of a blender are contaminated, the contaminant would probably not be uniformly dispersed in the placebo; it would most likely be concentrated in the initial discharge portion of the batch. Additionally, if the contaminant or residue is of a larger particle size, it may not be uniformly dispersed in the placebo.

Some firms have made the assumption that a residual contaminant would be worn off the equipment surface uniformly; this is also an invalid conclusion. Finally, the analytical power may be greatly reduced by dilution of the contaminant. Because of such problems, rinse and/or swab samples should be used in conjunction with the placebo method.

b. Detergent

If a detergent or soap is used for cleaning, determine and consider the difficulty that may arise when attempting to test for residues. A common problem associated with detergent use is its composition. Many detergent suppliers will not provide specific composition, which makes it difficult for the user to evaluate residues. As with product residues, it is important and it is expected that the manufacturer evaluate the efficiency of the cleaning process for the removal of residues. However, unlike product residues, it is expected that no (or for ultra-sensitive analytical test methods - very low) detergent levels remain after cleaning. Detergents are not part of the manufacturing process and are only added to facilitate cleaning during the cleaning process. Thus, they should be easily removable. Otherwise, a different detergent should be selected.

c. Test Until Clean

Examine and evaluate the level of testing and the retest results since testing until clean is a concept utilized by some manufacturers. They test, resample, and retest equipment or systems until an "acceptable" residue level is attained. For the system or equipment with a validated cleaning process, this practice of resampling should not be utilized and is acceptable only in rare cases. Constant retesting and resampling can show that the cleaning process is not validated since these retests actually document the presence of unacceptable residue and contaminants from an ineffective cleaning process.

References

1. J. Rodehamel, "Cleaning and Maintenance," Pgs 82–87, University of Wisconsin's Control Procedures in Drug Production Seminar, July 17–22, 1966, William Blockstein, Editor, Published by the University of Wisconsin, L.O.C.#66–64234.
2. J.A. Constance, "Why Some Dust Control Exhaust Systems Don't Work," Pharm. Eng., January-February, 24–26 (1983).
3. S.W. Harder, "The Validation of Cleaning Procedures," *Pharm. Technol.* **8** (5), 29–34 (1984).
4. W.J. Mead, "Maintenance: Its Interrelationship with Drug Quality," *Pharm. Eng.* **7**(3), 29–33 (1987).
5. J.A. Smith, "A Modified Swabbing Technique for Validation of Detergent Residues in Clean-in-Place Systems," *Pharm. Technol.* **16**(1), 60–66 (1992).
6. Fourman, G.L. and Mullen, M.V., "Determining Cleaning Validation Acceptance Limits for Pharmaceutical Manufacturing Operations," *Pharm. Technol.* **17**(4), 54–60 (1993).
7. McCormick, P.Y. and Cullen, L.F., in Pharmaceutical Process Validation, 2nd Ed., edited by I.R. Berry and R.A. Nash, 319–349 (1993).

Appendix F

Guide to Inspections of Dosage Form Drug Manufacturer's—CGMPR's

The following is a US Food and Drug Administration (FDA) training guide used by the FDA in the training of their inspectors. Such training guides are used to prepare and provide FDA inspectors with a general knowledge and understanding of what it is they should be considering during inspection of an FDA-regulated manufacturing facility.

In many cases these are general considerations due in large part to the many and varied drugs and the proprietary processes required to manufacture them. These guides are intended to augment the education, experience, and accumulated knowledge each inspector already possesses. The inspection process and procedures by which they perform it leave room for on-site judgment consideration and assessment by the inspectors.

The purpose of including these FDA guides in this book is to provide the reader, designer, and engineer involved with the design and engineering of a bioprocessing facility some insight as to what the expectations of an FDA inspector might be. What expectations an FDA inspector might possess can most likely be attributed to their training and these guidelines. Providing these guidelines for the readers of this book, those of you designing, constructing, and operating FDA-regulated manufacturing facilities, will hopefully lessen the gap of understanding between you and the FDA inspector. Having some insight into the inspector's understanding and training helps one to be better prepared to meet those expectations when it's time for the inspection.

These inspection guidelines are reproduced here through the courtesy of the FDA. These and others also reside on the FDA web site at:

http://www.fda.gov/ICECI/Inspections/InspectionGuides/default.htm

Bioprocessing Piping and Equipment Design: A Companion Guide for the ASME BPE Standard, First Edition. William M. (Bill) Huitt.

Guide to Inspections of Dosage form Drug Manufacturer's - Cgmpr's

Note: This document is reference material for investigators and other FDA personnel. The document does not bind FDA, and does no confer any rights, privileges, benefits, or immunities for or on any person(s).

I. Introduction

This document is intended to be a general guide to inspections of drug manufacturers to determine their compliance with the drug CGMPR's. This guide should be used with instructions in the IOM, other drug inspection guides, and compliance programs. A list of the inspection guides is referenced in Chapter 10 of the IOM. Some of these guides are:

- Guide to Inspections of Bulk Pharmaceutical Chemicals.
- Guide to Inspections of High Purity Water Systems.
- Guide to Inspections of Pharmaceutical Quality Control Laboratories.
- Guide to Inspections of Microbiological Pharmaceutical Quality Control Laboratories.
- Guide to Inspections of Lyophilization of Parenterals.
- Guide to Inspections of Validation of Cleaning Processes.
- Guide to Inspections of Computerized Systems in Drug Processing.
- Guideline on General Principles of Process Validation.

II. Current Good Manufacturing Practice Regulations

Prescription vs. Non-prescription

All drugs must be manufactured in accordance with the current good manufacturing practice regulations otherwise they are considered to be adulterated within the meaning of the FD&C Act, Section 501(a)(2)(B). Records relating to prescription drugs must be readily available for review in accordance with Sec. 704(a)(1)(B) of the FD&C Act. If the product is an OTC drug which is covered by an NDA or ANDA, FDA may review, copy and verify the records under Sec. 505(k)(2) of the FD&C Act. However, if the product is an OTC drug for which there is no application filed with FDA, the firm is not legally required to show these records to the investigator during an inspection being conducted under Section 704 of the FD&C Act. Nonetheless, all manufacturers of prescription and OTC drugs must comply with the drug CGMPR requirements, including those involving records. The investigator should review these records as part of the inspection in determining the firm's compliance with the CGMP regulations. On rare occasions, a firm may refuse to allow review of OTC records stating they are not legally required to. While the firm may be under no legal obligation to permit review of such records, this does not relieve the firm of its statutory requirement to comply with the good manufacturing practices under section 501(a)(2)(B) of the Food Drug and Cosmetic Act, including the requirements for maintaining records.

If a firm refuses review of OTC records, the investigator should determine by other inspectional means the extent of the firm's compliance with CGMPR's. Inspectional observations and findings that CGMPR's are not being followed are to be cited on a List of Inspectional Observations, FDA-483, for both prescription and non-prescription drugs.

Organization and Personnel [21 CFR 211 Subpart B]

The firm must have a quality control department that has the responsibility and authority as described in the referenced CFR. The quality control department must maintain its independence from the production department, and its responsibilities must be in writing.

Obtain the name, title and individual responsibilities of corporate officers and other key employees as indicated in the IOM.

In the drug industry, an employee's education and training for their position has a significant impact on the production of a quality product. Report whether the firm has a formalized training program, and describe the type of training received. The training received by an employee should be documented.

Quality control must do product annual review on each drug manufactured, and have written annual review procedures. Review these reports in detail. This report will quickly let you know if the manufacturing process is under control. The report should provide a summary all lots that failed in-process or finished product testing, and other critical factors. Investigate any failures.

Quality control must validate the manufacturing process for each drug manufactured. Review and evaluate this data.

Buildings and Facilities [21 CFR 211 Subpart C]

Review the construction, size, and location of plant in relation to surroundings. There must be adequate lighting, ventilation, screening, and proper physical barriers for all operations including dust, temperature, humidity, and bacteriological controls. There must be adequate blueprints which describe the high purity water, HEPA, and compressed air systems. The site must have adequate locker, toilet, and hand washing facilities.

The firm must provide adequate space for the placement of equipment and materials to prevent mix-ups in the following operations:

- receiving, sampling, and storage of raw materials;
- manufacturing or processing;
- packaging and labeling;
- storage for containers, packaging materials, labeling, and finished products;
- production and control laboratories.

Equipment [21 CFR 211 Subpart D]

Review the design, capacity, construction, and location of equipment used in the manufacturing, processing, packaging, labeling, and laboratories. Describe the manufacturing equipment including brief descriptions of operating principles. Consider the use of photographs, flow charts, and diagrams to supplement written descriptions.

New equipment must be properly installed, and operate as designed. Determine if the equipment change would require FDA pre-approval and/or revalidation of the manufacturing process. The equipment must be cleaned before use according to written procedures. The cleaning must be documented and validated.

The equipment should not adversely affect the identity, strength, quality, or purity of the drug. The material used to manufacture the equipment must not react with the drug. Also, lubricants or coolants must not contaminate the drug.

The equipment should be constructed and located to ease cleaning, adjustments, and maintenance. Also, it should prevent contamination from other or previous manufacturing operations. Equipment must be identified as to its cleaning status and content. The cleaning and maintenance of the equipment are usually documented in a log book maintained in the immediate area. Determine if the equipment is of suitable capacity and accuracy for use in measuring, weighing, or mixing operations. If the equipment requires calibration, they must have a written procedure for calibrating the equipment and document the calibration.

Components and Product Containers [21 CFR 211 Subpart E]

Inspect the warehouse and determine how components, drug product containers, and closures are received, identified, stored, handled, sampled, tested, and approved or rejected. They must have written procedures which describe how these operations are done. Challenge the system to decide if it is functioning correctly. If the handling and storage of components are computer controlled, the program must be validated.

The receiving records must provide traceability to the component manufacturer and supplier. The receiving records for components should contain the name of the component, manufacturer, supplier if different from the manufacturer, and carrier. In addition, it should include the receiving date, manufacturer's lot number, quantity received, and control number assigned by the firm.

Check sanitary conditions in the storage area, stock rotation practices, retest dates, and special storage conditions (protection from light, moisture, temperature, air, etc.). Inspect glandular and botanical components for insect infestation.

Components or finished product adulterated by rodents, insects, or chemicals must be documented and submitted for seizure.

Collect the evidence even if the firm plans to voluntarily destroy the product. Be alert for components, colors, and food additives that may be new drug substances, appear to have no use in the plant or appear to be from an unknown supplier. Check the colors against the Color Additives Status List in the IOM

Determine if the color is approved for its intended use, and required statements are declared on the drug label.

Components might be received at more than one location. Components must be handled in accordance with the drug CGMP's including components used in the research and development lab. Determine how components are identified after receipt and quarantined until released. Components must be identified so the status (quarantine, approved, or rejected) is known. Review the criteria for removing components from quarantine and challenge the system. Determine what records are maintained in the storage area to document the movement of components to other areas, and how rejected components handled. The component container has an identification code affixed to it. This unique code provides traceability from the component manufacturer to its use in the finished product.

Review the sampling and testing procedures for components, and the process by which approved materials are released for use. Decide if these practices are adequate and followed.

Determine the validity, and accuracy of the firm's inventory system for drug components, containers, closures and labeling. Challenge the component inventory records by weighing a lot and comparing the results against the quantity remaining on the inventory record. Significant discrepancies in these records should be investigated.

Evaluate the following to determine whether the firm has shown that the containers and closures are compatible with the product, will provide adequate protection for the drug against deterioration or contamination, are not additive or absorptive, and are suitable for use:

- Specifications for containers, closures, cotton filler, and desiccant, etc.
- What tests or checks are made (cracks, glass particles, durability of material, metal particles in ointment tubes, compliance with compendium specifications, etc.).
- Cleaning procedures and how containers are stored.
- Handling of preprinted containers. Are these controlled as labeling, or as containers? The firm must review the labeling for accuracy.

Production and Process Controls [21 CFR Subpart F]

1. Critical Manufacturing Steps [21 CFR 211.101]

Each critical step in the manufacturing process shall be done by a responsible individual and checked by a second responsible individual. If such steps in the processing are controlled by automatic mechanical or electronic equipment, its performance should be verified.

Critical manufacturing steps include the selection, weighing, measuring and identifying of components, and addition of components during processing. It includes the recording of deviations from the batch record, mixing time and testing of in-process material, and the determination of actual yield and percent of theoretical yield. These manufacturing steps are documented when done, and not before or after the fact.

2. Equipment Identification [21 CFR 211.105]

All containers and equipment used in to manufacture a drug should be labeled at all times. The label should identify the contents of the container or equipment including the batch number, and stage of processing. Previous identification labels should be removed. The batch should be handled and stored to prevent mixups or contamination.

3. In-Line and Bulk Testing [21 CFR 211.110]

To ensure the uniformity and integrity of products, there shall be adequate in-process controls, such as checking the weights and disintegration time of tablets, the fill of liquids, the adequacy of mixing, the homogeneity of suspensions, and the clarity of solutions.

Determine if in-process test equipment is on site and the specified tests are done. Be alert for prerecording of test results such as tablet weight determinations.

The bulk drug is usually held in quarantine until all tests are completed before it is released to the packaging and labeling department. However, the testing might be done after packaging product.

4. Actual Yield [21 CFR 211.103]

Determine if personnel check the actual against the theoretical yield of each batch of drug manufactured. In the event of any significant unexplained discrepancies, determine if there is a procedure to prevent distribution of the batch in question, and related batches.

5. Personnel Habits

Observe the work habits of plant personnel. Determine:

- Their attitudes and actions involving the jobs they perform. (Careless, lackadaisical, disgruntled, etc.).
- Their dress. (Clean dresses, coats, shirts and pants, head coverings, etc.
- If proper equipment is used for a given job or whether short cuts are taken (i.e. use of hands and arms to mix or empty trays of drug components).
- If there are significant written or verbal language barriers that could affect their job performance.

Tablet and Capsule Products

Become familiar with the type of equipment and its location in the tableting operation. The equipment may include rotary tableting machines, coating and polishing pans, punches and dies, etc. The equipment should be constructed and located to facilitate maintenance and cleaning at the end of each batch or at suitable intervals in the case of a continuous batch operation. If possible, observe the cleaning and determine if the cleaning procedure is followed.

The ingredients in a tablet are the active ingredient, binders, disintegrators, bases, and lubricants. The binder is added to the batch to keep the tablet together. Excess binder will make the tablet too hard for use. The disintegrator is used to help disintegration of the tablet after administration. The base should be an inert substance which is compatible with the active ingredient and is added to provide size and weight. The lubricant helps in the flow of granulated material, prevents adhesion of the tablet material to the surface of punches and dies, and helps in tablet ejection from the machine.

Tablets and capsules are susceptible to airborne contamination because of the manipulation of large quantities of dry ingredients. To prevent cross-contamination in the tableting department, pay close attention to the maintenance, cleaning, and location of equipment, and the storage of granulations and tablets. To prevent cross-contamination, the mixing, granulation, drying and/or tableting operation should be segregated in enclosed areas with its own air handling system. Determine what precautions are taken to prevent cross-contamination. When cross-contamination is suspect, investigate the problem and collect in-line samples (INV) and official samples of the suspect product. Determine what temperature, humidity, and dust collecting controls are used by the firm in manufacturing operations. Lack of temperature and humidity controls can affect the quality of the tablet.

Observe the actual operation of the equipment and determine whether powders or granulations are processed according to the firm's specifications. The mixing process must be validated. The drying ovens should have their own air handling system which will prevent cross-contamination. Does the firm record drying time/temperature and maintain recording charts including loss on drying test results? Review the in-line tests performed by production and/or quality control. Some in-process tests are tablet weight, thickness, hardness, disintegration, and friability. Evaluate the disposition of in-process samples.

Capsules may be either hard, or soft type. They are filled with powder, beads, or liquid by machine. The manufacturing operation of powders for capsules should follow the same practice as for tablets. Determine manufacturing controls used, in-line testing, and basis for evaluating test results for the filling operations.

Sterile Products

Typically, a sterile drug contains no viable microorganisms and is non-pyrogenic. Drugs for intravenous injection, irrigation, and as ophthalmic preparations, etc., meet this criteria. In addition, other dosage forms might be labeled as sterile. For instance, an ointment applied to a puncture wound or skin abrasion.

Parenteral drugs must be non-pyrogenic, because the presence of pyrogens can cause a febrile reaction in human beings. Pyrogens are the products of the growth of microorganisms. Therefore, any condition that permits bacterial growth should be avoided in the manufacturing process. Pyrogens may develop in water located in stills, storage tanks, dead legs, and piping, or from surface contamination of containers, closures, or other equipment. Parenterals may also contain chemical contaminants that will produce a pyretic response in humans or animals although there are no pyrogens present.

There are many excellent reference materials which should be reviewed before the inspection. Some of these are the "Guideline on Sterile Drug Products Produced by Aseptic Processing," and chapter 84 on pyrogens in the Remington's Pharmaceutical Sciences.

Determine and evaluate the procedures used to minimize the hazard of contamination with microorganisms and particulates of sterile drugs.

- Personnel
  - Review the training program to ensure that personnel performing production and control procedures have experience and training commensurate with their intended duties. It is important that personnel be trained in aseptic procedures. The employees must be properly gowned and use good aseptic techniques.
- Buildings
  - The non-sterile preparation areas for sterile drugs should be controlled. Refer to Subpart C of the proposed CGMPR's for LVP's; however, deviations from these proposed regulations are not necessarily deviations from the CGMPR's. Evaluate the air cleanliness classification of the area. For guidance in this area, review Federal Standard #209E entitled "Airborne Particulate Cleanliness Classes in Cleanrooms and Clean Zones." Observe the formulation practices or procedures used in the preparation areas. Be alert for routes of contamination. Determine how the firm minimizes traffic and unnecessary activity in the preparation area. Determine if filling rooms and other aseptic areas are constructed to eliminate possible areas for microbiological/ particulate contamination. For instance, dust-collecting ledges, porous surfaces, etc. Determine how aseptic areas are cleaned and maintained.

1. Air

Air supplied to the non-sterile preparation or formulation area for manufacturing solutions prior to sterilization should be filtered as necessary to control particulates. Air being supplied to product exposure areas where sterile drugs are processed and handled should be high efficiency particulate air (HEPA) filtered under positive pressure.

Review the firm's system for HEPA filters, determine if they are certified and/ or Dioctyl Phthalate (DOP) tested and frequency of testing.

Review the compressed air system and determine if it is filtered at the point of use to control particulates. Diagrams of the HEPA filtered and compressed air systems should be reviewed and evaluated.

2. Environmental Controls

Specifications for viable and non-viable particulates must be established. Specifications for viable particulates must include provisions for both air and surface sampling of aseptic processing areas and equipment. Review the firm's environmental control program, specifications, and test data. Determine if the firm follows its procedure for reviewing out-of-limit test results. Also, determine if review of environmental test data is included as a part of the firm's release procedures.

Note: In the preparation of media for environmental air and surface sampling, suitable inactivating agents should be added. For example, the addition of penicillinase to media used for monitoring sterile penicillin operations and cephalosporin products.

3. Equipment

Determine how the equipment operates including the cleaning and maintenance practices. Determine how equipment used in the filling room is sterilized, and if the sterilization cycle has been validated. Determine the practice of re-sterilizing equipment if sterility has been compromised.

Determine the type of filters used. Determine the purpose of the filters, how they are assembled, cleaned, and inspected for damage. Determine if a microbial retentive filter, and integrity testing is required.

4. Water for Injection

Water used in the production of sterile drugs must be controlled to assure that it meets U.S.P. specifications. Review the firm's water for injection production, storage, and delivery system. Determine that the stills, filters, storage tanks, and pipes are installed and operated in a manner that will not contaminate the water. Evaluate the firm's procedures and specifications that assure the quality of the water for injection. As reference material, review the "FDA Guide to Inspections of High Purity Water Systems" before initiating an inspection.

5. Containers and Closures

Determine how containers and closures are handled and stored. Decide if the cleaning, sterilization, and depyrogenization are adequate, and have been validated.

6. Sterilization

- Methods
  - Determine what method of sterilization is used. A good source of reference material on validation of various sterilization processes is the Parenteral Drug Association Technical Reports. For instance, Technical Report #1 covers "Validation of Steam Sterilization Cycles." Review and evaluate the validation data whatever the method employed.
  - If steam under pressure is used, an essential control is a mercury thermometer and a recording thermometer installed in the exhaust line. The time required to heat the center of the largest container to the desired temperature must be known. Steam must expel all air from the sterilizer chamber to eliminate cold

spots. The drain lines should be connected to the sewer by means of an air break to prevent back siphoning. The use of paper layers or liners and other practices which might block the flow of steam should be avoided. Charts of time, temperature, and pressure should be filed for each sterilizer load.

- If sterile filtration is used, determine the firm's criteria for selecting the filter and the frequency of changing. Review the filter validation data. Determine if the firm knows the bioburden of the drug, and examine their procedures for filter integrity testing. Filters might not be changed after each batch is sterilized. Determine if there is data to justify the integrity of the filters for the time used and that "grow through" has not occurred.
- If ethylene oxide sterilization is used, determine what tests are made for residues and degradation. Review the ETO sterilization cycle including preconditioning of the product, ETO concentration, gas exposure time, chamber and product temperature, and chamber humidity.

- Indicators
  - Determine the type of indicator used to assure sterility. Such as, lag thermometers, peak controls, Steam Klox, test cultures, biological indicators, etc.
  - Caution: When spore test strips are used to test the effectiveness of ethylene oxide sterilization, be aware that refrigeration may cause condensation on removal to room temperature. Moisture on the strips converts the spore to the more susceptible vegetative forms of the organism which may affect the reliability of the sterilization test. The spore strips should not be stored where they could be exposed to low levels of ethylene oxide.
  - If biological indicators are used, review the current U.S.P. on sterilization and biological indicators. In some cases, testing biological indicators may become all or part of the sterility testing.
  - Biological indicators are of two forms, each of which incorporates a viable culture of a single species of microorganism. In one form, the culture is added to representative units of the lot to be sterilized or to a simulated product that offers no less resistance to sterilization than the product to be sterilized. The second form is used when the first form is not practical as in the case of solids. In the second form, the culture is added to disks or strips of filter paper, or metal, glass, or plastic beads.
  - During the inspection of a firm which relies on biological indicators, review background data compiled by the firm to include:
  - Surveys of the types and numbers of organisms in the product before sterilization.
  - Data on the resistance of the organism to the specific sterilization process.
  - Data used for selecting the most resistant organism and its form (spore or vegetative cell).
  - Studies of the stability and resistance of the selected organism to the specific sterilization process.
  - Studies on the recovery of the organism used to inoculate the product.
  - If a simulated product or surface similar to the solid product is used, validation of the simulation or similarity. The simulated product or similar surface must not affect the recovery of the numbers of indicator organisms applied.

- Validation of the number of organisms used to inoculate the product, simulated product, or similar surface, to include stability of the inoculum during the sterilization process.

Since qualified personnel are crucial to the selection and application of these indicators, review their qualifications including experience dealing with the process, expected contaminants, testing of resistance of organisms, and technique.

Review the firm's instructions regarding use, control and testing, of the biological indicator by product including a description of the method used to demonstrate presence or absence of viable indicator in or on the product.

Review the data used to support the use of the indicator each time it is used. Include the counts of the inoculum used; recovery data to control the method used to demonstrate the sterilization of the indicator organism; counts on unprocessed, inoculated material to indicate the stability of the inoculum for the process time; and results of sterility testing specifically designed to demonstrate the presence or absence of the indicator organism for each batch or filling operation. In using indicators, you must assure yourself that the organisms are handled so they don't contaminate the drug manufacturing area and product.

7. Filled Containers

Evaluate how the filled vials or ampules leave the filling room. Is the capping or sealing done in the sterile fill area? If not, how is sterility maintained until capped?

Review the tests done on finished vials, ampules, or other containers, to assure proper fill and seal. For instance, leak and torque tests.

Review examinations made for particulate contamination. You can quickly check for suspected particulate matter by using a polariscope. Employees doing visual examinations on line must be properly trained. If particle counts are done by machine, this operation must be validated.

8. Personnel Practices

Check how the employees sterilize and operate the equipment used in the filling area.

Observe filling room personnel practices. Are the employees properly dressed in sterile gowns, masks, caps, and shoe coverings? Observe and evaluate the gowning procedures, and determine if good aseptic technique is maintained in the dressing and filling rooms.

Check on the practices after lunch and other absences. Is fresh sterile garb supplied, or are soiled garments reused?

Determine if the dressing room is next to the filling area and how employees and supplies enter the sterile area.

9. Laboratory Controls

For guidance on how to inspect micro and chemistry labs, review the "FDA Guide to Inspections of Pharmaceutical Quality Control Laboratories" and "FDA Guide to Inspections of Microbiological Pharmaceutical Quality Control Laboratories."

10. Retesting for Sterility

See the USP for guidance on sterility testing. Sterility retesting is acceptable provided the cause of the initial non-sterility is known, and thereby invalidates the original results. It cannot be assumed that the initial sterility test failure is a false positive. This conclusion must be justified by sufficient documented investigation. Additionally, spotty or low level contamination may not be identified by repeated sampling and testing.

Review sterility test failures and determine the incidence, procedures for handling, and final disposition of the batches involved.

11. Retesting for Pyrogens

As with sterility, pyrogen retesting can be performed provided it is known that the test system was compromised. It cannot be assumed that the failure is a false positive without documented justification.

Review any initial pyrogen test failures and determine the firm's justification for retesting.

12. Particulate Matter Testing

Particulate matter consists of extraneous, mobile, undissolved substances, other than gas bubbles, unintentionally present in parenteral solutions.

Cleanliness specifications or levels of non-viable particulate contamination must be established. Limits are usually based on the history of the process. The particulate matter test procedure and limits for LVP's in the U.S.P. can be used as a general guideline. However, the levels of particulate contamination in sterile powders are generally greater than in LVP's. LVP solutions are filtered during the filling operation. However, sterile powders, except powders lyophilized in vials, cannot include filtration as a part of the filling operation. Considerable particulate contamination is also present in sterile powders which are spray dried due to charring during the process.

Review the particulate matter test procedure and release criteria. Review production and control records of any batches for which complaints of particulate matter have been received.

13 Production Records

Production records should be similar to those for other dosage forms. Critical steps, such as integrity testing of filters, should be signed and dated by a second responsible person.

Review production records to ensure that directions for significant manufacturing steps are included and reflect a complete history of production.

Ointments, Liquids, and Lotions

Major factors in the preparation of these drugs are the selection of raw materials, manufacturing practices, equipment, controls, and laboratory testing.

Following the basic drug inspection fundamentals, fully evaluate the production procedures. In addition, evaluate specific information regarding:

- The selection and compatibility of ingredients.
- Whether the drug is a homogeneous preparation free of extraneous matter.
- The possibility of decomposition, separation, or crystallization of ingredients.
- The adequacy of ultimate containers to hold and dispense contents.
- Procedure for cleaning the containers before filling.
- Maintenance of homogeneity during manufacturing and filling operations.

The most common problem associated with the production of these dosage forms is microbiological contamination caused by faulty design and/or control of purified water systems. During inspections, evaluate the adequacy of the water system. Review and evaluate the micro/chemistry test results on the routine monitoring of the water system including validation of the water system. Review any microbiological tests done on the finished drug including in-process testing.

Some of these drugs have preservatives added which protect them from microbial contamination. The preservatives are used primarily in multiple-dose containers to inhibit the growth of microorganisms introduced inadvertently during or after manufacturing. Evaluate the adequacy of preservative system. Preservative effectiveness testing for these products should be reviewed. For additional information, review the "Antimicrobial Preservatives-Effectiveness" section of the U.S.P.

Equipment employed for manufacturing topical drugs is sometimes difficult to clean. This is especially true for those which contain insoluble active ingredients, such as the sulfa drugs. The firm's equipment cleaning procedures including cleaning validation data should be reviewed and evaluated.

Packaging and Labeling [21 CFR Subpart G]

Packaging and labeling operations must be controlled so only those drugs which meet the specifications established in the master formula records are distributed. Review in detail the packaging and labeling operations to decide if the system will prevent drug and label mix-ups. Approximately 25% of all drug recalls originate in this area.

Evaluate what controls or procedures the firm has to provide positive assurance that all labels are correct. Determine if packaging and labeling operations include:

- Adequate physical separation of labeling and packaging operations from manufacturing process.
- Review of:
- Label copy before delivery to the printer.

- Printer's copy.
  - Whether firm's representative inspects the printer.
  - Whether or not gang printing is prohibited.
  - Whether labels are checked against the master label before released to stock. Determine who is responsible for label review prior to release of the labels to production. Also, whether the labels are identical to the labeling specified in the batch production records.
- Separate storage of each label (including package inserts) to avoid mixups.
- Inventory of label stocks. Determine if the printer's count is accepted or if labels are counted upon receipt.
- Designation of one individual to be responsible for storage and issuance of all labels.
- Receipt by the packaging and labeling department of a batch record, or other record, showing the quantity of labels needed for a batch. Determine if the batch record is retained by the packaging supervisor or accompanies the labels to the actual packaging and labeling line.
- Adequate controls of the quantities of labeling issued, used, and returned. Determine if excess labels are accounted for and if excess labels bearing specific control codes, and obsolete or changed labels are destroyed.
- Inspection of the facilities before labeling to ensure that all previously used labeling and drugs have been removed.
- Assurance that batch identification is maintained during packaging.
- Control procedures to follow if a significant unexplained discrepancy occurs between quantity of drug packaged and the quantity of labeling issued
- Segregated facilities for labeling one batch of the drug at a time. If this is not practiced, determine what steps are taken to prevent mix-ups.
- Methods for checking similar type labels of different drugs or potencies to prevent mixing.
- Quarantine of finished packaged products to permit adequate examination or testing of a representative sample to safeguard against errors. Also, to prevent distribution of any batch until all specified tests have been met.
- An individual who makes the final decision that the drug should go to the warehouse, or the shipping department.
- Utilization of any outside firms, such as contract packers, and what controls are exercised over such operations. Special attention should be devoted to firms using "rolls" of pressure sensitive labels. Investigators have found instances where:
- Paper chips cut from label backing to help running the labels through a coder interfered with the code printer causing digits in the lot number to be blocked out.
- Some rolls contained spliced sections resulting in label changes in the roll.
- Some labels shifted on the roll when the labels were printed resulting in omitting required information.

The use of cut labels can cause a significant problem and should be evaluated in detail. Most firms are replacing their cut labels with roll labels.

Review prescription drugs for which full disclosure information may be lacking. If such products are found, submit labels and other labeling as exhibits with the EIR See 21 CFR 201.56 for the recommended sequence in which full disclosure information should be presented.

Review labels of OTC products for warnings required by 21 CFR 369.

A control code must be used to identify the finished product with a lot, or control number that permits determination of the complete history of the manufacture and control of the batch.

Determine:

- The complete key (breakdown) to the code.
- Whether the batch number is the same as the control number on the finished package. If not, determine how the finished package control number relates, and how it is used to find the identity of the original batch.

Beginning August 3, 1994 the following new requirements will become effective:

- Use of gang-printed labels will be prohibited unless they are adequately differentiated by size, shape or color. (211.122(f))
- If cut labels are used one of the following special control procedures shall be used (211.122(g)):
  - Dedication of packaging lines.
  - Use of electronic or electromechanical equipment to conduct a 100-percent examination of finished product.
  - Use of visual inspection to examine 100-percent of the finished product for hand applied labeling. The visual examination will be conducted by one person and independently verified by a second person.
- Labeling reconciliation required by 211.125 is waived for cut or roll labeling if a 100-percent examination is performed according to 211.22(g)(2).

Holding and Distribution [21 CFR subpart H]

Check the finished product storage and shipping areas for sanitary condition, stock rotation, and special storage conditions needed for specific drugs. Evaluate any drugs that have been rejected, or are on hold for other than routine reasons.

Laboratory Controls [21 CFR Subpart I]

Laboratory controls should include adequate specifications and test procedures to assure that components, in-process and finished products conform to appropriate standards of identity, strength, quality, and purity.

In order to permit proper evaluation of the firm's laboratory controls, determine:

- Whether the firm has established a master file of specifications for all raw materials used in drug manufacture. This master file should include sampling procedures, sample size, number of containers to be sampled, manner in which samples will be identified, tests to be performed, and retest dates for components subject to deterioration.

- The firm's policies about protocols of assay. These reports are often furnished by raw material suppliers; however, the manufacturer is responsible for verifying the validity of the protocols by periodically performing their own complete testing and routinely conducting identity tests on all raw materials received
- Laboratory procedure for releasing raw materials, finished bulk drugs or packaged drugs from quarantine. Determine who is responsible for this decision. Raw material specifications should include approved suppliers. For NDA or ANDA drugs, the approved suppliers listed in their specifications should be the same as those approved in the NDA or ANDA.
- If the laboratory is staffed and equipped to do all raw material, in-process, and finished product testing that is claimed.
- Whether drug preparations are tested during processing. If so, determine what type of tests are made and whether a representative sample is obtained from various stages of processing.
- Specifications and description of laboratory testing procedures for finished products.
- Procedures for checking the identity and strength of all active ingredients including pyrogen and sterility testing, if applicable.
- If the laboratory conducts pyrogen tests, safety tests, or bioassays; determine the number of laboratory animals and if they are adequately fed and housed. Determine what care is provided on weekends and holidays.
- Sterility testing procedures.

Entries should be permanently recorded and show all results, both positive and negative. Examine representative samples being tested and their records. When checking the sterility testing procedures, determine:

- Physical conditions of testing room. The facility used to conduct sterility testing should be similar to those used for manufacturing products.
- Laboratory procedures for handling sterile sample.
- Use of ultra-violet lights.
- Number of units tested per batch.
- Procedure for identifying test media with specific batches.
- Test media's ability to support growth of organisms.
- Length of incubation period.
- Procedure for diluting products to offset the effects of bacteriostatic agents.
- Pyrogen testing procedures

Determine if animals involved in positive pyrogen tests are withdrawn from use for the required period.

If the L.A.L. Test is used, review the FDA "Guideline on Validation of the Limulus Amebocyte Lysate Test ***."

- If any tests are made by outside laboratories, report the names of the laboratories and the tests they perform. Determine what precautions the firm takes to insure that the laboratories' work is bona fide.
- Methods used to check the reliability, accuracy, and precision of laboratory test procedures and instrumentation.
- How final acceptance or rejection of raw materials, intermediates, and finished products is determined. Review recent rejections and disposition of affected items.
- The provisions for complete records of all data concerning laboratory tests performed, including dates and endorsements of individuals performing the tests, and traceability.
- For components and finished product, the reserve sample program and procedures should be evaluated. Challenge the system and determine if the samples are maintained and can be retrieved. The storage container must maintain the integrity of the product.
- Whether stability tests are performed on:
  - The drug product in the container and closure system in which marketed.
  - Solutions prepared as directed in the labeling at the time of dispensing. Determine if expiration dates, based on appropriate stability studies, are placed on labels.
- If penicillin and non-penicillin products are manufactured on the same premises, whether non-penicillin products are tested for penicillin contamination.

Obtain copies of laboratory records, batch records, and any other documents that show errors or other deficiencies.

Control Records [21 CFR Subpart J]

1. Master Production and Control Records [21 CFR 211.186]

The various master production and control records are important because all phases of production and control are governed by them. Master records, if erroneous, may adversely affect the product. These records must be prepared according to the drug CGMPR's outlined in 21 CFR 211.186. These records might not be in one location, but should be readily available for review.

2. Batch Production and Control Records [21 CFR 211.188]

The batch production and control records must document each significant step in the manufacture, labeling, packaging, and control of specific batches of drugs. 21 CFR 211.188 provides the basic information the batch records must provide. A complete production and control record may consist of several separate records which should be readily available to the investigator.

Routinely check the batch record calculations against the master formula record. Give special attention to those products on which there have been complaints.

Be alert for transcription errors from the master formula record to the batch record. Be alert for transcription or photocopying errors involving misinterpretation of symbols, abbreviations, and decimal points, etc.

It is important that batch production records be specific in terms of equipment (v-blender vs. ribbon blender) and processing times (mixing time and speed). The equipment should have its own unique identification number. The manufacturing process for these products must be standardized, controlled, and validated.

3. Distribution [21 CFR 211.196]

Complete distribution records should be maintained per 21 CFR 211.196. Be alert for suspicious shipments of products subject to abuse or which have been targeted for high priority investigation by the agency. These include steroids, counterfeits, diverted drugs (i.e.; physician samples, clinical packs, etc.).

Determine and evaluate if the firm checks on the authenticity of orders received. What references are used, e.g. current editions of the AMA Directory, Hays Directory, etc.

4. Complaint Files [21 CFR 211.198]

21 CFR 211.198 requires that records of all written and oral complaints be maintained. Although FDA has no authority to require a drug firm, except for prescription drugs, to open its complaint files, attempt to review the firm's files.

The complaint files should be readily available for review. Do a follow-up investigation on all applicable consumer complaints in the firm's district factory jacket. Review and evaluate the firm's procedures for handling complaints. Determine if all complaints are handled as complaints and not inappropriately excluded.

Review the complaints and determine if they were fully investigated. Evaluate the firm's conclusions of the investigation, and determine if appropriate corrective action was taken. Determine if the product should be recalled, or warrant a comprehensive investigation by FDA.

Returned Drug Products [21 CFR Subpart K]

Returned drugs often serve as an indication that products may have decomposed during storage, are being recalled or discontinued.

Determine how returned drug items are handled. For example, are they quarantined, destroyed after credit, or returned to storage?

If an abnormally large amount of a specific drug item is on hand, determine why. Check if returned drug items are examined in the laboratory, and who makes the ultimate decision as to the use of the returned drugs.

Note: Dumping salvage drugs in the trash is a potentially dangerous practice. Advise management to properly dispose of the drugs to preclude salvage. Drugs should be disposed of in accordance with E.P.A. regulations.

Appendix G

Guide to Inspections Oral Solutions and Suspensions

The following is a US Food and Drug Administration (FDA) training guide used by the FDA in the training of their inspectors. Such training guides are used to prepare and provide FDA inspectors with a general knowledge and understanding of what it is they should be considering during inspection of an FDA-regulated manufacturing facility.

In many cases these are general considerations due in large part to the many and varied drugs and the proprietary processes required to manufacture them. These guides are intended to augment the education, experience, and accumulated knowledge each inspector already possesses. The inspection process and procedures by which they perform it leave room for on-site judgment consideration and assessment by the inspectors.

The purpose of including these FDA guides in this book is to provide the reader, designer, and engineer involved with the design and engineering of a bioprocessing facility some insight as to what the expectations of an FDA inspector might be. What expectations an FDA inspector might possess can most likely be attributed to their training and these guidelines. Providing these guidelines for the readers of this book, those of you designing, constructing, and operating FDA-regulated manufacturing facilities, will hopefully lessen the gap of understanding between you and the FDA inspector. Having some insight into the inspector's understanding and training helps one to be better prepared to meet those expectations when it's time for the inspection.

These inspection guidelines are reproduced here through the courtesy of the FDA. These and others also reside on the FDA web site at:

http://www.fda.gov/ICECI/Inspections/InspectionGuides/default.htm

Bioprocessing Piping and Equipment Design: A Companion Guide for the ASME BPE Standard, First Edition. William M. (Bill) Huitt.

Guide to Inspections Oral Solutions and Suspensions

Note: This document is reference material for investigators and other FDA personnel. The document does not bind FDA, and does no confer any rights, privileges, benefits, or immunities for or on any person(s).

I. Introduction

The manufacture and control of oral solutions and oral suspensions has presented some problems to the industry. While bio-equivalency concerns are minimal (except for the antiseptic products such as phenytoin suspension), there are other issues which have led to recalls. These include microbiological, potency and stability problems. Additionally, because the population using these oral dosage forms includes newborns, pediatrics and geriatrics who may not be able to take oral solid dosage forms and may be compromised, defective dosage forms can pose a greater risk because of the population being dosed. Thus, this guide will review some of the significant potential problem areas and provide direction to the investigator when giving inspectional coverage.

II. Facilities

The design of the facilities are largely dependent upon the type of products manufactured and the potential for cross-contamination and microbiological contamination. For example, the facilities used for the manufacture of OTC oral products might not require the isolation that a steroid or sulfa product would require.

Review the products manufactured and the procedures used by the firm for the isolation of processes to minimize contamination. Observe the addition of drug substance and powdered excipients to manufacturing vessels to determine if operations generate dust. Observe the systems and the efficiency of the dust removal system.

The firm's HVAC (Heating Ventilation and Air Conditioning) system may also warrant coverage particularly where potent or highly sensitizing drugs are processed. Some manufacturers recirculate air without adequate filtration. Where air is recirculated, review the firm's data which demonstrates the efficiency of air filtration such should include surface and/or air sampling.

III. Equipment

Equipment should be of sanitary design. This includes sanitary pumps, valves, flow meters and other equipment which can be easily sanitized. Ball valves, packing in pumps and pockets in flow meters have been identified as sources of contamination.

In order to facilitate cleaning and sanitization, manufacturing and filling lines should be identified and detailed in drawings and SOPs. In some cases, long delivery lines between manufacturing areas and filling areas have been a source of contamination.

Also, SOPs, particularly with regard to time limitations between batches and for cleaning have been found deficient in many manufacturers. Review cleaning SOPs, including drawings and validation data with regard to cleaning and sanitization.

Equipment used for batching and mixing of oral solutions and suspensions is relatively basic. Generally, these products are formulated on a weight basis with the batching tank on load cells so that a final Q.S. can be made by weight. Volumetric means, such as using a dip stick or line on a tank, have been found to be inaccurate.

In most cases, manufacturers will assay samples of the bulk solution or suspension prior to filling. A much greater variability has been found with batches that have been manufactured volumetrically rather than by weight. For example, one manufacturer had to adjust approximately 8% of the batches manufactured after the final Q.S. because of failure to comply with potency specifications. Unfortunately, the manufacturer relied solely on the bulk assay. After readjustment of the potency based on the assay, batches occasionally were found out of specification because of analytical errors.

The design of the batching tank with regard to the location of the bottom discharge valve has also presented problems. Ideally, the bottom discharge valve is flush with the bottom of the tank. In some cases valves, including undesirable ball valves, have been found to be several inches to a foot below the bottom of the tank. In others, drug or preservative was not completely dissolved and was lying in the "dead leg" below the tank with initial samples being found to be sub-potent. For the manufacture of suspensions, valves should be flush. Review and observe the batching equipment and transfer lines.

With regard to transfer lines, they are generally hard piped and easily cleaned and sanitized. In some cases manufacturers have used flexible hoses to transfer product. It is not unusual to see flexible hoses lying on the floor, thus significantly increasing the potential for contamination. Such contamination can occur by operators picking up or handling hoses, and possibly even placing them in transfer or batching tanks after they had been lying on the floor. It is also a good practice to store hoses in a way that allows them to drain rather than be coiled which may allow moisture to collect and be a potential source of microbial contamination. Observe manufacturing areas and operator practices, particularly when flexible hose connection are employed.

Another common problem occurs when a manifold or common connections are used, especially in water supply, premix or raw material supply tanks. Such common connections have been shown to be a source of contamination.

IV. Raw Materials

The physical characteristics, particularly the particle size of the drug substance, are very important for suspensions. As with topical products in which the drug is suspended, particles are usually very fine to micronized (less than 25 microns).

For syrups, elixir or solution dosage forms in which there is nothing suspended, particle size and physical characteristics of raw materials are not that important. However, they can affect the rate of dissolution of such raw materials in the manufacturing process. Raw materials of a finer particle size may dissolve faster than those of a larger particle size when the product is compounded.

Examples of a few of the oral suspensions in which a specific and well defined particle size specification for the drug substance is important include phenytoin suspension, carbamazepine suspension, trimethoprim and sulfamethoxazole suspension, and hydrocortisone suspension. Review the physical specifications for any drug substance which is suspended in the dosage form.

V. Compounding

In addition to a determination of the final volume (Q.S.) as previously discussed, there are microbiological concerns. For oral suspensions, there is the additional concern with uniformity, particularly because of the potential for segregation during manufacture and storage of the bulk suspension, during transfer to the filling line and during filling. Review the firm's data that support storage times and transfer operations. There should be established procedures and time limits for such operations to address the potential for segregation or settling as well as other unexpected effects that may be caused by extended holding or stirring.

For oral solutions and suspensions, the amount and control of temperature is important from a microbiological as well as a potency aspect. For those products in which temperature is identified as a critical part of the operation, the firm's documentation of temperature, such as by control charts, should be reviewed.

There are some manufacturers that rely on heat during compounding to control the microbiological levels in product. For such products, the addition of purified water to final Q.S., the batch, and the temperatures during processing should be reviewed.

In addition to drug substances, some additives, such as the parabens are difficult to dissolve and require heat. The control and assurance of their dissolution during the compounding stage should be reviewed. From a potency aspect, the storage of product at high temperatures may increase the level of degradants. Storage limitations (time and temperature) should be justified by the firm and evaluated during your inspection.

There are also some oral liquids which are sensitive to oxygen and have been known to undergo degradation. This is particularly true of the phenothiazine class of drugs, such as perphenazine and chlorpromazine. The manufacture of such products might require the removal of oxygen such as by nitrogen purging. Additionally, such products might also require storage in sealed tanks, rather than those with loose lids. Manufacturing directions for these products should be reviewed.

VI. Microbiological Quality

There are some oral liquids in which microbiological contamination can present significant health hazards. For example, some oral liquids, such as nystatin suspension are used in infants and immuno-compromised patients, and microbiological contamination with organisms, such as Gram-negative organisms, is objectionable. There are other oral liquid preparations such as antacids in which Pseudomonas sp. contamination is also objectionable. For other oral liquids such as cough preparations, the contamination with Pseudomonas sp. might not present the same health hazard. Obviously, the contamination of any preparation with Gram-negative organisms is not desirable.

In addition to the specific contaminant being objectionable, such contamination would be indicative of a deficient process as well as an inadequate preservative system. The presence of a specific Pseudomonas sp. may also indicate that other plant or raw material contaminants could survive the process. For example, the fact that a Pseudomonas putida contaminant is present could also indicate that Pseudomonas aeruginosa, a similar source organism, could also be present.

Both the topical and microbiological inspection guides discuss the methods and limitations of microbiological testing. Similar microbiological testing concepts discussed apply to the testing of oral liquids for microbiological contamination. Review the microbiological testing of raw materials, including purified water, as well as the microbiological testing of finished products. Since FDA laboratories typically utilize more sensitive test methods than industry, consider sampling any oral liquids in which manufacturers have found microbiological counts, no matter how low. Submit samples for testing for objectionable microorganisms.

VII. Oral Suspensions Uniformity

Those liquid products in which the drug is suspended (and not in solution) present manufacturer and control problems.

Those liquid products in which the drug is suspended (and not in solution) present manufacture and control problems. Depending upon the viscosity, many suspensions require continuous or periodic agitation during the filling process. If delivery lines are used between the bulk storage tank and the filling equipment, some segregation may occur, particularly if the product is not viscous. Review the firm's procedures for filling and diagrams for line set-up prior to the filling equipment.

Good manufacturing practice would warrant testing bottles from the beginning, middle and end to assure that segregation has not occurred. Such samples should not be composited.

In-process testing for suspensions might also include an assay of a sample from the bulk tank. More important, however, may be testing for viscosity.

VIII. Product Specifications

Important specifications for the manufacture of all solutions include assay and microbial limits. Additional important specifications for suspensions include particle size of the suspended drug, viscosity, pH, and in some cases dissolution. Viscosity can be important from a processing aspect to minimize segregation. Additionally, viscosity has also been shown to be associated with bio-equivalency. pH may also have some meaning regarding effectiveness of preservative systems and may even have an effect on the amount of drug in solution. With regard to dissolution, there are at least three products which have dissolution specifications. These products include phenytoin suspension, carbamazepine suspension, and sulfamethoxazole and trimethoprim suspension. Particle size is also important and at this point it would seem that any suspension should have some type of particle size specification. As with other dosage forms, the underlying data to support specifications should be reviewed.

IX. Process Validation

As with other products, the amount of data needed to support the manufacturing process will vary from product to product. Development (data) should have identified critical phases of the operation, including the predetermined specifications that should be monitored during process validation.

For example, for solutions the key aspects that should be addressed during validation include assurance that the drug substance and preservatives are dissolved. Parameters, such as heat and time should be measured. Also, in-process assay of the bulk solution during and/or after compounding according to predetermined limits are also an important aspects of process validation. For solutions that are sensitive to oxygen and/or light, dissolved oxygen levels would also be an important test. Again, the development data and the protocol should provide limits. Review firm's development data and/or documentation for their justification of the process.

As discussed, the manufacture of suspensions presents additional problems, particularly in the area of uniformity. Again, development data should have addressed the key compounding and filling steps that assure uniformity. The protocol should provide for the key in-process and finished product tests, along with their specifications. For oral solutions, bio-equivalency studies may not always be needed. However, oral suspensions, with the possible exception of some of the antacids, OTC products, usually require a bio-equivalency or clinical study to demonstrate effectiveness. As with oral solid dosage forms, comparison to the bio-batch is an important part of validation of the process.

Review the firm's protocol and process validation report and, if appropriate, compare data for full scale batches to bio-batch, data and manufacturing processes.

X. Stability

One area that has presented a number of problems includes the assurance of stability of oral liquid products throughout their expiry period. For example, there have been a number of recalls of the vitamins with fluoride oral liquid products because of vitamin degradation. Drugs in the phenothiazine class, such as perphenazine, chlorpromazine and promethazine have also shown evidence of instability. Good practice for this class of drug products would include quantitation of both the active and primary degradant. Dosage form manufacturers should know and have specifications for the primary degradant. Review the firm's data and validation data for methods used to quantitate both the active drug and degradant.

Because interactions of products with closure systems are possible, liquids and suspensions undergoing stability studies should be stored on their side or inverted in order to determine whether contact of the drug product with the closure system affects product integrity.

Moisture loss which can cause the remaining contents to become super-potent and microbiological contamination are other problems associated with inadequate closure systems.

XI. Packaging

Problems in the packaging of oral liquids have included potency (fill) of unit dose products, accurate calibration of measuring devices such as droppers that are often provided. The USP does not provide for dose uniformity testing for oral solutions. Thus, for unit dose solution products, they should deliver the label claim within the limits described in the USP. Review the firm's data to assure uniformity of fill and test procedures to assure that unit dose samples are being tested.

Another problem in the packaging of Oral Liquids is the lack of cleanliness of containers prior to filling. Fibers and even insects have been identified as debris in containers, and particularly plastic containers used for these products. Many manufacturers receive containers shrink-wrapped in plastic to minimize contamination from fiberboard cartons. Many manufacturers utilize compressed air to clean containers. Vapors, such as oil vapors, from the compressed air have occasionally been found to present problems. Review the firm's systems for the cleaning of containers.

Appendix H

Guide to Inspections of Sterile Drug Substance Manufacturers

The following is a US Food and Drug Administration (FDA) training guide used by the FDA in the training of their inspectors. Such training guides are used to prepare and provide FDA inspectors with a general knowledge and understanding of what it is they should be considering during inspection of an FDA-regulated manufacturing facility.

In many cases these are general considerations due in large part to the many and varied drugs and the proprietary processes required to manufacture them. These guides are intended to augment the education, experience, and accumulated knowledge each inspector already possesses. The inspection process and procedures by which they perform it leave room for on-site judgment consideration and assessment by the inspectors.

The purpose of including these FDA guides in this book is to provide the reader, designer, and engineer involved with the design and engineering of a bioprocessing facility some insight as to what the expectations of an FDA inspector might be. What expectations an FDA inspector might possess can most likely be attributed to their training and these guidelines. Providing these guidelines for the readers of this book, those of you designing, constructing, and operating FDA-regulated manufacturing facilities, will hopefully lessen the gap of understanding between you and the FDA inspector. Having some insight into the inspector's understanding and training helps one to be better prepared to meet those expectations when it's time for the inspection.

These inspection guidelines are reproduced here through the courtesy of the FDA. These and others also reside on the FDA web site at:

http://www.fda.gov/ICECI/Inspections/InspectionGuides/default.htm

Bioprocessing Piping and Equipment Design: A Companion Guide for the ASME BPE Standard,
First Edition. William M. (Bill) Huitt.

Guide to Inspections of Sterile Drug Substance Manufacturers

Note: This document is reference material for investigators and other FDA personnel. The document does not bind FDA, and does no confer any rights, privileges, benefits, or immunities for or on any person(s).

One of the more difficult processes to inspect and one which has presented considerable problems over the years is that of the manufacture of sterile bulk drug substances. Within the past several years, there have been a number of batches of sterile bulk drug substances from different manufacturers which exhibited microbiological contamination. One manufacturer had approximately 100 batches contaminated in a 6 month time period. Another had approximately 25 batches contaminated in a similar period. Other manufacturers have had recalls due to the lack of assurance of sterility. Although the Inspection Guide for Bulk Drug Substances provides some direction for the inspection of the sterile bulk drug substance, it does not provide the detailed direction needed.

I. Introduction

In the manufacture of the sterile bulk powders, it is important to recognize that there is no further processing of the finished sterile bulk powder to remove contaminants or impurities such as particulates, endotoxins and degradants.

As with other inspections, any rejected batches, along with the various reasons for rejection, should be identified early in the inspection to provide direction for the investigator. For example, lists of batches rejected and/or retested over a period of time should be obtained from the manufacturer to provide direction for coverage to be given to specific processes or systems. Because some of the actual sterile bulk operations may not be seen, and because of the complexity of the process, it is particularly important to review reports and summaries, such as validation studies, reject lists, Environmental Monitoring Summary Reports, QA Investigation Logs, etc. These systems and others are discussed in the Basic Inspection Guide. This is particularly important for the foreign sterile bulk drug substance manufacturer where time is limited.

In the preparation for a sterile bulk drug substance inspection, a flow chart with the major processing steps should be obtained. Generally, the manufacture of a sterile bulk substance usually includes the following steps:

1. Conversion of the non-sterile drug substance to the sterile form by dissolving in a solvent, sterilization of the solution by filtration and collection in a sterilized reactor (crystallizer).
2. Aseptic precipitation or crystallization of the sterile drug substance in the sterile reactor.
3. Aseptic isolation of the sterile substance by centrifugation or filtration.
4. Aseptic drying, milling and blending of the sterile substance.
5. Aseptic sampling and packaging the drug substance.

These operations should be performed in closed systems, with minimal operator handling. Any aseptic operations performed by an operator(s) other than in a closed system should be identified and carefully reviewed.

II. Components

In addition to the impurity concerns for the manufacture of bulk drug substances, there is a concern with endotoxins in the manufacture of the sterile bulk drug substances. The validation report, which demonstrates the removal, if present, of endotoxins to acceptable levels, should be reviewed. Some manufacturers have commented that since an organic solvent is typically used for the conversion of the non-sterile bulk drug substance to the sterile bulk drug substance, that endotoxins will be reduced at this stage. As with any operation, this may or may not be correct. For example, in an inspection of a manufacturer who conducted extensive studies of the conversion (crystallization) of the non-sterile substance to the sterile drug substance, they found no change from the initial endotoxin level. Organic solvents were used in this conversion. Thus, it is important to review and assess this aspect of the validation report.

In the validation of this conversion (non-sterile to sterile) from an endotoxin perspective, challenge studies can be carried out on a laboratory or pilot scale to determine the efficiency of the step. Once it is established that the process will result in acceptable endotoxin levels, some monitoring of the production batches would be appropriate. As with any validation process, the purpose and efficiency of each step should be evaluated. For example, if the conversion (crystallization) from the non-sterile to the sterile substance is to reduce endotoxins by one log, then data should support this step.

Since endotoxins may not be uniformly distributed, it is also important to monitor the bioburden of the non-sterile substance(s) being sterilized. For example, gram negative contaminants in a non-sterile bulk drug substance prior to sterilization are of concern, particularly if the sterilization (filtration) and crystallization steps do not reduce the endotoxins to acceptable levels. Therefore, microbiological, as well as endotoxin data on the critical components and operational steps should be reviewed.

III. Facility

Facility design for the aseptic processing of sterile bulk drug substances should have the same design features as an SVP aseptic processing facility. These would include temperature, humidity and pressure control. Because sterile bulk aseptic facilities are usually larger, problems with pressure differentials and sanitization have been encountered. For example, a manufacturer was found to have the gowning area under greater pressure than the adjacent aseptic areas. The need to remove solvent vapors may also impact on area pressurization.

Unnecessary equipment and/or equipment that cannot be adequately sanitized, such as wooden skids and forklift trucks, should be identified. Inquire about the movement of large quantities of sterile drug substance and the location of pass-through areas between the sterile core and non-sterile areas. Observe these areas, review environmental monitoring results and sanitization procedures.

The CGMP Regulations prohibit the use of asbestos filters in the final filtration of solutions. At present, it would be difficult for a manufacturer to justify the use of asbestos filters for filtration of air or solutions. Inquire about the use of asbestos filters.

Facilities used for the charge or addition of non-sterile components, such as the non-sterile drug substance, should be similar to those used for the compounding of parenteral solutions prior to sterilization. The concern is soluble extraneous contaminants, including endotoxins that may be carried through the process. Observe this area and review the environmental controls and specifications to determine the viable and non-viable particulate levels allowed in this area.

IV. Processing

Sterile powders are usually produced by dissolving the non-sterile substance or reactants in an organic solvent and then filtering the solution through a sterilizing filter. After filtration, the sterile bulk material is separated from the solvent by crystallization or precipitation. Other methods include dissolution in an aqueous solution, filtration sterilization and separation by crystallization/filtration. Aqueous solutions can also be sterile filtered and spray dried or lyophilized.

In the handling of aqueous solutions, prior to solvent evaporation (either by spray drying or lyophilization), check the adequacy of the system and controls to minimize endotoxin contamination. In some instances, piping systems for aqueous solutions have been shown to be the source of endotoxin contamination in sterile powders. There should be a print available of the piping system. Trace the actual piping, compare it with the print and assure that there are no “dead legs” in the system.

The validation data for the filtration (sterilization) process should also be reviewed. Determine the firm’s criteria for selection of the filter and the frequency of changing filters. Determine if the firm knows the bioburden and examine their procedures for integrity testing filters.

Filters might not be changed after each batch is sterilized. Determine if there is data to justify the integrity of the filters for the time periods utilized and that “grow through” has not occurred.

In the spray drying of sterile powders, there are some concerns. These include the sterilization of the spray dryer, the source of air and its quality, the chamber temperatures and the particle residence or contact time. In some cases, charring and product degradation have been found for small portions of a batch.

With regard to bulk lyophilization, concerns include air classification and aseptic barriers for loading and unloading the unit, partial meltback, uneven freezing and heat transfer throughout the powder bed, and the additional aseptic manipulations required to break up the large cake. For bulk lyophilization, unlike other sterile bulk operations, media challenges can be performed. At this point in time, with today's level of technology, it would seem that it would be difficult to justify the bulk lyophilization of sterile powders (from a microbiological aspect). Refer to the Guide for the Inspection of a Lyophilization Process for additional direction regarding this process.

Seek to determine the number and frequency of process changes made to a specific process or step. This can be an indicator of a problem experienced in a number of batches. A number of changes in a short period of time can be an indicator that the firm is experiencing problems. Review the Process Change SOP and the log for process changes, including the reason for such changes.

V. Equipment

Equipment used in the processing of sterile bulk drug substances should be sterile and capable of being sterilized. This includes the crystallizer, centrifuge and dryer. The sanitization, rather than sterilization of this equipment, is unacceptable. Sterilization procedures and the validation of the sterilization of suspect pieces of equipment and transfer lines should be reviewed.

The method of choice for the sterilization of equipment and transfer lines is saturated clean steam under pressure. In the validation of the sterilization of equipment and of transfer systems, Biological Indicators (BIs), as well as temperature sensors (Thermocouple (TC) or Resistance Thermal Device (RTD)) should be strategically located in cold spots where condensate may accumulate. These include the point of steam injection and steam discharge, as well as cold spots, which are usually low spots. For example, in a recent inspection, a manufacturer utilized a Sterilize-In-Place (SIP) system and only monitored the temperature at the point of discharge and not in low spots in the system where condensate can accumulate.

The use of formaldehyde is a much less desirable method of sterilization of equipment. It is not used in the United States, primarily because of residue levels in both the environment and in the product. A major problem with formaldehyde is its removal from piping and surfaces. In the inspection of a facility utilizing formaldehyde as a sterilant, pay particular attention to the validation of the cleaning process. The indirect testing of product or drug substance to demonstrate the absence of formaldehyde levels in a system is unacceptable. As discussed in the Cleaning Validation Guide, there should be some direct measure or determination of the absence of formaldehyde. Since contamination in a system and in a substance is not going to be uniform, merely testing the substance as a means of validating the absence of formaldehyde is unacceptable. Key surfaces should be sampled directly for residual formaldehyde.

One large foreign drug substance manufacturer, after formaldehyde sterilization of the system, had to reject the initial batches coming through the system because of formaldehyde contamination. Unfortunately, they relied on end product testing of the product and not on direct sampling to determine the absence of formaldehyde residues on equipment.

SIP systems for the bulk drug substance industry require considerable maintenance, and their malfunction has directly led to considerable product contamination and recall. The corrosive nature of the sterilant, whether it is clean steam, formaldehyde, peroxide or ethylene oxide, has caused problems with gaskets and seals. In two cases, inadequate operating procedures have led to even weld failure. For example, tower or pond water was inadvertently allowed to remain in a jacket and was valved shut. Clean steam applied to the tank resulted in pressure as high as 1,000 lbs., causing pinhole formation and contamination. Review the equipment maintenance logs. Review non-schedule equipment maintenance and the possible impact on product quality. Identify those suspect batches manufactured and released prior to the repair of the equipment.

Another potential problem with SIP systems is condensate removal from the environment. Condensate and excessive moisture can result in increased humidity and increases in levels of microorganisms on surfaces of equipment. Therefore, it is particularly important to review environmental monitoring after sterilization of the system.

The sterile bulk industry, as the non-sterile bulk industry, typically manufactures batches on a campaign basis. While this may be efficient with regard to system sterilization, it can present problems when a batch is found contaminated in the middle of a campaign. Frequently, all batches processed in a campaign in which a contaminated batch is identified are suspect. Review the failure investigation reports and the logic for the release of any batches in a campaign. Some of the more significant recalls have occurred because of the failure of a manufacturer to conclusively identify and isolate the source of a contaminant.

VI. Environmental Monitoring

The environmental monitoring program for the sterile bulk drug substance manufacturer should be similar to the programs employed by the SVP industry. This includes the daily use of surface plates and the monitoring of personnel. As with the SVP industry, alert or action limits should be established and appropriate follow-up action taken when they are reached.

There are some bulk drug substance manufacturers that utilize UV lights in operating areas. Such lights are of limited value. They may mask a contaminant on a settling or aerobic plate. They may even contribute to the generation of a resistant (flora) organism. Thus, the use of Rodac or surface plates will provide more information on levels of contamination.

There are some manufacturers that set alert/action levels on averages of plates. For the sampling of critical surfaces, such as operators' gloves, the average of results on plates is unacceptable. The primary concern is any incidence of objectionable levels of contamination that may result in a non-sterile product.

As previously discussed, it is not unusual to see the highest level of contamination on the surfaces of equipment shortly after systems are steamed. If this occurs, the cause is usually the inadequate removal of condensate.

Since processing of the sterile bulk drug substance usually occurs around the clock, monitoring surfaces and personnel during the second and third shifts should be routine.

In the management of a sterile bulk operation, periodic (weekly/monthly/quarterly) summary reports of environmental monitoring are generated. Review these reports to obtain those situations in which alert/action limits were exceeded. Review the firm's investigation report and the disposition of batches processed when objectionable environmental conditions existed.

VII. Validation

The validation of the sterilization of some of the equipment and delivery systems and the validation of the process from an endotoxin perspective have been discussed.

In addition to these parameters, demonstration of the adequacy of the process to control other physicochemical aspects should also be addressed in a validation report. Depending upon the particular substance, these include potency, impurities, particulate matter, particle size, solvent residues, moisture content, and blend uniformity. For example, if the bulk substance is a blend of two active substances or an active substance and excipient, then there should be some discussion/evaluation of the process for assuring uniformity. The process validation report for such a blend would include documentation for the evaluation and assurance of uniformity. A list of validation reports and process variables evaluated should be reviewed.

As with a non-sterile bulk drug substance, there should be an impurity profile and specific, validated analytical methods. Those should also be reviewed.

Manufacturers are expected to validate the aseptic processing of sterile BPCs. Such validation must encompass all parts, phases, steps, and activities of any process where components, fluid pathways, in-process fluids, etc., are expected to remain sterile. Furthermore, such validation must include all probable potentials for loss of sterility as a result of processing. Validation must also account for all potential avenues of microbial ingress associated with the routine use of the process.

The validation procedure should approximate as closely as possible all those processing steps, activities, conditions, and characteristics that may have a bearing on the possibility of microbial ingress into the system during routine production. In this regard, it is essential that validation runs are as representative as possible of routine production to ensure that the results obtained from validation are generalizable to routine production.

Validation must include the 100% assessment of sterility of an appropriate material that is subjected to the validation procedure. Culture media is the material of choice, whenever feasible. Where not feasible, non-media alternatives would be acceptable. Where necessary, different materials can be used in series for different phases of a composite aseptic process incapable of accommodating a single material. In any event, some material simulating the sterile BPC, or the sterile BPC itself, must pass through the entire system that is intended to be sterile. Any material used for process validation must be microbiologically inert.

Environmental and personnel monitoring must be performed during validation, in a manner and amount sufficient to establish appropriate monitoring limits for routine production.

At least three consecutive, successful validation runs are necessary before an aseptic process can be considered to be validated.

Alternative proposals for the validation of the aseptic processing of bulk pharmaceuticals will be considered by FDA on a case-by-case basis. For example, it may be acceptable to exclude from the aseptic processing validation procedure certain stages of the post-sterilization bulk process that take place in a totally closed system. Such closed systems should be sterilized in place by a validated procedure, integrity tested for each lot, and should not be subject to any intrusions whereby there may be the likelihood of microbial ingress. Suitable continuous system pressurization would be considered an appropriate means for ensuring system integrity.

VIII. Water for Injection

Although water may not be a component of the sterile drug substance, water that comes in contact with the equipment or that enters into the reaction can be a source of impurities (e.g., endotoxins). Therefore, only water for injection should be utilized.

Some manufacturers have attempted to utilize marginal systems, such as single pass Reverse Osmosis (RO) systems. For example, a foreign drug substance manufacturer was using a single pass RO system with post RO sterilizing filters to minimize microbiological contamination. This system was found to be unacceptable. RO filters are not absolute and should therefore be in series. Also, the use of sterilizing filters in a Water for Injection system to mask a microbiological (endotoxin) problem has also been unacceptable. As with environmental monitoring, periodic reports should be reviewed.

If any questionable conditions are found, refer to the Inspection Guide for High Purity Water Systems.

IX. Terminal Sterilization

There are some manufacturers who sterilize bulk powders after processing, by the use of ethylene oxide or dry heat. Some sterile bulk powders can withstand the lengthy times and high temperatures necessary for dry heat sterilization. In the process validation for a dry heat cycle for a sterile powder, important aspects that should be reviewed include: heat penetration and heat distribution, times, temperatures, stability (in relation to the amount of heat received), and particulates.

With regard to ethylene oxide, a substantial part of the sterile bulk drug industry has discontinued the use of ethylene oxide as a "sterilizing" agent. Because of employee safety considerations, ethylene oxide residues in product and the inability to validate ethylene oxide sterilization, its use is on the decline. As a primary means of sterilization, its utilization is questionable because of lack of assurance of penetration into the crystal core of a sterile powder.

Ethylene oxide has also been utilized in the treatment of sterile powders. Its principal use has been for surface sterilization of powders as a precaution against potential microbiological contamination of the sterile powder during aseptic handling.

There are some manufacturers of ophthalmics that continue to use it as a sterilant for the drug used in the formulation of sterile ophthalmic ointments and suspensions. If used as a primary sterilant, validation data should be reviewed. Refer to the Inspection Guide for Topical Products for further discussion.

X. Rework/Reprocessing/Reclamation

As with the principal manufacturing process, reprocessing procedures should also be validated. Additionally, these procedures must be approved in filings.

Review reprocessed batches and data that were used to validate the process. Detailed investigation reports, including the description, cause, and corrective action should be available for the batch to be reprocessed.

XI. Laboratory Testing and Specifications

The sterility testing of sterile bulk substances should be observed. Additionally, any examples of initial sterility test failures should be investigated. The release of a batch, particularly of a sterile bulk drug substance, which fails an initial sterility test and passes a retest is very difficult to justify. Refer to the Microbiological Guide and Laboratory Guide for additional direction.

Particulate matter is another major concern with sterile powders. Specifications for particulate matter should be tighter than the compendial limits established for sterile dosage forms. The subsequent handling, transfer and filling of sterile powders increases the level of particulates. It is also important to identify particulates so that their source can be determined. Review the firm's program for performing particulate matter testing. If there are no official limits established, review their release criteria for particulates, and the basis of their limit.

With regard to residues, since some sterile powders are crystallized out of organic solvents, low levels of these solvents may be unavoidable. In addition to evaluation of the process to assure that minimal levels are established, data used by the firm to establish a residue level should be reviewed. Obviously, each batch should be tested for conformance with the residue specification. Refer to the Inspection Guide for Bulk Drug Substances for additional direction regarding limits for impurities.

XII. Packaging

Sterile bulk drug substances are filled into different type containers which include sterile plastic bags and sterile cans. With regard to sterile bags, sterilization by irradiation is the method of choice because of the absence of residues. There are some manufacturers, particularly foreign, which utilize formaldehyde. A major disadvantage is that formaldehyde residues may and frequently do appear in the sterile drug substance. Consequently, we have reservations about the acceptability of the use of formaldehyde for, container sterilization because of the possibility of product contamination with formaldehyde residues.

If multiple sterile bags are used, operations should be performed in aseptic processing areas. Since the dosage form manufacturer expects all inner bags to be sterile, outer bags should be applied over the primary bag containing the sterile drug in an aseptic processing area. One large manufacturer of a sterile powder only applied the immediate or primary bag in an aseptic processing area. Thus, the outer portion of this primary bag was contaminated when the other bags were applied over this bag in non-sterile processing areas.

With regard to sterile cans, a concern is particulates, which can be generated due to banging and movement. Because of some with trace quantities of aluminum, companies have moved to stainless steel cans.

The firm's validation data for the packaging system should be reviewed. Important aspects of the sterile bag system include residues, pinholes, foreign matter (particulates), sterility and endotoxins. Important aspects of the rigid container systems include moisture, particulates and sterility.

Appendix J

Guide to Inspections of Topical Drug Products

The following is a US Food and Drug Administration (FDA) training guide used by the FDA in the training of their inspectors. Such training guides are used to prepare and provide FDA inspectors with a general knowledge and understanding of what it is they should be considering during inspection of an FDA-regulated manufacturing facility.

In many cases these are general considerations due in large part to the many and varied drugs and the proprietary processes required to manufacture them. These guides are intended to augment the education, experience, and accumulated knowledge each inspector already possesses. The inspection process and procedures by which they perform it leave room for on-site judgment consideration and assessment by the inspectors.

The purpose of including these FDA guides in this book is to provide the reader, designer, and engineer involved with the design and engineering of a bioprocessing facility some insight as to what the expectations of an FDA inspector might be. What expectations an FDA inspector might possess can most likely be attributed to their training and these guidelines. Providing these guidelines for the readers of this book, those of you designing, constructing, and operating FDA-regulated manufacturing facilities, will hopefully lessen the gap of understanding between you and the FDA inspector. Having some insight into the inspector's understanding and training helps one to be better prepared to meet those expectations when it's time for the inspection.

These inspection guidelines are reproduced here through the courtesy of the FDA. These and others also reside on the FDA web site at:

http://www.fda.gov/ICECI/Inspections/InspectionGuides/default.htm

Bioprocessing Piping and Equipment Design: A Companion Guide for the ASME BPE Standard, First Edition. William M. (Bill) Huitt.

Guide to Inspections of Topical Drug Products

Note: This document is reference material for investigators and other FDA personnel. The document does not bind FDA, and does no confer any rights, privileges, benefits, or immunities for or on any person(s).

I. Purpose

The purpose of this guide is to provide field investigators, who are familiar with the provisions of the Current Good Manufacturing Practice (CGMP) regulations for pharmaceuticals, with guidance on inspecting selected facets of topical drug product production. The subjects covered in the guide are generally applicable to all forms of topical drug products, including those that are intended to be sterile. However, this guide does not address every problem area that the investigator may encounter, nor every policy that pertains to topical drug products.

II. Introduction

This inspectional guide addresses several problem areas that may be encountered in the production of topical drug products potency, active ingredient uniformity, physical characteristics, microbial purity and chemical purity. The guide also addresses problems relating to the growing number of transdermal products. If a new drug pre-approval inspection is being conducted, then an examination of the filed manufacturing and control data, and correspondence should be accomplished early in the inspection. As with other pre-approval inspections, the manufacturing and controls information filed in the relevant application should be compared with the data used for clinical batches and for production (validation) batches. Filed production control data should be specific and complete.

III. Potency Uniformity

Active ingredient solubility and particle size are generally important ingredient characteristics that need to be controlled to assure potency uniformity in many topical drug products such as emulsions, creams and ointments. Crystalline form is also important where the active ingredient is dispersed as a solid phase in either the oil or water phase of an emulsion, cream, or ointment.

It is important that active ingredient solubility in the carrier vehicle be known and quantified at the manufacturing step in which the ingredient is added to the liquid phase. The inspection should determine if the manufacturer has data on such solubility and how that data was considered by the firm in validating the process.

Substances which are very soluble, as is frequently the case with ointments, would be expected to present less of a problem than if the drug substance were to be suspended, as is the case with creams. If the drug substance is soluble, then potency uniformity would be based largely upon adequate distribution of the component throughout the mix.

If the active ingredient is insoluble in the vehicle, then in addition to assuring uniformity of distribution in the mix, potency uniformity depends upon control of particle size, and use of a validated mixing process. Particle size can also affect the activity of the drug substance because the smaller the particle size the greater its surface area, which may influence its activity. Particle size also affects the degree to which the product may be physically irritating when applied; generally, smaller particles are less irritating.

Production controls should be implemented that account for the solubility characteristics of the drug substance; inadequate controls can adversely affect product potency, efficacy and safety. For example, in one instance, residual water remaining in the manufacturing vessel, used to produce an ophthalmic ointment, resulted in partial solubilization and subsequent recrystallization of the drug substance; the substance recrystallized in a larger particle size than expected and thereby raised questions about the product efficacy.

In addition to ingredient solubility/particle size, the inspection should include a review of other physical characteristics and specifications for both ingredients and finished products.

IV. Equipment and Production Control

Mixers

There are many different kinds of mixers used in the manufacture of topical products. It is important that the design of a given mixer is appropriate for the type of topical product being mixed. One important aspect of mixer design is how well the internal walls of the mixer are scraped during the mixing process. This can present some problems with stainless steel mixers because scraper blades should be flexible enough to remove interior material, yet not rigid enough to damage the mixer itself. Generally, good design of a stainless steel mixer includes blades which are made of some hard plastic, such as Teflon, which facilitates scrapping of the mixer walls without damaging the mixer.

If the internal walls of the mixer are not adequately scraped during mixing, and the residual material becomes part of the batch, the result may be non-uniformity. Such non-uniformity may occur, for example, if operators use hand held spatulas to scrape the walls of the mixer.

Another mixer design concern is the presence of "dead spots" where quantities of the formula are stationary and not subject to mixing. Where such "dead spots" exist, there should be adequate procedures for recirculation or non-use of the cream or ointment removed from the dead spots in the tank. Ideally, during the inspection, mixers should be observed under operating conditions.

Filling and Packaging

Suspension products often require constant mixing of the bulk suspension during filling to maintain uniformity. When inspecting a suspension manufacturing process determine how the firm assures that the product remains homogeneous during the filling process and audit the data that supports the adequacy of the firm's process. When the batch size is large and the bulk suspension is in large tanks, determine how the firm deals with low levels of bulk suspension near the end of the filling process. Does the bulk suspension drop below a level where it can be adequately mixed? Is residual material transferred to a smaller tank? Does the firm rely upon hand mixing of the residual material? The firm should have demonstrated the adequacy of the process for dealing with residual material.

Process Temperature Control

Typically, heat is applied in the manufacture of topicals to facilitate mixing and/or filling operations. Heat may also be generated by the action of high energy mixers. It is important to control the temperature within specified parameters, not only to facilitate those operations, but also to assure that product stability is not adversely affected. Excessive temperatures may cause physical and/or chemical degradation of the drug product, vehicle, the active ingredient(s), and/or preservatives. Furthermore, excessive temperatures may cause insoluble ingredients to dissolve, re-precipitate, or change particle size or crystalline form.

Temperature control is also important where microbial quality of the product is a concern. The processing of topicals at higher temperatures can destroy some of the objectionable microorganisms that may be present. However, elevated temperatures may also promote incubation of microorganisms.

Temperature uniformity within a mixer should be controlled. In addressing temperature uniformity, firms should consider the complex interaction among vat size, mixer speed, blade design, viscosity of the contents and the rate of heat transfer. Where temperature control is critical, use of recording thermometers to continuously monitor/document temperature measurements is preferred to frequent manual checks. Where temperature control is not critical, it may be adequate to manually monitor/document temperatures periodically by use of hand held thermometers.

V. Cleaning Validation

It is CGMP for a manufacturer to establish and follow written SOPs to clean production equipment in a manner that precludes contamination of current and future batches. This is especially critical where contamination may present direct safety concerns, as with a potent drug, such as a steroid (e.g., cortisone, and estrogen), antibiotic, or a sulfa drug where there are hypersensitivity concerns.

The insolubility of some excipients and active substances used in the manufacture of topicals makes some equipment, such as mixing vessels, pipes and plastic hoses, difficult to clean. Often, piping and transfer lines are inaccessible to direct physical cleaning. Some firms address this problem by dedicating lines and hoses to specific products or product classes.

It is therefore important that the following considerations be adequately addressed in a firm's cleaning validation protocol and in the procedures that are established for production batches.

Detailed Cleaning Procedures

Cleaning procedures should be detailed and provide specific understandable instructions. The procedure should identify equipment, cleaning method(s), solvents/detergents approved for use, inspection/release mechanisms, and documentation. For some of the more complex systems, such as clean-in-place (CIP) systems, it is usually necessary to provide a level of detail that includes drawings, and provision to label valves. The time that may elapse from completion of a manufacturing operation to initiation of equipment cleaning should also be stated where excessive delay may affect the adequacy of the established cleaning procedure. For example, residual product may dry and become more difficult to clean.

Sampling Plan For Contaminants

As part of the validation of the cleaning method, the cleaned surface is sampled for the presence of residues. Sampling should be by an appropriate method, selected based on factors such as equipment and solubility of residues. For example, representative swabbing of surfaces is often used, especially in hard to clean areas and/or where the residue is relatively insoluble. Analysis of rinse solutions for residues has also been shown to be of value where the residue is soluble and/or difficult to access for direct swabbing. Both methods are useful when there is a direct measurement of the residual substance. However, it is unacceptable to test rinse solutions (such as purified water) for conformance to the purity specifications for those solutions, instead of testing directly for the presence of possible residues.

Equipment Residue Limits

Because of improved technology, analytical methods are becoming much more sensitive and capable of determining very low levels of residues. Thus, it is important that a firm establish appropriate limits on levels of post-equipment cleaning residues. Such limits must be safe, practical, achievable, verifiable, and must ensure that residues remaining in the equipment will not cause the quality of subsequent batches to be altered beyond established product specifications. During inspections, the rationale for residue limits should be reviewed.

Because surface residues will not be uniform, it should be recognized that a detected residue level may not represent the maximum amount that may be present. This is particularly true when surface sampling by swabs is performed on equipment.

VI. Microbiological

Controls (Non-sterile Topicals)

The extent of microbiological controls needed for a given topical product will depend upon the nature of the product, the use of the product, and the potential hazard to users posed by microbial contamination. This concept is reflected in the Current Good Manufacturing (CGMP) regulations at 21 Code of Federal Regulations (CFR) 211.113(a) (Control of microbiological contamination), and in the U.S. Pharmacopeia (USP). It is therefore vital that manufacturers assess the health hazard of all organisms isolated from the product.

Deionized Water Systems for Purified Water

Inspectional coverage should extend to microbiological control of deionized water systems used to produce purified water. Deionizers are usually excellent breeding areas for microorganisms. The microbial population tends to increase as the length of time between deionizer service periods increases. Other factors which influence microbial growth include flow rates, temperature, surface area of resin beds and, and of course, the microbial quality of the feed water. These factors should be considered in assessing the suitability of deionizing systems where microbial integrity of the product incorporating the purified water is significant. From this assessment, a firm should be able to design a suitable routine water monitoring program and a program of other controls as necessary.

It would be inappropriate for a firm to assess and monitor the suitability of a deionizer by relying solely upon representations of the deionizer manufacturer. Specifically, product quality could be compromised if a firm had a deionizer serviced at intervals based not on validation studies, but rather on the "recharge" indicator built into the unit. Unfortunately, such indicators are not triggered by microbial population, but rather they are typically triggered by measures of electrical conductivity or resistance. If a unit is infrequently used, sufficient time could elapse between recharging/sanitizing to allow the microbial population to increase significantly.

Pre-use validation of deionizing systems used to produce purified water should include consideration of such factors as microbial quality of feed water (and residual chlorine levels of feed water where applicable), surface area of ion-exchange resin beds, temperature range of water during processing, operational range of flow rates, recirculation systems to minimize intermittent use and low flow, frequency of use, quality of regenerant chemicals, and frequency and method of sanitization.

A monitoring program used to control deionizing systems should include established water quality and conductivity monitoring intervals, measurement of conditions and quality at significant stages through the deionizer (influent, post cation, post anion, post mixed-bed, etc.), microbial conditions of the bed, and specific methods of microbial testing. Frequency of monitoring should be based upon the firm's experience with the systems.

Other methods of controlling deionizing systems include establishment of water quality specifications and corresponding action levels, remedial action when microbial levels are exceeded, documentation of regeneration and a description of sanitization/ sterilization procedures for piping, filters, etc.

Microbiological Specifications and Test Methods

During inspections it is important to audit the microbiological specifications and microbial test methods used for each topical product to assure that they are consistent with any described in the relevant application, or U.S.P. It is often helpful for the inspection to include an FDA microbiologist.

Generally, product specifications should cover the total number of organisms permitted, as well as specific organisms that must not be present. These specifications must be based on use of specified sampling and analytical procedures. Where appropriate, the specifications should describe action levels where additional sampling and/or speciation of organisms is necessary.

Manufacturers must demonstrate that the test methods and specifications are appropriate for their intended purpose. Where possible, firms should utilize methods that isolate and identify organisms that may present a hazard to the user under the intended use. It should be noted that the USP does not state methods that are specific for water insoluble topical products.

One test deficiency to be aware of during inspections is inadequate dispersement of a cream or ointment on microbial test plates. Firms may claim to follow USP procedures, yet in actual practice may not disperse product over the test plate, resulting in inhibited growth due to concentrated preservative in the non- dispersed inoculate. The spread technique is critical and the firm should have documentation that the personnel performing the technique have been adequately trained and are capable of performing the task. Validation of the spread plate technique is particularly important where the product has a potential antimicrobial affect.

In assessing the significance of microbial contamination of a topical product, both the identification of the isolated organisms and the number of organisms found are significant. For example, the presence of a high number of organisms may indicate that the manufacturing process, component quality, and/or container integrity may be deficient. Although high numbers of non-pathogenic organisms may not pose a health hazard, they may affect product efficacy and/or physical/ chemical stability. Inconsistent batch to batch microbial levels may indicate some process or control failure in the batch. The batch release evaluation should extend to both organism identification and numbers and, if limits are exceeded, there should be an investigation into the cause.

Preservative Activity

Manufacturing controls necessary to maintain the anti- microbiological effectiveness of preservatives should be evaluated by the firm. For example, For those products that separate on standing, the firm should have data that show the continued effectiveness of the preservative throughout the product's shelf-life.

For preservative-containing products, finished product testing must ensure that the specified level of preservative is present prior to release. In addition, preservative effectiveness must be monitored as part of the final on-going stability program. This can be accomplished through analysis for the level of preservative previously shown to be effective and/or through appropriate microbiological challenge at testing intervals.

For concepts relating to sterility assurance and bioburden controls on the manufacture of sterile topicals see the Guideline on Sterile Drug Products Produced by Aseptic Processing.

VII. Change Control

As with other dosage forms, it is important for the firm to carefully control how changes are made in the production of topical products. Firms should be able to support changes which represent departures from approved and validated manufacturing processes.

Firms should have written change control procedures that have been reviewed and approved by the quality control unit. The procedures should provide for full description of the proposed change, the purpose of the change, and controls to assure that the change will not adversely alter product safety and efficacy. Factors to consider include potency and/or bioactivity, uniformity, particle size (if the active ingredient is suspended), viscosity, chemical and physical stability, and microbiological quality.

Of particular concern are the effects that formulation and process changes may have on the therapeutic activity and uniformity of the product. For example, changes in vehicle can affect absorption, and processing changes can alter the solubility and microbiological quality of the product.

VIII. Transdermal Topical Products

Inspections of topical transdermal products (patches) have identified many problems in scale-up and validation. Problems analogous to production of topical creams or ointments include uniformity of the drug substance and particle size in the bulk gel or ointment. Uniformity and particle size are particularly significant where the drug substance is suspended or partially suspended in the vehicle. Viscosity also needs control because it can affect the absorption of the drug; the dissolution test is important in this regard.

Other areas that need special inspectional attention are assembly and packaging of the patch, including adhesion, package integrity (regarding pinholes) and controls to assure that a dose is present in each unit.

Because of the many quality parameters that must be considered in the manufacture and control of a transdermal dosage form, scale- up may be considerably more difficult than for many other dosage forms. Therefore, special attention should be given to evaluating the adequacy of the process validation efforts. As with other dosage forms, process validation must be based on multiple lots, typically at least three consecutive successful batches. Inspection of summary data should be augmented by comparison to selected data contained in supporting batch records, particularly where the data appear unusually uniform or disparate. Given the complexities associated with this dosage form, you may encounter tolerances and/or variances broader than for other dosage forms. In addition, batches may not be entirely problem-free. Nevertheless, the firm should have adequate rationale for the tolerances and production experiences, based on appropriate developmental efforts and investigation of problems.

IX. Other References

Other relevant inspection guides that should be used in conjunction with this guide include:

- Guide to Inspections of Validation of Cleaning Processes.
- Guide to Inspections of High Purity Water Systems

Appendix K

BPE History—Letters and Notes

As referenced from Section B "Early History and Development of the ASME BPE Standard" of the Preface, the following documents attest to the determination and methodical work that went into creating the Bioprocessing Equipment (BPE) Standard. These are first hand evidentiary accounts of communications, memorandums, notifications, and Newsletter clippings pulled from dust covered file boxes and yellowed file folders that were thought to be long gone.

It was recognized in 2013 that with the upcoming 25 year anniversary of the founding of the BPE Standard this would be an opportune time, if ever it were to happen, to document the history of the development of the BPE Standard. I had no idea how opportune it was.

This turned out to be a point in time in which many of the original founding members of the Standard were still very active in the Standard. It turns out too that a number of these members, once they located and begin digging around their packed up and forgotten file boxes, discovered they still had documents that they had stuck away and forgotten about. I have trouble locating work I did last week, much less something I would have thought, in all likelihood, I might never need again from 25 years ago.

This detailed account with first hand recollections and hard copy documents telling the story of how the BPE Standard was created is analogous to the history of how other codes and standards were created, but the details lost, forgotten, or resting in a dusting and decaying file box in some basement.

Bioprocessing Piping and Equipment Design: A Companion Guide for the ASME BPE Standard, First Edition. William M. (Bill) Huitt.

Codes and Standards

The American Society of Mechanical Engineers

Pressure Technology
212-705-7087

345 East 47th Street
New York, NY 10017

December 6, 1989

David Smith; Ld PV Engr
United Engrs & Constructors
Stearns Catalytic Div
30 South 17th St
PO Box 8223
Philadelphia, PA 19101

Dear Mr. Smith:

It gives me sincere pleasure to advise you of your appointment to serve as Chairman and Member of the Ad Hoc Committee on Bioprocessing Equipment (BPE) (FN10120000) effective December 6, 1989 for a term expiring June 30, 1992 in the Interest Category of Designer/Constructor (AB).

We understand this appointment is agreeable to you and that you will be able to allocate the time necessary for this activity. We are very pleased that you can take on this assignment and I express ASME's appreciation for your efforts in this regard. Would you please forward to ASME Headquarters a signed copy of the attached form indicating your acceptance.

Sincerely,

Walter R. Mikesell Jr.

Walter R. Mikesell, Chmn, Board on PTCS
Vice Pres, Pressure Technology Codes and Standards

Reference 01 – ASME Notification to Dave Smith Approving His Appointment as Chair of the Ad Hoc Committee on BPE_1 Page

Codes and Standards

The American Society of Mechanical Engineers

Pressure Technology
212-705-7087

345 East 47th Street
New York, NY 10017

December, 1989

This letter will explain the current status of the formation of the ASME's Standards Writing Committee for Biopharmaceutical Process Equipment. We will be holding an organizational meeting sometime after the Bioprocess Engineering Program to be held during ASME's Winter Annual Meeting, December 10-15 in San Francisco. If you are not already planning to attend, we encourage your participation, especially in the session on Wednesday, December 13, 1989 on "Codes and Standards in the Bioprocess Industry." For more information on that program, call Julie T. Lee, ASME Headquarters, 212-705-7797.

On June 20, 1989, the ASME Codes and Standards Council approved the Biopharmaceutical Process Engineering Standards project(BPE) and assigned it to the Board on Pressure Technology Codes and Standards, with the following tentative scope: "This standard is intended for design, materials, construction, inspection, and testing of vessels, piping, and related accessories; such as pumps, valves, and fittings for use in the biopharmaceutical industry. The rules provide for adoption of other ASME and related national standards, and when so referenced become a part of the standard." At this time an Ad Hoc Committee is being formed to:

1) develop a recommendation with respect to the formation of a Main Committee to develop and administer Codes for Bioprocessing Equipment;

2) develop a proposed charter for the Committee, if recommended; and

3) develop proposed preliminary operating procedures for the Committee, if recommended.

In order to comply with the item above, please complete the enclosed Personnel Form, including a resume. Be sure to indicate what interest category you fall under, or what category you think you fall under. In your transmittal, include a statement of your company's commitment of support of your participation on the Committee. If possible, please return your response to us in the enclosed envelope by November 24th.

The American Society of Mechanical Engineers requires that all of its Codes and Standards Committees be balanced in relation to member interests.

Reference 02-01 – ASME Solicitation for Prospective Members for the Ad Hoc Committee on BPE_3 Pages

BPE Membership Letter
December, 1989
Page 2 of 2

Anticipated membership categories include:

- **Design/Constructor (AC):** An organization performing design, facility design and erection, or design related services;
- **General Interest (AF):** Consultants, educators, research and development organization personnel, and public interest persons;
- **Insurance/Inspection (AH):** An organization which insures equipment and/or provides required independent inspection of the manufacture and installation of components, parts, and items;
- **Manufacturer (AK):** An organization producing components or assemblies;
- **Materials Manufacturer (AM):** An organization producing materials or ancillary materials related accessories or component parts;
- **Regulatory (AT):** An agency or organization which regulates the design, manufacture, installation, or operation of the products(s) covered by the applicable standard(s); and
- **User (AW):** An organization which uses a product(s) covered by the applicable standard(s).

Membership on ASME Committees requires a commitment from both members and their employers. Members will be appointed to the Committee for a maximum term of five years, during which time they will be required to attend 3-4 meetings per year. Meeting locations will be decided upon by the Committee, when formed. In addition to travel and attendance commitments, members must be prepared to devote time to committee ballots and correspondence.

Note that the initial Ad Hoc Committee membership will be small for ease in organizing the structure. Once the Committee has been established, membership will be increased. Therefore, please be patient if you are not initially appointed as a member of the Biopharmaceutical Process Equipment Standards Committee.

All meetings are open to the public, except when discussing personnel, accreditation, or legal matters. We urge your continued contact with us to keep abreast of scheduled meetings. Feel free to contact us if you have any questions, and we look forward to your participation.

Arlene A. Spadafino; Secretary, BPTCS
Director, Pressure Technology Codes and Standards
(212) 705-7030 FAX (212) 705-7533

Reference 02-02 – ASME Solicitation for Prospective Members for the Ad Hoc Committee on BPE_3 Pages

Codes and Standards

The American Society of Mechanical Engineers

345 East 47th Street
New York, NY 10017

ASME CODES & STANDARDS COMMITTEE PERSONNEL FORM (PF-1 5/89)

Name: ____________ Date: ____________

Title: ____________ Committee: ____________

Company: ____________ ____________

Address: ____________ ____________

____________ [ ] Member

____________ [ ] Alt. to: ____________

Tel No : ____________ [ ] Officer: ____________

FAX No: ____________

ASME Membership Grade: [ ] Fellow [ ] Associate Member [ ] Applicant
[ ] Member [ ] Executive Affiliate [ ] None

Professional Engineers License: [ ] No [ ] Yes States: ____________

Degrees: ____________ Date of Birth (YY/MM/DD): ____________

Current ASME Committee membership: [ ] None [ ] Listed below:

Membership in other technical organizations: [ ] None [ ] Listed below:

Experience and qualifications for appointment (use attachment if necessary):

Interest Class: ____________
Designated Liaison
(if applicable): ____________

Reference 02-03 – ASME Solicitation for Prospective Members for the Ad Hoc Committee on BPE_3 Pages

Codes and Standards

The American Society of Mechanical Engineers

Pressure Technology
212-705-7087

345 East 47th Street
New York, NY 10017

February 23, 1990

Frank J. Manning; Vice Pres.
TEK Supply, Inc.
9 Walkup Drive
Westboro, MA 01581

Dear Mr. Manning:

It gives me sincere pleasure to advise you of your appointment to serve as a Member on the Ad Hoc Committee on Bioprocessing Equipment (BPE) (FN10120000) effective February 18, 1990 for a term expiring June 30, 1992 in the Interest Category of Material Manufacturer (AM).

We understand this appointment is agreeable to you and that you will be able to allocate the time necessary for this activity. We are very pleased that you can take on this assignment and I express ASME's appreciation for your efforts in this regard. Would you please forward to ASME Headquarters a signed copy of the attached form indicating your acceptance.

Walter R. Mikesell Jr

Walter R. Mikesell; Vice President, PTCS
Chairman, Board on Pressure Technology
Codes and Standards

enc.

Reference 03-01 – ASME Notice of Membership Appointment to the Ad Hoc BPE Committee_2 Pages

Codes and Standards

Pressure Technology
212-705-7087

The American Society of Mechanical Engineers

345 East 47th Street
New York, NY 10017

February 23, 1990

Lloyd J. Peterman; Mgr.
Tri-Clover, Inc.
Intl. Sales Engrg.
9201 Wilmot Road
Kenosha, WI 53141

Dear Mr. Peterman:

It gives me sincere pleasure to advise you of your appointment to serve as a Member on the Ad Hoc Committee on Bioprocessing Equipment (BPE) (FN10120000) effective February 18, 1990 for a term expiring June 30, 1992 in the Interest Category of Material Manufacturer (AM).

We understand this appointment is agreeable to you and that you will be able to allocate the time necessary for this activity. We are very pleased that you can take on this assignment and I express ASME's appreciation for your efforts in this regard. Would you please forward to ASME Headquarters a signed copy of the attached form indicating your acceptance.

Walter R. Mikesell; Vice President, PTCS
Chairman, Board on Pressure Technology
Codes and Standards

enc.

Reference 03-02 – ASME Notice of Membership Appointment to the Ad Hoc BPE Committee_2 Pages

Notice: These Minutes are subject to approval and are for committee use only. They are not to be duplicated or quoted for other than committee business.

MINUTES

ASME
Ad Hoc Committee on Bioprocessing Equipment
Meeting of
March 20, 1990
Room 125
United Engineering Center
New York City, NY

90-3-1 Call to Order

Chairman David Smith called the meeting to order at 10:10 AM.

Attendance

Members Present

David Smith, Chairman
William H. Cagney, Vice C'man
Mark E. Sheehan, Secretary
Hans Koning-Bastiaan
Ivy Logsdon
Frank J. Manning
Richard E. Markovitz
Theodore Mehalko
Lloyd J. Peterman
Joseph VanHouten
Frederick D. Zikas

Members Absent

None.

Visitors

Pat Banes	Oakley Services Co., Inc.
David Baram	Vanasyl Valves, Ltd.
Nigel Brooks	Fluor Daniel
Robert Daggett	Allegheny Bradford Corp.
William M. Dodson	Precision Stainless
Randolph Greasham	Merck & Co., Inc.
Barbara Henon	Arc Machines, Inc.
Peter Leavesly	Membrex, Inc.
Julie Lee	ASME BPEP Manager
Tom Ransohoff	Dorr-Oliver
Arlene Spadafino	ASME Codes and Standards
Rick Zinkowski	ITT Engineered Valves

Reference 04-01 – Minutes of the March 20, 1990 Meeting of the Ad Hoc Committee on BPE_13 Pages

MINUTES ASME Ad Hoc Committee
on Bioprocessing Equipment
March 20, 1990

90-3-2 Introductions

The Chairman asked the members and guests to stand and identify themselves and their affiliations.

90-3-3 Chairman's Opening Statement

The Chairman announced that some of the members did not receive the correct letters of appointment and that the mixup is being corrected. He then recapped activities since the ASME Winter Annual Meeting noting that since that time, a decided interest in this activity has been expressed by many people. Events of note were:

- a decision to operate as an ASME Committee
- a call for participation based on BPEP mailing lists
- delivery of a charge from the ASME Council on Codes and Standards to the Board on Pressure Technology to initiate this project with a specific statement of scope
- formation of this Ad Hoc Committee by ASME appointment procedures
- charge to this Ad Hoc Committee to decide whether to go ahead with development of a standard.

The Chairman then stated that the purpose of this meeting is to determine the need for a standard and to initiate preparation of a presentation to the Board on PTCS describing said need. Since these meetings are to be conducted under ASME policies, all meetings are open to the public and guests are invited to participate in discussions, but not in voted actions.

90-3-4 Adoption of Agenda

The Agenda for the March 20, 1990 meeting of the Ad Hoc Bioprocessing Equipment Committee was unanimously adopted as distributed with the Secretary's February 21, 1990 transmittals.

90-3-5 Review of Board on PTCS Charge

The Chairman returned the discussion to the charge to this committee as handed down from the Council on Codes and Standards via the Board on Pressure Technology Codes and Standards. This Ad Hoc Committee has been charged with the development of a recommendation to the Board on PTCS with respect to the formation of a Main Committee to develop and administer Standards for bioprocessing equipment. The scope of such standards was stated by the Council as follows:

"This standard is intended for design, materials, construction, inspection, and testing of vessels, piping, and related accessories; such as pumps, valves, and fittings for use in the biopharmaceutical industry. The rules provide for adoption of other ASME and related national standards and when so referenced become a part of the standard."

Reference 04-02 – Minutes of the March 20, 1990 Meeting of the Ad Hoc Committee on BPE_13 Pages

MINUTES ASME Ad Hoc Committee
on Bioprocessing Equipment
March 20, 1990

Assuming that this committee determines that there is a need for these standards and that the Board on PTCS approves the formation of a new Main COmmittee, the next task for this froup would be to develop a proposed charter for the committee and subsequently develop preliminary operating procedures for the Main Committee. The development of the charter may involve some fine tuning of the scope of the standards to be developed.

90-3-6 <u>Ad Hoc Committee Operating Procedures</u>

The Secretary presented a Handout showing some procedural guidelines by which the Ad Hoc Committee will operate, and this is shown on pages 7 - 12 of these Minutes. Included in this package is a description of balloting procedures for letter ballots. This was included to give the members a feel for what is involved in a letter ballot and what their responsibilities would be if one were issued. The Chairman noted that for the time being, he only envisions taking such actions as could be voice-voted at committee meetings. The members should, however, read these procedures and familiarize themselves with the mechanics of ballots.

A discussion began on the Ad Hoc nature of this committee and whether that designation implies that this is a preliminary group. Ms. Spadafino gave an overview of the formation of this group and its specific charge, which translates into a justification of the need for a Main Committee. When that is accomplished, the charge to this group will be fulfilled and the committee will evolve from there, based on the decision of the Board.

90-3-7 <u>Formulation of Activities</u>

The Chairman began this discussion by directing attention to the BPEP Newsletter, in which the topics of concern in the bioprocessing industry were delineated. A copy of an excerpt from that newsletter is shown on page 13 of these Minutes. Several members emphasized the importance of cleanability of equipment as essential to the operation of a bioprocess. Other members pointed out deficiencies in existing standards, including ASME and EPA standards. In the area of welding, bioprocessing applications require a greater emphasis on the quality of the surface of the weld, assuming that the needed strength is present. This leads to a conflict in philosophy; most ASME standards are primarily concerned with pressure integrity and safety, while the needs of the bioprocessing industry go beyond that, to the point where equipment construction specifications are largely determined by the process requirements (cleanability, economics, etc.). This also causes difficulty in creating a standard for an industry where there are so many varied processes and applications. There was general agreement that standardized definitions and terminology are needed as a benchmark for further development. As a correlary, standards need to be developed to improve communication between the User and the Designer/Fabricator. Since the needs of the industry apply equally to vessels, piping, appurtenances and other equipment covered by existing individual standards, some kind of standard to bring

- 3 -

Reference 04-03 – Minutes of the March 20, 1990 Meeting of the Ad Hoc Committee on BPE_13 Pages

MINUTES ASME Ad Hoc Committee
on Bioprocessing Equipment
March 20, 1990

them all together as a system is needed, without infringing on the scopes of the existing standards. This led to discussion on what exists presently in the way of standards and what areas need to be improved. When that discussion wound down, the Chairman began to delineate specific needs as follows:

1) Existing pressure vessel, component and piping codes are inadequate or incomplete in covering critical areas of design, fabrication and installation of bioprocessing equipment, such as:

   a) selection of materials and additional requirements for material manufacture
   b) specialized welding specifications
   c) requirements for establishment and verification of surface conditions
   d) permissible standards for fittings, valves, closures, etc.
   e) special rules for containment of toxics
   f) more stringent rules for dimensional tolerances and measurement
   g) establishment of standardized methods of communication between Users and Designer/Fabricators.

2) There is a need to establish a standard which would oversee all aspects of design, fabrication and installation of the entire bioprocessing equipment <u>system</u>, including vessels, piping, appurtenances, pressure relief, final inspection and verification.

3) There is an immediate need throughout the bioprocessing industry for standardized language encompassing the entire spectrum of bioprocessing operations, including equipment, processes and products.

At this point in the discussion, the Chairman stated that he would entertain a motion on how the committee should proceed. It was moved, seconded and unanimously

<u>VOTED</u> to proceed to petition the Board on Pressure Technology Codes and Standards for establishment of a Main Committee for the development and administration of standards for bioprocessing equipment.

90-3-8 <u>Formation of Task Groups</u>

The Chairman called for volunteers to be members of a Task Group assigned with development of the text of the presentation which the Chairman will make to the Board on PTCS. Mr. Cagney agreed to Chair the TG with Mr. Koning-Bastiaan and Mr. Mehalko as members. The Secretary agreed to send them printouts of the needs described in the previous section of these Minutes by FAX within 24 hours and to include a copy of the scope as described by the Council. Mr. Cagney was also directed to include in the text of the presentation a petition to appoint the existing Ad Hoc Committee members as the first members of the Main Committee, keeping the current officers, as well.

- 4 -

Reference 04-04 – Minutes of the March 20, 1990 Meeting of the Ad Hoc Committee on BPE_13 Pages

MINUTES ASME Ad Hoc Committee
on Bioprocessing Equipment
March 20, 1990

90-3-9 Long Range Schedule of Activities

The Chairman realized that it could be possible to complete the necessary preparations to allow him to make the presentation to the Board on PTCS at their June 6, 1990 meeting. Working back from that date, he developed the following schedule of activities:

3/21/90 Printout of needs transmitted to TG members.

4/15/90 Text of presentation to the Board on PTCS completed by the TG and submitted to the Secretary for distribution to the Ad Hoc Committee members for review and approval.

5/1/90 Review period completed and any comments received by the Secretary are distributed to the TG for resolution (copies sent to all members).

5/15/90 Resolution of comments completed and final version of the presentation submitted to the Secretary of the Board on PTCS (A. Spadafino) for inclusion in the Agenda for their June 6, 1990 meeting.

6/6/90 Chairman D. Smith makes presentation to the Board on PTCS and requests formation of a new Bioprocessing Equipment Main Committee with membership based on that of the Ad Hoc Committee.

6/15/90 (Assuming affirmative action by the Board) Submittal of additional candidates to the Board on PTCS for approval as members of the new Main Committee, filling out the roster to 25 members.

90-3-10 New Business/Closing Remarks

There was no new business to be discussed at this time.

90-3-11 Next Meeting Date

The Chairman announced that it is his intention to hold three meetings a year for starters with a basic format consisting of one day of Subcommittee meetings followed by a second day when the Main Committee would meet. Anticipating timely action by the Board on PTCS, it was decided to hold a meeting as sson as possible after the Board takes action. Therefore, the next meeting of the Bioprocessing Equipment Committee (Ad Hoc or otherwise) was scheduled for Wednesday, June 27, 1990, at the United Engineering Center in New York. Meeting time would be 10 AM to 4 PM and meeting announcements and an Agenda will be issued sometime before the end of May.

- 5 -

Reference 04-05 – Minutes of the March 20, 1990 Meeting of the Ad Hoc Committee on BPE_13 Pages

MINUTES ASME Ad Hoc Committee
on Bioprocessing Equipment
March 20, 1990

90-3-12 Adjournment

The meeting was adjourned at 3:25 PM.

Respectfully submitted,

Mark E. Sheehan
Secretary, Ad Hoc Committee
on Bioprocessing Equipment

MES/mes

- 6 -

Reference 04-06 – Minutes of the March 20, 1990 Meeting of the Ad Hoc Committee on BPE_13 Pages

Meeting Copy

PROCEDURAL GUIDELINES

RELATED TO THE ACTIVITIES OF

THE AD HOC BIOPROCESSING EQUIPMENT COMMITTEE

Section 2.0 ORGANIZATION (pg 1)

Gives an overall look at the make-up of the BPE committee and describes the mechanism for establishment of Task Groups.

Section 6.0 AD HOC COMMITTEE ACTIONS (pgs 2 - 5)

Describes the mechanism for voted actions of the Ad Hoc Committee, including voting categories and steps to be taken when negatives and comments are received. Also describes how ballots are initiated. Note that some of the passages refer to standards approval, which this committee is not concerned with at the present time. These passages are included to give the members a taste of standards approval procedures.

- 7 -

Reference 04-07 – Minutes of the March 20, 1990 Meeting of the Ad Hoc Committee on BPE_13 Pages

2.0 ORGANIZATION

2.1 The ASME Ad Hoc Bioprocessing Equipment Committee (BPE) is responsible to the Board on Pressure Technology Codes & Standards (BPTCS).

The Board on Pressure Technology Codes & Standards is a Supervisory Board responsible for the management of all ASME activities related to codes, standards, guidelines, and accreditation programs directly applicable to non-nuclear pressure containing equipment.

2.2 Ad Hoc BPE Committee

This Committee is made up of a number of members to be determined by the Chairman based on the needs of the group. Appointment will be for a term not to exceed three years, subject to the approval of the Board on PTCS. A balance of interest will be maintained such that no interest category has membership which exceeds 50% of the total membership.

→ 33%

2.3 Task Groups

When a need arises for a specialized review group, the Chairman will appoint such Task Groups from the membership of the Ad Hoc Committee. A Task Group Chairman will be assigned who will be responsible for reporting the results of the TG's review to the Ad Hoc Committee.

(- 1 -)

Reference 04-08 – Minutes of the March 20, 1990 Meeting of the Ad Hoc Committee on BPE_13 Pages

6.0 AD HOC COMMITTEE ACTIONS

6.1 At Ad Hoc Committee meetings, a quorum shall consist of more than fifty percent of the eligible voting members. All meetings called shall commence and continue at the discretion of the Chairman regardless of the attendance at that meeting. Actions taken at a meeting where a quorum is not present are not official until approved by more than fifty percent of the members, by letter ballot, or at a subsequent meeting.

6.2 Under normal circumstances, the Ad Hoc Committee will take action only on items which have cleared a Task Group. Any new business, unless it is extremely simple and clear cut, shall be referred to the appropriate Task Group for study and recommendations. In addition, an item which has been submitted to the Chairman or Secretary or to one of the Task Group Chairmen prior to a committee meeting, which has not been discussed by an appropriate Task Group, shall not be brought before the Ad Hoc Committee until the Task Group has acted upon it.

6.3 On question of Parliamentary Procedure not covered in these operating procedures, "Roberts Rules of Order" shall prevail.

6.4 Balloting/Voting Procedures

Committee actions involving balloting/voting procedures are of two types:

6.4.1 Approval of standards actions (approvals, new standards, revisions, withdrawals, reaffirmations, etc.) or review of documents submitted under the canvass method, by letter ballot.

6.4.2 Approval of personnel and administrative items or actions relating to policy or ASME position.

6.5 Authorization of Ballots

6.5.1 A letter ballot may be authorized by any of the following:

(1) Committee Officer
(2) BPTCS
(3) A majority vote of those members present in Ad Hoc Committee meeting
(4) Petition of five members of the Ad Hoc Committee

6.5.2 Voting Obligations. Each member shall exercise his voting privilege within prescribed time limits. When a member fails repeatedly to return ballots when due, or consistently abstains from voting, the member shall be subject to termination by the Committee. The individual or organization may appeal such action.

(- 2 -)

Reference 04-09 – Minutes of the March 20, 1990 Meeting of the Ad Hoc Committee on BPE_13 Pages

6.6 Letter Ballot Procedure

6.6.1 The approved form of letter ballot (Appendix B) contains four forms of response: "approved", "no objection", "disapproved with reason", and "objection with reason". All ballots shall also give the option of "abstaining with reason" or "not voting with reason".

"Not Voting with reason", to be used for conflict or apparent conflict of interest, signifies neither approval nor disapproval, but the total committee voting membership is reduced by one. "Abstain with reason" signifies neither approval nor disapproval, but the total committee voting membership count remains unaffected.

"Approved", "abstain with reason", "not voting with reason", and "disapproved with reason" are to be used as a set when voting on any issue other than ASME position on a standard submitted through the ANSI canvass method.

6.6.2 Letter ballots shall be closed upon receipt of all ballots but not later than six weeks of the date of issue. Members of the Committee are encouraged to take action on letter ballots as soon as possible.

6.6.3 After two thirds of the balloting period has elapsed, a follow-up will be sent to those who have not returned the ballot. Ballots received by the Secretary after the close of the voting period will not be considered in the evaluation of the first ballot, unless an extension of the balloting period has been established by the Chairman. At the close of the voting period, the Secretary shall submit a complete voting tally to the Chairman of the Ad Hoc Committee who shall determine the most expeditious method of resolving any outstanding negative votes (see para. 6.6.6).

6.6.4 One negative ballot shall be sufficient to require reconsideration of the question. All negative ballots shall be accompanied by the reasons for the objection.

(- 3 -)

– 10 –

Reference 04-10 – Minutes of the March 20, 1990 Meeting of the Ad Hoc Committee on BPE_13 Pages

6.6.5 All comments and objections to proposals shall be referred to the concerned Task Group for consideration. However, the Task Group is responsible for development of a resolution of comments and objections. Persons making adverse comments shall be contacted by the Task Group and invited to work with the Task Group to resolve adverse comments. A response shall be made to each commentator for indicating the action taken and the disposition of their comments.

6.6.6 Comments and objections that are resolved or withdrawn without change or with editorial change require no further action. Any member who wishes to change his vote, shall so indicate within thirty days. Resolutions accomplished by substantive change require ballot of changes. When comments or objections are not resolved, the subcommittee must report to the Ad Hoc Committee that the issue is either withdrawn or submitted for an overriding vote. Reasons for either position must accompany the report.

6.6.7 The Ad Hoc Committee shall consider the Task Group report at its meeting subsequent to receiving the report. No action is required of the Ad Hoc Committee on items withdrawn. Editorial changes must be considered by the Ad Hoc Committee.

Items on which a substantive change has been recommended shall be reballoted in accordance with para. 6.6.4. Items recommended to be overriden shall be discussed after which a vote shall be taken. Two thirds of the total membership, either at the meeting or by letter ballot, shall be required to affirm the original position. Not voting ballots shall be deducted from the membership in determining the two thirds majority.

6.6.8 When a proposal of the type described in para. 6.4.1 has been approved (see para. 6.7) by the Ad Hoc Committee, the Secretary shall submit it to the BPTCS. Such submittals shall be accompanied by the voting record in the Ad Hoc Committee and details of any unresolved negative ballots accumulated to that point.

6.7 Proposals of the type described in para. 6.4.1 require at least two thirds affirmative letter ballot vote of the committee membership for submittal to the BPTCS.

(- 4 -)

Reference 04-11 – Minutes of the March 20, 1990 Meeting of the Ad Hoc Committee on BPE_13 Pages

6.8 Proposals of the type described in para. 6.4.2 may be approved by majority letter ballot or at a Ad Hoc Committee meeting without letter ballot provided a quorum is present. Approval of motions shall be by at least a majority vote of the members voting; however, the Chairman may rule that a motion has not passed even if a majority vote has been cast. The reason for such ruling shall be the closeness of the vote, abstention of some members, or a combination of these reasons.

(- 5 -)

Reference 04-12 – Minutes of the March 20, 1990 Meeting of the Ad Hoc Committee on BPE_13 Pages

Bioprocess Engrg Program Newsletter 3/[illegible]

Advanced Programs at University of Virginia

(continued from page 1)

(1) Mammalian-Cell-Bioreactor Kinetics October 12-14, 1990

Biology and biochemistry of mammalian cells, metabolism, and synthesis of biopharmaceuticals. Reaction kinetics. Oxygenation and gas transfer. Cell retention and immobilization. Nutrients and culture conditions. Economics, safety regulatory aspect and reliability. Impact on down-stream processing.

- Cell Productivity and Yield
- Bioreactor Performance and Improvement
- Batch and Continuous Operation
- Damaging Agents

(2) Bioprocess Equipment Technology October 15-17, 1990

Technology of equipment for biopharmaceutical manufacturing. Process development engineers will learn about the physical and chemical phenomena governing their process equipment performance; about how those phenomena can be exploited most economically; and about what could be achieved with better engineered process equipment and manufacturing systems.

- Mammalian-Cell-Bioreactor Mass Transfer and Fluid Mechanics
- Chromatographic Separation of Proteins
- Fouling and Cleaning—Surface Chemistry
- Design of a Typical Production Plant for Therapeutic Proteins
- Lessons from the Production Floor

(3) Bioprocess Equipment Design October 17-19, 1990

Information required for practicing engineers to create systems designs, equipment specification, choose materials and fabrication techniques, predict system performance, integrate facility designs and qualify process systems. Safety, reliability, FDA.

- Design of a Typical Production Plant for Therapeutic Proteins
- Lessons from the Production Floor
- Plant System Design
- Sterilization
- Clean-in-Place
- Reactor Design
- Sealing

Hands-on participation at the '89 UVA Program.

Bioprocess Equipment Standards Status Report

This annual session at the ASME Winter Annual Meeting has become a popular venue for informal public discussions between all interested parties involved with bioprocess equipment. A panel-facilitated dialogue culminated in a consensus that further pursuit of standards and references for bioprocess equipment is essential to the industry.

TOPICS OF CONCERN

Suggested areas (among others) emerged to be investigated initially:

- Welding
  - Materials
  - Inspection
  - Dimensional Tolerances
- Cleanability
- Passivation
- Surface finish
- Seals
  - Static
  - Dynamic
  - Dimensional Tolerances

This session generated much enthusiasm indicating that a very active committee is forming, which ultimately will result in new standards for bioprocess equipment.

ORGANIZATION

The ASME Ad Hoc Standards Writing Committee

Arlene Spadafino, Director, ASME Pressure Technology Codes and Standards is the ASME staff support to the Ad Hoc Committee for Bioprocess Equipment. She introduced Dave Smith from Stearns Catalytic Division of United Engineers & Constructors. He is the Chairman of the Ad Hoc Committee for Bioprocess Equipment. Bill Cagney, who recently left Genentech for a position at Biogen, is the Vice Chairman of the committee.

During this session, Arlene and Dave outlined the protocol involved to participate on the Committee. Several individuals have already formally applied for committee membership. A core group of about 12 people is expected to be approved by the ASME Pressure Technology Board soon.

APPLICATION TO JOIN COMMITTEE WELCOME AT ANY TIME

Active participation will be expected from the committee members. Many areas will be addressed simultaneously and input will be required on a continuous basis. Several meetings will be held throughout the year. Travel, time and fund commitments should be obtained from employers before application to the committee is made. Standards may take years to establish.

IF YOU WANT TO BE KEPT INFORMED

All individuals who submit their application to Arlene Spadafino will be maintained on a mailing list to receive meeting dates. These meetings will always be open to the public. The first one is anticipated to be scheduled in March 1990.

Write to: Arlene Spadafino
Director
Pressure Technology
Codes and Standards
ASME MS/8E
345 E. 47th St., NY, NY 10017
201-705-7030 FAX 201-705-7674

THANKS

We wish to acknowledge the active participation and guidance of Dave DeLucia from Verax, Cas Perkowski from Stearns Catalytic, Arlene Spadafino and Julie Lee from ASME, and the Stearns Catalytic Division management for their support.

Dave Smith, *Chairman*,
ASME Ad Hoc Committee BPE
Bill Cagney, *Vice Chairman*,
ASME Ad Hoc Committee BPE

Markets for Drugs to Treat and Prevent Strokes Caused by Blood Clots to Reach $1.8 Billion Worldwide by 1999

by Technology Management Group, New Haven, CT

Products that can be used to help prevent strokes include drugs that control high blood pressure and clot-inhibiting agents. At least 41 companies are working on products for the multi-billion dollar hypertension market. Drugs that will likely be used to prevent strokes in at-risk patients include platelet anti-aggregants and calcium channel blockers.

As second generation tPA products become available, a substantial price reduction is likely to occur. These products are now used for heart attack treatment. Strokes are their next large application area. As understanding of tPA and other clot dissolving agents increases, each drug is likely to become the first choice for certain types of strokes.

Seven companies now have drugs in clinical trials to determine effectiveness in stroke treatment. These include three trials for clot-dissolvers. The preliminary results from these trials appear encouraging.

Over 101 companies and 61 other organizations are involved in research, development or production of products for strokes.

To My Fellow Department Heads

From:
Robert M. Hochmuth
Chairman of Mechanical Engineering & Materials Science Department
Duke University, Durham, NC

Scholarship Information

Your faculty may apply for a scholarship from ASME to attend the Process Equipment Seminars to be held at the University of Virginia, Charlottesville, VA, October 12-19, 1990.

Five scholarships will be awarded to mechanical engineering faculty to attend the seminar(s) with the intention of incorporating bioprocessing technology into their engineering curricula at their university.

The National Science Foundation (NSF) is expected to support these scholarships. Send written request (no application form is necessary) on university letterhead to:

Julie Lee
ASME MS 5A
345 E. 47th St.,
New York, NY 10017.

Reference 04-13 – Minutes of the March 20, 1990 Meeting of the Ad Hoc Committee on BPE_13 Pages

COMMITTEE CORRESPONDENCE

committee: Board on Pressure Technology Codes and Standards (BPTCS)

address writer care of:
ASME PTCS Dept. M/S 8E
345 East 47th Street
New York, NY 10017

subject: Establishment of the Bioprocessing Equipment Committee (BPE)

date: June 26, 1990

copy to:
M.E. Sheehan; Secy, BPE
BPTCS Officers

to: David Smith; Chmn, Ad Hoc BPE

Based on the presentation prepared by the Ad Hoc Bioprocessing Equipment Committee and presented by yourself as Chairman, the Board on Pressure Technology Codes and Standards (BPTCS), at its 6/4/90 meeting, took action to:

(1) to recommend to CCS the formation of the Bioprocessing Equipment Main Committee (BPE) with a scope to read: "Design, materials, construction, inspection, and testing of vessels, piping, and related accessories such as pumps, valves, and fittings for use in the biopharmaceutical industry";

(2) to approve the current Ad Hoc BPE Committee Officers and Members as Officers for a three year term ending June 30, 1993 and Members for a five year term ending June 30, 1995 on the new BPE Main Committee when approved by CCS;

At its 6/5/90 meeting, the Council on Codes and Standards (CCS) approved the recommendation of the BPTCS.

Therefore this letter is to advise that your Chairmanship is reaffirmed with an extension of term until June 30, 1993 and membership is reaffirmed with an extension of term until June 30, 1995 on the newly approved BPE Main Committee. Your previous acceptance to the Ad Hoc Committee is being considered as your acceptance to the new BPE Main Committee for the extend term unless we hear from you to the contrary.

Again, we are very pleased that you can take on this assignment and I express ASME's appreciation for your efforts in this regard.

Arlene A. Spadafino; Secy, BPTCS
Director, PTCS Dept; M/S 8E
(212) 705-7030 - FAX (212) 705-7533

The American Society of Mechanical Engineers
345 East 47th Street, New York, NY 10017

Keep ASME Codes and Standards Department Informed

Reference 05-01 – Board on PTCS Reaffirming Membership to the Newly Approved BPE Main Committee_4 Pages

COMMITTEE CORRESPONDENCE

committee: Board on Pressure Technology Codes and Standards (BPTCS)

address writer care of:

ASME PTCS Dept. M/S 8E
345 East 47th Street
New York, NY 10017

subject: Establishment of the Bioprocessing Equipment Committee (BPE)

date: June 26, 1990

copy to:

M.E. Sheehan; Secy, BPE
BPTCS Officers

to: William H. Cagney; Vice Chmn, Ad Hoc BPE

Based on the presentation prepared by the Ad Hoc Bioprocessing Equipment Committee and presented by its Chairman, David Smith, the Board on Pressure Technology Codes and Standards (BPTCS), at its 6/4/90 meeting, took action to:

(1) to recommend to CCS the formation of the Bioprocessing Equipment Main Committee (BPE) with a scope to read: "Design, materials, construction, inspection, and testing of vessels, piping, and related accessories such as pumps, valves, and fittings for use in the biopharmaceutical industry";

(2) to approve the current Ad Hoc BPE Committee Officers and Members as Officers for a three year term ending June 30, 1993 and Members for a five year term ending June 30, 1995 on the new BPE Main Committee when approved by CCS;

At its 6/5/90 meeting, the Council on Codes and Standards (CCS) approved the recommendation of the BPTCS.

Therefore this letter is to advise that your Vice Chairmanship is reaffirmed with an extension of term until June 30, 1993 and membership is reaffirmed with an extension of term until June 30, 1995 on the newly approved BPE Main Committee. Your previous acceptance to the Ad Hoc Committee is being considered as your acceptance to the new BPE Main Committee for the extend term unless we hear from you to the contrary.

Again, we are very pleased that you can take on this assignment and I express ASME's appreciation for your efforts in this regard.

Arlene A. Spadafino; Secy, BPTCS
Director, PTCS Dept; M/S 8E
(212) 705-7030 - FAX (212) 705-7533

The American Society of Mechanical Engineers

345 East 47th Street,
New York, NY 10017

Keep ASME Codes and Standards Department Informed

Reference 05-02 – Board on PTCS Reaffirming Membership to the Newly Approved BPE Main Committee_4 Pages

COMMITTEE CORRESPONDENCE

committee: Board on Pressure Technology Codes and Standards (BPTCS)

subject: Establishment of the Bioprocessing Equipment Committee (BPE)

date: June 26, 1990

to: Frank J. Manning

address writer care of:

ASME PTCS Dept. M/S 8E
345 East 47th Street
New York, NY 10017

copy to:

M.E. Sheehan; Secy, BPE
BPTCS Officers

Based on the presentation prepared by the Ad Hoc Bioprocessing Equipment Committee and presented by its Chairman, David Smith, the Board on Pressure Technology Codes and Standards (BPTCS), at its 6/4/90 meeting, took action to:

(1) to recommend to CCS the formation of the Bioprocessing Equipment Main Committee (BPE) with a scope to read: "Design, materials, construction, inspection, and testing of vessels, piping, and related accessories such as pumps, valves, and fittings for use in the biopharmaceutical industry";

(2) to approve the current Ad Hoc BPE Committee Officers and Members as Officers for a three year term ending June 30, 1993 and Members for a five year term ending June 30, 1995 on the new BPE Main Committee when approved by CCS;

At its 6/5/90 meeting, the Council on Codes and Standards (CCS) approved the recommendation of the BPTCS.

Therefore this letter is to advise that your membership is reaffirmed on the newly approved BPE Main Committee with an extension of term until June 30, 1995. Your previous acceptance to the Ad Hoc Committee is being considered as your acceptance to the new BPE Main Committee for the extend term unless we hear from you to the contrary.

Again, we are very pleased that you can take on this assignment and I express ASME's appreciation for your efforts in this regard.

Arlene A. Spadafino; Secy, BPTCS
Director, PTCS Dept; M/S 8E
(212) 705-7030 - FAX (212) 705-7533

The American Society of Mechanical Engineers

345 East 47th Street
New York, NY 10017

Keep ASME Codes and Standards Department Informed

Reference 05-03 – Board on PTCS Reaffirming Membership to the Newly Approved BPE Main Committee_4 Pages

COMMITTEE CORRESPONDENCE

committee: Board on Pressure Technology Codes and Standards (BPTCS)

address writer care of:

ASME PTCS Dept. M/S 8E
345 East 47th Street
New York, NY 10017

subject: Establishment of the Bioprocessing Equipment Committee (BPE)

date: June 26, 1990

copy to:

M.E. Sheehan; Secy, BPE
BPTCS Officers

to: Lloyd J. Peterman

Based on the presentation prepared by the Ad Hoc Bioprocessing Equipment Committee and presented by its Chairman, David Smith, the Board on Pressure Technology Codes and Standards (BPTCS), at its 6/4/90 meeting, took action to:

(1) to recommend to CCS the formation of the Bioprocessing Equipment Main Committee (BPE) with a scope to read: "Design, materials, construction, inspection, and testing of vessels, piping, and related accessories such as pumps, valves, and fittings for use in the biopharmaceutical industry";

(2) to approve the current Ad Hoc BPE Committee Officers and Members as Officers for a three year term ending June 30, 1993 and Members for a five year term ending June 30, 1995 on the new BPE Main Committee when approved by CCS;

At its 6/5/90 meeting, the Council on Codes and Standards (CCS) approved the recommendation of the BPTCS.

Therefore this letter is to advise that your membership is reaffirmed on the newly approved BPE Main Committee with an extension of term until June 30, 1995. Your previous acceptance to the Ad Hoc Committee is being considered as your acceptance to the new BPE Main Committee for the extend term unless we hear from you to the contrary.

Again, we are very pleased that you can take on this assignment and I express ASME's appreciation for your efforts in this regard.

Arlene A. Spadafino; Secy, BPTCS
Director, PTCS Dept; M/S 8E
(212) 705-7030 - FAX (212) 705-7533

The American Society of Mechanical Engineers

345 East 47th Street
New York, NY 10017

Keep ASME Codes and Standards Department Informed

Reference 05-04 – Board on PTCS Reaffirming Membership to the Newly Approved BPE Main Committee_4 Pages

345 East 47th Street New York NY 10017

Bioprocess Engineering Program Newsletter

May 1991

Robert C. Dean, Jr., Editor

Barbara B. Green, Associate Editor

Three Seminars of Instruction Developed Specifically for The Biotechnology and Pharmaceutical Industry

Offered by the ASME and the Center for Bioprocess/Product Development of the University of Virginia

Bioreactor Technology
October 17–20, 1991

Bioprocess Equipment Technology
October 20–23, 1991

Bioprocess Equipment Design
October 23–25, 1991

Hardware Presentations and Activities
Bioreactor Equipment
Chromatography and Filtration Equipment
Materials and Fabrication
Visit to UVA Laboratories
Receptions and Banquets

Mark Your Calendar

Phone or Fax to Reserve Space Early!

Three seminars will be offered at the University of Virginia from October 17–25, 1991. Leading practitioners and technical experts will instruct. The fee for the 3-day programs is $1,175. Program 2 is expanded to 4 days and the fee is $1,550. As in the past, discounts will be offered to individuals who attend multiple seminars, and groups of individuals who attend from a single group.

NSF will again support scholarships for mechanical engineering faculty to obtain information on bioprocessing technology for incorporation into their teaching.

The brochure for the seminars will be mailed in June. For information, scholarship application, or early registration call Julie Lee/Raj Manchanda, ASME, 212-705-7284. FAX 212-705-7674.

ASME Bioprocess Equipment Standards

The ASME BPE Standards Writing Committee continues to expand and formalize its subcommittees.

There are currently 23 members on the main committee divided into the following 6 subcommittees: Dimensions and Tolerances; Material Joining; Gaskets and Seals; Surface Finish; General Requirements; and the newly formed subcommittee, Equipment Design for Sterility and Cleanability.

There are 7 interest categories from which a consensus main committee is composed. Membership application is accepted at any time.

Overall Scope

To develop a standard covering design, materials, construction, inspection and

(continued on page 5)

Chairman's Message

Petition for the Establishment of a Technical Subdivision

R. M. Hochmuth

I am now just a few months from the end of my two-year tenure as Chairman of the Bioprocess Engineering Program so I thought it would be worthwhile to reflect on our program and future.

BPEP, for five years now, has been a major participant at the ASME Winter Annual Meeting. We began at WAM in Boston in 1986 and continued on to Anaheim, Chicago, San Francisco and Dallas. Typically, we organized about a

(continued on page 3)

"Solution preparation area—Dr. Karl Thomae, GmbH, Biberach an der Riss, Germany" (Courtesy of Fluor Daniel)

Reference 06-01 – May 1991 Issue of BPEP Newsletter Discussing the BPE Standard_4 Pages

try, the use of Bioprocessing to solve environmental problems. This work is exceptionally important; it could be the most economical way to clean up environmental disasters.

BPEP-3 Panel on Bioprocess Equipment Standards

This session, as in the past, was an excellent opportunity for the audience to exchange ideas with a panel of experts. The session was well-attended with a lot of input and expressions of concern from the audience. The most vocal concern revolved around building quality into bioprocess products, whether an individual component or a completed plant. Many new ideas and opinions were expressed. It was a provocative interchange.

BPEP-4 & 5 Bioseparations

Because both of these sessions dealt with types of separation or purification, it was appropriate that these were held consecutively. The importance of these types of sessions for engineers in our industry cannot be understated. As more and more exotic bio-products are developed, the separation and purification portion of our projects become more complicated and costly. These sessions provided very diverse outlooks at these problems and also provided a good basis for further research.

BPEP-6 Process Control/Validation

This session provided first an overview of validation of a control system and then an inside look at designing software so that it can be validated. The highlight for me was a comment from one individual, "Now I see what we've been doing wrong." Ron Nekula was a last minute replacement for P. DeSantis; his presentation was excellent.

BPEP-7 Panel on Sanitary Valves

This session provided an in-depth view of several valves from different suppliers. The attributes of these and the companies future developments were presented. This session was a good overview of the valve industry. Happily, this session remained very non-commercial.

BPEP-8A Panel on Opportunities for Mechanical Engineers in Bioprocess Engineering

This session as in other years gave an excellent viewpoint of the wide varieties of opportunities available to Mechanical Engineers in our industry. The moderator was Duane Bruley who was later elected Chairman of the Bioprocess Engineering Program.

BPEP-8B Passivation and Surface Finishes of Type 316L Austenitic Stainless Steel

The highlight of this session was the variety of opinions and data presented. The Scanning Electron Microscope slides shown by Drew Coleman and Bob Evans were spectacular. The pass around samples of rouging were also big hits. The variety of opinions is healthy and should stimulate additional papers for later sessions. This was, in my opinion, not only well done from the authors' viewpoint, but the information presented is very definitely needed.

BPEP-9 Sulfur and Other Issues for Stainless Steel Piping Systems

This also was a very stimulating session. Again the diversity of data and opinions was noteworthy. This variety, continuing through the last paper, is our real strength—a variety of experts sharing their experiences and talents.

All sessions were well crafted; my congratulations to all presenters, chairmen and co-chairmen. Thank you for your time and talents in support of our program. We are beginning to break down the parochial attitudes exhibited in some industries. This makes our program an even better forum for sharing ideas, problems and solutions. Until WAM '91, I wish each of you the best. See you then!

Equipment Standards

(continued from page 1)

testing of vessels, piping and related accessories (components), such as pumps, valves, and fittings for use in the Bioprocessing industry. The Standard may provide for adoption of other ASME and related national standards and when so referenced become a part of the Standard.

All meetings are open to the public and are announced in *ME Magazine*. Since there is no real mailing list for the BPE Standards Writing Committee a personnel form can be requested from Steve Weinman at ASME and sent back to be kept on file.

The next ASME BPE Standards Writing Committee meeting is scheduled for September 12, 1991, in Chicago.

Dimensions & Tolerances Subcommittee

1. General Comments & Task Group Accomplishments:

a) Contacted eight of the major manufacturers of Sanitary Fittings & Tubing. Survey all with respect to: Tolerances, Dimensional Differences, Finish, Chemical/Physical Min/Max, Pressure Ratings

b) Manufacturers Contacted: Fittings Manufacturers and Tube Manufacturers.

c) Overall Consensus & Opinions:

- Most all agree that something must be done to create a more detailed and workable specification.
- Many people want to get involved with our committee.
- Almost all feel that they are somewhat industry driven and would change to meet the requirements.
- Most manufacturers contacted think by reviewing existing specifications we can add supplemental changes to meet demands
- Concerns for a mandatory change lie in tooling cost, production time increase, inventory now built, cost increases to products, increased QC/QA.

d) Specification Reviews: Specifications currently used in Biotechnology, Pharmaceutical and semi-conductor applications are:

- ASTM A-269—seamless and welded austenitic stainless tubing for general service.
- ASTM A-270—same as above except for sanitary tubing.
- ASTM A-213—seamless ferritic and austenitic alloy—steel boiler, super heater, and heat exchanger tubes.
- ASTM A-450—general requirements for carbon, ferritic and austenitic alloy steel tubes. MSS-SP43—stainless steel butt weld fittings (now only for schedule S/10 piping)
- 3A standard #08-17—Sanitary standards for fittings used on milk and milk products. 3A standard #33-00 Sanitary standards for polished metal tubing for dairy products.

Currently all manufacturers of sanitary tube and fittings start with tube made from ASTM A269/270. Concerns not only lie in making sure dimensions are met, but if in fact the finished tube and fittings still meet the applicable specification. In the process of machining, bending, annealing, grinding, polishing and electropolishing, will the fitting and tube meet minimum standard? This, of course, is our charge to recognize, decipher and come up with a new or added specification or to change to existing ones.

e) Specification outline (proposed)
1) A269/270 wall thickness allows too much variation for a smooth fit-up on tube and fittings. A269—allows for ±10%. A270—allows ±12 1/2%
PROPOSAL: Come up with agreeable and workable min/max wall tolerance. Strip on welded tube can be ordered to exact thickness. Currently most manufacturers order strip to minimums to keep costs low. This makes fitting manufacturers who remove more metal by processing to end up with still lower tolerances on same heat lots of tube. Recommend start-

Reference 06-02 – May 1991 Issue of BPEP Newsletter Discussing the BPE Standard_4 Pages

ing with nominal wall thickness required. It is possible to create a minimum wall specification similar to A213.

• A269/270—Outside Diameter—again this is too open with regard to nominal diameter.
PROPOSAL: Reestablish min./max. tolerances for the above.

• B16.9—Addresses dimensions and tolerances for wrought stainless steel butt welding fittings.
PROPOSAL: Review formulas used in creating dimensional guidelines, wall thickness established by pressure and temperature parameters.

MSS SP-43—Manufacturers' standardization society non-pressure related standard for light weight type fittings.
PROPOSAL: Again review and consider using a formula similar to create new specification.

3A standards—Polished metal tubing #33-00 and fittings used on sanitary lines #08-17.
PROPOSAL: Much too loose for our work. Does not address dimensional/tolerance areas. Bevel seat fittings well addressed but not used in our industry.

f) FINAL SUMMARY:

Dimensions—On-like fittings need to be a top priority.

Designs in tight areas around fermentors, tanks and other equipment require fitting to fitting construction. Would allow the use of any brand fitting if all have the same dimensions.

• Tolerances—On wall thickness present big fit up problems and poor drainability in piping systems. The general consensus is that we can tighten up the wall thickness dimensional tolerance. We must consider auto-welding machine needs with regard to this area. Fittings and tubing must match in order to hold needed tolerances.

• Ovality/Out Of Roundness—This also must be much tighter to maintain a good fit-up. Fittings are sometimes sized after manufacture on tangents to bring tube back to a closer tolerance. Tube is sized during final finishing on weld mill, but during shipments the tolerance is sometimes distorted. Larger sized tubing must be closer if fit-up is to be within the prescribed limits.

• Recommendations: Create supplementary requirements to A269/A270 for minimum wall thickness similar to ASTM A213; Tighten up ovality for outside diameters on tube sizes; Dimensions should be the same for autoweld fittings and clamp type fittings. We should discuss a reasonable and least costly way of coming up with a formula, using existing specifications now out, with some supplementary requirements to create a whole new specification.

Clearly this subject is very important to our final goals. We must move to come up with an unbiased decision with regard to selection of specification, consideration of directs costs to the manufacturer and the end user. How much additional cost will we put into products and will it be accepted by our industry?

Material Joining Subcommittee

The aim of this Task Group Statement is to establish an outline guide for the subcommittee on the items to be addressed in covering the area of material joining (welding) for the new proposed standard.

Our aim, at this point, is to survey, identify and assemble the commercial available data on the materials and processes that have been developed and are applicable; to incorporate them into a workable standard which can be utilized to resolve problems related to this area facing the bio-pharmaceutical industry. It is not the purpose of this subcommittee to research and develop new novel material and processes. However, this does not mean that this standard, when developed, will not inspire some novel techniques or processes to be developed. If this does happen, and it is found to be applicable and beneficial, then these items may be added later to the standard through an issue of addendum, etc.

This standard shall apply to all major types of material joining in all areas of biopharmaceutical processing, that is all areas of distribution piping, component parts, vessels and tankage that has an effect on the purity and reliability of the product. This standard shall directly address areas of high purity and process control in systems of high purity water, water for injections, high purity steam, product lines, gases, etc.

This standard is not limited to the items listed below, but shall serve as a guide or starting point for the Subcommittee to commence. The subcommittee shall have the option to add, remove or amend as they determine necessary to meet their objective.

I. Material
A. Establish specification that details control requirements of chemistry, surface finishes, dimensional control, cleanliness and packaging, etc., which shall promote improved joining procedures. This includes the following items:

1. Stainless steel piping, tubing, fittings, valves and other component parts.
2. Stainless steel plate, sheet, forgings, casting, etc. related to vessels, tankage, etc.
3. Alloy steel pipe, tubing, fittings, valves and other component parts.
4. Copper and copper alloys for tubing, fittings, valves, etc.
5. Non-metallic items, such as pvc, polypropylene, PVDF, etc., piping, tubing, fittings, valves and other component parts.

II. Joining Process and Procedure
A. Welding Processes—Tungsten inert gas; Shielded metal arc; Metal inert gas; Submerged arc
B. Brazing
C. Mechanical—Screw and threads; Clamping
D. Joining of non-metallics—Chemical/Solvent bonding; Thermal welding (socket joints, forged thermal fusion, others)
E. Methods—Manual; Semi-automatic; Automatic

III. Weld Joint Design
Joint Design for: Tubing; Piping; Fittings; Components; Copper systems; Non-metallics; Mechanical.

IV. Weld Filler Metal
Tungsten inert gas; Shielded metal arc; Metal inert gas; Submerged arc; Brazing; Non-metallic

V. Weld Acceptance Criteria
A. Fit-up and mismatch; concavity; convexity; weld geometry; oxidation; discoloration; meandering; dross and surface slag; defects (undercuts, lack of fusion, surface porosity, cracks, etc.)
B. Weld repairs
C. Weldment surface conditioning

VI. Inspection
A. Visual: Borescope/Videoscope
B. Radiographic
C. Surface inspection: Dye-penetrant; Profilator; Magna-flux; etc.

VII. Procedure Qualification
Evaluated for: Structural reliability; Cleanliness; Interior surface contour.

VIII. Performance Qualification
A. A test to illustrate the ability of the welder to make an acceptable weld.
B. Criteria for acceptance.

IX. Documentation Requirements
A. Establish what documentation is required.
B. Develop an acceptable format.

Seals Subcommittee

General synopsis: There are two main concerns regarding pump seals and gaskets for the Bioprocessing Industry.

First, to prevent contamination to the product or, inversely, to prevent leakage of potentially dangerous or costly products to the atmosphere.

Reference 06-03 – May 1991 Issue of BPEP Newsletter Discussing the BPE Standard_4 Pages

Secondly, compatibility to insure that the seal components and elastomers will not be "attacked" with the resulting residues introduced into the products.

General scope: Pump seals, valve seals and seats; Accessory seals, gaskets, diaphragms, etc.; Agitator shaft seals; Reactor nozzles and manways; Fermenter apertures; Pipeline gaskets.

General concerns: Design to withstand product/sterilization, pressure/temperature; Product compatibility; FDA et al. compliance; Elastomers used; Flexibility to allow seal movement between shaft and housing; Installation; Energy consumption; Fluid barrier design and buffer fluid; Shaft/housing matching tolerances; CIP and sterilizing ability; Corrosion resistance; Cost; Equipment design modifications; Life expectancy; Ease of inspection; Monitoring ability; Carbon black and rouge; Thermal expansion; Maintenance.

General mechanical seal types used in bioprocessing: Single seal; Double seal; Rotary seal; Stationary seal; Cartridge seal; Tandem seal.

Common seal face materials: Silicon carbide vs. silicon carbide; Silicon carbide vs. tungsten carbide; Carbon; Ceramic; Chrome oxide.

Common elastomers used with valves, pumps, accessories: steam resistance viton; Silicone; Buna; Teflon; EPDM.

Recommendations: Review existing documents, standards, and proposed standards for the above type seals and gasket. For example, ASME-ANSI B73 and 3A Standards.

Prepare survey questionnaire to gather information and advice as to demand or need for such standards.

Proceed with specialized task study groups or disband.

Surface Finish Subcommittee

The aim of this Task Group statement is to establish an outline for the subcommittee on the items to be addressed in this section of the standard covering the areas of surface finishes, passivation and cleaning. Our aim at this point is to survey, search, identify and assemble the commercial, technical, available data on the material surfaces, cleaning, passivation, sterilization and their related processes that have been developed and are applicable, and incorporate them into a workable standard which can be utilized to resolve the related and respective cleanliness and purity problems facing the biopharmaceutical industry. It is not the purpose of this subcommittee to do research and develop new novel techniques and processes. However, this does not mean that this standard, when finally developed, will not inspire some unique and novel standard to be developed. If this does happen, and found to be applicable and beneficial, then these new founded techniques and/or processes may be added later to the standard through the issuance of an addendum, etc. This standard shall apply to all surfaces of the various materials and items that make up the systems which come in contact with the high purity media and product. This shall include all distribution piping, component parts, vessels tankage and miscellaneous items that become a wetted or exposed surface. This standard shall directly address areas of high purity and process control systems of high purity water, water for injection, high purity steam, product line, tankage, gases, etc.

This standard is not limited to the items listed below, but shall serve as a guide or starting point for the subcommittee to commence. This subcommittee shall have the option to add, detract, remove or amend as they determine necessary to meet their objective.

I. Materials and Their Related Surfaces: Stainless steel; Alloy steels; Copper; Nonmetallic systems; Gaskets, seals, seats, etc.

II. Components and Items: Pipe, tubing and fittings; Valves, controls and other components; Tankage, mixer and vessels; Nonmetallics; Seals, seats and gaskets.

III. Type of Finishes: Mill finishes; Bright annealed; Mechanical polishing; Mechanical buffing; Chemical polish; Electropolish; Grid blast; Machine; Misc.

IV. Identification and Classification of Typical Surface Anomalies: Mechanical defects (scratches, crevices); Corrosion; Surface porosity; Oxidation; Discoloration; Weld defects; Misc.

V. Measurements Procedures and Classification of Surface Finishes: Machine finish roughness scale; Grit finish measurements; Surface profilator; Optical comparitor; Optical microscope (macro, micro); Scanning electron microscope; Spectrographic (EDX, auger, others).

VI. Precleaning Processes: Chemical wash; Solvent wash; Acid wash, D.I. water wash; Detergent (surfactant, chemical chelating agents); Mechanical and friction cleaning.

VII. Passivation: Available commercial solutions (Nitric acid solution, Phosphoric/mixed acid solutions; Citric acid solutions, Ammonium citrate solutions); Process techniques (Pre-Clean, Passivation, Neutralization, Rinse); Testing technique.

VIII. Sterilization Procedure: Solution; Temperature; Testing; Maintenance of systems.

ABC Applauds Findings of White House Report on National Biotech Policies

The nation's largest biotechnology trade group sees signs of increasing support from the President's Council of Competitiveness after reading its "Report on National Biotechnology Policy." Dr. Pamela Bridgen, executive director of the Association of Biotechnology Companies, is encouraged by several points that will have a positive impact on small, entrepreneurial companies—the backbone of the industry and bulk of ABC membership.

Bridgen, in reacting to the White House report about biotechnology, made the following points:

"The competitiveness council has recommended we continue to examine tax issues. Of long concern to ABC members is the research and experimentation tax credit. Many small, research-intensive companies focusing on biotechnology are not making profits yet, so the R&E tax credit in its current form is of no value to them. ABC has worked hard to bring this issue before policymakers, and we are extremely pleased to see the White House wants to revisit this issue.

"ABC would support appropriate legislation where necessary to ensure biotechnology's competitiveness in world markets. Although we do not support duplicative, cumbersome legislation for the sake of legislation, we would support initiatives if they proved beneficial to the entire industry that includes academia, small and large corporations, and related services."

"The White House is supportive of biotechnology. This report shows it recognizes our industry has certain needs that must be addressed, such as training, and today's report indicates they'll address them."

"The competitiveness council has identified action areas that need follow up. ABC particularly lauds White House recommendations to publish final biotechnology oversight to environmental releases and, related to that, the preparation of an oversight policy of products and research not included in the 1986 Coordinated Framework of Biotechnology Regulations."

For copies of the report call the White House Press Office, 202-456-1414.

7

Reference 06-04 – May 1991 Issue of BPEP Newsletter Discussing the BPE Standard_4 Pages

The ASME BPE Standards Writing Committee continues to expand and formalize its subcommittees. There are currently 23 members on the main committee divided into the following six subcommittees: Dimensions and Tolerances; Material Joining; Gaskets and Seals; Surface Finish; General Requirements; and the newly formed subcommittee, Equipment Design for Sterility and Cleanability.

Chairman

David Smith	**215-422-4486
United Engineers	215-422-4871

Vice-Chairman

Bill Cagney	508-650-9237
Abbott Biotech	

Subcommittee Chairmen

F. Manning, dimensional tolerances	508-366-0710
TEK Supply	508-898-2349
R. Cotter, welding	617-938-1505
Cotter Corp.	617-938-3703
L. Peterman, Seals	414-697-2416
Tri-Clover, Inc.	414-694-7104
T. Mehalko, surface finishes and cleanliness	408-727-7740
Kinetic Systems	408-727-9774
C. Perkowski, general requirements	215-422-4293
United Engineers	215-422-4871
N. Brooks, equipment design for sterility and cleanability. *Fluor Daniel*	803-281-4907
	803-281-6913

**Telephone Number is listed first. If a FAX Number is available it is listed second.

Committee members represent a vast cross section of areas of interest, as shown below. No more than one-third of the total membership of the main committee may be derived from any given area of interest.

Design / construction:
Organizations performing design, facility design and erection, or design-related services.

General Interest:
Consultants, educators, research and development organizations personnel, and public interest persons.

Manufacturers:
Organizations producing assemblies.

Insurance / Inspections:
Organizations that insure equipment and / or provide independent inspection of the manufacture and installation of components, parts, and items.

Regulatory:
Agencies and organizations that regulate the design, manufacture, installation or operation of the products covered by the applicable standards.

Materials Manufacturers:
Organizations producing materials or ancillary materials, related accessories, or parts.

Users:
Organizations that use products covered by the applicable standards.

We have had six meetings since our establishment, the most recent on 12 September 1991 in Chicago Illinois. We anticipate holding four meetings a year during the formation of this Code or Standard. Persons interetsed in participating in the committee's activities should contact a BPE officier, a subcommittee chairman, or the committee staff secretary.

The next main committee and subcommittee meetings will be held in conjunction with:

ASME
Winter Annual Meeting
Atlanta, Georgia
December 2nd to the 4th, 1991

For Future Meeting Dates
Contact
Steve Weinman
Secretary, BPE Main Committee
ASME
345 - East 17th Street
New York, New York 10017
212-705-7025
FAX 212-705-7533

All meetings are open to the public and are announced in "ME" Magazine. A personnel form can be requested from Steve Weinman at ASME and sent back to be kept on file.

Help Set Standards for Bioprocessing Equipment

The ASME Committee of Bio-Processing Equipment (BPE) is actively soliciting participation of all interested parties involved in bioprocessing. The BPE committee was formally established and approved by the ASME Board of Pressure Technology Codes and Standards (BPTCS) in June 1990 to create a standard "for design, materials, construction, inspection, and testing of vessels, piping, and related accessories such as pumps, valves, and fittings for use in the bioprocessing industry." The standards may provide for adaptation of other ASME and related national standards, and when so referenced become a part of the standard.

There are seven interest categories from which a consensus main committee is composed. Membership application is accepted at any time.

Reference 07 – 1991 Pamphlet Distributed by ASME to Solicit Membership_Back and Front

Codes and Standards
Tel: 212-705-8500
Fax: 212-705-8501

2/10

April 16, 1996

To: All BPE Committee Members

Subject: Draft of Bioprocessing Equipment Standard

Dear Members:

Enclosed you will find the latest complete draft of the Bioprocessing Equipment Standard, consisting of all six "Parts".

This draft contains of all the changes made as a result of the January, 1996 meeting.

Please review this document for content, clarity, etc. Any comments, concerns and/or questions are to be saved for the April 25, 1996 meeting.

The purpose of this meeting will be to take a final look at the complete draft of the proposed standard prior to submittal to the BPTCS for approval.

Sincerely,

Paul D. Stumpf

Paul D. Stumpf
Secretary, BPE
(212) 705-8536

The American Society of Mechanical Engineers

Reference 08 – Initial 1996 Draft of the BPE Standard_1 Page

SENT BY: 5-21-97 ; 3:36PM ; ANSI→ 7058501:# 2/ 2

ANSI
Procedures and Standards Administration
American National Standards Institute 11 WEST 42ND STREET, NEW YORK, NEW YORK 10036
TEL. 212.642.4976
212.642.4912
212.642.4914
FAX 212-398.0023

Today's Date: May 21, 1997
Our Reference: BSR LB 3673

Ms. Silvana Rodriguez-Bhatti
Manager, Standards Administration
American Society of Mechanical Engineers
345 East 47th Street
M/S 10B
New York, NY 10017

<u>Notification of Approval of Standard</u>

The Board of Standards Review (BSR) of the American National Standards Institute (ANSI) has approved as an American National Standard the following:

ANSI/ASME BPE-1997
Bioprocessing Equipment
Create new standard:
Approval Date: **May 20, 1997**

It will be appreciated if you would print the standard in accordance with the ANSI Style Manual, copies of which are available from our Publications Department. If it is not possible to do this, the official ANSI designation, which is shown above the title, should appear on the cover of the approved standard along with the date of ANSI approval, preferably in the upper right corner.

As soon as printed copies are available, the Institute would appreciate receiving one (1) complimentary copy, addressed to my attention, for ANSI administrative purposes.

Please inform appropriate personnel of this approval – officers of standards committee, members of the balloting group, other secretariat, the organization which has the responsibility for printing the standard, etc. The Institute will insert a notice of approval in a forthcoming issue of *Standards Action*.

In accordance with ANSI Procedures, those objecting to this approval **who have completed the appeals process at the Standards Developer**, are hereby notified of their right to file a procedural appeal. If your name is listed below, but you have not completed the appeals process at the standards developer, you are normally not eligible to appeal to the BSR.

Your appeal and all supporting documentation must be filed in writing with the office of the undersigned **within 15 working days** after receipt of this official notification. The appeal must be based on procedural or substantive criteria, or both, and include a statement as to why the BSR action should be modified. The BSR will not render decisions on the relative merits of technical matters, but it shall consider whether due process was afforded technical concerns.

The appeal must be accompanied by a $150.00 filing fee, payable to ANSI.

Sincerely,

Anne Caldas, Secretary
Board of Standards Review

CB/AC/B
Copy to: SB Chairman, SB Secretary

Reference 09 ANSI Letter of Approval_1 Page

For Immediate Release

Contact: Anne Buckley
(212) 705-8157
buckleya@asme.org
Download: http//www.asme.org

ASME PUBLISHES STANDARD TO BE USED BY PHARMACEUTICAL AND BIOTECHNOLOGY INDUSTRIES

NEW YORK, MARCH 14, 1998 -- ASME International (The American Society of Mechanical Engineers) recently published a standard to be used in the biopharmaceutical industry. The ASME bioprocessing equipment standard (ASME BPE-1997) includes and addresses such concerns as:

- equipment designs that can be cleaned and sterilized;
- the quality of weld surfaces;
- material joining;
- sealing;
- standardized definitions that can be used by material suppliers, designers, fabricators and users;
- dimensions and tolerances for stainless steel automatic welding and hygienic clamp tube fittings;
- the need to integrate existing standards relating to vessels, piping, appurtenances and other equipment necessary for the biopharmaceutical industry without infringing on those standards.

The ASME BPE 1997 standard applies to all parts that contact either the product, raw materials or product intermediates during the development and manufacturing processes.

-more-

Reference 10-01 – ASME Press Release on the Publication of the ASME BPE Standard_2 Pages

ASME BPE-1997/Page 2

The price of the publication, *ASME BPE-1997 Bioprocessing Equipment*, is $165. To order contact ASME Information Central at 1-800-843-2763 or via email at infocentral@asme.

The 125,000-member ASME is a worldwide engineering society focused on technical, educational and research issues. It conducts one of the world's largest technical publishing operations, holds some 30 technical conferences and 200 professional development courses each year, and sets many industrial and manufacturing standards.

- # -

Reference 10-02 – ASME Press Release on the Publication of the ASME BPE Standard_2 Pages

ASME Ad Hoc Committee on Bioprocessing Equipment

Presentation of Need for Standards
Covering
Bioprocessing Equipment

1. **General**

1.1 The current direction for equipment design and fabrication in the Bioprocessing industry focuses on assuring the ability to clean, sterilize and maintain environmental containment for both product and personnel protection. The emphasis in design is on obtaining high surface quality, elimination of pockets or crevices, full drainability, and system seal integrity to assure a contained environment. Achievement of sanitary design criteria can require compromises that have a potential impact on structural integrity, as interpreted from currently available codes and standards.

1.2 The lack of standards specific to the field has caused problems for all parties concerned due to confusion and uncertainty over the interpretation of systems requirements. The knowledge and capability to establish baseline standards for the industry exist. These new standards should be comprised of a Systems Standard along with amendments to existing codes and standards. The Systems Standard would tie together the systems performance criteria with the appropriately amended design, materials, fabrication, and construction codes and standards under a standard definition of terminology.

2. **Currently Identified Needs**

2.1 Existing pressure vessel and piping codes are inadequate or incomplete in covering critical areas of high purity for design, fabrication, and installation of bioprocessing equipment, such as:

2.1.1 Selection of materials and additional requirements for material manufacture.

2.1.1.1 The current material of choice is 316 or other stainless steels purchased to the ASTM specification. The ASTM specification is not tight enough on the control of residual or trace elements which can have an effect on the weldability, surface quality and corrosion resistance.

1

Reference 11-01 – Presentation Given to the Board on PTCS_5 Pages

ASME Ad Hoc Committee on Bioprocessing Equipment

Presentation of Need for Standards
Covering
Bioprocessing Equipment

2.1.2 More stringent rules for dimensional tolerances and measurement.

2.1.2.1 The current allowable tolerances for variation in wall thickness, ovality, internal diameter, and outer diameter of tubing and fittings are not appropriate for the quality of welds required to meet the industry's performance criteria. A new or amended ASTM standard is called for to meet this need.

2.1.2.2 The fabricators of sanitary fittings do not follow dimensional standards of length, radius, and postfabrication ID, OD, ovality, and wall thickness. This leads to field failure to meet design criteria when "equivalent" fittings are purchased for installation.

2.1.3 Specialized welding specifications

2.1.3.1 ASME Code Section IX addresses the structural integrity of welds but does not address the surface quality. There is a need to provide specifications covering high purity purging, oxidation, discoloration, and surface contour. Standardization of inspection and acceptance criteria need to be developed. The design emphasis on smooth surfaces can conflict with the existing welding and fabrication practice according to current codes and standards. The ideal weld should just achieve complete penetration without any weld concavity, normally associated with a strong root pass, for providing assurance of structural integrity.

2.1.3.2 Control of dimensional tolerances and surface preparation for welded components is critical for obtaining welds with acceptable surface characteristics.

2.1.4 Requirements for establishment and verification of surface conditions.

2

ASME Ad Hoc Committee on Bioprocessing Equipment

Presentation of Need for Standards
Covering
Bioprocessing Equipment

2.1.4.1 Standard terminology for specifying surface conditions need to be established. The specifications should outline the surface finish treatment to be utilized (ie. mechanical polish, electro-polish, chemical polish, etc.), the inspection technique, acceptance criteria, and the tolerances required.

2.1.4.2 Surface treatments to inhibit corrosion need to be specified with consideration given to impact on surface quality, efficiency, and environmental impact of disposal.

2.1.5 Permissible standards for fittings, valves, closures, etc.

2.1.5.1 The performance criteria for systems design often call for fittings that do not clearly meet existing codes and standards for Pressure Vessels. May vessel fabricators limit their boundaries for testing to exclude these components.

2.1.5.1.1 Single bolt closures are often utilized on the first fittings attached to pressure vessels. Standards need to be established to define the conditions under which this practice can be acceptable.

2.1.5.2 Components from "equivalent" suppliers can fit together but do not especially obtain equivalent capability for pressure containment.

2.1.6 Special rules for containment of toxic and hazardous systems.

2.1.6.1 Standards need to be established outlining the special systems requirements for different levels of containment.

2.1.6.1.1 Protecting the personnel from the product.

2.1.6.1.2 Protecting the product from the personnel.

3

Reference 11-03 – Presentation Given to the Board on PTCS_5 Pages

ASME Ad Hoc Committee on Bioprocessing Equipment

Presentation of Need for Standards
Covering
Bioprocessing Equipment

2.1.6.1.3 Conditions requiring double containment and disposal.

2.1.7 Standardization in the responsibilities in communication between Users, Designers, Fabricators, and Inspectors.

2.1.7.1 A standard outline identifying the specific communications responsibilities between user, designer, fabricator, and inspector would help to assure that the completed system will meet the user's performance criteria, be fabricated according to design, and will be inspected in a consistent manner with the design standards.

3. Requirement for a Systems Standard

3.1 There is a need to establish a standard which would oversee all aspects of design, fabrication, installation, and inspection of the entire bioprocessing equipment system, including vessels, piping, appurtenances, pressure relief, final inspection and verification. The performance criteria for the ability of the processing equipment to be cleaned, sterilized, and maintain containment are global to all system components.

3.1.1 The intent of this standard would be to coordinate the appropriate standards as established and amended by other committees to meet the overall system requirements. The identity and specifics of amendments would be established by the Bio-Process Equipment Committee and recommended to the appropriate vessel code, piping code, or other committees.

4. Requirement for Standardized Language

4.1 There is an immediate need throughout the bioprocessing industry for standardized language encompassing the entire spectrum of bioprocessing operations, including equipment, processes, and products. New technology consistently creates its own terminology and confusion can exist until clear definitions are established. A lexicon of terms with their definitions would improve the communications between user, designer, fabricator, and inspector. This would be a cardinal requirement of this standard.

4

ASME Ad Hoc Committee on Bioprocessing Equipment

Presentation of Need for Standards
Covering
Bioprocessing Equipment

5. Conclusions

5.1 Therefore, the Ad Hoc Committee on Bioprocessing Equipment requests that the Board on PTCS approve the establishment of a Main Committee to develop and administer codes or standards for Bioprocessing Equipment, and to formally adopt the proposed scope as previously established by the Board.

5.1.1 We further request the currently approved officers and members of the Ad Hoc Committee, be approved as members and officers respectively of the Main Committee Bioprocessing Equipment (BPE-1). (See attached list.)

5

Reference 11-05 – Presentation Given to the Board on PTCS_5 Pages

Tri-Clover ... Interoffice Correspondence

June 28, 1990

TO: Geyer

cc: DenGitzlaff

FR: Peterman

SUBJECT: American Society of Mechanical Engineers (ASME)
Bio-processing Equipment Committee

The second meeting of the subject group was held June 27, 1990 in New York City at the United Engineering Center of ASME.

The minutes of the meeting will be available in approx. three weeks which, of course, I will pass on to you, but this is just a recap of the meeting in general, which was called to order at 10 A.M.

Ten of the twelve original committee members were present, and we were immediately informed that this is no longer an ad hoc committe inasmuch as this comm. was accepted by the board of ASME to become a full-fledged comm. which will probably last a lifetime. Our official designation within ASME is BPE-1. Further, the original 12 members (writer included) had our terms approved for a five-year period ending June 30, 1995. (See ltr. attached.)

After the normal formalities of adoption of general approval of minutes, etc., we spent approx. 4 hrs. reviewing the operating procedures manual of the bio-processing comm. The manual as revised per our discussion today will be available in approx. four wks.

We will have a minimum of four meetings per year, hopefully, one to be in conjunction with Interphex in April and one with the ASME winter meeting which is traditionally held in November. In fact, the next three meeting dates were set as follows:

Reference 12-01 – Notes on June 27, 1990 Meeting_3 Pages

-2-

Sept. 18, 1990	New York City	ASME Building
Nov. 28, 1990	Dallas, TX	In conjunction w/ASME winter meeting and in April 1991 w/Interphex.

It shd. be noted, however, that there may be an additional meeting between November and April, depending on the actions of the committee as a whole.

Apparently, as the result of the letter written by Manning to 3-A, there was a huge supply of 3-A material on hand for the meeting. This was sent directly to ASME, and I was surprised that many of the members grabbed this information as if they had never seen it before.

One of the visitors, but who also was elected to the permanent committee, Randy Greashman, Director, Bio-Process Research, Merck & Co. in Rahway, NJ, is really pushing franctically for quick results of this comm. He actually stated he wd. prefer a document by the end of 1990; however, the group as a whole, especially those with more experience with these types of meetings, said they wd. be quite happy with a document by December 1991.

A task group was formed to specifically outline exactly what items shd. be researched for the committee and as to set the scope of work. Nigel Brooks, Fluor-Daniel, will be Chairman of this new group and immediately had many volunteers. Being a task group (differentiated from a sub-committee), the participation is informal, and all comments are welcome and shd. be directed to the Chairman.

Bill Cagney, the Vice Chairman and a director of Biogen, Inc. of Cambridge, MA, indicated during the meeting that they would prefer to have more industry users on the committee and mentioned a name or two, one of which was Dr. Grimes of Upjohn. He said that he had contacted Dr. Grimes more than 1-1/2 years ago but that Grimes refused, saying that he did not want to spend a lot of time on something that will probably be squashed by the FDA. Scott, except for myself due to your involvement, I was the only one in the room who had even an inkling of knowledge that this possibility either did exist or could possibly exist in

Reference 12-02 – Notes on June 27, 1990 Meeting_3 Pages

-2-

the future. They are planning to invite someone from the PDA to be a committee member and I personally thought we might be inviting the enemy into the camp, but on the other hand, perhaps it's good to involve them from the start. It would seem to me that ISPE in the beginning was a lot weaker than ASME as a whole today, and, therefore, the ASME group probably has a much better chance of succeeding.

I suppose every group has at least one or two problem people, and this group seems to have David Baram of Vanasyl Valves Limited. Perhaps you know this individual -- in any case, he is quite outspoken and is apparently representing a European standards committee, trying to do the same thing, except that he has held them off pending the results of this group's work. He has an opinion on everything, and Manning and I at an executive session of the same meeting, suggested that perhaps this individual takes up too much time and something will have to be done about it.

The formal meeting was adjourned at 3:30 and an exec. meeting followed with the ten permanent committee members present to discuss the addition of eight new members to the permanent committee, making the total now 20. The operating procedure allows for a maximum of 25.

The executive committee meeting ended at approx. 4 P.M.

As stated, I will send you a copy of the formal minutes when I receive them.

LP 6/28

L. J. Peterman

/rt

Attach.

Reference 12-03 – Notes on June 27, 1990 Meeting_3 Pages

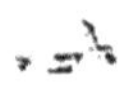

Tri-Clover ... Interoffice Correspondence

December 10, 1990

TO: Geyer

cc: Riemer/ DenGitzlaff/ Kish/ Schroyer/ Becker

FR: Peterman

SUBJECT: ASME Winter Meeting
Dallas, TX
Nov. 28-30, 1990

This report is a synopsis of the writer's activities at the subject meeting.

It is divided into three segments: 1) Participation in the Bio-Processing Equipment Main Comm. Meeting, 2) Panel discussion on Sanitary Valves for the Pharmaceutical/ Bio-Tech Industry, 3) the writer's general observations/ comments.

Bio-Processing Equipment Main Committee

This meeting was called to order on Wed., Nov. 28, 1990, and we immediately began discussing the operating procedure manual for this committee. As is typical with this type of a group, discussions on operating procedures for the committee itself seemed to occupy the majority of time spent. This meeting was no different, and after approx. 3 hrs. of discussion, we were able to agree on the final language which will be compiled and sent to the ASME Board.

As usual, the official minutes of this meeting will be published, and I will distribute to those wishing to keep informed.

At this meeting, the (4) Task Groups assigned at the previous meeting were scheduled to give a brief report on their activities during the last two months.

Chip Manning reported on his group's findings on dimensional tolerances. Scott, I gave you separately a copy of all correspondence Chip sent and/or accumulated for his report.

Reference 13-01 – Notes on November 28-30, 1990 Meeting_4 Pages

-2-

The writer charged with the gasket and seals Task Group gave a report which appeared to be well received.

All these (4) Task Groups were voted upon to become full-fledged sub-committees. The writer was selected as Chairman for the sub-comm. on gaskets and seals.

Although I will be looking for additional volunteers, my two co-chairmen will be Donald Parker, Sr. Engr. of Triad Technologies, Inc., New Castle, DE, and Nigel Brooks, Principal Process Engr. of Fluor Daniel, Greenville, SC.

The next scheduled meeting of this committee was set for January 23, 1991 in Scottsdale, AZ. The reason this date was chosen is because most of the ASME staff will be meeting during that week for other activities and this would be less costly for them. However, during the next few days, many of the comm. members discussed the possibility that the meeting would be too premature, but due to the Christmas holidays, we may not be able to have a meaningful meeting. It was left that we would get direction from ASME as soon as possible. Whether or not this meeting comes through, the next meeting, or the meeting after will be in conjunction with Interphex in NYC in April.

Sanitary Valves for Bio-Processing

This was something new to the writer but is quite common with the ASME group in that instead of having full-blown lectures, they have what is called a panel session. These sessions are chaired by an ASME selected chairman with the individuals selected from the manufacturing/ processing industries and in some cases, regulatory.

This particular panel discussion was chaired by Harry Adey, Life Sciences International, and the Co-Chairman was Rick Zinkowski, ITT Engineered Valves. The panelists were Ivy Logsdon, Eli Lilly Co.; Mike Miles, Worcester Controls; Bob Peoples, P.B.M., Inc.; and the writer.

Harry Adey spent about 30 minutes going from the ground floor pharmaceutical process design, and as far as I'm concerned, bored most of the audience to death.

Reference 13-02 – Notes on November 28-30, 1990 Meeting_4 Pages

-3-

There were approx. (65) people in the audience of which (9) were actual pharmaceutical manufacturers. After his presentation, each of us (except Zinkowski, that is) gave a presentation approx. 10-15 minutes in length. My subject was the 650-type valve and its place in this industry.

I and everyone more or less conformed to the non-commercial aspect of the presentation.

As Co-Chairman, Rick forbade the other panelists to have our product (I and everyone else had a product sample) sitting on the table in front of us. Although he did not have any product sample, his slide presentation was probably the next best thing, if not better.

Observations/ Comments

Ivy Logsdon, Engineering Associate of Eli Lilly & Co., was involved in every meeting/ panel discussion that I was, and at each occasion his favorite story was to tell the audience that, "The FDA reports that 65% of all recalls by the FDA are traced back to the water." He, therefore, stressed careful attention to all aspects of the WFI/ pure water lines. If there were people at these meetings who had not intended to take notes, each time Logsdon made this remark, the participants grappled for their pencil and something to write on.

There seemed to be four themes that were echoed constantly throughout the various meetings and discussions: 1) That all stainless steel components in the various water/ processing lines should be obtained from the same manufacturer with all from the same heat. 2) Each component should have its own

Reference 13-03 – Notes on November 28-30, 1990 Meeting_4 Pages

-4-

certificate of quality control/ surface preparation/ validation. 3) Using the word "grit" is taboo. 4) Containment is also a very big issue and was brought up often.

During some of our discussions, some of the pharmaceutical manufacturing participants also seemed to insist upon manufacturers to provide them with factory-assembled "sections," such as use points, or use points containing valves (such as ITT Engineering valve use point). This is because they feel the factory can have much better control and provide standardized sections for their WFI lines.

During my valve panel discussion, I was asked if Tri-Clover had an ASME-rated pressure relief valve. I wasn't aware that we had any, and according to a fellow from Bristol-Myers and our friend Logsdon, no such valve exists today (sanitary ASME related relief valve) and, therefore, there would be a big demand. This consensus was murmured throughout the audience.

Attending this special session on sanitary valves was Darold Alderman of Cherry-Burrell, Delavan, WI, and Richard Klemp of Advanced Products.

There were, of course, many other meetings. Some of the interesting ones were "Passivation and Surface Finishes of Type 316L SS," "Sulphur and Other Issues for Stainless Steel Piping Systems." I assume Ralph Morrison will be reporting on these meetings and other observations that he, himself, had.

LP 12/10

L. J. Peterman
/rt

Reference 13-04 – Notes on November 28-30, 1990 Meeting_4 Pages

Appendix L

Component Dimensions

The dimensional tables that follow include both BPE-listed components as well as frequently used components not listed in the BPE Standard. Such unlisted components are considered nonstandard fittings[1] and comply with Part DT under Para. DT-4.2 nonstandard fitting dimensions. Refer to Chapter 3 for further information.

[1] All fittings—not BPE listed, Courtesy of MaxPure.

Bioprocessing Piping and Equipment Design: A Companion Guide for the ASME BPE Standard, First Edition. William M. (Bill) Huitt.

ELBOWS - 45°

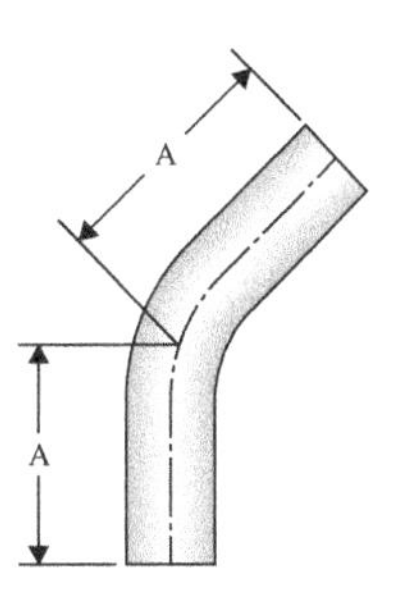

45° ELBOW

Nominal Size in.	Dimensions	
	A in.	A mm
¼	2.000	50.8
⅜	2.000	50.8
½	2.250	57.2
¾	2.250	57.2
1	2.250	57.2
1½	2.500	63.5
2	3.000	76.2
2½	3.375	85.7
3	3.625	92.1
4	4.500	114.3
6	6.250	158.8

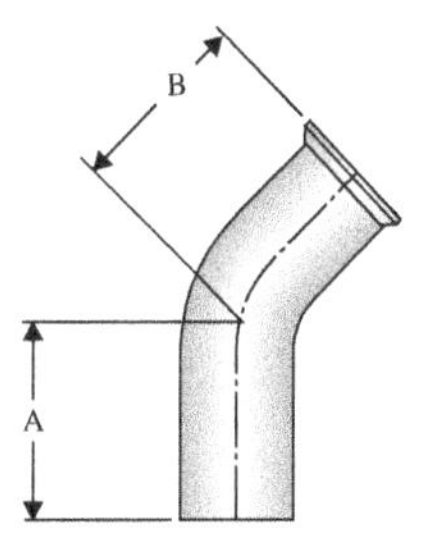

45° ELBOW CLAMP × WELD

Nominal Size in.	Dimensions			
	A in.	A mm	B in.	B mm
¼	2.000	50.8	1.000	25.4
⅜	2.000	50.8	1.000	25.4
½	2.250	57.2	1.000	25.4
¾	2.250	57.2	1.000	25.4
1	2.250	57.2	1.125	28.6
1½	2.500	63.5	1.438	36.5
2	3.000	76.2	1.750	44.5
2½	3.375	85.7	2.063	52.4
3	3.625	92.1	2.375	60.3
4	4.500	114.3	3.125	79.4
6	6.250	158.8	5.250	133.4

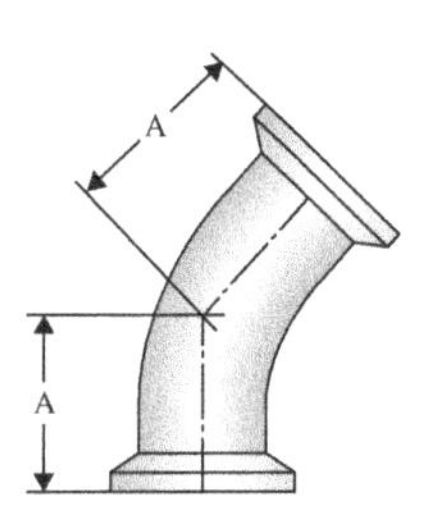

45° CLAMP ELBOW

Nominal Size in.	Dimensions	
	A in.	A mm
¼	1.000	25.4
⅜	1.000	25.4
½	1.000	25.4
¾	1.000	25.4
1	1.125	28.6
1½	1.438	36.5
2	1.750	44.5
2½	2.063	52.4
3	2.375	60.3
4	3.125	79.4
6	5.250	133.4

THESE FITTINGS – NOT BPE LISTED

ELBOWS - 88°

TE2S - 88° ELBOW

Nominal Size in.	Dimensions					
	T		$\alpha = 88°$			
	in.	mm	A in.	A mm	B in.	B mm
½	0.65	1.65	3.06	77.72	2.96	75.18
¾	0.65	1.65	3.06	77.72	2.96	75.18
1	0.65	1.65	3.43	87.12	2.95	74.93
1½	0.65	1.65	3.80	96.52	3.67	93.22
2	0.65	1.65	4.81	122.17	4.64	117.86
2½	0.65	1.65	5.56	141.22	5.37	139.40
3	0.65	1.65	6.31	160.27	6.09	154.69
4	0.83	2.115	8.07	204.98	7.79	197.87

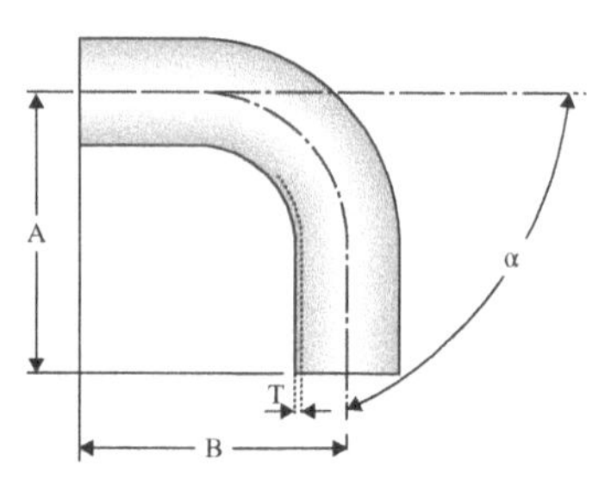

TEG2C - 88° ELBOW

Nominal Size in.	Dimensions					
	T		$\alpha = 88°$			
	in.	mm	A in.	A mm	B in.	B mm
½	0.65	1.65	1.64	41.66	1.59	40.39
¾	0.65	1.65	1.64	41.66	1.59	40.39
1	0.65	1.65	2.02	51.31	1.95	49.53
1½	0.65	1.65	2.77	70.36	2.67	67.82
2	0.65	1.65	3.52	89.41	3.40	86.36
2½	0.65	1.65	4.26	108.20	4.12	104.65
3	0.65	1.65	5.02	127.51	4.84	122.94
4	0.83	2.115	6.64	168.66	6.42	163.07

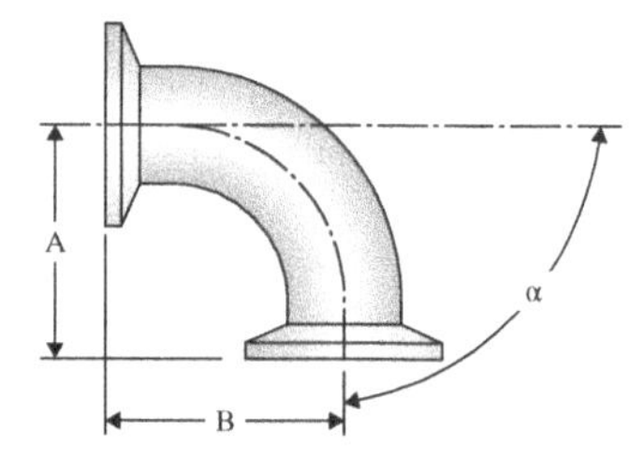

TE2C - 88° ELBOW

Nominal Size in.	Dimensions					
	T		$\alpha = 88°$			
	in.	mm	A in.	A mm	B in.	B mm
½	0.65	1.65	3.02	76.71	1.59	40.39
¾	0.65	1.65	3.02	76.71	1.59	40.39
1	0.65	1.65	3.02	76.71	1.95	49.53
1½	0.65	1.65	3.77	95.76	2.67	67.82
2	0.65	1.65	4.77	121.16	3.40	86.36
2½	0.65	1.65	5.52	140.21	4.12	104.65
3	0.65	1.65	6.27	159.26	4.84	122.94
4	0.83	2.115	8.02	203.71	6.42	163.07

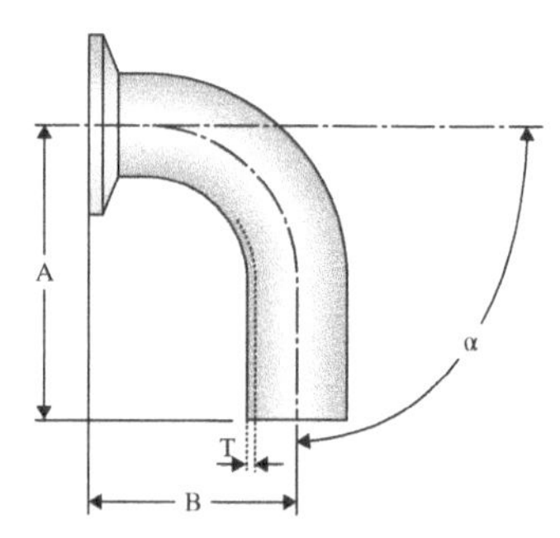

ELBOWS - 90°

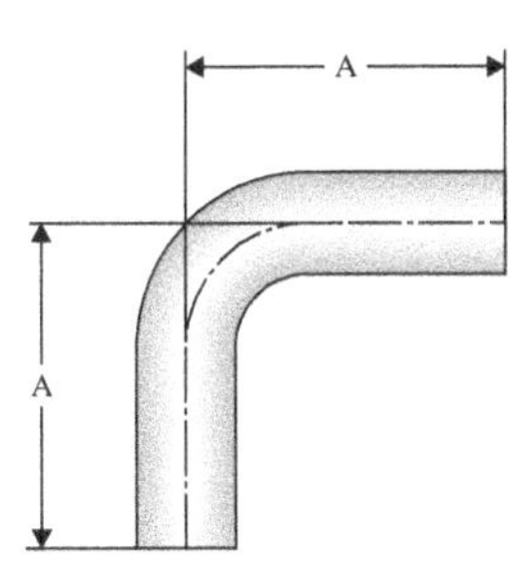

90° ELBOW

Nominal Size in.	Dimensions	
	A in.	A mm
¼	2.625	66.7
⅜	2.625	66.7
½	3.000	76.2
¾	3.000	76.2
1	3.000	76.2
1½	3.750	95.3
2	4.750	120.7
2½	5.500	139.7
3	6.250	158.8
4	8.000	203.2
6	11.500	292.1

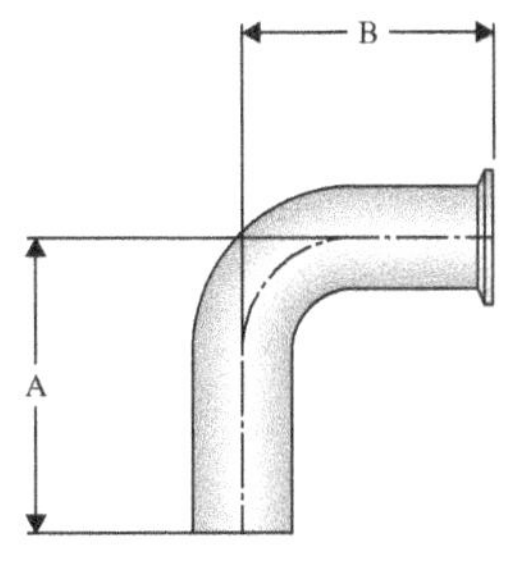

90° ELBOW CLAMP × WELD

Nominal Size in.	Dimensions			
	A in.	A mm	B in.	B mm
¼	2.625	66.7	1.625	41.3
⅜	2.625	66.7	1.625	41.3
½	3.000	76.2	1.625	41.3
¾	3.000	76.2	1.625	41.3
1	3.000	76.2	2.000	50.8
1½	3.750	95.3	2.750	69.9
2	4.750	120.7	3.500	88.9
2½	5.500	139.7	4.250	108.0
3	6.250	158.8	5.000	127.0
4	8.000	203.2	6.625	168.3
6	11.500	292.1	10.500	266.7

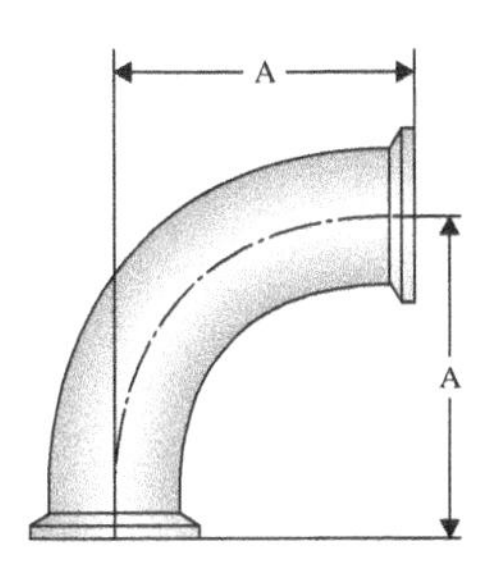

90° CLAMP ELBOW

Nominal Size in.	Dimensions	
	A in.	A mm
¼	1.625	41.3
⅜	1.625	41.3
½	1.625	41.3
¾	1.625	41.3
1	2.000	50.8
1½	2.750	69.9
2	3.500	88.9
2½	4.250	108.0
3	5.000	127.0
4	6.625	168.3
6	10.500	266.7

THESE FITTINGS – NOT BPE LISTED

ELBOWS - 92°

TE2S - 92° ELBOW

Nominal Size in.	Dimensions					
	T		α = 92°			
	in.	mm	A in.	A mm	B in.	B mm
½	0.65	1.65	2.94	74.68	3.04	77.22
¾	0.65	1.65	2.95	74.93	3.04	77.22
1	0.65	1.65	2.95	74.93	3.05	77.22
1½	0.65	1.65	3.74	95.00	3.83	97.28
2	0.65	1.65	4.73	120.14	4.85	123.19
2½	0.65	1.65	5.44	138.18	5.63	143.00
3	0.65	1.65	6.19	157.23	6.41	162.81
4	0.83	2.115	7.93	201.42	8.21	208.53

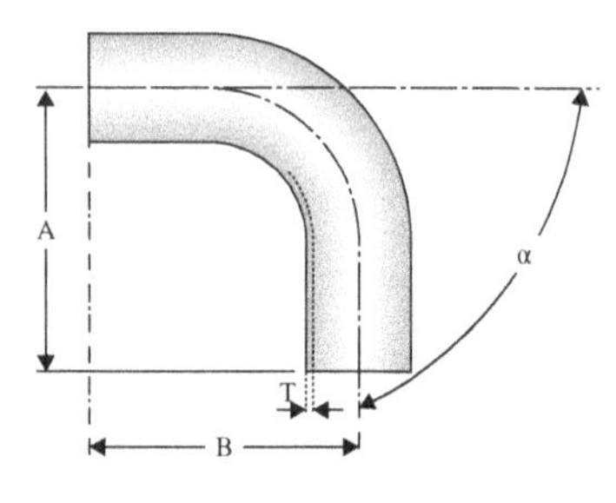

TEG2C - 92° ELBOW

Nominal Size in.	Dimensions					
	T		α = 92°			
	in.	mm	A in.	A mm	B in.	B mm
½	0.65	1.65	1.63	41.40	1.66	42.16
¾	0.65	1.65	1.63	41.40	1.66	42.16
1	0.65	1.65	2.00	50.80	2.50	52.07
1½	0.65	1.65	2.75	69.85	2.83	71.88
2	0.65	1.65	3.50	88.90	3.60	91.44
2½	0.65	1.65	4.25	107.95	4.38	122.68
3	0.65	1.65	5.00	127.00	5.16	131.06
4	0.83	2.115	6.62	168.15	6.83	173.48

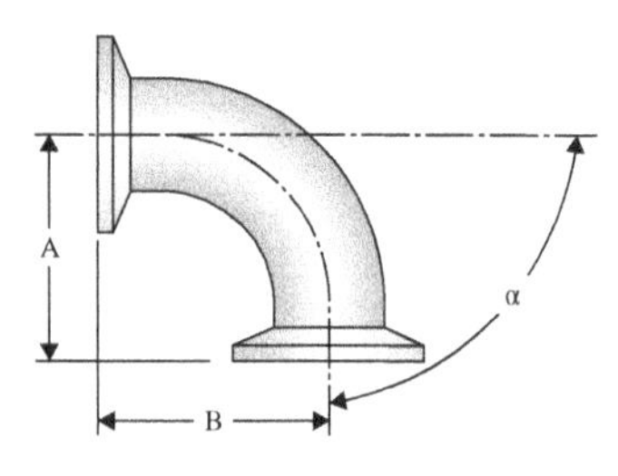

TE2C - 92° ELBOW

Nominal Size in.	Dimensions					
	T		α = 92°			
	in.	mm	A in.	A mm	B in.	B mm
½	0.65	1.65	2.98	75.69	1.66	42.16
¾	0.65	1.65	2.98	75.69	1.66	42.16
1	0.65	1.65	2.98	75.69	2.50	52.07
1½	0.65	1.65	3.73	94.74	2.83	71.88
2	0.65	1.65	4.73	120.14	3.60	91.44
2½	0.65	1.65	5.48	139.19	4.38	122.68
3	0.65	1.65	6.25	158.75	5.16	131.06
4	0.83	2.115	7.98	202.69	6.83	173.48

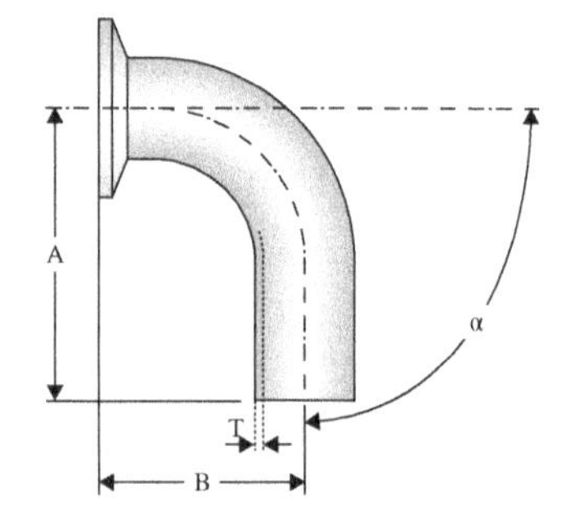

TEES - EQUAL

TEE

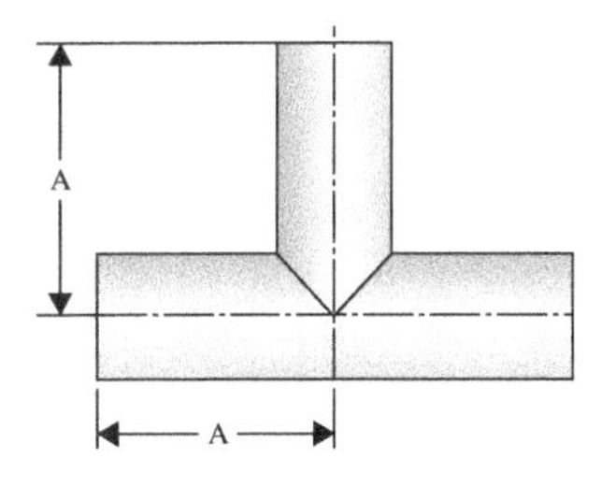

Nominal Size in.	Dimensions	
	A in.	A mm
¼	1.750	44.5
⅜	1.750	44.5
½	1.875	47.6
¾	2.000	50.8
1	2.125	54.0
1½	2.375	60.3
2	2.875	73.0
2½	3.125	79.4
3	3.375	85.7
4	4.125	104.8
6	5.625	142.9

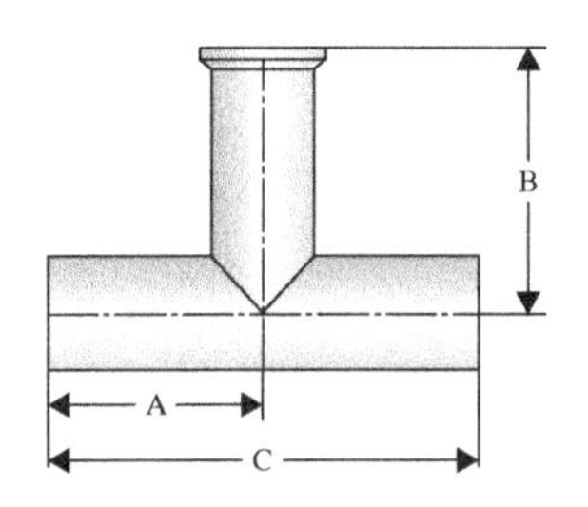

Nominal Size in.	Dimensions					
	A in.	A mm	B in.	B mm	C in.	C mm
½	1.875	47.6	2.250	57.20	3.750	95.2
¾	2.000	50.8	2.375	60.30	4.000	101.6
1	2.125	54.0	2.625	66.68	4.250	108.0
1½	2.375	60.3	2.875	73.03	4.750	120.6
2	2.875	73.0	3.375	85.70	5.750	146.0
2½	3.125	79.4	3.625	92.08	6.250	158.8
3	3.375	85.7	3.875	98.43	6.750	171.4
4	4.125	104.8	4.750	120.65	8.250	209.6
6	5.625	142.9	7.125	181.00	11.250	285.8

TEE CLAMP ENDS

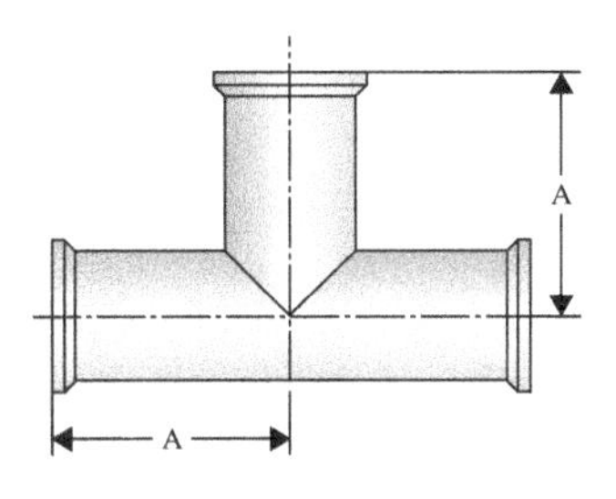

Nominal Size in.	Dimensions	
	A in.	A mm
¼	2.250	57.2
⅜	2.250	57.2
½	2.250	57.2
¾	2.375	60.3
1	2.625	66.7
1½	2.875	73.0
2	3.375	85.7
2½	3.625	92.1
3	3.875	98.4
4	4.750	120.7
6	7.125	181.0

THESE FITTINGS – NOT BPE LISTED

TEES - 88°

TE7WWW886L - BRANCH TEE 88°

Nominal Size in.	Dimensions	
	A in.	A mm
1	2.125	54
1½	2.375	60.3
2	2.875	73
2½	3.125	79.4
3	3.375	85.7
4	4.125	104.8

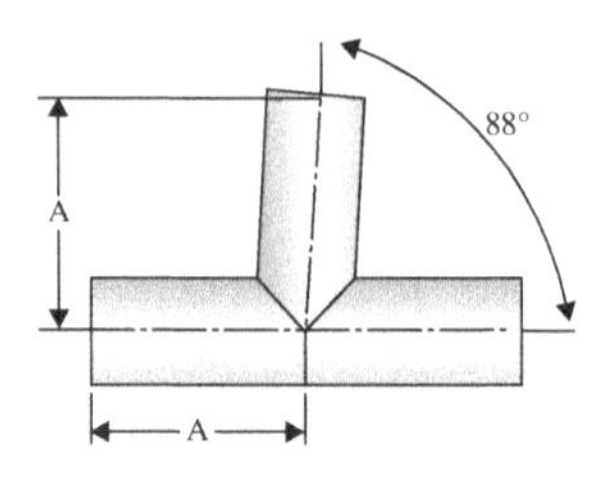

TEES - 176°

TE7WWW17667 - 176° RUN TEE

Nominal Size in.	Dimensions			
	A in.	A mm	B in.	B mm
1	2.2	55.88	2.125	53.848
1½	2.46	62.484	2.375	60.198
2	2.98	75.692	2.875	72.898
2½	3.24	82.296	3.125	79.248
3	3.5	88.9	3.735	94.742
4	4.27	108.458	4.125	104.648

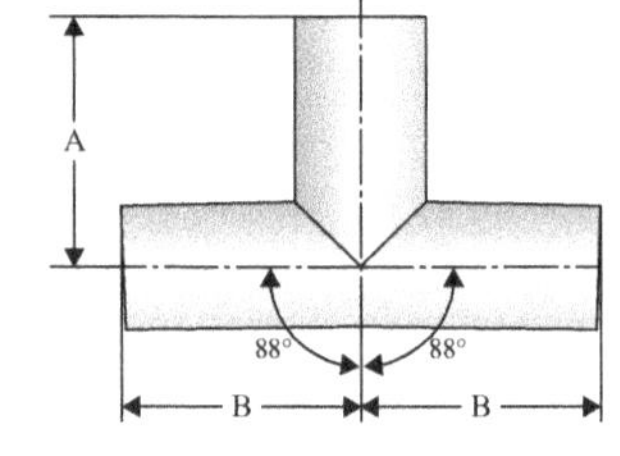

TEES - SHORTOUTLET

SHORT OUTLET TEE – CLAMP ON BRANCH

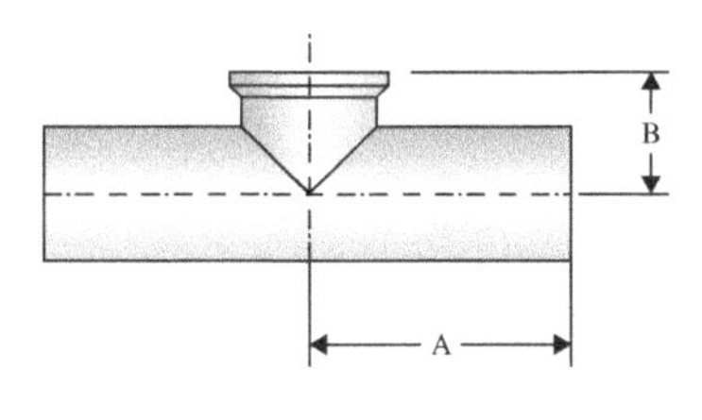

Nominal Size in.	Dimensions			
	A in.	A mm	B in.	B mm
¼	1.750	44.5	1.000	25.4
⅜	1.750	44.5	1.000	25.4
½	1.875	47.6	1.000	25.4
¾	2.000	50.8	1.125	28.6
1	2.125	54.0	1.125	28.6
1½	2.375	60.3	1.375	34.9
2	2.875	73.0	1.625	41.3
2½	3.125	79.4	1.875	47.6
3	3.375	85.7	2.125	54.0
4	4.125	104.8	2.750	69.9
6	5.625	142.9	4.625	117.5

SHORT OUTLET TEE CLAMP ENDS

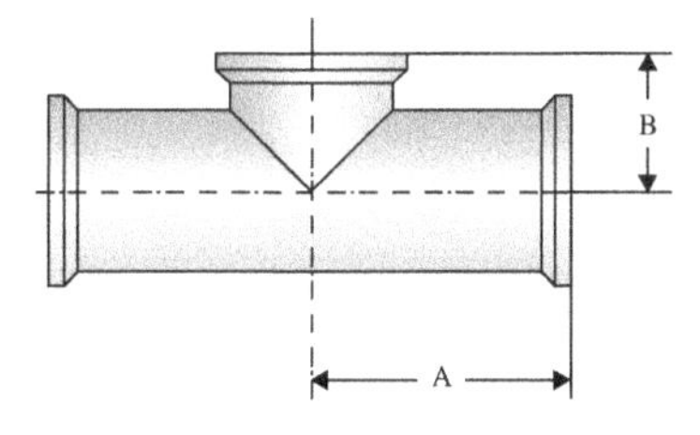

Nominal Size in.	Dimensions			
	A in.	A mm	B in.	B mm
½	2.250	57.2	1.000	25.4
¾	2.375	60.3	1.125	28.6
1	2.625	66.7	1.125	28.6
1½	2.875	73.0	1.375	34.9
2	3.375	85.7	1.625	41.3
2½	3.625	92.1	1.875	47.6
3	3.875	98.4	2.125	54.0
4	4.750	120.7	2.750	69.9
6	7.125	181.0	4.625	117.5

SHORT OUTLET RUN TEE CLAMP ON RUN

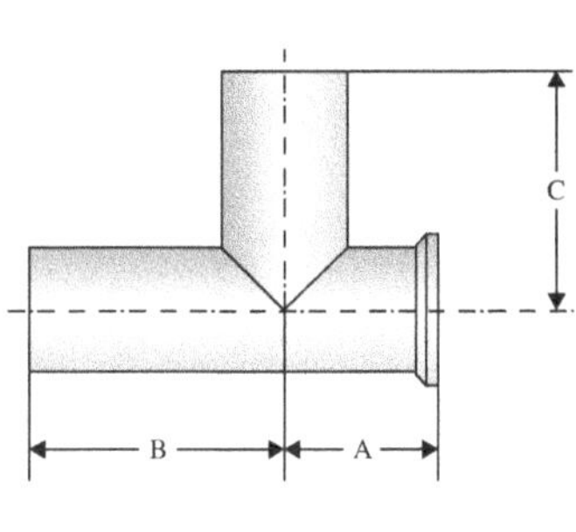

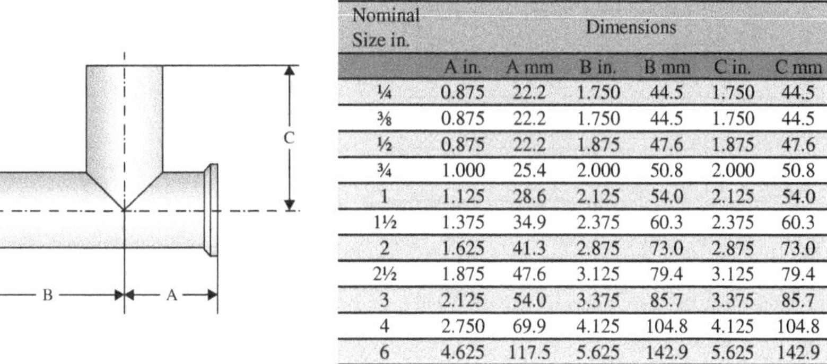

Nominal Size in.	Dimensions					
	A in.	A mm	B in.	B mm	C in.	C mm
¼	0.875	22.2	1.750	44.5	1.750	44.5
⅜	0.875	22.2	1.750	44.5	1.750	44.5
½	0.875	22.2	1.875	47.6	1.875	47.6
¾	1.000	25.4	2.000	50.8	2.000	50.8
1	1.125	28.6	2.125	54.0	2.125	54.0
1½	1.375	34.9	2.375	60.3	2.375	60.3
2	1.625	41.3	2.875	73.0	2.875	73.0
2½	1.875	47.6	3.125	79.4	3.125	79.4
3	2.125	54.0	3.375	85.7	3.375	85.7
4	2.750	69.9	4.125	104.8	4.125	104.8
6	4.625	117.5	5.625	142.9	5.625	142.9

TEES - REDUCING

TE7RWWW - REDUCING TEE

Nominal Size in.	Dimensions			
	A in.	A mm	B in.	B mm
⅜ × ¼	1.750	44.5	1.750	44.5
½ × ¼	1.875	47.6	1.875	47.6
½ × ⅜	1.875	47.6	1.875	47.6
¾ × ¼	2.000	50.8	2.000	50.8
¾ × ⅜	2.000	50.8	2.000	50.8
¾ × ½	2.000	50.8	2.000	50.8
1 × ¼	2.125	54.0	2.125	54.0
1 × ⅜	2.125	54.0	2.125	54.0
1 × ½	2.125	54.0	2.125	54.0
1 × ¾	2.125	54.0	2.125	54.0
1½ × ½	2.375	60.3	2.375	60.3
1½ × ¾	2.375	60.3	2.375	60.3
1 ½ × 1	2.375	60.3	2.375	60.3
2 × ½	2.875	73.0	2.625	66.7
2 × ¾	2.875	73.0	2.625	66.7
2 × 1	2.875	73.0	2.625	66.7
2 × 1½	2.875	73.0	2.625	66.7
2½ × ½	3.125	79.4	2.875	73.0
2½ × ¾	3.125	79.4	2.875	73.0
2½ × 1	3.125	79.4	2.875	73.0
2½ × 1½	3.125	79.4	2.875	73.0
2½ × 2	3.125	79.4	2.875	73.0
3 × ½	3.375	85.7	3.125	79.4
3 × ¾	3.375	85.7	3.125	79.4
3 × 1	3.375	85.7	3.125	79.4
3 × 1½	3.375	85.7	3.125	79.4
3 × 2	3.375	85.7	3.125	79.4
3 × 2½	3.375	85.7	3.125	79.4
4 × ½	4.125	104.8	3.625	92.1
4 × ¾	4.125	104.8	3.625	92.1
4 × 1	4.125	104.8	3.625	92.1
4 × 1½	4.125	104.8	3.625	92.1
4 × 2	4.125	104.8	3.875	98.4
4 × 2½	4.125	104.8	3.875	98.4
4 × 3	4.125	104.8	3.875	98.4
6 × 3	5.625	142.9	4.875	123.8
6 × 4	5.625	142.9	5.125	130.2

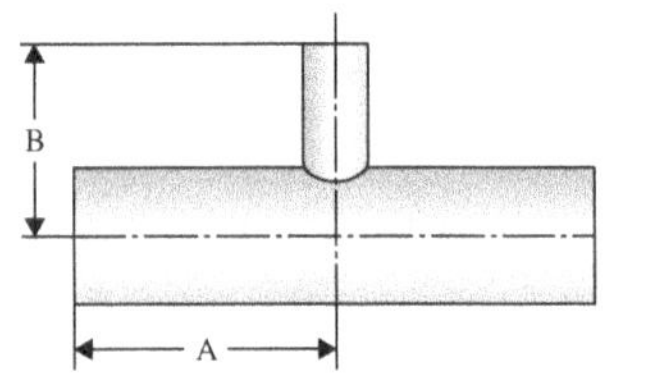

TEES - REDUCING

TEG7R - REDUCING TEE

Nominal Size in.	Dimensions			
	A in.	A mm	B in.	B mm
⅜ × ¾	2.250	57.2	2.250	57.2
½ × ¼	2.375	60.3	2.375	60.3
½ × ⅜	2.375	60.3	2.375	60.3
¾ × ¼	2.500	63.5	2.500	63.5
¾ × ⅜	2.500	63.5	2.500	63.5
¾ × ½	2.500	63.5	2.500	63.5
1 × ¼	2.625	66.7	2.625	66.7
1 × ⅜	2.625	66.7	2.625	66.7
1 × ½	2.625	66.7	2.625	66.7
1 × ¾	2.625	66.7	2.625	66.7
1½ × ½	2.875	73.0	2.875	73.0
1½ × ¾	2.875	73.0	2.875	73.0
1½ × 1	2.875	73.0	2.875	73.0
2 × ½	3.375	85.7	3.125	79.4
2 × ¾	3.375	85.7	3.125	79.4
2 × 1	3.375	85.7	3.125	79.4
2 × 1½	3.375	85.7	3.375	79.4
2½ × ½	3.625	92.1	3.375	85.7
2½ × ¾	3.625	92.1	3.375	85.7
2½ × 1	3.625	92.1	3.375	85.7
2½ × 1½	3.625	92.1	3.375	85.7
2½ × 2	3.625	92.1	3.375	85.7
3 × ½	3.875	98.4	3.625	92.1
3 × ¾	3.875	98.4	3.625	92.1
3 × 1	3.875	98.4	3.625	92.1
3 × 1½	3.875	98.4	3.625	92.1
3 × 2	3.875	98.4	3.625	92.1
3 × 2½	3.875	98.4	3.625	92.1
4 × ½	4.750	120.7	4.125	104.8
4 × ¾	4.750	120.7	4.125	104.8
4 × 1	4.750	120.7	4.125	104.8
4 × 1½	4.750	120.7	4.125	104.8
4 × 2	4.750	120.7	4.375	111.1
4 × 2½	4.750	120.7	4.375	111.1
4 × 3	4.750	120.7	4.375	111.1
6 × 3	7.125	181.0	5.375	136.5
6 × 4	7.125	181.0	5.750	146.1

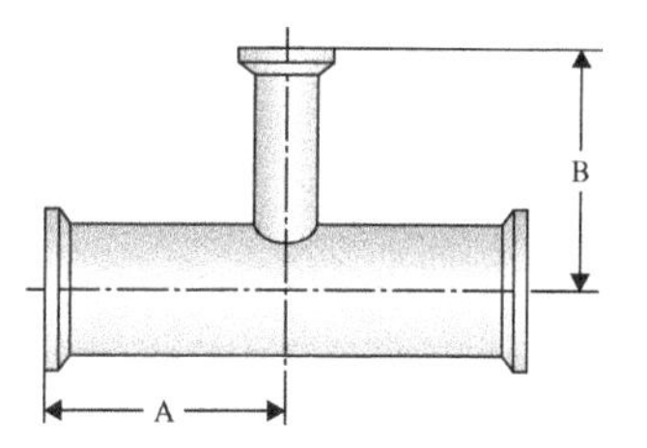

TEES - REDUCING

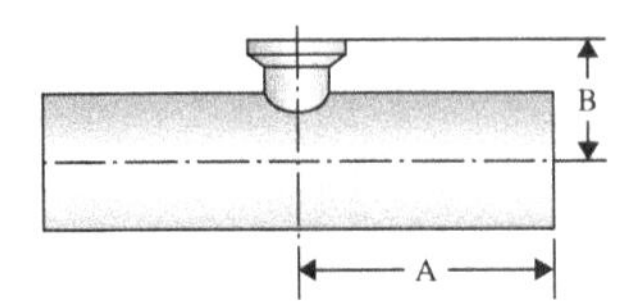

SHORT OUTLET REDUCING TEE
CLAMP ON BRANCH

Nominal Size in.	Dimensions			
	A in.	A mm	B in.	B mm
⅜ × ¼	1.750	44.5	1.000	25.4
½ × ¼	1.875	47.6	1.000	25.4
½ × ⅜	1.875	47.6	1.000	25.4
¾ × ¼	2.000	50.8	1.000	25.4
¾ × ⅜	2.000	50.8	1.000	25.4
¾ × ½	2.000	50.8	1.000	25.4
1 × ¼	2.125	54.0	1.125	28.6
1 × ⅜	2.125	54.0	1.125	28.6
1 × ½	2.125	54.0	1.125	28.6
1 × ¾	2.125	54.0	1.125	28.6
1½ × ½	2.375	60.3	1.375	34.9
1½ × ¾	2.375	60.3	1.375	34.9
1½ × 1	2.375	60.3	1.375	34.9
2 × ½	2.875	73.0	1.625	41.3
2 × ¾	2.875	73.0	1.625	41.3
2 × 1	2.875	73.0	1.625	41.3
2 × 1½	2.875	73.0	1.625	41.3
2½ × ½	3.125	79.4	1.875	47.6
2½ × ¾	3.125	79.4	1.875	47.6
2½ × 1	3.125	79.4	1.875	47.6
2½ × 1½	3.125	79.4	1.875	47.6
2½ × 2	3.125	79.4	1.875	47.6
3 × ½	3.375	85.7	2.125	54.0
3 × ¾	3.375	85.7	2.125	54.0
3 × 1	3.375	85.7	2.125	54.0
3 × 1½	3.375	85.7	2.125	54.0
3 × 2	3.375	85.7	2.125	54.0
3 × 2½	3.375	85.7	2.125	54.0
4 × ½	4.125	104.8	2.625	66.7
4 × ¾	4.125	104.8	2.625	66.7
4 × 1	4.125	104.8	2.625	66.7
4 × 1½	4.125	104.8	2.625	66.7
4 × 2	4.125	104.8	2.625	66.7
4 × 2½	4.125	104.8	2.625	66.7
4 × 3	4.125	104.8	2.625	66.7
6 × ½	5.625	142.9	3.625	92.1
6 × ¾	5.625	142.9	3.625	92.1
6 × 1	5.625	142.9	3.625	92.1
6 × 1½	5.625	142.9	3.625	92.1
6 × 2	5.625	142.9	3.625	92.1
6 × 2½	5.625	142.9	3.625	92.1
6 × 3	5.625	142.9	3.625	92.1
6 × 4	5.625	142.9	3.750	95.3

TEES - REDUCING

SHORT OUTLET REDUCING TEE – CLAMP ENDS

Nominal Size in.	Dimensions			
	A in.	A mm	B in.	B mm
⅜ × ¼	2.250	57.2	1.000	25.4
½ × ¼	2.375	60.3	1.000	25.4
½ × ⅜	2.375	60.3	1.000	25.4
¾ × ¼	2.500	63.5	1.000	25.4
¾ × ⅜	2.500	63.5	1.000	25.4
¾ × ½	2.500	63.5	1.000	25.4
1 × ¼	2.625	66.7	1.125	28.6
1 × ⅜	2.625	66.7	1.125	28.6
1 × ½	2.625	66.7	1.125	28.6
1 × ¾	2.625	66.7	1.125	28.6
1½ × ½	2.875	73.0	1.375	34.9
1½ × ¾	2.875	73.0	1.375	34.9
1½ × 1	2.875	73.0	1.375	34.9
2 × ½	3.375	85.7	1.625	41.3
2 × ¾	3.375	85.7	1.625	41.3
2 × 1	3.375	85.7	1.625	41.3
2 × 1½	3.375	85.7	1.625	41.3
2½ × ½	3.625	92.1	1.875	47.6
2½ × ¾	3.625	92.1	1.875	47.6
2½ × 1	3.625	92.1	1.875	47.6
2½ × 1½	3.625	92.1	1.875	47.6
2 ½ × 2	3.625	92.1	1.875	47.6
3 × ½	3.875	98.4	2.125	54.0
3 × ¾	3.875	98.4	2.125	54.0
3 × 1	3.875	98.4	2.125	54.0
3 × 1½	3.875	98.4	2.125	54.0
3 × 2	3.875	98.4	2.125	54.0
3 × 2½	3.875	98.4	2.125	54.0
4 × ½	4.750	120.7	2.625	66.7
4 × ¾	4.750	120.7	2.625	66.7
4 × 1	4.750	120.7	2.625	66.7
4 × 1½	4.750	120.7	2.625	66.7
4 × 2	4.750	120.7	2.625	66.7
4 × 2½	4.750	120.7	2.625	66.7
4 × 3	4.750	120.7	2.625	66.7
6 × ½	7.125	181.0	3.625	92.1
6 × ¾	7.125	181.0	3.625	92.1
6 × 1	7.125	181.0	3.625	92.1
6 × 1½	7.125	181.0	3.625	92.1
6 × 2	7.125	181.0	3.625	92.1
6 × 2½	7.125	181.0	3.625	92.1
6 × 3	7.125	181.0	3.625	92.1
6 × 4	7.125	181.0	3.750	95.3

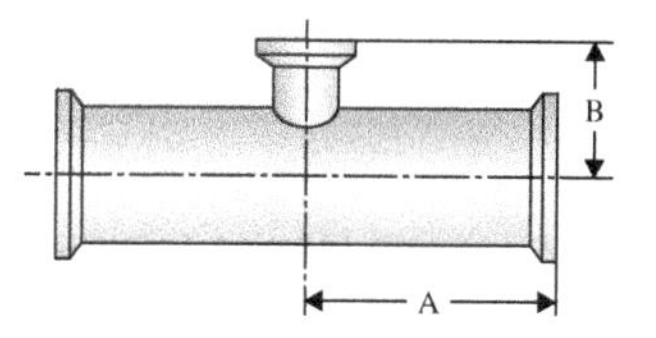

TEES - INSTRUMENT

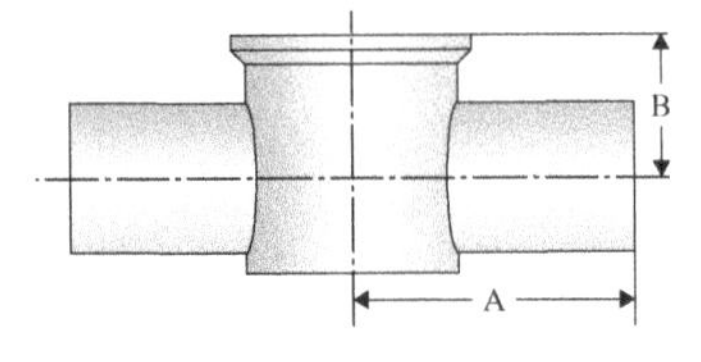

INSTRUMENT TEE – CLAMP ON BRANCH

Nominal Size in.	Dimensions			
	A in.	A mm	B in.	B mm
½ × 1	2.250	57.15	1.000	25.40
¾ × 1	2.250	57.15	1.000	25.40
½ × 1½	2.500	63.5	0.875	22.2
¾ × 1½	2.500	63.5	1.000	25.4
1 × 1½	2.500	63.5	1.125	28.6
½ × 2	2.750	69.9	1.000	25.4
¾ × 2	2.750	69.9	1.125	28.6
1 × 2	2.750	69.9	1.250	31.8
1½ × 2	2.750	69.9	1.500	38.1

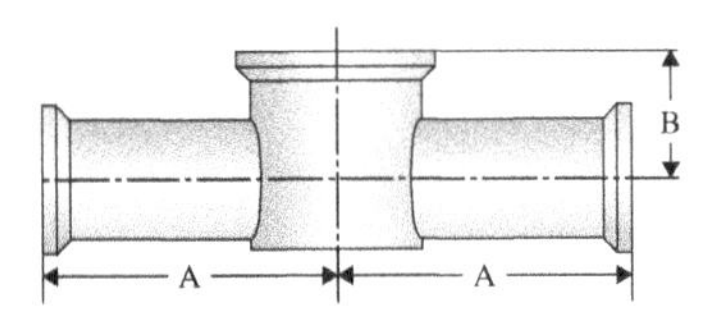

INSTRUMENT TEE CLAMP ENDS

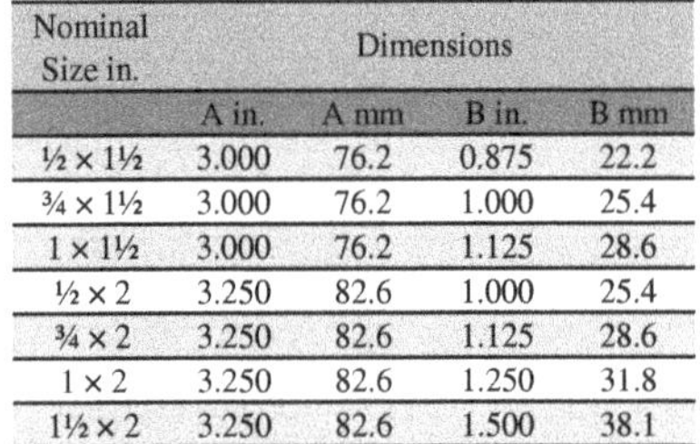

Nominal Size in.	Dimensions			
	A in.	A mm	B in.	B mm
½ × 1½	3.000	76.2	0.875	22.2
¾ × 1½	3.000	76.2	1.000	25.4
1 × 1½	3.000	76.2	1.125	28.6
½ × 2	3.250	82.6	1.000	25.4
¾ × 2	3.250	82.6	1.125	28.6
1 × 2	3.250	82.6	1.250	31.8
1½ × 2	3.250	82.6	1.500	38.1

REDUCERS - CONCENTRIC

TE31SWW - SHORT CONCENTRIC REDUCER

Nominal Size in.	Dimensions					
	Overall Length, A, in	Overall Length, A, mm	Minimum O.D Tangent, Small End, L3, in.	Minimum O.D Tangent, Small End, L3, mm	Minimum O.D Tangent, Large End, L4, in.	Minimum O.D Tangent, Large End, L4, mm
⅜ × ¼	1.625	41.275	0.750	19.05	0.750	19.05
½ × ¼	1.875	47.625	0.750	19.05	1.000	25.4
½ × ⅜	1.875	47.625	0.750	19.05	1.000	25.4
¾ × ⅜	2.000	508	0.750	19.05	1.000	25.4
¾ × ½	2.125	53.975	1.000	25.4	1.000	25.4
1 × ½	2.500	63.5	1.000	25.4	1.000	25.4
1 × ¾	2.125	53.975	1.000	25.4	1.000	25.4
1½ × ¾	3.000	76.2	1.000	25.4	1.000	25.4
1½ × 1	2.500	63.5	1.000	25.4	1.000	25.4
2 × 1	3.375	85.725	1.000	25.4	1.000	25.4
2 × 1½	2.500	63.5	1.000	25.4	1.000	25.4
2½ × 1½	3.375	85.725	1.000	25.4	1.000	25.4
2½ × 2	2.500	63.5	1.000	25.4	1.000	25.4
3 × 1½	4.250	107.95	1.000	25.4	1.500	38.1
3 × 2	3.375	85.725	1.000	25.4	1.500	38.1
3 × 2½	2.625	66.675	1.000	25.4	1.500	38.1
4 × 2	5.125	130.175	1.000	25.4	1.500	38.1
4 × 2½	4.250	107.95	1.000	25.4	1.500	38.1
4 × 3	3.875	98.425	1.500	38.1	1.500	38.1
6 × 3	7.250	184.15	1.500	38.1	2.000	50.8
6 × 4	5.625	142.875	1.500	38.1	2.000	50.8

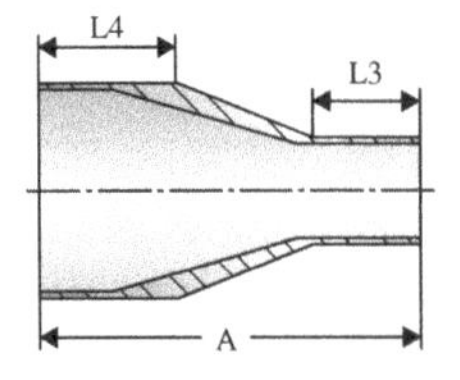

THESE FITTINGS – NOT BPE LISTED

REDUCERS - CONICAL

TEG31L - SHORT CONICAL INSTRUMENT REDUCER

Nominal Size in.	Dimensions					
	A in.	A mm	B in.	B mm	C in.	C mm
¾ × ½	1.250	31.75	0.620	15.748	0.370	9.398
1 × ½	1.250	31.75	0.870	22.098	0.370	9.398
1 × ¾	1.250	31.75	0.870	22.098	0.620	15.748
1½ × ½	1.250	31.75	1.370	34.798	0.370	9.398
1½ × ¾	1.250	31.75	1.370	34.798	0.620	15.748
1½ × 1	1.250	31.75	1.370	34.798	0.870	22.098
2 × ½	1.250	31.75	1.870	47.498	0.370	9.398
2 × ¾	1.250	31.75	1.870	47.498	0.620	15.748
2 × 1	1.250	31.75	1.870	47.498	0.870	22.098
2 × 1½	1.250	31.75	1.870	47.498	1.370	34.798
2½ × ½	1.250	31.75	2.370	60.198	0.370	9.398
2½ × ¾	1.250	31.75	2.370	60.198	0.620	15.748
2½ × 1	1.250	31.75	2.370	60.198	0.870	22.098
2½ × 1½	1.250	31.75	2.370	60.198	1.370	34.798
2½ × 2	1.250	31.75	2.370	60.198	1.870	47.498
3 × ½	1.250	31.75	2.870	72.898	0.370	9.398
3 × ¾	1.250	31.75	2.870	72.898	0.620	15.748
3 × 1	1.250	31.75	2.870	72.898	0.870	22.098
3 × 1½	1.250	31.75	2.870	72.898	1.370	34.798
3 × 2	1.250	31.75	2.870	72.898	1.870	47.498
3 × 2½	1.250	31.75	2.870	72.898	2.370	60.198
4 × ½	1.250	31.75	3.387	97.384	0.370	9.398
4 × ¾	1.250	31.75	3.834	97.384	0.620	15.748
4 × 1	1.250	31.75	3.834	97.384	0.870	22.098
4 × 1½	1.250	31.75	3.834	97.384	1.370	34.798
4 × 2	1.250	31.75	3.834	97.384	1.870	47.498
4 × 2½	1.250	31.75	3.834	97.384	2.370	60.198
4 × 3	1.250	31.75	3.834	97.384	2.870	72.898

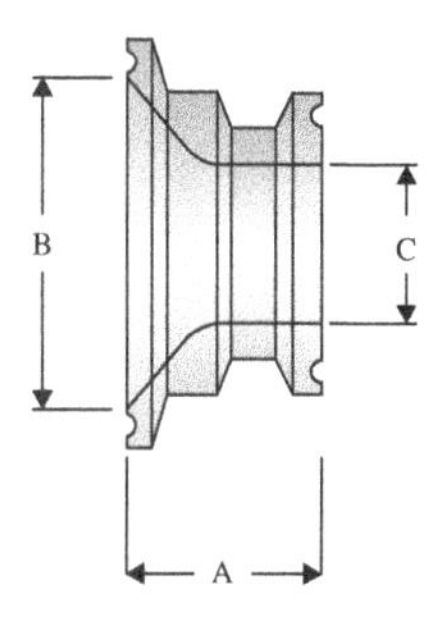

REDUCERS - ECCENTRIC

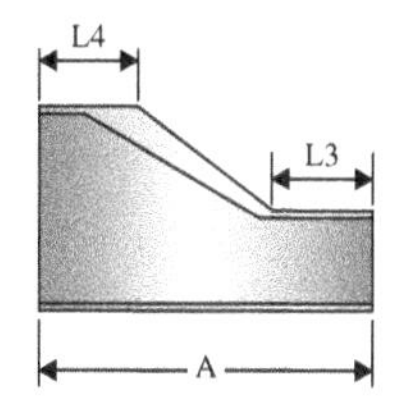

SHORT ECCENTRIC REDUCER

Nominal Size in.	Dimensions					
	Overall Length, A, in	Overall Length, A, mm	Minimum O.D. Tangent, Small End, L3, in.	Minimum O.D. Tangent, Small End, L3,mm	Minimum O.D. Tangent, Large End, L4, in.	Minimum O.D. Tangent, Large End, L4, mm
⅜ × ¼	1.625	41.275	0.750	19.05	0.750	19.05
½ × ¼	1.875	47.625	0.750	19.05	1.000	25.4
½ × ⅜	1.875	47.625	0.750	19.05	1.000	25.4
¾ × ⅜	2.000	50.8	0.750	19.05	1.000	25.4
¾ × ½	2.125	53.975	1.000	25.4	1.000	25.4
1 × ½	2.500	63.5	1.000	25.4	1.000	25.4
1 × ¾	2.125	53.975	1.000	25.4	1.000	25.4
1½ × ¾	3.000	76.2	1.000	25.4	1.000	25.4
1½ × 1	2.500	63.5	1.000	25.4	1.000	25.4
2 × 1	3.375	85.725	1.000	25.4	1.000	25.4
2 × 1½	2.500	63.5	1.000	25.4	1.000	25.4
2½ × 1½	3.375	85.725	1.000	25.4	1.000	25.4
2½ × 2	2.500	63.5	1.000	25.4	1.000	25.4
3 × 1½	4.250	107.95	1.000	25.4	1.500	38.1
3 × 2	3.375	85.725	1.000	25.4	1.500	38.1
3 × 2½	2.625	66.675	1.000	25.4	1.500	38.1
4 × 2	5.125	130.175	1.000	25.4	1.500	38.1
4 × 2½	4.250	107.95	1.000	25.4	1.500	38.1
4 × 3	3.875	98.425	1.500	38.1	1.500	38.1
6 × 3	7.250	184.15	1.500	38.1	2.000	50.8
6 × 4	5.625	142.875	1.500	38.1	2.000	50.8

REDUCERS - CONCENTRIC

TE32SCW - SHORT ECCENTRIC REDUCER

Nominal Size in.	Dimensions			
	Overall Length, A, in	Overall Length, A, mm	Minimum O.D Tangent, Small End, L3, in.	Minimum O.D Tangent, Small End, L3, mm
⅜ × ¼	2.125	53.975	0.750	19.05
½ × ¼	2.375	60.325	0.750	19.05
½ × ⅜	2.375	60.325	0.750	19.05
¾ × ⅜	2.500	63.5	0.750	19.05
¾ × ½	2.625	66.675	1.000	25.4
1 × 1½	3.000	76.2	1.000	25.4
1 × ¾	2.625	66.675	1.000	25.4
1½ × ¾	3.500	88.9	1.000	25.4
1½ × 1	3.000	76.2	1.000	25.4
2 × 1	3.875	98.425	1.000	25.4
2 × 1½	3.000	76.2	1.000	25.4
2½ × 1½	3.875	98.425	1.000	25.4
2½ × 2	3.000	76.2	1.000	25.4
3 × 1½	4.750	120.65	1.000	25.4
3 × 2	3.875	98.425	1.000	25.4
3 × 2½	3.125	79.375	1.000	25.4
4 × 2	5.750	146.05	1.000	25.4
4 × 2½	4.875	123.825	1.000	25.4
4 × 3	4.500	114.3	1.500	38.1
6 × 3	8.000	203.2	1.500	38.1
6 × 4	6.375	161.925	1.500	38.1

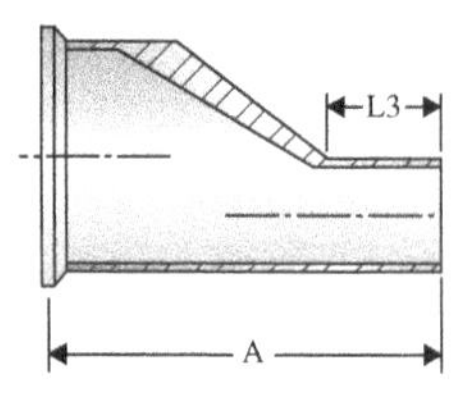

REDUCERS - ECCENTRIC

SHORT ECCENTRIC REDUCER
CLAMP ENDS

Nominal Size in.	Dimensions	
	Overall Length, A, in	Overall Length, A, mm
⅜ × ¼	2.625	66.675
½ × ¼	2.875	73.025
½ × ⅜	2.875	73.025
¾ × ⅜	3.000	76.2
¾ × ½	3.125	79.375
1 × 1½	3.500	88.9
1 × ¾	3.125	79.375
1½ × ¾	4.000	101.6
1½ × 1	3.500	88.9
2 × 1	4.375	111.125
2 × 1½	3.500	88.9
2½ × 1½	4.375	111.125
2½ × 2	3.500	88.9
3 × 1½	5.250	133.35
3 × 2	4.375	111.125
3 × 2½	3.625	92.075
4 × 2	6.250	158.75
4 × 2½	5.375	136.525
4 × 3	5.000	127
6 × 3	8.500	215.9
6 × 4	7.000	177.8

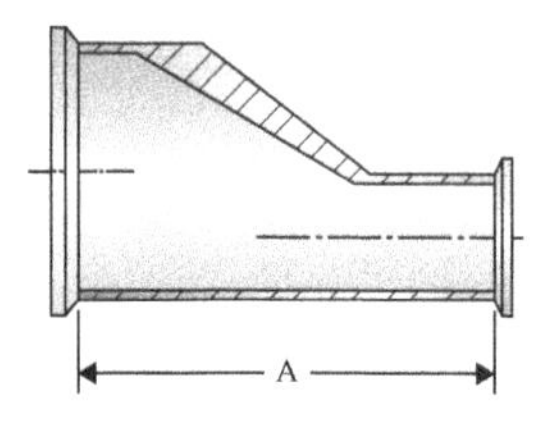

THESE FITTINGS – NOT BPE LISTED

USE POINTS

TE2UBWWW - 180° BOTTOM OUTLET
WELD USE POINT

Normal Size in.	Dimensions					
	A in.	A mm	B in.	B mm	C in.	C mm
¾ × ½	4.500	114.3	3.000	76.2	1.875	47.6
¾ × ¾	4.500	114.3	3.000	76.2	1.875	47.6
1 × ½	3.000	76.2	3.000	76.2	2.062	52.4
1½ × ½	4.500	114.3	4.500	114.3	2.312	58.7
2 × ½	6.000	152.4	5.000	127.0	2.562	65.1
2½ × ½	7.500	190.5	5.750	146.1	2.812	71.4
3 × ½	9.000	228.6	6.500	165.1	3.062	77.8
4 × /½	12.000	304.8	8.000	215.9	3.562	90.5

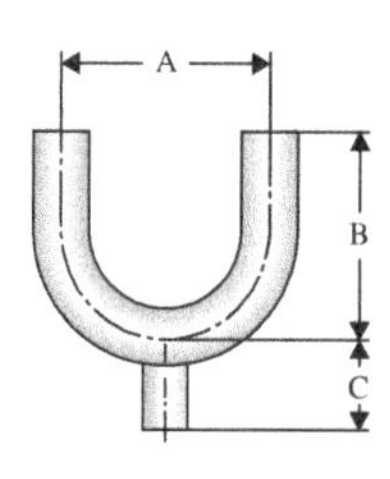

TE2UBWWW - 180° BOTTOM OUTLET
CLAMP USE POINT

Normal Size in.	Dimensions					
	A in.	A mm	B in.	B mm	C in.	C mm
¾ × ½	4.500	114.3	3.000	76.2	0.875	22.2
¾ × ¾	4.500	114.3	3.000	76.2	0.875	22.2
1 × ½	3.000	76.2	3.000	76.2	1.062	27.0
1½ × ½	4.500	114.3	4.500	114.3	1.312	33.3
2 × ½	6.000	152.4	5.000	127.0	1.562	39.7
2½ × ½	7.500	190.5	5.750	146.1	1.812	46
3 × ½	9.000	228.6	6.500	165.1	2.062	52.4
4 × /½	12.000	304.8	8.000	215.9	2.562	65.1

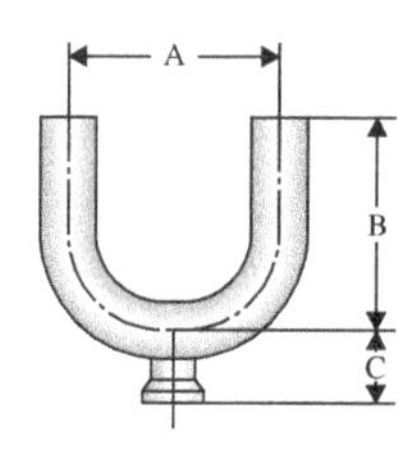

TE2UBWWW - 180° SIDE OUTLET
CLAMP USE POINT

Normal Size in.	Dimensions					
	A in.	A mm	B in.	B mm	C in.	C mm
¾ × ½	4.500	114.3	3.000	76.2	0.875	22.2
¾ × ¾	4.500	114.3	3.000	76.2	0.875	22.2
1 × ½	3.000	76.2	3.000	76.2	1.062	26.9
1½ × ½	4.500	114.3	4.500	114.3	1.312	33.3
2 × ½	6.000	152.4	5.000	127.0	1.562	39.7
2½ × ½	7.500	190.5	5.750	146.1	1.812	46
3 × ½	9.000	228.6	6.500	165.1	2.062	52.4
4 × ½	12.000	304.8	8.000	215.9	2.562	65

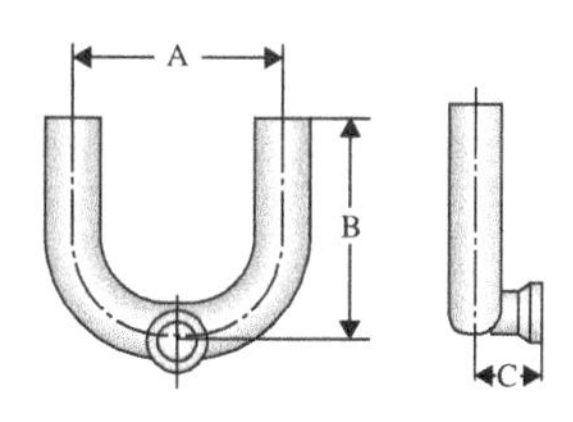

THESE FITTINGS – NOT BPE LISTED

TRUE Y'S & LATERALS

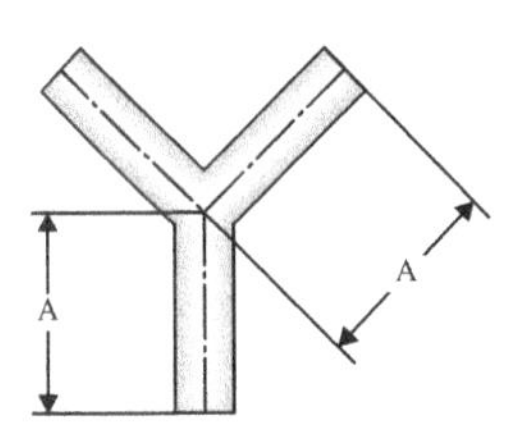

TRUE Y

Nominal Size in.	Dimensions		
	A in.	A mm	Nom. Wall
1	3.000	76.2	0.065
1½	3.000	76.2	0.065
2	4.000	101.6	0.065
2½	5.000	127.0	0.065
3	6.000	152.4	0.065
4	8.000	302.2	0.065
6	8.000	302.2	0.065

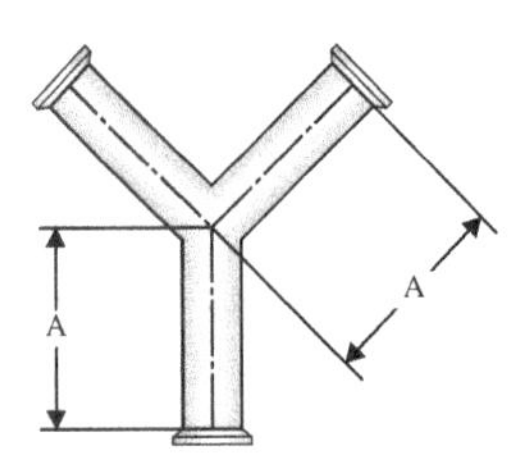

CLAMP TRUE Y

Nominal Size in.	Dimensions	
	A in.	A mm
1	3.500	88.9
1½	3.500	88.9
2	4.500	114.9
2½	5.500	139.7
3	6.500	165.1
4	8.625	219.1
6	8.875	225.4

TRUE Y'S AND LATERALS

TE28 - 45° LATERAL

Nominal Size in.	Dimensions							
	A in.	A mm	B in.	B mm	C in.	C mm	D in.	D mm
1	6.000	152.4	5.000	127.0	1.000	25.4	1.000	25.4
1½	7.380	187.5	6.190	157.2	1.190	30.2	1.500	38.1
2	8.750	222.3	7.120	181.0	1.630	41.4	2.000	50.8
2½	10.000	254.0	8.500	215.9	1.500	38.1	2.500	63.50
3	10.750	270.1	8.870	225.4	1.870	47.5	3.000	76.2
4	12.810	325.4	10.750	273.1	2.060	52.4	4.000	101.6
6	16.500	419.1	12.500	317.5	4.000	101.6	6.000	152.4

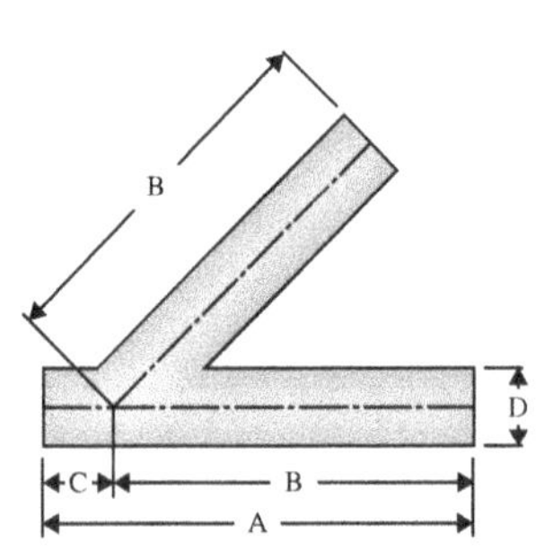

TEG28A - 45° CLAMP LATERAL

Nominal Size in.	Dimensions					
	A in.	A mm	B in.	B mm	C in.	C mm
1	7.000	177.8	5.500	139.7	1.500	38.1
1½	8.375	212.7	6.687	169.9	1.687	42.9
2	9.750	247.7	7.625	193.7	2.125	54.0
2½	11.000	279.4	9.000	228.6	2.000	50.8
3	11.750	298.5	9.375	238.1	2.375	60.3
4	14.062	357.2	11.375	288.9	2.687	68.3
6	18.250	479.4	15.375	390.5	4.875	111.1

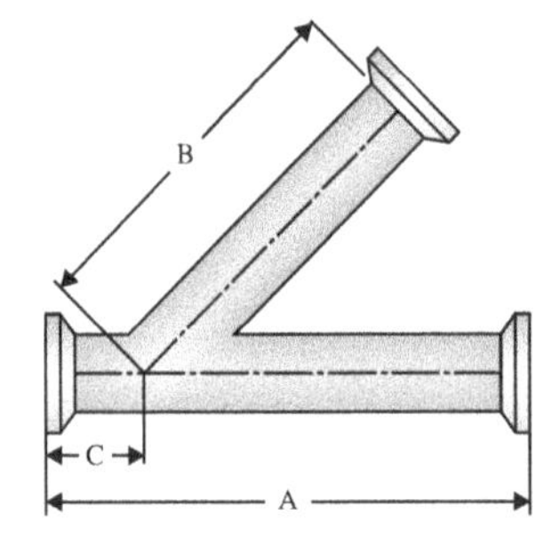

CROSSES

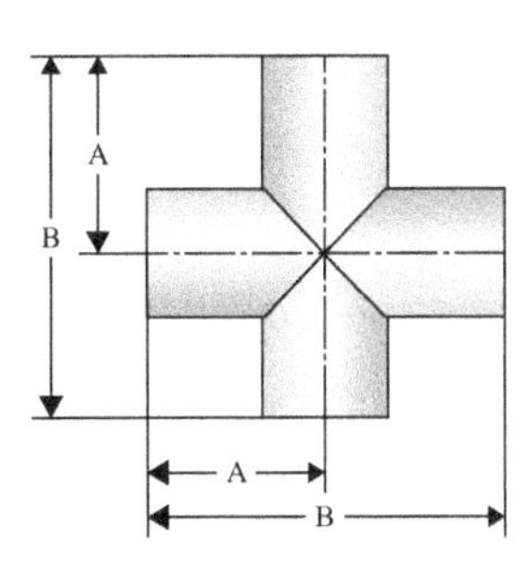

CROSS

Nominal Size in.	Dimensions			
	A in.	A mm	B in.	B mm
¼	1.750	44.5	3.500	89.0
⅜	1.750	44.5	3.500	89.0
½	1.875	47.6	3.750	95.2
¾	2.000	50.8	4.000	101.6
1	2.125	54	4.250	108.0
1½	2.375	60.3	4.750	120.6
2	2.875	73	5.750	146.0
2½	3.125	79.4	6.250	158.8
3	3.375	85.7	6.750	171.4
4	4.125	104.8	8.250	209.6
6	5.625	142.9	11.250	285.8

CROSS CLAMP ENDS

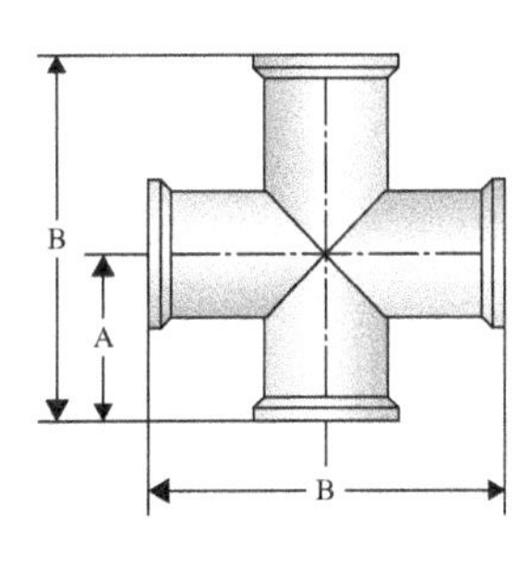

Nominal Size in.	Dimensions			
	A in.	A mm	B in.	B mm
¼	2.250	57.2	4.500	114.4
⅜	2.250	57.2	4.500	114.4
½	2.250	57.2	4.500	114.4
¾	2.375	60.3	4.750	120.6
1	2.625	66.7	5.250	133.4
1½	2.875	73.0	5.750	146.0
2	3.375	85.7	6.750	171.4
2½	3.625	92.1	7.250	184.2
3	3.875	98.4	7.750	196.8
4	4.750	120.7	9.500	241.4
6	7.125	181.0	14.250	362.0

THESE FITTINGS – NOT BPE LISTED

CROSSES

TE9RWWW - REDUCING CROSS

Nominal Size in.	Dimensions			
	A in.	A mm	B in.	B mm
⅜ × ¼	1.750	44.5	1.750	44.5
½ × ¼	1.875	47.6	1.875	47.6
½ × ⅜	1.875	47.6	1.875	47.6
¾ × ¼	2.000	50.8	2.000	50.8
¾ × ⅜	2.000	50.8	2.000	50.8
¾ × ½	2.000	50.8	2.000	50.8
1 × ¼	2.125	54.0	2.125	54.0
1 × ⅜	2.125	54.0	2.125	54.0
1 × ½	2.125	54.0	2.125	54.0
1 × ¾	2.125	54.0	2.125	54.0
1½ × ½	2.375	60.3	2.375	60.3
1½ × ¾	2.375	60.3	2.375	60.3
1½ × ½	2.375	60.3	2.375	60.3
2 × ½	2.875	73.0	2.625	66.7
2 × ¾	2.875	73.0	2.625	66.7
2 × 1	2.875	73.0	2.625	66.7
2 × 1½	2.875	73.0	2.625	66.7
2½ × ½	3.125	79.4	2.875	73.0
2½ × ¾	3.125	79.4	2.875	73.0
2½ × 1	3.125	79.4	2.875	73.0
2½ × 1½	3.125	79.4	2.875	73.0
2½ × 2	3.125	79.4	2.875	73.0
3 × ½	3.375	85.7	3.125	79.4
3 × ¾	3.375	85.7	3.125	79.4
3 × 1	3.375	85.7	3.125	79.4
3 × 1½	3.375	85.7	3.125	79.4
3 × 2	3.375	85.7	3.125	79.4
3 × 2½	3.375	85.7	3.125	79.4
4 × ½	4.125	104.8	3.625	92.1
4 × ¾	4.125	104.8	3.625	92.1
4 × 1	4.125	104.8	3.625	92.1
4 × 1½	4.125	104.8	3.625	92.1
4 × 2	4.125	104.8	3.625	92.1
4 × 2½	4.125	104.8	3.625	92.1
4 × 3	4.125	104.8	3.625	92.1
6 × 3	5.625	142.9	4.875	123.8
6 × 4	5.625	142.9	5.125	130.2

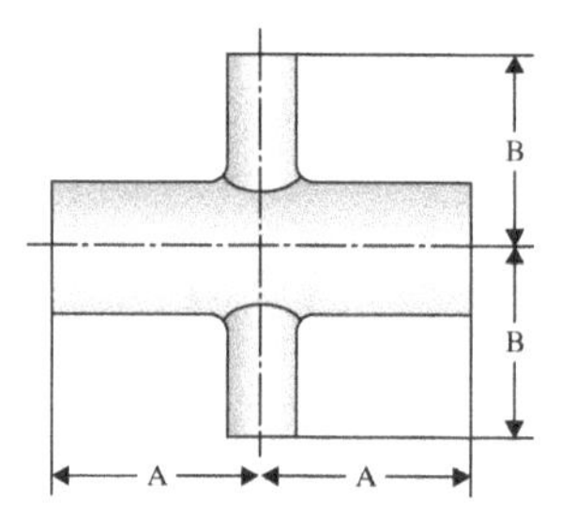

FERRULES

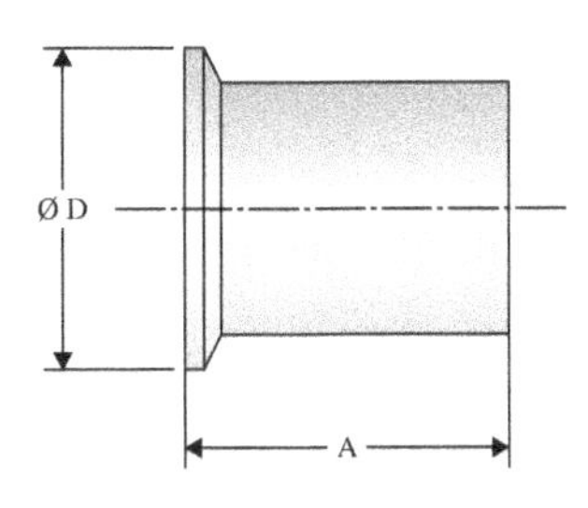

CLAMP FERRULE LONG

Nominal Size in.	Dimensions				
	Type	A in.	A mm	D in.	D mm
¼	A	1.750	44.5	0.984	24.9
⅜	A	1.750	44.5	0.984	24.9
½	A	1.750	44.5	0.984	24.9
¾	A	1.750	44.5	0.984	24.9
1	A	1.750	44.5	1.339	34.0
1	B	1.750	44.5	1.984	50.3
1½	B	1.750	44.5	1.984	50.3
2	B	2.250	57.2	2.516	63.9
2½	B	2.250	57.2	3.047	77.3
3	B	2.250	57.2	3.579	90.9
4	B	2.250	57.2	4.682	118.9
6	B	3.000	76.2	6.570	166.8

CLAMP FERRULE MEDIUM

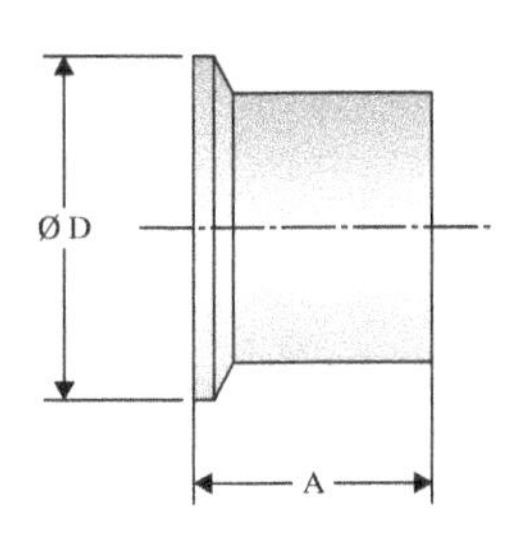

Nominal Size in.	Dimensions				
	Type	A in.	A mm	D in.	D mm
¼	A	1.130	28.7	0.984	24.9
⅜	A	1.130	28.7	0.984	24.9
½	A	1.130	28.7	0.984	24.9
¾	A	1.130	28.7	0.984	24.9
1	A	1.130	28.7	1.339	34.0
1	B	1.130	28.7	1.984	50.3
1½	B	1.130	28.7	1.984	50.3
2	B	1.130	28.7	2.516	63.9
2½	B	1.130	28.7	3.047	77.3
3	B	1.130	28.7	3.579	90.9
4	B	1.130	28.7	4.682	118.9
6	B	1.500	38.1	6.570	166.8

CLAMP FERRULE SHORT

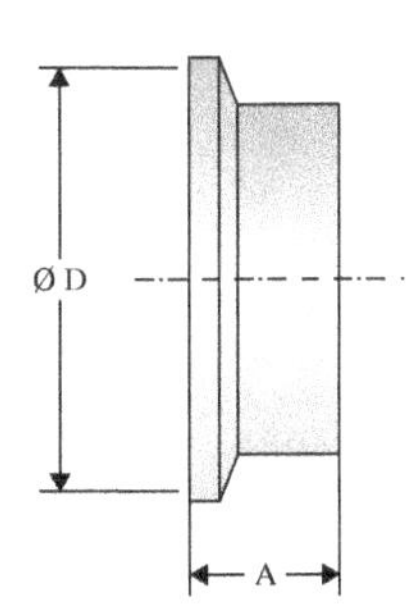

Nominal Size in.	Dimensions				
	Type	A in.	A mm	D in.	D mm
¼	A	0.500	12.7	0.984	24.9
⅜	A	0.500	12.7	0.984	24.9
½	A	0.500	12.7	0.984	24.9
¾	A	0.500	12.7	0.984	24.9
1	A	0.500	12.7	1.339	34.0
1	B	0.500	12.7	1.984	50.3
1½	B	0.500	12.7	1.984	50.3
2	B	0.500	12.7	2.516	63.9
2½	B	0.500	12.7	3.047	77.3
3	B	0.500	12.7	3.579	90.9
4	B	0.625	15.9	4.682	118.9
6	B	0.750	19.1	6.570	166.8

Further Reading

Alloy Casting Institute (ACI) has changed to the Steel Founders Society of America (SFSA), details available at: https://www.sfsa.org/ (accessed on May 10, 2016).

Arnold, J. W. and Bailey, G. W., "Surface Finishes on Stainless Steel Reduce Bacterial Attachment and Early Biofilm Formation: Scanning Electron and Atomic Force Microscopy Study," Poultry Science, Vol. **79**(12) (2000).

Arnold, J. W., Boothe, D. H., and Bailey, G. W, "Parameters of Treated Stainless Steel Surfaces Important for Resistance to Bacterial Contamination," American Society of Agricultural Engineers, Vol. **44**(2) (2001).

ASME B31.3 "Process Piping," New York: American Society of Mechanical Engineers, 2014.

ASME Bioprocessing Equipment (BPE) Standard, New York: American Society of Mechanical Engineers, 2014.

ASME Boiler and Pressure Vessel Code, Section VIII, Division 1, "Pressure Vessels," New York: American Society of Mechanical Engineers, 2013.

ASME Boiler and Pressure Vessel Code, Section VIII, Division 2, "Pressure Vessels," New York: American Society of Mechanical Engineers, 2013.

ASME Boiler and Pressure Vessel Code, Section IX, "Welding and Brazing Qualifications," New York: American Society of Mechanical Engineers, 2013.

ASTM, A213/A213M-03b, "Standard Specification for Seamless Ferritic and Austenitic Alloy-Steel Boiler, Superheater, and Heat-Exchanger Tubes," New York: American Society for Testing and Materials.

ASTM, A270-03, "Standard Specification for Seamless and Welded Austenitic and Ferritic/Austenitic Stainless Steel Sanitary Tubing," New York: American Society for Testing and Materials.

ASTM G48 "Standard Test Methods for Pitting and Crevice Corrosion Resistance of Stainless Steels and Related Alloys by Use of Ferric Chloride Solution," New York: American Society for Testing and Materials, 2011.

Blessman, E. R., "Processing Requirements for Corrosion Resistance Optimization of High Performance Stainless Steel Tubulars," Presentation at IDA World Congress, Dubai, UAE (November 2009).

Bioprocessing Piping and Equipment Design: A Companion Guide for the ASME BPE Standard, First Edition. William M. (Bill) Huitt.

Campbell, R. D., Dvorscek, J., Elkins, C., and Roth, W., "High Purity Welding in the Biotechnology and Pharmaceutical Industries," Welding Journal, Vol. **93**(11) (November 2014).

Charles, J., "Duplex Stainless Steels: A Review after DSS '07 held in Grado," Steel Research International, Vol. **79**(6) (June 2008).

Cobb, H. M., "Development of the Unified Numbering System," *ASTM Standard News* (May 2002). Available at: http://www.astm.org/SNEWS/MAY_2002/cobb_may02.html (accessed on June 9, 2016).

Cutler, P. and Coats, G., "The Nickel Advantage in Duplex Stainless Steels," Presented at the Nickel Institute, Taiyuan, China (November 25, 2009).

DeBold, T. A. and Martin, J. W., "How to Passivate Stainless Steel Parts: Modern Machine Shop," details available at: http://www.mmsonline.com/articles/how-to-passivate-stainless-steel-parts (accessed on May 17, 2016) (October 2003).

European Committee for Standardization (CEN), a European Standard (EN), details available at: https://www.cen.eu/Pages/default.aspx (accessed on May 10, 2016).

European Committee for Electrotechnical Standardization (CENELEC), a European Standard (EN), details available at: http://www.cenelec.eu/ (accessed on May 10, 2016).

European Telecommunications Standards Institute (ETSI), a European Standard (EN). Available at: http://www.etsi.org/eloper (accessed on June 9, 2016).

Gonzales, M. M., "Stainless Steel Tubing in the Biotechnology Industry," Pharmaceutical Engineering, Vol. **21** (April 2001).

Gonzales, M. M., "Materials of Construction for Biopharmaceutical Water Systems, Part 2," American Pharmaceutical Review, Vol. **9**(3) (March/April 2006).

Gonzales, M. M., "Rouge: The Intrinsic Phenomenon in 316L Stainless Steel – A Key Material for Biopharmaceutical Facilities," Pharmaceutical Engineering Magazine, Vol. **32**(4) (July/August 2012).

Haynes International, Inc., "Fabrication of Hastelloy® Corrosion-Resistant Alloys," *Corporate Brochure* (2003).

Henon, B. K., "Orbital Welding of 316L Stainless Steel," presented at the 12th Annual World Tube Congress, October 7–10, 1996, sponsored by the Tube and Pipe Association International (TPA).

Henon, B. K., "High Purity Process Piping: Harmonization of ASME Codes and Standards," Pharmaceutical Engineering, Vol. **34**(2) (March/April 2014).

Henon, B. K. and Tan, W. K., "Autogenous Orbital GTAW of Large, High Purity Tubes," The Fabricator Magazine (July 2010). Available at: http://www.thefabricator.com/article/arcwelding/autogenous-orbital-gtaw-of-large-high-purity-tubes (accessed on June 9, 2016).

Huitt, W. M., "Piping for Process Plants: Part 1," *Chemical Engineering Magazine* (February 2007).

Huitt, W. M., "Piping Design, Part 3 – Design Elements," *Chemical Engineering Magazine* (July 2007).

Huitt, W. M., "Piping Design Part 5: Installation and Cleaning," *Chemical Engineering Magazine* (April 2008).

Huitt, W. M., "Piping for Process Plants Part 6: Testing and Verification," *Chemical Engineering Magazine* (June 2008).

Huitt, W. M., "Crossover Applications for the ASME-Bioprocessing Equipment Standard," *Chemical Engineering Magazine* (October 2010).

Huitt, W. M., "ASTM and ASME-BPE Standards – Complying with the Needs of the Pharmaceutical Industry," PDA Journal, Vol. **65**(1) (January/February 2011).

Huitt, W. M., Henon, B. K., Molina, V. B., III, "New Piping Code for High Purity Processes," *Chemical Engineering Magazine* (July 2011).

Huitt, W. M., "FDA Form 483: Minimizing FDA Inspection Citations," *Chemical Engineering Magazine* (March 2014).

Iannuzzi, M., "Maximum Allowable Service Temperatures and the ASTM International G48 Test" (November 2013).

Kapp, T., Boehm, J., Chase, J., Craig, J., Davis, K., Gupta, V., Stover, J., Wilkowski, S., Montgomery, A., and Ott, K., "Road Map to Implementation of Single-Use Systems," Bioprocess International (April 2010).

Lewis, D. B., "Current FDA Perspective on Leachable Impurities in Parenteral and Ophthalmic Drug Products," AAPS Workshop on Pharmaceutical Stability – Scientific and Regulatory Considerations for Global Drug Development and Commercialization, Washington, DC, USA (October 22–23, 2011).
Mannion, B. and Heinzman, J. III, "Setting Up and Determining Parameters for Orbital Tube Welding," *The Fabricator Magazine* (May 1999).
Mannion, B. and Heinzman, J. III, "Orbital Welding," *Flow Control Magazine* (December 1999).
Moffatt. F., "Extractables and Leachables in Pharma – A Serious Issue," Solvias White Paper (2010).
National Board (National Board of Boiler and Pressure Vessel Inspectors), https://www.nationalboard.org/ (accessed on May 10, 2016).
Riedewald, F., "Microbial Biofilms – Are They a Problem in the Pharmaceutical Industry?" *ASME BPE Symposium*, Cork, Ireland (June 2004).
Riedewald, F., "Bacterial Adhesion to Surfaces: The Influence of Surface Roughness," PDA Journal of Pharmaceutical Science and Technology, Vol. **60**(3) (May–June 2006).
Roth, W. L., Campbell, R. D., Henon, B. K., and Avery, R. E., "High-Purity Welding for Hygienic Applications," Welding Journal, Vol. **93**(11) (November 2014).
SAE International, https://www.sae.org/ (accessed on May 10, 2016).
Schmitz, J., "Best Practices in Adopting Single-Use Systems," BioPharm International, Vol. **27**(1) (January 2014).
Sperko, Walter J., P. E., "Rust on Stainless Steel," A White Paper (December 2014).
Trotter, A. M., "Adoption of Single-Use Disposable Technology in Biopharma Industries – Manufacturing, Economic and Regulatory Issues to Consider," American Pharmaceutical Review, Vol. **15**(2) (March 2012).
Tuthill, A. H. and Avery, R. E., "Specifying Stainless Steel Surface Treatments," Advanced Material Processes, Vol. **142**(6) (1992).
Tverberg, J. C., "Rouging of Stainless Steel in WFI and High Purity Water Systems," presented at the Institute for International Research Conference in San Francisco, CA, USA (October 27–29, 1999).
Tverberg, J. C., "The Role of Alloying Elements on the Fabricability of Austenitic Stainless Steel." White paper presented on June 30, 2011, to Metals and Materials Consulting Engineers, Mukwonago, WI.
Vacher, H. C. and Bechtoldt, C. J., "Delta Ferrite-Austenite Reactions and the Formation of Carbide, Sigma, and Chi Phases in 18 Chromium–8 Nickel–3.5 Molybdenum Steels," Journal of Research of the National Bureau of Standards, Vol. **53**(2), Research Paper 2517 (August 1954).
Vogel, J. D. "The Maturation of Single-Use Applications," Bioprocess International, Vol. **10**(Suppl. 5) (May 2012).
Vogel, J. D., "Single-Use and Stainless Steel Technologies: Comparison, Contrast – Where are they going?," Bioprocess International, Vol. **13** (December 2015).
Vogel, J. D. and Eustis, M., "The Single-Use Watering Hole: Where Innovation Needs Harmonization, Collaboration, and Standardization," Bioprocess International (January 2015). Available at: http://www.bioprocessintl.com/manufacturing/single-use/special-report-single-use-watering-hole/ (accessed on June 9, 2016).

Referenced Organizations

Within the narration of this book, various organizations such as governmental groups, industry-based organizations, and privately held companies are referenced. A list of those organizations is provided here for the benefit of you, the reader, and your ongoing interest in learning with respect to the BPE Standard and the bioprocessing industry in general:

API (American Petroleum Industry), http://www.americanpetroleuminstitute.com/
ASME (American Society of Mechanical Engineers), https://www.asme.org/
ASTM International (American Society for Testing and Materials), http://www.astm.org/
CDC (Centers for Disease Control and Prevention), http://www.cdc.gov/
CFR (Code of Federal Regulations) or e-CFR (Electronic Code of Federal Regulations), http://www.ecfr.gov/cgi-bin/ECFR?page=browse

CSB (US Chemical Safety Board), http://www.csb.gov/
DOT (Department of Transportation), https://www.transportation.gov/
EHEDG (European Hygienic Engineering and Design Group), http://www.ehedg.org/
EPA (Environmental Protection Agency), http://www3.epa.gov/
EMA (European Medicines Agency), http://www.ema.europa.eu/ema/
FDA (Food and Drug Administration), http://www.fda.gov/
OSHA (Occupational Safety and Health Administration), https://www.osha.gov/
NIH (National Institute of Health), http://www.nih.gov/

Attributions

Some of the graphics and photos that make up the figures in this book have been provided as a courtesy by various companies and organizations. This book has been made the better for their courtesy, and my deep appreciation therefore goes out to the following:

3-A Sanitary Standards, Inc., http://www.3-a.org/, Figure 7.5-7.
Ace Sanitary, http://acesanitary.com/, Figure 7.5-5.
American Society of Mechanical Engineers (ASME) Bioprocessing Equipment (BPE) Standard, https://www.asme.org/products/codes-standards/bpe-2014-bioprocessing-equipment, Figures 2.2-1, 2.3-3, 2.3-4, 2.3-5, 2.3-6, 2.6-1, 2.6-2, 2.6-3, 2.6-4, 3.2-1, 3.2-2, 3.2-3, 3.3-1, 3.3-2, 3.3-3, 3.3-4, 3.3-5, 3.3-6, 3.3-7, 3.3-8, 3.3-9a, 3.3-9b, 3.3-10, 3.3-11, 3.3-12a, 3.3-12b, 3.3-13, 3.3-14, 3.3-15, 3.3-16, 3.3-17, 3.3-18, 3.3-19, 3.3-20, 3.3-21, 3.3-22, 3.3-23, 3.3-24, 3.3-25, 4.3-4, 4.4-1, 4.4-2, 4.4-3, 4.4-4, 4.4-5, 4.4-6, 4.4-7, 4.4-8, 4.5-8, 5.1-1, 5.1-2, 6.1-2, 6.1-3, 6.2-1, 7.5-1, 7.5-2, 7.5-3, 7.5-4, 7.5-6, 7.5-10. 7.5-11, 7.5-12.
CELLS alive!, https://www.cellsalive.com, Figures 7.3-2 through 7.3-10.
Centers for Disease Control and Prevention (CDC), http://www.cdc.gov/, Figure 7.3-12
Central States Industrial Equipment and Service, Inc., http://www.csidesigns.com, Figures 7.5-8A and 7.5-8B.
US Chemical Safety Board (CSB), http://www.csb.gov/, Figures 1.7-1, 1.7-2. 1.7-3.
Cotter Brothers, Corp., http://cotterbrother.com, Figures 4.6-4, 4.6-5, and 7.5-9.
Dockweiler, http://www.dockweiler.com/en/, Figure 7.3-14.
FDA (Food and Drug Administration), http://www.fda.gov/, Appendices B, C, D, E, F, G, H, and J.
National Board of Boiler and Pressure Vessel Inspectors (National Board), https://www.nationalboard.org/, Figure b1.
UltraClean Electropolish, Inc, http://ultracleanep.com/, Figures 2.5-4 and 2.5-5.
US Chemical Safety Board, http://www.csb.gov/, Figures 1.7-1, 1.7-2, and 1.7-3.
VNE, member of the NEUMO-Ehrenberg Group, http://www.vnestainless.com/, Appendix L.

Index

Bioprocessing Piping and Equipment Design: A Companion Guide for the ASME BPE Standard,
First Edition. William M. (Bill) Huitt.